INTERNATIONAL ASTRONOMICAL UNION
UNION ASTRONOMIQUE INTERNATIONALE

SYMPOSIUM No. 102

HELD IN ZURICH, SWITZERLAND, AUGUST 2–6, 1982

SOLAR AND STELLAR MAGNETIC FIELDS: ORIGINS AND CORONAL EFFECTS

EDITED BY

J. O. STENFLO
Institute of Astronomy, Zurich, Switzerland

D. REIDEL PUBLISHING COMPANY
A MEMBER OF THE KLUWER ACADEMIC PUBLISHERS GROUP
DORDRECHT / BOSTON / LANCASTER

Library of Congress Cataloging in Publication Data
Main entry under title:

Solar and stellar magnetic fields: origins and coronal effects

At head of title: International Astronomical Union.
Includes indexes.
1. Solar magnetic field–Congresses. 2. Stars–Magnetic fields–Congresses. 3. Sun–Corona–Magnetic fields–Congresses. 4. Stars–Corona–Magnetic fields–Congresses. I. Stenflo, Jan Olof, 1942– II. International Astronomical Union.
QB539.M23S64 1983 523.8'2 83-9571
ISBN 90-277-1619-6
ISBN 90-277-1621-8 (pbk.)

Published on behalf of
the International Astronomical Union
by
D. Reidel Publishing Company, P. O. Box 17, 3300 AA Dordrecht, Holland

Sold and distributed in the U.S.A. and Canada
by Kluwer Academic Publishers,
190 Old Derby Street, Hingham, MA 02043, U.S.A.

In all other countries, sold and distributed
by Kluwer Academic Publishers Group,
P. O. Box 322, 3300 AH Dordrecht, Holland

Printed in The Netherlands

TABLE OF CONTENTS

FOREWORD

Traditionally, solar and stellar physics have been two separate branches of astronomy, which independently of each other have developed their own scientific goals and methods. During the last decade, however, we have witnessed a gradual convergence of these two areas: The solar physicists realize more and more that the sun has to be seen as a special case in a large family of stars of various properties. A more complete understanding of the sun can only be achieved by considering it in this broader context. The stellar physicists on the other hand have become aware that the detailed knowledge of the physical processes that the solar physicists have reached has a more general significance and can be applied to a variety of other astrophysical objects. Observational techniques developed in solar work can frequently be adapted for other stars as well. This unified approach to solar and stellar physics is often called the "solar–stellar connection".

One main goal of this approach has been to understand the general nature and causes of stellar activity. The pioneering and visionary program to search for activity cycles on other stars started by Olin Wilson at the Mount Wilson Observatory 16 years ago has born fruit: in his sample of 91 stars, cyclic behaviour similar to that of the sun is found to be quite common, but many stars also show irregular activity fluctuations of large amplitude. For some stars, no activity is seen. It has thus become feasible to search for empirical "scaling laws" for stellar activity, to find how the activity depends on, for instance, stellar mass, age, and rotation. In particular, differential rotation may now be observed on other stars as an extension of Olin Wilson's original program.

On the theoretical side, a new class of unified theories have emerged, which seek to explain the origin of all cosmic magnetic fields, like that of the Earth, the Sun, or the Galaxy. The solar magnetic cycle has been a well used testing ground for these dynamo theories, but too many free or unknown parameters have prevented these theories to be sufficiently constrained. As the unified theories can now be applied to a wide variety of stars of various properties, we may achieve a much deeper understanding of stellar activity and magnetism.

The central role played by magnetic fields in the activity of a star has been investigated in great detail on the sun. Using the Zeeman effect, the magnetic field in the solar photosphere can be mapped with high spatial resolution. One unexpected discovery made a decade ago was the extreme fine structure of the solar magnetic field: most of the magnetic flux in the photosphere occurs in strong–field, kG form, in isolated flux fragments or fluxtubes occupying a fraction of a percent of the surface area. This discovery has had important implications for the attempts to understand the origins of stellar magnetism and its influence on the outer atmospheres of the stars. Recent observations have demonstrated that this intermittent nature of the magnetic field occurs on other stars as well.

The magnetic fields not only cause stellar activity, but also have a decisive influence on the dynamics and energy balance of the outer stellar atmospheres, in particular their coronae. It is in fact being debated whether the very existence of a corona is a consequence of stellar magnetism. This question can now be penetrated in depth using the wealth of

J. O. Stenflo (ed.), Solar and Magnetic Fields: Origins and Coronal Effects, ix–x.

observations from the Einstein satellite, which records the X-ray emission arising in the hot stellar coronae. Coronal expansion leads to stellar winds, which transport angular momentum away from the stars. The evolution of stellar rotation — the spindown problem — is directly linked to stellar magnetic fields and coronae.

The title of the symposium, "Solar and Stellar Magnetic Fields: Origins and Coronal Effects", tries to express these interconnections in our synergistic approach to solar and stellar physics: the central and unifying role played by the magnetic field, and the link between the problem of magnetic-field origin in the stellar interiors and the problem of the structure and dynamics of the outer stellar atmospheres.

The symposium was held at the Swiss Federal Institute of Technology Zurich (ETHZ), August 2-6, 1982. It was attended by about 170 scientists from 21 countries. 110 papers were presented: 17 invited reviews, 36 orally presented contributed papers, and 56 poster papers. Due to lack of space, we list the poster papers by title only in the present volume.

The Scientific Organization Committee consisted of: J.O. Stenflo (Chairman), D. Mihalas, R.W. Noyes, A.B. Severny, M. Stix, Y. Uchida, G. Vaiana, N.O. Weiss, O.C. Wilson, C. Zwaan The members of the Local Organizing Committee were: M.C.E. Huber (Chairman), A.O. Benz, C.V.H. Huber, H. Nussbaumer, U. Pauls, H. Rüf, D.S. Spicer, J.O. Stenflo, A.-M. Thalmann, S. Weber. Using his extraordinary organizational talent and experience, the chairman Martin Huber elevated his task to a "fine art". He thereby greatly contributed to the smooth operation and the pleasant atmosphere of the symposium. As Session Chairmen during the meeting served: A.B. Severny and R.G. Giovanelli (Session I), E.A. Müller and D.J. Mullan (Session II), N.O. Weiss and M. Stix (Session III), R.M. Bonnet and Y. Uchida (Session IV), A.K. Dupree (Session V), and J.O. Stenflo (Session VI).

The symposium was sponsored by IAU Commission 12 and cosponsored by IAU Commissions 10, 29, 35, 36, and 44. Financial support was received from IAU and ETH Zurich, from the Swiss National Academy of Sciences (SNG) and the Swiss National Science Foundation, as well as from Swiss industry. Thanks to the efforts of H. Rüf, who tended with great aptitude the liaison to industry during an economically unfavourable period, we obtained donations from 22 private companies and corporations. We are very grateful for all the support given.

The proceedings have been prepared from camera-ready typescripts, but it was often necessary to retype portions of the manuscripts to bring them as close as possible to a standard format. The printed text of the discussions was edited. It is based on a combination of written questions and comments, handed in by those who participated in the discussions, and a tape recording of the proceedings. Unfortunately the review on Solar Magnetic Fields is missing in this volume, since no manuscript was available in February 1983, when the proceedings had to be sent to the Publisher. The interested reader is referred to four relatively recent and still relevant reviews on this topic: by J.W. Harvey (1977, *Highlights of Astronomy* **4**, pp. 223-239), and by J.O. Stenflo (1976, *IAU Symp.* **71**, pp. 69-99; 1977, *IAU Coll.* **36**, pp. 143-188; 1978, *Rep. Prog. Phys.* **41**, pp. 865-907).

The secretary of the Institute of Astronomy at ETH, Mrs. S. Weber, has with her efficiency, skill, and dedication played a central role in all phases of the organization of the symposium and in the preparation of the proceedings. To her and all the other members of the Institute of Astronomy who have so greatly contributed to the success of this undertaking, I express my sincere thanks.

Zurich, February 1983 Jan Olof Stenflo

I. MAGNETIC FIELDS IN STELLAR PHOTOSPHERES

OBSERVATIONS OF MAGNETIC FIELDS ON LATE-TYPE STARS

Geoffrey W. Marcy
Lick Observatory, Board of Studies in Astronomy and Astrophysics, University of California, Santa Cruz

ABSTRACT

The "Robinson" method for measuring magnetic fields on solar- and late-type stars is reviewed. The results of such measurements for a sample of 29 G and K main-sequence stars are presented. The area covering-factors of magnetic regions are greater in the K dwarfs than in the G dwarfs, but no spectral-type dependence is found for the field strengths, contrary to expectations of some flux-tube models. The dependence of Ca II H and K emission on magnetic fields and T_{eff} is consistent with theoretical expectations for "slow-mode" mhd wave-generation rates, but inconsistent with those of other mhd modes. Coronal soft X-ray fluxes correlate well with the magnetic fields, and it is argued that Alfvén waves are the likely energy-transport mechanism. Surface magnetic fluxes vary with rotation as $V_{rot}^{0.5-1.0}$, depending on spectral type.

1. INTRODUCTION

Cool stars are observed to exhibit a variety of phenomena such as chromospheres, coronae, starspots, flares, and activity cycles, that appear qualitatively similar to features associated with the magnetic fields on the sun. (For recent reviews, see Linsky 1980, Golub 1981, Hartmann 1981, Noyes 1981, Skumanich and Eddy 1981, Zwaan 1981.) The correlation of all of these telltale signs of fields with stellar rotation has led to the widespread belief that the Russel-Vogt theorem must be amended to include rotation as another fundamental parameter, with magnetic fields constituting its most visible effect. To further understand the processes by which rotation is coupled to magnetic fields, and in turn determine the influence of those fields on the surfaces of solar-type stars, we must observe how such effects change on stars with different masses and rotation rates.

In this review I will try to provide a summary of the several attempts made to detect and measure magnetic fields on late-type stars directly. The written history of such trials is brief; however, I have

J. O. Stenflo (ed.), Solar and Magnetic Fields: Origins and Coronal Effects, 3–22.

been told that various exploratory attempts by such farsighted observers as G. E. Hale, H. W. Babcock, G. W. Preston and R. P. Kraft were made, but proved impossible to pursue with existing detectors. More recently, the so-called "Babcock technique," in which one measures the circular polarization induced by the Zeeman effect in absorption-line wings, was used by Boesgaard (1974) and Boesgaard, Chesley, and Preston (1975), resulting in marginal field detections for XI Boo A and 70 Oph A. A similar effort by Vogt (1980) with a Reticon detector yielded upper limits of about 100 gauss for the "effective" field strength on active late-type stars. Recently, Brown and Landstreet (1981) and Borra and Mayor (personal communication) have modified "Griffin-type" radial velocity spectrometers to perform a multi-line Zeeman analysis. These efforts have further reduced the upper limit to about 5 gauss.

These null results are thought to have been due to two effects: (1) small magnetic-field area coverages on the stellar surfaces; (2) the small-scale, bipolar character of the field regions, similar to those on the sun, so that the opposing field polarities result in cancellation of the circular polarization expected from the Zeeman effect (Robinson 1980). Such bipolar regions, however, could produce net transverse fields if, for example, a star is viewed equator-on and has radial field lines located near the limb. The Zeeman-induced linear polarization from such transverse fields is expected to be observable both in absorption-line wings, and with broad-band polarimetry (e.g. Landi Degl'Innocenti 1982). The latter is possible because of the relative saturation of the central π-component in the Zeeman pattern, thereby suppressing its state of polarization. Unfortunately, such broadband polarimetric observations apparently require accuracy of at least 0.01%. Careful polarimetric studies by Pettersen and Hsu (1981) and Clayton and Martin (1981) on selected stars have yielded no positive detections, though the stated errors tend to be somewhat higher than 0.01%. However, Tinbergen and Zwaan (1981) have reported intrinsic linear polarization in a large sample of late-type stars based on the statistically higher polarization in the bandpass containing the greater density of lines. Though the levels of polarization are too low to be interpreted in detail, the results suggest that future measurements of linear polarization in the line wings may prove fruitful in extracting not only field strengths but also geometry information about photospheric magnetic fields. E. Borra has begun such a program.

Perhaps the most promising method for detecting and measuring magnetic fields on late-type stars is the so-called "Robinson technique" (Robinson 1980). This approach was motivated by recent solar line-profile analyses that have proven eminently successful in measuring field strengths in spatially unresolved magnetic structures in the solar photosphere (e.g., Tarbell and Title 1976). The first attempts on stars with Robinson's method were quite encouraging, and this success has led several investigators, including myself, to attempt similar observations and analyses.

2. THE ROBINSON METHOD

Robinson (1980) revived an old idea, exployed successfully on Ap stars as recently as 1971 (Preston 1971), that the Zeeman effect on lines may be detectable by the excess "broadening" of a profile observed in unpolarized light. Robinson suggested this be done on solar-type stars by comparing line profiles that have very different Zeeman sensitivities but which otherwise have similar transition characteristics. Any difference between two such profiles would be attributable to the presence of magnetic fields in the line-forming region.

The analysis of the two profiles is based on the premise that a Zeeman-broadened line formed in a photospheric region with a uniform magnetic field may be represented by the sum of three displaced components (a Zeeman triplet), the shapes of which are, to first order, given by the profile of a Zeeman-insensitive line. We know, from atomic physics, the expected ratio of the strength of the outer "sigma" components to that of the central "pi" components in a given Zeeman triplet. Therefore, any observed enhancement of the central component, over that expected, must have arisen from the nonmagnetic areas where no splitting occurs. This enhancement yields the area-coverage of the nonmagnetic and hence magnetic regions. The splitting of the components is proportional to the field strength: $\Delta\lambda = 4.67 \times 10^{-13} \lambda^2 gB$ (Å), where λ is the wavelength, g the Landé factor, and B the field strength in gauss.

Even for lines with Landé factors as large as 2.5, the Zeeman triplet is spectroscopically unresolved for solar field strengths of 1 to 3 kilogauss, necessitating some deconvolution procedure. Robinson (1980) has advocated a general Fourier approach in which a continuum of field strengths is permitted, and the area coverage for each field strength is deduced. The method I have employed is a numerical chi-square fit to the Zeeman-sensitive profile, to solve for the strength and separation of the underlying Zeeman-triplet components. This approach yields some ill-defined average of the surface field strength and area coverage, which is also the ultimate result of the Fourier technique. This is because currently available spectral resolution and signal-to-noise ratios, as well as ubiquitous weak blends, permit at most three or four degrees of freedom to characterize a profile. A direct comparison of Robinson's Fourier technique with my chi-square profile-fitting approach yielded the same result to within 10%, for the one data set (generously supplied by R. Robinson) to which both were applied.

A third approach is possible, <u>viz</u>., to interpret the Fourier transform of the Zeeman-broadened profile in terms of a two-component stellar surface -- a nonmagnetic component and a magnetic component of some characteristic field strength (M. Smith, personal communication). Such a technique might allow the user to more readily acquire an understanding of the analysis, as with the profile-fitting approach, while retaining the advantages of Fourier methods in handling noise in the data.

Unfortunately, all of the versions of the Robinson technique require two significant assumptions. First, the implicit assumption is made that both the energy removed from the continuum in the line and the shape of the line are the same in and out of magnetic regions. The assumption of similar profile shapes has been used successfully in solar profile analyses (e.g., Stenflo 1973). However, it is well known that continuum brightness increases and line strength decreases in the small-scale magnetic elements of which solar faculae and network regions are comprised (Stenflo 1975, Chapman and Sheeley 1968). Second, the field geometry modulates the expected relative strengths of the sigma and pi components from the magnetic regions (Seares 1913, Beckers 1969). I have chosen to assume that the field lines lie predominantly normal to the surface and that they are uniformly distributed over a spherical surface, which implies an average field-line to line-of-sight angle of 34° for a (solar) limb-darkened hemisphere.

Both the assumption of constant line strength and the field-geometry assumption will affect only the deduced area-coverage of the fields since the interpretation of the splitting of the Zeeman-components remains model-independent. Tests with theoretical profiles (Unno 1956) indicate that errors of about 20% in magnetic area-coverage may accrue from either of the above assumptions, and larger errors are possible if active regions on other stars are dramatically different in character than solar. Note also that starspot umbrae, if similar to their solar counterparts, will have surface brightnesses roughly 20% of the surrounding photosphere and hence will contribute little to the observed integrated line profiles. Starspot penumbrae, however, would contribute to the deduced magnetic field measurements (both strength and area-coverage) because the locally enhanced strength of the neutral metal lines would compensate for the slightly reduced continuum brightness.

In addition to the model-dependent errors discussed above, random errors from noise in the data demand that spectra be obtained with high signal-to-noise ratio (S/N > 100), owing to the small effect of Zeeman broadening compared with other sources. A good estimate of the magnitude of the Zeeman-induced broadening may be gained from the theoretical line profiles shown in fig. 1, derived from Unno's (1956) relations. The small vertical discrepancy between the Zeeman-sensitive line (assumed to have Landé g = 2.5) and the insensitive line (g = 1.0) shows the necessity of low-noise data. Tests of the random errors in the field measurements have been made by myself and Robinson (1980) by introducing random fluctuations in the theoretical profiles. Such measurement errors in field strength and area-coverage are neither gaussian nor independent, and they depend on the strength and coverage themselves. For fields of solar strength, about 1500 gauss, covering 30% of a stellar surface, spectra with a signal-to-noise ratio of 100 will result in random errors of about 25% in strength and flux (strength multiplied by area) and about 35% in area-coverage.

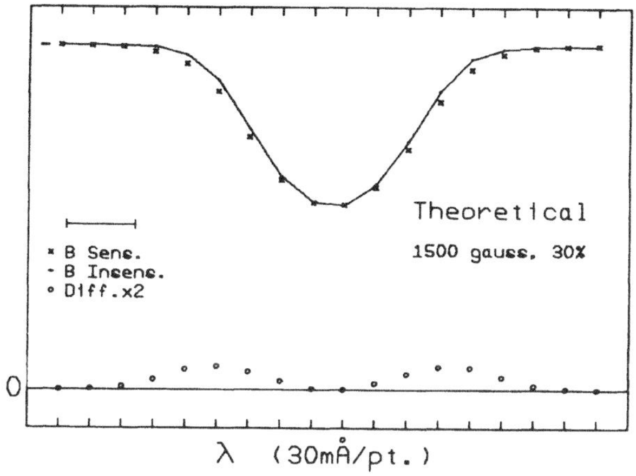

Figure 1. Comparison of a theoretical, Zeeman-sensitive profile (upper points) with that of a Zeeman-insensitive line (continuous line), for a star with magnetic fields of 1500 gauss, covering 30% of its surface. The points at the bottom show the difference of the profiles multiplied by 2. The bar indicates the assumed resolution of 60 mÅ.

Finally, tests similar to those above indicate that the Robinson method, applied in its pure form, can detect fields of solar strength only if the area coverage is greater than about 5%. The magnetic flux tubes on the sun cover only about 1% of the solar photosphere. Therefore, we can successfully detect fields only on stars which have 5 times the magnetic flux of the sun. One ray of hope for significant future improvements is to obtain high-resolution spectra of hundreds of line profiles simultaneously, thereby minimizing the effect of ubiquitous weak blends in late-type stellar spectra.

3. RESULTS OF MAGNETIC-FIELD MEASUREMENTS

The first measurements of magnetic fields with the Robinson technique were made of XI Boo A (G8V) and 70 Oph A (K0V) by Robinson, Worden and Harvey (1980). Their deduced field strength of 2500 gauss covering 20-45% of XI Boo A implies a magnetic flux 50 times greater than that of the sun! Later failures to measure the field of XI Boo A (Marcy 1981), followed by positive detections by myself and G. Timothy, C. Joseph and J. Linsky at Boulder, Colorado, have confirmed the validity of the original detection and suggested that the fields on active stars may be quite variable. In addition, Timothy, Joseph and Linsky (1981) have reported detections of fields on several other late-type stars, and I am aware to date, via personal communications, of similar detections by M. Smith at Sacramento Peak Observatory, and by J. Harvey at Kitt Peak National Observatory.

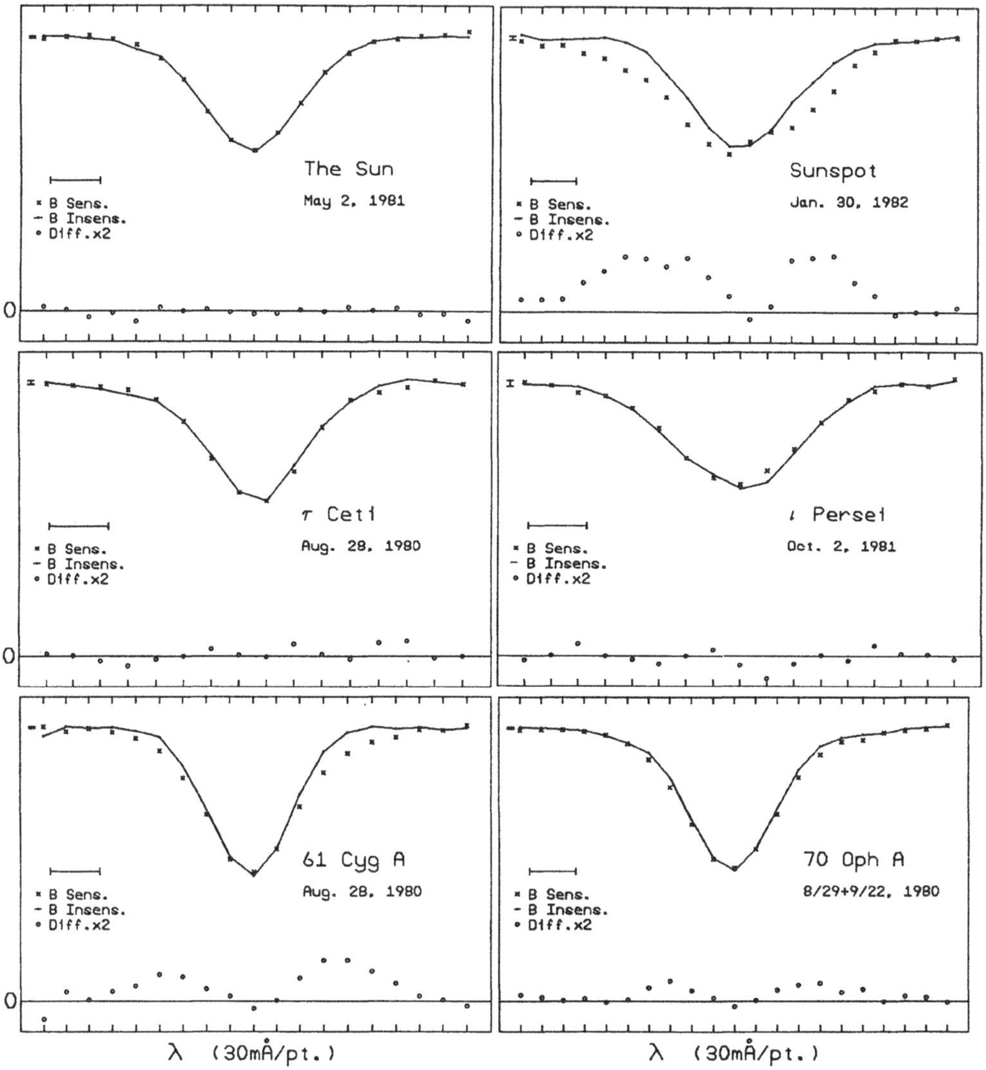

Figure 2. Examples of observed line profile pairs for the daytime sky, a sunspot, and four stars, as in figure 1. The difference in the Zeeman-sensitive and insensitive profiles is shown multiplied by 2 at the bottom. Tau Ceti and iota Per do not show evidence of split Zeeman components in their difference profiles, but the sigma components are visible in 61 Cyg A and 70 Oph A.

During the past two years I have been engaged in a program of measuring and monitoring fields on 29 G and K main-sequence stars, and the remainder of this review paper will contain a description of the results obtained thus far. The observations have been made with the coudé double-pass echelle spectrograph (Soderblom et al. 1978) and the image-dissector scanner (Robinson and Wampler 1972) at Lick Observatory. The Zeeman-sensitive line employed was λ6173.34 (lower EP = 2.21 eV, g = 2.5) and the insensitive line was λ6240.65 (lower EP = 2.21 eV, g = 1.0), both arising from the same lower state of Fe I.

Examples of some of the data are shown in figure 2. Profiles from the daytime sky (labeled "sun") show no evidence of Zeeman broadening because of its low area-coverage of fields (1%). However, the sunspot data show clear evidence of fields and the deduced field strength of 2000 gauss is comparable with the value obtained at Mt. Wilson Observatory (kindly provided by Dr. Robert Howard) of 2300 gauss. Examination of the "difference profile" shown at the bottom of the panels reveals no sigma components on either tau Ceti (G8V) or iota Per (G4V), but prominent Zeeman components are seen for 61 Cyg A (K5V) and 70 Oph A (K0V). To date, 19 of the 29 surveyed stars have shown fields. The sample, though not complete to a given magnitude, is not biased in rotational velocity. However, most of the stars on which fields have not yet been detected are G dwarfs.

Indeed, the histogram of area-coverage (or "filling factor") shown in figure 3 shows that many of the G dwarfs have less than 10% of their surfaces covered by fields while the K dwarfs have filling factors ranging from 20% to 80%. In order to show an unbiased histogram, the nondetections have been included at half of their upper-limit values. The very large area-coverages seen on some stars may indicate that a model-dependent systematic error is being made in our interpretation of the line profiles; however, a year of searching for such an error source has uncovered nothing save the possible 20% errors, mentioned previously, from field geometry and line-strength changes. Further, the independent measurements of area-coverages for XI Boo A of 20-45% by Robinson, Worden and Harvey (1980), and of 40-60% by G. Timothy and C. Joseph (private communciation) confirm the large area-coverages found here for many active stars. Possibly related are the reports that the observed amplitudes of light variation on some spotted stars demand that the spots cover up to 40% of the stellar surface (e.g., Vogt 1981). Similarly, analyses of chromospheric lines from active stars suggest the need for much more extensive active regions than are found on the sun (Kelch, Linsky and Worden 1979; Giampapa et al. 1982. The apparently larger median value of the magnetic covering-factors (indicated by the arrows in figure 3) for the K dwarfs, compared with that of the G dwarfs, will be addressed in section 6 in conjunction with stellar rotation effects.

The accompanying histogram in figure 3 for the magnetic field strengths shows no difference in field strength between the G and K dwarfs. This is remarkable because the characteristic field strength

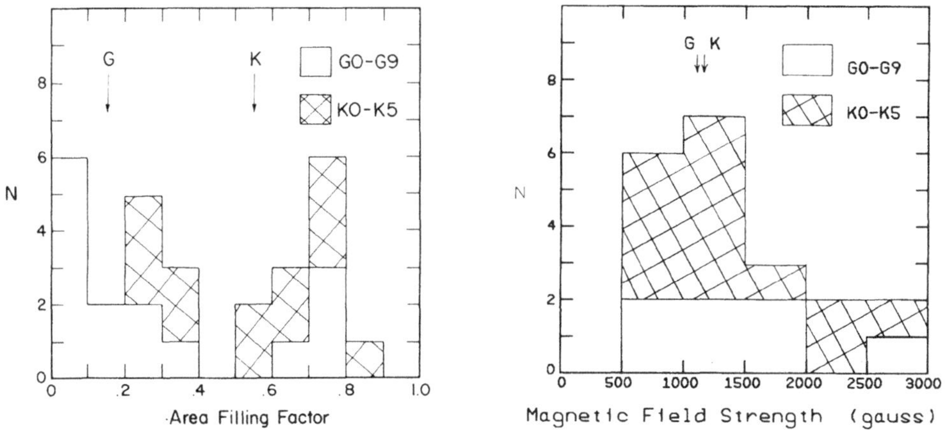

Figure 3. Left: a histogram of magnetic area-coverage for all observed dwarfs with nondetections included as 1/2 their upper limits to minimize bias between G and K stars. Right: a histogram of field strength. Nondetections are left out because no upper-limit estimates are possible. Both panels display G dwarfs as open rectangles and K dwarfs as cross-hatched. The arrows represent the median values.

on the sun of about 1500 gauss has been attributed by some as the result of gas-pressure confinement of the thin flux tubes ($B^2/8\pi = P_g$), i.e., tubes with stronger fields would disperse owing to high internal magnetic pressure (Galloway and Weiss 1981; Parker 1981). Since photospheric gas pressure is expected to increase toward later spectral types approximately as, $P_g \propto T_{eff}^{-3}$, the thin flux-tube models would predict fields of about 2400 gauss on all K2 dwarfs. This expectation is in disagreement with, for example, the well-observed stars epsilon Eridani (K2V) and 70 Oph A (K0V) that usually show field strengths under 1500 gauss. (Robinson, Worden and Harvey found 1880 gauss for 70 Oph A for their single observation.) It may be that the concept of isolated, stable, thin flux-tubes, confined by gas pressure, will not be useful in describing either the nature or the dynamics of the fields on the more active stars.

An alternative possibility, that magnetic-field strength is determined by confinement from turbulent pressure, ρV_{turb}^2, has led some to suggest that flux ropes of lower field strengths (600 gauss) might rise to the surface (Zwaan 1978). Though this might explain the 1 kilogauss fields observed on some stars, a further refinement is necessary to account for the wide variation in field stregths observed at a given spectral type. In view of the short time scale variations of magnetic fields (1 day) seen by Timothy, Joseph, and Linsky (1981), and also found in the Lick data, a more dynamic and stochastic model of rising and dissipating flux ropes may be needed to describe the surface fields on active stars.

4. MAGNETIC FIELDS AND CHROMOSPHERES

A longstanding problem in stellar astronomy has been the unknown source of heating in solar- and late-type stellar chromospheres. Until recently, pure acoustic waves generated by convective turbulence at the base of the photosphere was perceived as the most likely energy-transport mechanism (e.g., Renzini et al. 1977; Ulmschneider and Bohn 1981). Observations of velocity fluctuations in solar optical lines by Deubner (1976) seemed to support their existence. However, a variety of solar and stellar observations are in conflict with the notion that pure acoustic waves supply much of the chromospheric energy. These arguments have been summarized elsewhere (e.g. Linsky 1980), but the most compelling are: (1) spatial coincidence on the sun of chromospheric and transition-region emission with photospheric magnetic fields; (2) the dependences of stellar chromospheric emission on effective temperature and gravity contradict those expected theoretically for pure acoustic-wave heating; and (3) the wide range of chromospheric emission at a given spectral type conflicts with expectations that acoustic-wave generation depend only on T_{eff} and gravity.

Alternatively, mhd waves have been proposed as the dominant energy transport mechanism (for a review, see Stein and Leibacher 1980), and three modes can satisfy the mhd equations: Alfvén waves, "slow"-mode, and "fast"-mode. Alfvén waves involve bending and twisting of the field lines, slow-mode waves are simply acoustic waves channeled by vertical field lines, and fast-mode waves involve fluid compression but depend on magnetic tension for the restoring force, and need not travel along field lines as is required for the former two wave types. The wave-flux generated by turbulent motions at the top of the convection zones of cool stars has been given by Stein (1981) and Ulmschneider and Stein (1982):

$$F(\mathrm{Acoust})/T_{eff}^{4} \propto g^{-1}\, T_{eff}^{10.6} \tag{1a}$$

$$F(\mathrm{Alfv\acute{e}n})/T_{eff}^{4} \propto g^{0.1}\, T_{eff}^{0.7}\, B^{-1} \tag{1b}$$

$$F(\mathrm{Slow})/T_{eff}^{4} \propto g^{-0.2}\, T_{eff}^{2.1} \tag{1c}$$

$$F(\mathrm{Fast})/T_{eff}^{4} \propto g^{0.5}\, T_{eff}^{3.4}\, B^{-5} \tag{1d}$$

where g is gravity and B is magnetic field strength.

We may use the Mt. Wilson measurements of Ca II H and K emission, as indicators of radiative loss rates in the lower chromosphere, by converting them to surface fluxes as done by Duncan (1981) and

Middelkoop (1982), and subtracting the contribution from the "photosphere" (Linsky et al. 1979). A three-parameter least-squares fit was made to these chromospheric Ca II H and K fluxes, $F'(H + K)$, to determine the observed dependence on T_{eff}, B and magnetic area-coverage. The result of the fit was:

$$F'(H + K)/\sigma\, T_{eff}^4 = 6.14 \times 10^{-14} B^{0.5}\, T_{eff}^{2.0}\, f^{0.6} \qquad (2)$$

where f is the area-coverage. The errors in the above exponents are about 1/2 the stated value in all three cases. A plot showing the relationship between $F'(H + K)$ and the above-determined dependences is given in the left panel of figure 4.

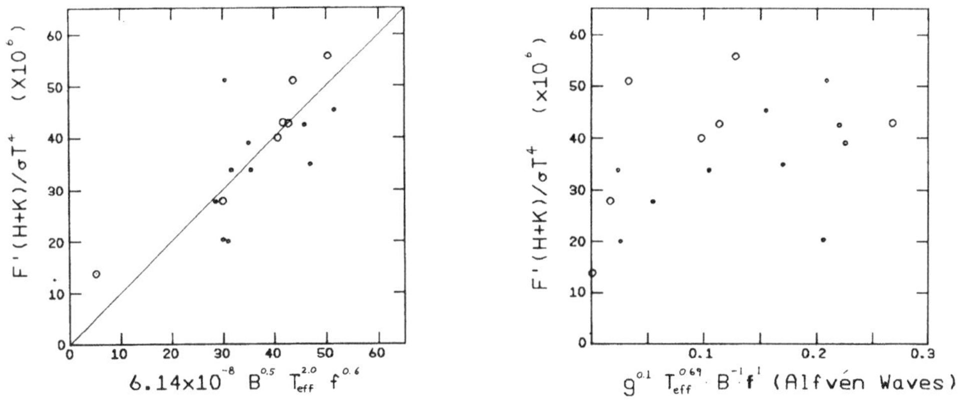

Figure 4. Left: The fraction of total stellar luminosity that is emitted in Ca II H and K from the chromosphere vs. the best-fit power-law dependences on field strength, B, effective temperature, and magnetic area-coverage. These exponents correspond closely with those expected for slow-mode mhd waves. Right: Same plot for the expected dependences assuming Alfvén waves heat lower chromospheres.

The power-law dependences on B and T_{eff} given above are in clear disagreement with those expected for pure acoustic waves and for fast-mode waves (eq. 1), and are many standard deviations away from those dependences expected for Alfvén waves. Indeed the right panel of figure 4 shows a plot of observed Ca II H and K emission vs. the function of B and T_{eff} expected if Alfvén waves provide the heating (eq. 1b). No correlation is apparent. However, the observed dependences of Ca II flux on T_{eff} and B are quite close to those expected if slow-mode waves (eq. 1c) heat stellar chromospheres. Indeed, magnetically channeled acoustic waves have been previously suspected as the energy transport mechanism to lower chromospheres on the basis of the variation of Mg II emission-line fluxes with effective temperature and gravity (Ulmschneider and Stein 1982; Basri and Linsky 1979).

A remaining uncertainty in the identification of slow-mode waves as the dominant mechanism derives from the poorly known variation in the radiative damping rate of such waves with changing T_{eff} and gravity, over the relevant range of G0V to K5V. Further theoretical work in this regard would certainly be most useful.

5. MAGNETIC FIELDS AND CORONAE

The detailed observations by Skylab of arch-like coronal structures having feet planted on bipolar active regions suggest that both coronal structure and heating are related to photospheric magnetic fields (e.g., Billings 1966; Vaiana et al. 1973; Rosner, Tucker and Vaiana 1978; and a recent review by Golub 1981). The measurements of soft X-ray fluxes from other G, K and M dwarfs have led a number of investigators (e.g., Rosner and Vaiana 1979; Ayres and Linsky 1980; Pallavicini et al. 1981; Walter 1982) to suspect that magnetic fields control stellar coronae as well. Though we do not know a priori how coronal development will depend on field strength, the above suspicion implies that stellar soft X-ray fluxes will be related in some fashion to the photospheric area-coverage of magnetic fields.

To test this, soft X-ray measurements from the Einstein Observatory have been taken from Johnson (1981), Pallavicini et al. (1981), Ayres et al. (1981), Walter (1982), Vaiana et al. (1981) and from Vogt, Walter and Marcy (unpublished, for XI Boo A, XI Boo B, and HR 5553). Figure 5 shows a plot of photospheric, magnetic area-coverage against $\log(f_x/l_{bol})$, the ratio of soft X-ray flux to bolometric flux. The correlation seen in Figure 5 provides strong confirmation of the notion that the coronae around all G and K main-sequence stars depend sensitively on the underlying, photospheric magnetic fields.

Additional information regarding the actual mechanism by which coronae are heated may be gained from determining how the soft X-ray fluxes depend on photospheric magnetic field strength. A least-squares fit to the meager data (there are only 7 points besides the sun) showed that

$$f_x/l_{bol} \propto B^{-1.5 \pm 1.0} \cdot f, \qquad (3)$$

where B is field strength and f is magnetic area-coverage. (Including T_{eff} in the fit results in a dependence of T_{eff}^{11} but this exponent is

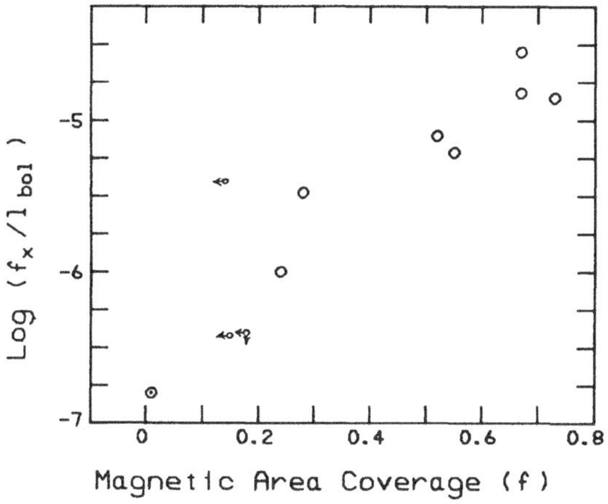

Figure 5. The ratio of soft X-ray flux to bolometric flux vs. photospheric magnetic area-coverage for G and K dwarfs.

uncertain by ±5, i.e., meaningless, because of the extremely narrow range of spectral types available, G0-K5, and the few stars.)

According to Stein (1981) and Ulmschneider and Stein (1982), we may apply equations 1a-1d, given here previously, to describe the generation rate of mhd waves which may eventually propagate to the coronae. Here again, we know little about the variation of deposition efficiency with spectral type for any of the waves. The expected generation rate for fast-mode waves (eq. 1d) disagrees, in field-strength dependence, with the $B^{-1.5}$ dependence observed here. And, of course, the good correlation between soft X-ray flux and magnetic area-coverage in figure 5 suggests that pure acoustic waves are not responsible for coronal heating. Unfortunately, the expected generation rates for both slow-mode and Alfvén waves given in equation 1 are consistent, within two standard deviations, with the observed dependence on B. However, the well known high radiative damping rate (Schmitz and Ulmschneider 1980) of all compression waves, such as slow-mode, in the photosphere and lower chromosphere has led some to choose Alfvén waves as the only surviving mhd wave in the corona (Leibacher and Stein 1981; Hartmann and MacGregor 1980).

Indeed it seems that in order to get Alfvén waves to dissipate at all, they must first couple to either the compressive slow or fast waves after traversing the lower chromosphere (Leibacher and Stein 1981). Certainly, further theoretical study of the dissipation mechanisms available for Alfvén waves are needed to clarify this process, and may lead to better understandings of the winds from cool giants and, perhaps, T Tauri stars (Ayres et al. 1981; Hartmann and MacGregor 1980).

6. MAGNETIC FIELDS AND STELLAR ROTATION

The observed correlations between stellar rotation and such characteristics as chromospheric and coronal emission, age, and the presence of starspots, flares and activity cycles have led many investigators to suggest that magnetic fields provide the missing link via dynamo generation (Skumanich 1972; Hartmann 1981; Pallavicini et al. 1981; Vaughan et al. 1981). Independently, most theoretical descriptions of the dynamo process require rotation and differential rotation for the production (by coriolis-affected convection) and enhancement (by longitudinal stretching of field lines) of magnetic fields (for a recent review, see Gilman 1981).

To investigate the dependence of stellar magnetic fields on rotation, equatorial velocities have been determined from Vaughan et al. (1981) and, when not available there, from Soderblom (1980) and Smith (1979). Unfortunately, few stars have both a measured field and rotation velocity. Those available are shown in the left panel of figure 6, where magnetic flux is plotted against equatorial velocity. The 3 data points representing G dwarfs (circles) are consistent with the

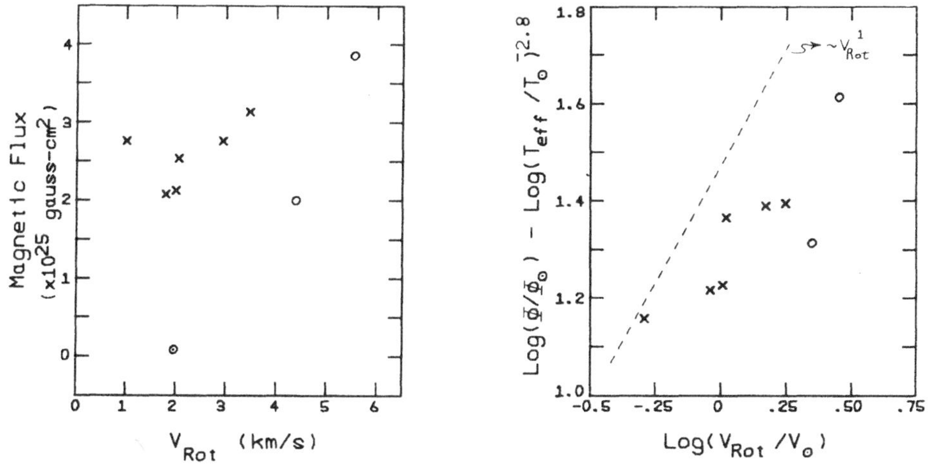

Figure 6. Left: Magnetic flux vs. equatorial velocity. Crosses are K dwarfs, circles are G dwarfs. Right: Same as left panel, but on a log scale and the magnetic flux, Φ, has been corrected for the T_{eff} dependence in observed flux. The dashed line indicates the slope for a linear dependence on V_{rot}.

possibility that magnetic flux increases with rotation. The K stars would show a similar relationship, except for the far left point in the diagram. This is 61 Cyg A and, at K5, has the lowest T_{eff} of the plotted stars.

This suggests (as does the visible difference between the K and G dwarfs in the plot) that magnetic field flux increases toward later spectral types. To quantify this, a least-squares fit to the data yields:

$$\Phi = 4.1 \times 10^{10}\, V_{rot}^{0.55 \pm 0.2}\, T_{eff}^{-2.8 \pm 1.1} , \qquad (4)$$

where Φ is magnetic flux in gauss-cm^2 and V_{rot} is equatorial velocity in km/s. The relationship between magnetic flux and rotation may be more graphically displayed by correcting the magnetic fluxes for the T_{eff} dependence, as has been done in the right panel of figure 6. There, the logarithm of the magnetic flux minus $-2.8 \log (T_{eff}/T_{\odot})$ is correlated with rotational velocity at the 1% confidence level. This deduced dependence of magnetic flux on $V_{rot}^{0.55}$ can be shown to be completely consistent with the rotational dependence of Ca II H and K emission found by Vaughan et al. 1981, for their sample of stars near K0 (B-V = 0.86 - 0.89, see their figure 4). However, their data indicate that at G0, Ca II H and K flux depends nearly linearly on rotation, suggesting that the same will be found true for magnetic flux when enough G dwarfs are measured. Skumanich and Eddy (1981) found corroborating evidence in the Ca II H and K emission from spectroscopic binaries that varies as $V_{rot}^{2/3}$.

Finally, it is difficult at this time to interpret the observed dependence of magnetic flux on V_{rot} and T_{eff} (equation 4) in terms of available dynamo models. Durney and Robinson (1981) have computed dynamo models in which the field generation at the bottom of the convection zone is determined by the rise time of a magnetic flux tube due to buoyancy. Their results suggest that observed magnetic flux should vary as $V_{rot}^{2.3}\, T_{eff}^{-15}$. These dependences are in the right direction, but are so discrepant both with those found here and with those implied by Vaughan et al. (1981), that some crucial physical process must have been overlooked.

A different approach is taken by Stix (1972) who argues that the critical field attained in the convection zone is that which suppresses the helicity of convection, i.e., $B_c \propto \Omega^{1/2}$, where Ω is angular velocity. It can be shown that the observed surface toroidal field will also vary as $\Omega^{1/2}$, assuming that the turbulent and magnetic Reynolds numbers depend only on spectral type (see Skumanich and Eddy 1981). The agreement between Stix's predicted square-root dependence on rotation and that observed suggests that the suppression of convective helicity by the fields may constitute an important process in late-type stellar dynamos.

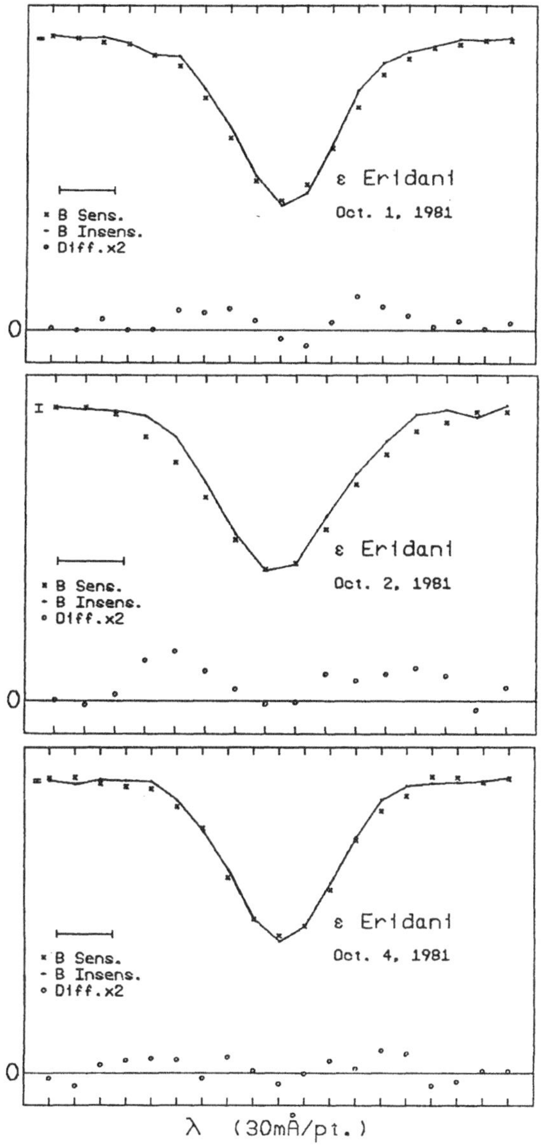

Figure 7. Magnetic field data for epsilon Eridani as in figure 2. Changes seen in the difference profile suggest that the magnetic fields are variable on time scales of one day.

7. MAGNETIC FIELD VARIABILITY

As first reported by Timothy, Joseph and Linsky (1981), the magnetic fields on epsilon Eridani seem to vary significantly on time scales as short as one day. The data taken at Lick have also suggested short time-scale variations for both epsilon Eridani and XI Boo A. Figure 7 shows data for epsilon Eridani from three closely-spaced nights. The difference profiles shown at the bottom of each panel suggest that the magnetic fields became somewhat more widespread from Oct. 1 to Oct. 2, 1981, and then nearly disappeared by Oct. 4. Since the rotation period of epsilon Eridani is 11 days (Vaughan et al. 1981), these magnetic field variations cannot be due to active regions rotating across the stellar disk. One possibility is that we are seeing large magnetic flux bundles, comparable in size to solar complexes, that are gently rising to the surface and then sinking below the photoshhere. Such processes normally take weeks on the solar surface.

In any case, these variations in magnetic flux must be more convincingly demonstrated and correlated with simultaneously obtained chromospheric diagnostics to better understand the dynamics and effects of magnetic fields on the more active stars.

I am indebted to E. Borra, A. Vaughan, R. Howard, R. Robinson, J. G. Timothy and C. Joseph for sending me unpublished data. I have also profited greatly from many conversations with T. Tarbell, S. Vogt and D. Penrod. I would like to extend my appreciation to G. H. Herbig for his continued support and guidance. Finally, I would like to thank Gayle Parlato for her careful editing and typing of the manuscript. This work was made possible by NASA grant 8407 and NSF grant AST 82-03115.

REFERENCES

Ayres, T.R. and Linsky, J.L.: 1980, Astrophys. J. 241, p. 279.
Ayres, T.R., Linsky, J.L., Vaiana, G.S., Golub, L., and Rosner, R.: 1981, Astrophys. J. 250, p. 293.
Basri, G.S. and Linsky, J.L.: 1979, Astrophys. J. 234, p. 1023.
Beckers, J.: 1969, AFCRL 471.
Billings, D.E.: 1966, "A Guide to the Solar Corona", Academic Press, New York.
Boesgaard, A.M.: 1974, Astrophys. J. 188, p. 567.
Boesgaard, A.M., Chesley, D., and Preston, G.W.: 1975, Publ. Astron. Soc. Pacific 87, p. 353.
Brown, D.N. and Landstreet, J.D.: 1981, Astrophys. J. 246, p. 899.
Chapman, G.A. and Sheeley, Jr., N.R.: 1968, Solar Phys. 5, p. 442.
Clayton, G.C. and Martin, P.G.: 1981, Astron. J. 86, p. 1518.
Deubner, F.-L.: 1976, Astron. Astrophys. 51, p. 189.
Duncan, D.K.: 1981, Astrophys. J. 248, p. 651.
Durney, B.R. and Robinson, R.D.: 1982, Astrophys. J. 253, p. 290.

Galloway, D.J. and Weiss, N.O.: 1981, Astrophys. J. 243, p. 945.
Giampapa, M.S., Worden, S.P., and Linsky, J.L.: 1982, Astrophys. J. 258, p. 740.
Gilman, P.A.: 1981, in "Second Cambridge Workshop on Cool Stars, Stellar Systems, and the Sun", ed. M.S. Giampapa and L. Golub (SAO Special Report No. 392).
Golub, L.: 1981, in "Second Cambridge Workshop on Cool Stars, Stellar Systems, and the Sun", ed. M.S. Giampapa and L. Golub (SAO Special Report No. 392).
Hartmann, L. and MacGregor, K.B.: 1980, Astrophys. J. 242, p. 260.
Hartmann, L.: 1981, in "Solar Phenomena in Stars and Stellar Systems", ed. R.M. Bonnet and A.K. Dupree (Dordrecht: D. Reidel).
Kelch, W.L., Linsky, J.L., and Worden, S.P.: 1978, Astrophys. J. 229, p. 700.
Johnson, H.M.: 1981, Astrophys. J. 243, p. 234.
Landi Degl'Innocenti, E.: 1982, Astron. Astrophys. 110, p. 25.
Leibacher, J. and Stein, R.F.: 1981, in "Second Cambridge Workshop on Cool Stars, Stellar Systems, and the Sun", ed. M.S. Giampapa and L. Golub (SAO Special Report No. 392).
Linsky, J.L.: 1980, Ann. Rev. Astron. Astrophys. 18, p. 439.
Linsky, J.L., Worden, S.P., McClintock, W., and Robertson, R.M.: 1978, Astrophys. J. 41, p. 47.
Marcy, G.W.: 1981, Astrophys. J. 245, p. 624.
Middelkoop, F.: 1982, preprint.
Noyes, R.W.: 1981, in "Solar Phenomena in Stars and Stellar Systems", ed. R.M. Bonnet and A.K. Dupree (Dordrecht: D. Reidel).
Pallavicini, R., Golub, L., Rosner, R., Vaiana, G.S., Ayres, T., and Linsky, J.L.: 1981, Astrophys. J. 248, p. 27.
Parker, E.N.: 1981, in "Solar Phenomena in Stars and Stellar Systems", ed. R.M. Bonnet and A.K. Dupree (Dordrecht: D. Reidel).
Pettersen, B.R. and Hsu, J.-C.: 1981, Astrophys. J. 247, p. 1013.
Preston, G.W.: 1971, Astrophys. J. 164, p. 309.
Renzini, A., Cacciari, C., Ulmschneider, P., and Schmitz, F.: 1977, Astron. Astrophys. 61, p. 39.
Robinson, L.B. and Wampler, E.J.: 1972, Publ. Astron. Soc. Pacific 84, p. 161.
Robinson, R.D.: 1980, Astrophys. J. 239, p. 961.
Robinson, R.D., Worden, S.P., and Harvey, J.W.: 1980, Astrophys. J. Letters 236, p. L155.
Rosner, R., Tucker, W.H., and Vaiana, G.S.: 1978, Astrophys. J. 220, p. 643.
Rosner, R. and Vaiana, G.S.: 1979, in "X-Ray Astronomy", ed. R. Giacconi and G. Setti (Dordrecht: D. Reidel).
Schmitz, F. and Ulmschneider, P.: 1980, Astron. Astrophys. 84, p. 93.
Seares, F.H.: 1913, Astrophys. J. 38, p. 99.
Skumanich, A.: 1972, Astrophys. J. 171, p. 565.
Skumanich, A. and Eddy, J.A.: 1981, in "Solar Phenomena in Stars and Stellar Systems", ed. R.M. Bonnet and A.K. Dupree (Dordrecht: D. Reidel).

Smith, M.A.: 1979, Publ. Astron. Soc. Pacific 91, p. 737.
Soderbloom, D.R., Hartoog, M.R., Herbig, G.H., Mueller, F.S., Robinson, L.B., and Wampler, E.J.: 1978, in "Proc. 4th Int. Coll. Astrophys.", ed. M. Hack (Trieste: Trieste Astrophys. Observ.), p. 449.
Soderblom, D.R.: 1980, Ph.D. thesis, Univ. of Calif., Santa Cruz.
Stein, R.F. and Leibacher, J.W.: 1980, in "Stellar Turbulence", ed. D.F. Gray and J.L. Linsky (New York: Springer-Verlag).
Stein, R.F.: 1981, Astrophys. J. 246, p. 966.
Stenflo, J.O.: 1973, Solar Phys. 32, p. 41.
Stenflo, J.O.: 1975, Solar Phys. 42, p. 79.
Stix, M.: 1972, Astron. Astrophys. 20, p. 9.
Tarbell, T.D. and Title, A.M.: 1976, Solar Phys. 47, p. 563.
Timothy, J.G., Joseph, C.L., and Linsky, J.L.: 1981, Bull. Amer. Astron. Soc. 13, p. 828.
Tinbergen, J. and Zwaan, C.: 1981, Astron. Astrophys. 101, p. 223.
Ulmschneider, P. and Bohn, H.U.: 1981, Astron. Astrophys. 99, p. 173.
Ulmschneider, P. and Stein, R.F.: 1982, Astron. Astrophys. 106, p. 9.
Unno, W.: 1956, Publ. Astron. Soc. Japan 8, p. 108.
Walter, F.M.: 1982, Astrophys. J. 253, p. 745.
Vaiana, G.S., Davis, J.M., Giacconi, R., Krieger, A.S., Silk, J.K., Timothy, A.F., and Zombeck, M.: 1973, Astrophys. J. Letters 185, p. L47.
Vaiana, G.S., et al.: 1981, Astrophys. J. 244, p. 163.
Vaughan, A.H., Baliunas, S.L., Middelkoop, F., Hartmann, L.W., Mihalas, D., Noyes, R.W., and Preston, G.W.: 1981, Astrophys. J. 250, p. 276.
Vogt, S.S.: 1980, Astrophys. J. 240, p. 567.
Vogt, S.S.: 1981, Astrophys. J. 250, p. 327.
Zwaan, C.: 1978, Solar Phys. 60, p. 213.
Zwaan, C.: 1981, in "Solar Phenomena in Stars and Stellar Systems", ed. R.M. Bonnet and A.K. Dupree (Dordrecht: D. Reidel).

DISCUSSION

HARTMANN: Are you not surprised that 61 Cyg A, which is not a very active star and is a slow rotator, has such a large measured magnetic field?

MARCY: The field *strength* is large (3000 G), but the area covering factor is about 20% , so the total surface magnetic flux is moderate. Also, 61 Cyg A, at K5, is the coolest star among the dwarfs in the survey, so it is consistent with the suspicion that magnetic fields increase towards later spectral types, for a given rotational velocity.

ZWAAN: Did you find a relation between covering factor and field strength?

MARCY: It seems that stars with exceptionally small field strengths have the larger covering factors. This may, however, be a selection effect since low strengths are not observable unless the fields cover a significant fraction of the surface.

IONSON: When you speak of "covering factor" it appears that you are referring to the percentage of the star covered by the maximum field. I would expect that the covering factor is a function of the B-field, which we know from solar studies can vary over quite a large range. Can you comment on this?

MARCY: The present data suggest that stars with the larger filling factors have smaller relative field strengths. Should a distribution of field strengths exist on a given star, the analysis of the Zeeman broadening given here will yield some ill-defined average of both the field strength and the covering factor.

SCHÜSSLER: The weak dependence of field strength on spectral type is no evidence against fluxtube models because the peak fields do not only depend on surface pressure but also on the efficiency of lateral heat influx and the depth of the Wilson depression, which determines the geometrical height where the observed lines are formed. Of course, thin fluxtube models may be irrelevant for a surface flux coverage $> 70\%$.

MARCY: I agree. Furthermore, both the Ca II H and K data and the magnetic-field data suggest that significant changes in the surface magnetic flux occur on time scales as short as one day. If so, the observed field strengths may be less related to surface conditions such as photospheric pressure, but more to the conditions and confinement processes at the top of the convection zone.

GIOVANELLI: We observe in magnetic regions on the sun two types of waves, propagating outwards from the feet of the tubes of force. One of these two types is a longitudinal wave. If you look at the center of the disk, for example, it is a wave that oscillates up and down. Call it an acoustic wave or a slow-mode wave, but at any rate it is one of these waves that is propagating along a flux tube. It is later in phase at greater heights. Therefore the wave is propagating outwards. We do not find any evidence of waves propagating inwards. If, however, 25% of the wave energy were propagating inwards, we probably would not see it. So you could have 25% reflection, but you would have 75% going out. And there is enough energy in these waves to cook up the corona. I am talking about periods of 3–5 min. These are the things that we see in Hα and in lines at lower levels.

The second type of wave is Alfvén waves propagating upwards along fibrils. (I am going to show a movie about this later during the Symposium.) Every fibril that you see on the sun in Hα carries these waves, which I at least interpret as Alfvén waves. They carry a sufficient amount of energy to cook up the corona. You never see them propagating backwards.

CRAM: Presumably, the $F(H+K)/T_{eff}^4$ ratio for a set of stars depends almost completely on the area covered by calcium emission. How was the area effect removed from the quoted relations between $F(H+K)/T_{eff}^4$ and $T_{eff}^n B^m$? After the area effect is removed, how much residual correlation remains to determine the indices n and m?

MARCY: The magnetic area covering-factor was in fact included in the least-squares solution to the dependence of $F'(H+K)/\sigma T_{eff}^4$ on T_{eff} and B, and the result was that $F'(H+K)/\sigma T_{eff}^4$ varies as covering factor to the 0.6 power. The somewhat weaker than linear dependence may be interpreted in terms of the overlapping of diverging, dense fluxtubes in the more active stars.

GRAY: I would like to ask you about the effects of line saturation. As I understand your analysis, no radiation transfer is taken into account. Saturation of the lines will systematically affect the values of the filling factor. Could you comment specifically on this?

MARCY: The lines used, λ6173 and λ6240, are weak, having equivalent widths of about 50 mÅ on the sun. However, "equivalent width broadening" is not negligible. Tests with theoretical profiles indicate that the effect of such saturation is significantly smaller than the Zeeman-broadening effect; however, I agree that future magnetic-field measurements should include line-transfer effects.

GRAY: Can you tell us (qualitatively) whether you expect your filling factors to be too large or too small if saturation is significant?

MARCY: The filling factors would be overestimated. This effect becomes important for later-type stars like K5, but should be essentially non-existent for K2.

LINSKY: There are two additional problems in the interpretation of the magnetic data using the Robinson technique. (1) Line blends can lead to false magnetic field strengths and filling factors. This may not be a severe problem for the G stars, but it becomes worse for the K and M stars. (2) The continuum in the magnetic regions could be darker (star spots) or brighter (stellar faculae) than in non-magnetic regions in late-type stars. In either case the derived magnetic-field filling factor would be in error. Since some of the derived filling factors are as large as 80% for the most active stars, I suspect that the continuum brightness in the most active stars is enhanced (stellar faculae), in which case the derived filling factors are upper limits. Could you comment on these two problems?

MARCY: The absorption lines were specifically chosen to be free of blends. But for stars later than K5, blends do become a problem, so we have observed no stars later than that. Regarding the second point, I have emphasized that indeed the derived filling factors are model dependent. Both the facular brightness and the field-geometry will affect the covering factors, while the derived field strengths are much less dependent on such assumptions.

LONG-TERM COHERENT PERIODICITIES IN THE MEAN MAGNETIC FIELD OF THE SUN

V.A. Kotov and L.S. Levitsky
Crimean Astrophysical Observatory
Nauchny, Crimea 334413, USSR

To investigate periodic variations of the magnetic field of the Sun as a star (SMMF; see Severny, 1969) we have used the mean field measurements made in Crimea, Mt. Wilson, and Stanford observatories; in total N = 5783 daily values were available for the time interval 1968 - 1981. In essence, these data offer a unique possibility to study the Sun as a variable magnetic star (Scherrer et al., 1977).

To combine the three data sets into a single time series we corrected the original measurements of each observatory for the systematic calibration error applying correction factors determined via inter-comparison of the three individual power spectra:

a(Cr.) = 0.77, a(M.W.) = 0.94, a(St.) = 1.57.

Then the common harmonic amplitude (A) spectrum was computed with the use of a simple least-squares technique to find the Fourier coefficients.

The power spectrum of the SMMF (Figure 1) exhibits a set of discrete lines within the range of (synodic) periods 27 to 29 d. Along with 5 dominant peaks: P = 26.95, 27.18, 27.34, 27.87, and 28.15 (± 0.08) d, there exist also several reliable peaks corresponding to fast rotating (P = 26.35 and 26.60 d) and slow rotating, higher-latitude structures (periods up to 30.00 d).

The signature of the discrete character of the SMMF spectrum is reinforced by the fact that almost all dominant peaks of the SMMF are well reproduced in the changes of the polarity of the interplanetary magnetic field (IMF, 1926 - 1981; in all 20332 daily values of the polarity (+ 1, - 1, or 0) were used in the analysis). The latter data were taken from Svalgaard (1972) and Mansurov et al. (1978) supplemented with new data now available from in situ measurements in space. A similar data set was used earlier by Svalgaard and Wilcox (1975) for the analysis of the long term evolution of the solar sector structure.

J. O. Stenflo (ed.), Solar and Magnetic Fields: Origins and Coronal Effects, 23–27.

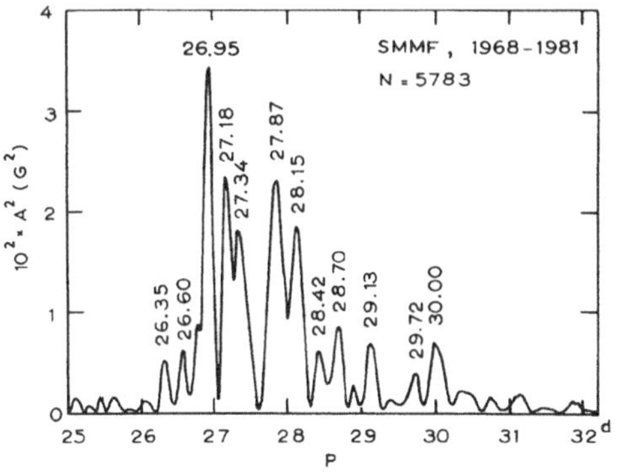

Figure 1. Power spectrum of the mean solar magnetic field.

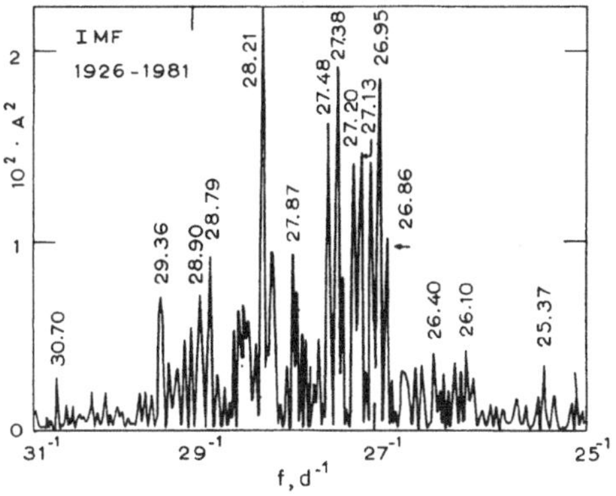

Figure 2. Power spectrum of the IMF polarity.

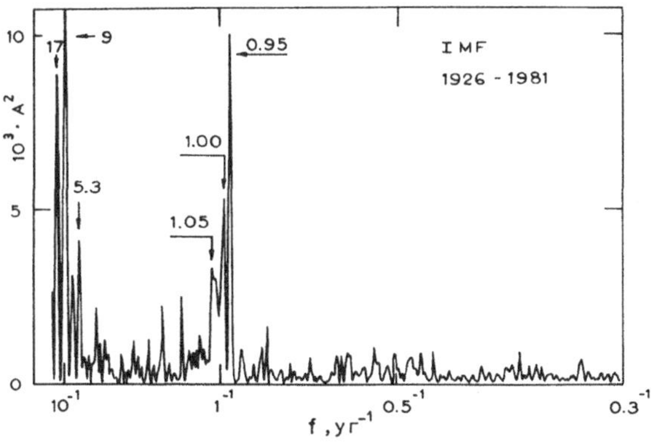

Figure 3. The low-frequency portion of the IMF spectrum.

The complete spectrum of the IMF polarity was computed in 0.11 nanoHz steps from 1.17 to 463 nanoHz, i.e. from 25 d to 27 yr in periods. A small part of the spectrum is shown in Figure 2 where we find nearly the same dominant periods as in the case of the SMMF: 26.95, 27.20, 27.38, and 28.21 (± 0.01) d, with addition of a few more peaks caused by splitting and possible phase-amplitude modulation of the IMF.

The detailed analysis (Kotov and Levitsky, 1982) has shown that the discrete character of the SMMF and IMF spectra cannot be ascribed to a sampling effect or modulation by the 11 yr cycle. The high confidence level of all peaks dominating both spectra (the ratio $2A/\sigma \approx 7$ to 11) rules out also the statistics as explaining the origin of the discrete lines.

It therefore appears that the mean field tends to be surprisingly stable at definite periods of solar rotation retaining the phase-coherency over 14 to 56 yr irrespective of the phase of the 11 yr cycle. This suggests a significant stratification of the long-lived magnetic structures in the solar atmosphere; the picture resembles the latitude stratification clearly seen, e.g., in the Jovian atmosphere. This solar discrete pattern might be directly related to the alternating latitude zones of slow and fast rotation found recently in Doppler measurements by LaBonte and Howard (1982). The mechanism involved to ensure the long-term phase-coherency of the field at some favourable latitudes seems to be a challenging problem for the theory of the solar magnetic field generation and evolution.

In the lower frequency portion of the IMF spectrum (Figure 3) the strongest peaks correspond to the triplet 0.95, 1.00, 1.05 (± 0.01) yr, which is produced mainly by the Rosenberg and Coleman's (1969) effect in the predominant IMF polarity, and 9 (± 0.4) and 17 (± 1) yr. The nature of the two latter peaks is not yet known due to their poor statistical significance.

The power spectrum of the SMMF (not shown here) computed for the period range 0.1 to 6 yr has the highest peak at exactly 1.00 (± 0.02) yr. The superposed epoch plot of the SMMF strength (H) for this period (Figure 4) indicates a remarkable annual modulation with an amplitude of about 0.1 G ($2A/\sigma \approx 6.5$). It was shown previously (Kotov et al., 1981) that this annual wave can be explained neither by instrumental effects and data processing nor by the annual change of the Earth's heliographic latitude. We suppose that some asymmetry of the Sun's global magnetic field in inertial space (with respect to the stars) might be responsible for the appearance of this variation as observed from the Earth orbiting the Sun.

The authors gratefully acknowledge valuable discussions with A.B. Severny and would like to thank also R. Howard, J.M. Wilcox, and

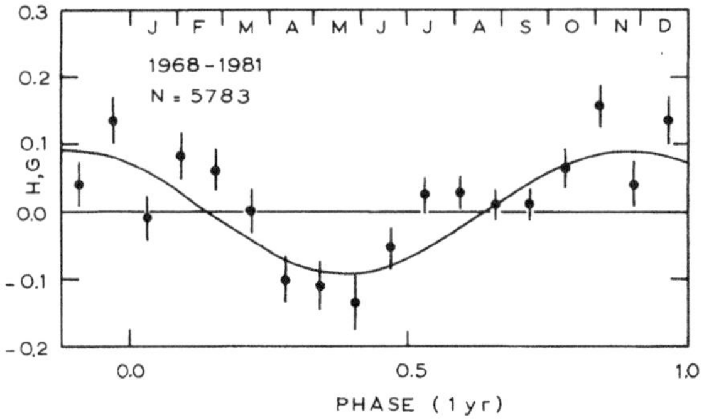

Figure 4. The annual variation of the SMMF. The solid line is the best fitted sinusoid; the vertical bars show double rms errors ($\pm\ \sigma$). The phase is defined to be 0 at 0 UT on 1968 January 1.

P.H. Scherrer for useful comments and for communicating to us their observations of the SMMF prior to publication.

REFERENCES

Kotov, V.A. and Levitsky, L.S.: 1982, Izv. Krymsk. Astrofiz. Obs. 68, in press.

Kotov, V.A., Levitsky, L.S., and Stepanyan, N.N.: 1981, Izv. Krymsk. Astrofiz. Obs. 63, pp. 3-14.

LaBonte, B.J. and Howard, R.: 1982, Solar Phys. 75, pp. 161-178.

Mansurov, S.M., Mansurova, L.G., and Okulova, L.S.: 1978, in "Antarktika. Doklady komissii", Moscow, Nauka, 17, pp. 268-270.

Rosenberg, R.L. and Coleman, P.J., Jr.: 1969, J. Geophys. Res. 74, pp. 5611-5622.

Scherrer, P.H., Wilcox, J.M., Kotov, V., Severny, A.B., and Howard, R.: 1977, Solar Phys. 52, pp. 3-12.

Severny, A.: 1969, Nature 224, pp. 53-54.

Svalgaard, L.: 1972, Dan. Meteorol Inst. Geophys. Pap. R-29, Charlottenlund, pp. 1-36.

Svalgaard, L. and Wilcox, J.M.: 1975, Solar Phys. 41, pp. 461-475.

DISCUSSION

LIVINGSTON: No doubt all of us view a solar periodicity of one year with suspicion. Recently, however, other evidence of annual periods have been found in the full disk Ca K data, namely in the Wilson-Bappu parameter and in the equivalent widths of photospheric lines. Also P. Simon, Meudon, has pointed out that the quiet sun component of the 10 cm radio flux can exhibit a one year period. So it may be that we are discovering here a new component of solar variability.

KUKLIN: I have a comment on the interpretation of the discrete spectrum of the oscillations. One can show that a triplet near 27.2 days may be explained as an amplitude-frequency modulation of the basic harmonic by an oscillation with 9 yr period, and a doublet near 28.3 days may be explained as a frequency modulation with 17.6 yr period. Therefore in the first case the background fields rotating rigidly change during the 11 yr cycle, and in the second case the magnetic fields of active regions are modulated during a magnetic cycle (Hale cycle).

KOTOV: The influence of long-term modulations by 9–22 yr cycles on the 27–29 day variations seems to be a rather complicated problem. For instance, if we attempt to ascribe several discrete lines to a certain phase-amplitude modulation, the question arises why there are no other discrete lines which would be produced by the modulation. In addition, the existence of almost the same discrete lines in the spectrum of the solar mean field data covering only 14 yr does not favour this explanation. However, I agree that there may be other reasonable sources of the spectral pattern observed, besides the latitude-zone hypothesis.

FOING: (1) Can the described data of the solar and interplanetary magnetic fields be used to derive a quantitative topology of the field from the sun to earth? (2) Why is the one year period privileged? Is there an offset in period due to the proper motion of the field lines rooted in the sun's surface?

KOTOV: (1) There is a good correlation between the sector structures observed in the SMMF and IMF polarity patterns. The topology of the heliospheric current sheet and, therefore, of the sector structure, computed on the basis of the observed photospheric field, agrees well with the IMF polarity observed at Earth, during about 80% of the days (cf. papers by Severny, Wilcox, and many others). (2) I think that the nature of an annual variation in the SMMF is still open to discussion. The effect of possible proper motion of the field lines in the photosphere, if any, is perhaps rather small and within the error of the period determination. The exact value of the period in the SMMF is (1.00 ± 0.02) yr.

METHODS AND RESULTS FOR DETECTING MAGNETIC FIELDS ON LATE-TYPE STARS

Mark S. Giampapa
Harvard-Smithsonian Center for Astrophysics

Simon P. Worden
United States Air Force, Headquarters, Space Division

ABSTRACT

We discuss our program to detect and measure magnetic flux on the surfaces of late-type stars. We adopt a novel technique to deconvolve magnetically insensitive lines from similar, magnetically sensitive lines to infer the degree of Zeeman splitting in the latter lines. These measurements yield values for the magnetic field strength and filling factor (flux). To illustrate our approach we present multiple observations of the RS CVn star λ And. At the epoch of observation 26 April 1981 we find a field strength of 1290 ± 50 gauss covering 48 ± 2 percent of this star's surface. Observations at other epochs clearly demonstrate magnetic flux variability on λ And.

1. INTRODUCTION

Quantitative stellar magnetic flux measurements provide essential inputs for theories which describe the origin of stellar magnetic fields and associated chromospheres and coronae. It is believed that magnetic fields arise from an interaction between existing magnetic fields, convection, and differential rotation, the "dynamo" process. To refine accutately nonlinear solar dynamo models, we require detailed comparisons with stellar dynamos. Stellar magnetic field strengths and extents (filling factors) are essential observational delineators of the dynamo process. Direct measures of stellar magnetic fields may also be compared with measures of chromospheric and coronal emission to study the mechanisms involved in the heating and evolution of these stellar atmospheric regions. Magnetic field detection methods, based on detailed analysis of line profiles, enable us to measure directly field strengths and filling factors for active chromosphere stars.

Standard polarization methods for measuring stellar magnetic fields are inappropriate for solar-type field topologies where the field polarities are tangled, and where polarization effects cancel.

J. O. Stenflo (ed.), Solar and Magnetic Fields: Origins and Coronal Effects, 29–34.

Recent results by our group and others have successfully deduced magnetic fields from the shapes of magnetically sensitive line profiles (Robinson, 1980). The Zeeman line splitting pattern is deconvolved from the line profiles by comparing magnetically sensitive lines with similar, but magnetically insensitive, lines. From this procedure we get both estimates of the actual field strengths and the fraction of visible stellar surface covered by the fields. This method has been verified by Robinson, Worden, and Harvey (1980). Recent work in this area is reviewed by Marcy in this volume. In this paper we illustrate applications of this method by presenting one of our results, flux measurements on the RS CVn star λ And.

2. OBSERVATIONS AND RESULTS

We use infrared spectral data in our analyses for two reasons. First, the interesting cool K-M stars have their energy maxima near 2μ. Second, magnetic sensitivity and therefore detectability is greater in the infrared based on the Zeeman splitting relation

$$\Delta_H = 4.7 \cdot 10^{-13}\ g \cdot H \cdot \lambda^2\ \text{Å}\ , \tag{2.1}$$

where H is the field strength in gauss, g is the Landé g factor, and Δ_H is the separation of each σ component from the cental π component. Although the splitting is proportional to λ^2, the intrinsic absorption line width is proportional to λ, so magnetic detectability is only proportional to λ.

Our data were obtained with the Kitt Peak National Observatory 4-m Mayall reflector and Fourier Transform Spectrometer (FTS). We record a bandpass of 0.32μ centered at 1.65μ (6178cm^{-1}). A large number of suitable spectral lines has been identified with effective g factors between 0 and 3. The spectral resolution of this data is 0.1 cm^{-1}.

We observed 19 late-type stars during 1981, many of them several times. Complete analysis for these objects is not yet complete, thus we provide here only our results for λ Andromedae to demonstrate our approach. This star is especially interesting since it is a G8 IV-III RS CVn-type system with variable chromospheric emission lines enhanced by factors of 10-100 over corresponding lines in the sun (Linsky *et al.*, 1979; Baliunas and Dupree, 1982). This star's X-ray luminosity is $L_x \sim 10^{30}$ ergs^{-1} or 10^3 times the solar value (Walter *et al.*, 1980; Swank *et al.* 1981). Periodic broadband optical fluctuations have been observed on λ And which are attributed to starspots covering 30% or more of the stellar surface (Eaton and Hall, 1979). λ And is bright in the infrared with an H magnitude of 2.0.

Figure 1 shows profiles of the magnetically sensitive (g=3.0) FeI line at 6388.65 cm^{-1} for λ And and for a quiet chromosphere star, α Tau (K III). The profile for the latter star was artificially broadened by convolution with a gaussian to account for differences in nonmagnetic broadening between the two stars. We estimated the necessary broadening by comparing absorption line profiles for the two stars which had $g \leqslant 1.0$. The FTS data provide a direct estimate of S/N since the spectrum is recorded simultaneously by two detectors. From the differences between the two signals we deduce that the λ And data has a normalized RMS noise of 0.023, and α Tau a RMS noise of 0.005.

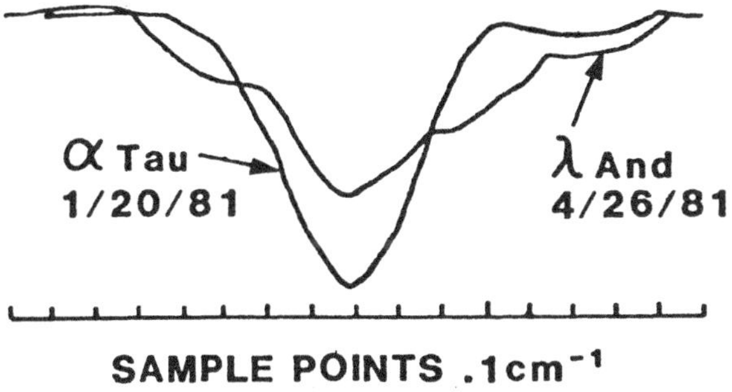

Figure 1. Plot of α Tau and λ And 6388.65 cm^{-1} line profiles. α Tau has been artificially broadened to a mean λ And value.

The line profile seen in the spectrum of λ And is broadened with respect to the comparison (α Tau) profile. The σ components are evident and partially blended with the central component, which is composed of the π component and a component profile from the nonmagnetic regions of the stellar surface.

3. ANALYSIS AND DISCUSSION

The intensity profile of a line produced in a region of uniform magnetic field may be represented by (Title and Tarbell, 1975)

$$I(\lambda) = A\cdot[Z(\lambda-\Delta_H) + Z(\lambda+\Delta_H)] + B\cdot Z(\lambda) , \qquad (3.1)$$

where $Z(\lambda)$ is the intensity profile of the unsplit line and A and B are constants given by Babcock (1949)

$$A = {}^1/_4\ (1 + \cos^2 \gamma\ ) \quad ; \qquad B = {}^1/_2\ \sin^2 \gamma\ , \qquad (3.2)$$

with γ the angle between the line of sight and the field orientation. Equation 3.1 is the convolution of the unsplit profile, $Z(\lambda)$, with a triple impulse function which parametizes the splitting. The magnetically sensitive stellar line profile is the combination the profiles from nonmagnetic (quiet) and magnetic (active) regions. The flux profile of such a two-component model is

$$M(\lambda) = (1-f)\cdot Q(\lambda) + A\cdot f\cdot[Z(\lambda-\Delta_H) + Z(\lambda+\Delta_H)] + B\cdot f\cdot Z(\lambda) \quad , \qquad (3.3)$$

where $Q(\lambda)$ is the flux profile of the unsplit line and f is the fraction of stellar surface covered by fields (filling factor). Upon

Fourier transforming with k as the transform variable, and applying the Fourier shift theorem

$$g(k) = (1-f) + {}^{M}/_{Q} \cdot f \cdot [B + 2A \cdot \cos(\Delta_H \cdot k)] \ , \tag{3.4}$$

where M/Q is the ratio of the relative depths of the central components of the unsplit profiles. For complex multi-component atmospheres, the profile is a summation of each component's profile, but we have characterized λ And by a two-component atmosphere. Averaging equation (3.4) over all possible line-of-sight angles yields the following:

$$g(k) = (1-f) + {}^{1}/_{4} \cdot {}^{M}/_{Q} \cdot f \cdot [1 + 3 \cdot \cos(\Delta_H \cdot k)] \ , \tag{3.5}$$

We performed the indicated Fourier deconvolution with the profiles shown in Figure 1, assuming the broadened α Tau profile to be a suitable comparison line. We used a χ^2 fitting routine to find best fit values of f, M/Q, and Δ_H in equation (3.5). From this fit we find a field strength of 1290 ± 50 G covering 48 ± 2 % of the visible hemisphere of λ And. We obtained multiple observations of λ And during 1981 and show the 6388 cm^{-1} line from these spectra along with the comparison α Tau data in Figure 2. We were able to derive magnetic field values for only the 26 Apr 1981 data. This result clearly suggests that the magnetic flux on λ And is variable.

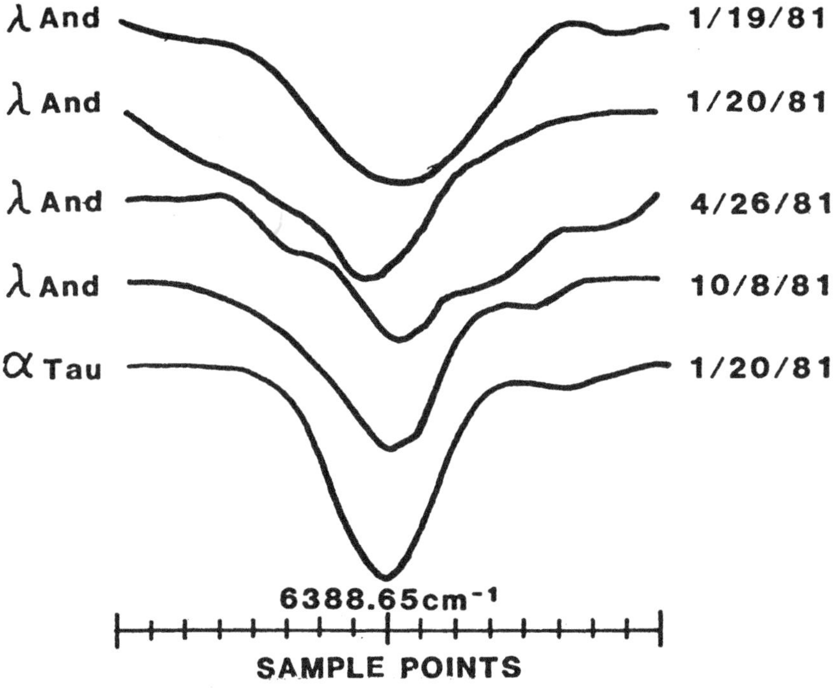

Figure 2. λ And and α Tau FeI 6388.65 cm^{-1} line profiles for 1981, the α Tau profile has been broadened to match λ And.

We can make a simple comparison of our result to magnetic flux estimates expected from equipartition arguments, which suggest that magnetic pressure ($H^2/8\pi$) should roughly equal the ambient gas pressure. If we adopt the photospheric model of Bell et al. (1976) with T_{eff} = 4500°K, log g = 3.0, and metallicity of - 0.5, we derive a gas pressure of Pg = 4.58 • 10^4 dynes cm^{-2} at a Rosseland mean optical depth τ_{ross} = 1. This gas pressure implies H = 1073G, in good agreement with our measurement.

An alternate estimate of magnetic flux for λ And may be based on scaling laws found in solar coronal studies (Golub et al., 1980) which relate magnetic field strength to coronal base pressures and loop lengths. By combining this information with scaling laws given by Rosner et al. (1978), we can predict magnetic field strengths if the X-ray luminosity and coronal temperature are known. Based on information supplied to us by Golub (1982) we estimate a mean coronal field of 650 G with a filling factor of unity, although the actual filling factor is probably larger than unity due to the large surface area of the loop structures. This estimate for the mean field does not include the effects of different turbulent velocities for λ And as compared to the sun, which would tend to increase the field strength estimate for λ And. Nonetheless, this estimate for coronal field strength is consistent with a photospheric field of 1290 G covering about 50% of the stellar surface.

REFERENCES

Babcock, H.W.: 1949, Astrophys. J. 110, pp. 126-142.
Baliunas, S.L. and Dupree, A.K.: 1982, Astrophys. J. 252, pp. 668-680.
Bell, R.A., Eriksson, K., Gustafsson, B., and Nordlund, A.: 1976, Astron. Astrophys. Suppl. 23, pp. 37-95.
Eaton, J.C. and Hall, D.S.: 1979, Astrophys. J. 227, pp. 907-922.
Golub, L.: 1982, private communication.
Golub, L., Maxson, C., Rosner, R., Serio, S., and Vaiana, G.S.: 1980, Astrophys. J. 238, pp. 343-348.
Marcy, G.W.: 1981, Astrophys. J. 245, pp. 624-629.
Robinson, R.D.: 1980, Astrophys. J. 239, pp. 961-967.
Robinson, R.D., Worden, S.P., and Harvey, J.W.: 1980, Astrophys. J. Letters 236, pp. L155-L158.
Rosner, R., Tucker, W.H., and Vaiana, G.S.: 1978, Astrophys. J. 220, pp. 643-655.
Swank, J.H., White, N.E., Holt, S.S., and Becker, R.H.: 1981, Astrophys. J. 246, pp. 208-214.
Title, A.M. and Tarbell, T.D.: 1975, Solar Phys. 41, pp. 255-269.
Walter, F.M., Cash, W., Charles, P.A., and Bowyer, C.S.: 1980, Astrophys. J. 236, pp. 212-218.

DISCUSSION

SCHÜSSLER: Could you comment on your determination of an equipartition field, and did you take into account the Wilson depression of a fluxtube?

WORDEN: When we spoke of "equipartition" we were comparing magnetic energy density ($H^2/8\pi$) with thermal energy density (nkT) rather than with a kinetic energy density based on turbulent motions within the fluxtubes ($\frac{1}{2}\rho v^2$). We did not take into account any peculiarities with the physics within the fluxtube. Once again, we used the pressure (or thermal energy density) at Rosseland mean optical depth unity in a representative photospheric model for λ And.

VOGT: (1) It seems that magnetic-field measures in the visible can go to much fainter objects, faster than your IR technique, even though your telescope aperture is much larger (4 meter vs. 24" for Marcy's work). Can you comment on this large discrepancy in detection efficiency? (2) Have you observed any of the stars for which Marcy measures fields, and are you in agreement with these results? (3) What are your blending problems with the lines selected?

WORDEN: (1) The IR Fourier Transform Spectrometer is a single channel device as well as a relatively wide spectral bandpass instrument. Thus, the efficiency on a single line is low. Moreover, the quantum efficiency of IR detectors is being improved to that of visible detectors. Our methods are greatly enhanced by using many lines, thus the broad spectral bandpass is required. But perhaps the best reason for suffering the lower efficiency in the infrared spectral region is that these spectra represent the energy maxima of the cooler K and M stars. For this reason the IR is much more efficient and indeed the only way to obtain adequate spectral resolution for the interesting dMe stars. Furthermore, due to increased magnetic sensitivity, lower spectral resolution is needed for the IR. Thus with improvements in IR detection, and keeping in mind the broad spectral bandpass we need, the IR is potentially more efficient than the visible for studying magnetic fields on later-type stars. (2) We have some stars in common with Marcy, e.g. the star for which results are presented here, λ And. I believe that Marcy's results for this star are similar within the limits of the errors. It is also important to make sure that the levels in the stellar atmosphere which the observations represent are comparable. Marcy's results, based on visible spectra, probably represent higher levels, especially for the cooler stars. This may explain why Marcy's results for the later-type stars tend to show lower field strengths than equipartition arguments suggest, because the field lines have spread with height. Clearly further intercomparison of IR and visible results are needed. (3) Blending problems in the IR are due primarily to terrestrial atmospheric lines. We have carefully avoided these for our line choices. The IR is considerably better than the visible for the later-type stars in avoiding blends. We have obtained extremely high resolution and high S/N solar data for quiet sun and sunspot to verify that our lines are unblended. In addition to proving that our lines are unblended, the Fourier reduction method is highly insensitive to blends except in the unlikely event that line blends are exactly evenly spaced in opposite wings of the absorption line being used.

SODERBLOM: In examining your profiles of λ And, I notice that the profiles are sampled at fairly wide intervals — maybe 5 or 6 points for the FWHM. This is a very broad-lined star; can one sample at much more closely spaced intervals needed for the slowly rotating late-type dwarfs?

WORDEN: Our resolution is 0.1 cm^{-1}, which is sufficient to oversample all but the narrowest stellar lines in the 5000–7000 cm^{-1} region. The 4-meter telescope Fourier Transform Spectrometer can be used to obtain resolutions of up to about 0.02 cm^{-1}, but at a corresponding penalty in integration time.

ZEEMAN EFFECT OBSERVATIONS IN THE RS CVn STAR HR 5110 AND IN LATE-TYPE STARS WITH STRONG Ca II EMISSION

Walter K. Bonsack and Theodore Simon
Institute for Astronomy
2680 Woodlawn Drive
Honolulu, Hawaii 96822 USA

1. INTRODUCTION

Although magnetic fields are presumed to be of substantial importance in the physics of stellar chromospheres (e.g., Vaiana et al. 1981), firm detections of magnetic fields in stars later than type A are rare, if not entirely absent. For example, Boesgaard (1974) made careful photographic measurements of longitudinal fields in a number of dwarf stars, and obtained 5 σ or greater detection of fields at the level of 100 G in the G8 V stars in ξ Boo A and in the KO V star 70 Oph A. Robinson et al. (1980) studied ξ Boo A by means of an analysis of the widths of magnetically sensitive lines, and obtained a mean field magnitude of 2600 G. However, Marcy (1981) used the same technique and found no significant field in this star. Brown and Landstreet (1981) looked for longitudinal fields in a variety of late type stars with an adaptation of a Griffin-type radial velocity spectrometer, which yielded standard errors near 10 G. They found no fields in any of the stars which they surveyed, including ξ Boo A, 70 Oph A, and a number of RS CVn stars and other stars with strong Ca II K and He I 10830 A emission.

Because of the wide range of possible field strengths in late stars suggested by indicators of chromospheric activity, further measurements remain of interest. The RS CVn stars are of particular interest in this context, since the evidence for "star spots" (e.g., Eaton & Hall 1979) suggests that their magnetic fields may be more simply organized than in the sun, and such evidence as flaring, polarized radio emission (Owen & Gibson 1978; Brown & Crane 1978), and strong X-ray (Walter et al. 1978) and UV emission (e.g., Simon et al. 1980a) suggest the possibility of quite strong magnetic fields. We report, therefore, the results of photographic measurements of the longitudinal Zeeman effect in the spectra of a number of late-type selected from the list of Bopp and Talcott (1978), plus a series of spectrograms of the RS CVn binary system HR 5110. The observations of Bopp and Talcott's stars were made with the 2.2-m telescope of the University of Hawaii at Mauna Kea Observatory in 1979; the plates of

J. O. Stenflo (ed.), Solar and Magnetic Fields: Origins and Coronal Effects, 35–39.

HR 5110 were obtained by P. S. Conti with the Lick Observatory 3-m telescope in 1967-68. We thank Dr. Conti for permitting us to use his plates. These observations have, at best, uncertainties three times larger than those attained by Brown and Landstreet (1981), but since there are no stars in common to the two programs, the present results constitute a significant addition to the available data.

2. OBSERVATIONS AND RESULTS

Table 1 lists the stars observed and the numerical results. The first three columns are self-explanatory. The fourth gives the heliocentric Julian Date of midexposure; the observations at HJD < 2,440,000 are Conti's Lick spectrograms; the remainder are Mauna Kea plates. The next two columns give the results: the effective longitudinal field and the radial velocity, each with its standard error. The last digit of the error is of the same significance as the last digit of the value to which it refers. The final column gives the phase for HR 5110 computed from the ephemeris of Burke et al. (1980). HR 5110 is the only RS CVn system observed; the remaining stars, except for the standard dwarf beta Vir, were selected by Bopp and Talcott (1978) as having strong Ca II emission.

Table 1. Observations and Results

HD	Name/HR	Spect.	HJD 2400000+	H_{eff}(s) gauss	V_{rad}(s) km/s	phase
118216	5110	F2 IV	39630.714	+4(61)	+11.6(1)	0.98
118216	5110	F2 IV	39636.756	+50(64	-1.5(2)	0.29
118216	5110	F2 IV	39874.006	-228(89)	+5.4(4)	0.08
118216	5110	F2 IV	39955.866	-40(22)	+2.1(2)	0.40
118216	5110	F2 IV	44029.893	+103(106)	-0.3(2)	0.44
95212	4280	K5 III	44030.787	-126(60	+2.7(6)	
102870	β Vir	F8 V	44028.769	-270(105)	+6.0(3)	
161797	μ Her	G5 IV	44028.902	-64(70)	-14.8(3)	
163770	θ Her	K1 IIa	44029.974	-1(62)	-35.3(6)	
173764	β Sct	G4 IIa	44028.950	-74(92)	-8.3(4)	

The equipment used for the observations was described by Preston and Pyper (1965) for Lick Observatory, and by Wolff and Bonsack (1972) for Mauna Kea Observatory. The measurement and data reduction were carried out as described by Wolff and Bonsack. The results given are means over all of the lines measured on each spectrogram, and the standard errors of the mean are based upon the differences between the results of individual lines. The line list contained principally Fe and other metals; strong, saturated lines and apparently blended lines were avoided. The lists for the Lick spectrograms contained approximately 120 lines in the range $0.38 < \lambda < 0.46$ μm, while that for the Mauna Kea plates contained 70 lines in the range $0.45 < \lambda < 0.55$ μm. The errors given in Table 1 are similar to those obtained in extensive previous measurements of sharp-line Ap stars.

It will be seen at once that none of the derived values of H_{eff} are as large as three times their standard error. We conclude, therefore, that we have made no reliable detections of magnetic fields, but we offer the data as representing upper limits to the possible net longitudinal fields in these stars at the epochs of the observations.

3. DISCUSSION

HR 5110. The spectroscopic orbit of this binary system was discussed most recently by Conti (1967), who found the system to consist of an F2 IV primary and a G-K secondary with a mass ratio near 0.3. Hall (1976) includes HR 5110 in his list of RS CVn variables, although HR 5110 in many ways resembles an Algol system, e.g., in its mass ratio and X-ray properties (Simon et al. 1980b). HR 5110 underwent a radio outburst in late 1979 May (Feldman 1979), but Simon et al. (1980b) observed that C IV and other high-temperature transition region lines were not enhanced, and these authors suggested that the radio outburst may have been the result of a binary mass-transfer event, rather than a magnetically-driven flare.

The Zeeman-effect observations presented in Table 1 cover reasonably well one-half of the orbit. Thus, all of the surface was sampled by these observations, except for an antipodal region which is permanently hidden as a result of the low inclination (Conti 1967). No significant magnetic field was detected; an upper limit of 100 G is suggested. The radial velocities from the 1967-68 Lick spectrograms agree well with the mean curve given by Conti based upon observations made in 1965-66. The Mauna Kea spectrogram, obtained in 1979, gives a somewhat deviant velocity, which would agree with the curve if the phase were reduced by 0.09. No large systematic difference has previously been noted between Mauna Kea and Lick coude velocities. The Mauna Kea spectrogram was obtained about one week after the onset of the radio flare; no magnetic effects were detected, which is in accord with the Simon et al. (1980b) suggestion regarding the origin of the flare. However, all such effects may have subsided in the week since the flare began.

Other Stars. Three of the stars observed are included in Wilson's (1976) catalog of Ca II emission line widths and strengths, in which he estimates strengths on a scale 0-5. μ Her rates 0 on this scale, β Sct, 2, and θ Her, 3. HR 4280 is included in Joy and Wilson's (1949) list of stars with Ca II emission, where its emision strength is given as 1 on a scale which apparently extends to 50. μ Her gives no other indications of an active chromosphere: Hartmann et al. (1982) find no detectable UV C IV emission, and Vianna et al. (1981) find a relatively weak X-ray luminosity compared to other G stars. θ Her has strong and variable emission at He I 10830 A as well as the K line (Zirin 1976) and Reimers (1982) find it to be a "hybrid" star, with significant chromospheric C IV emission, and a cool wind. However, recent IUE data shows no detectable C IV (Simon unpublished).

β Sct has apparently not been observed in the UV or X-ray regions. The remaining stars β Vir, is an F8 V standard which was observed to calibrate exposure times. It has not been noted to have Ca II emission, and a low Li abundance (Duncan 1981), low rotation (Smith 1978), and low X-ray luminosity (Pallavicini et al. 1981) are in accord in indicating advanced age.

The above stars cover a range of luminosities and a range of chromospheric activity. One active RS CVn system has been well sampled. None show significant net longitudinal magnetic fields greater than 100 G. It is clear that it is not possible to distinguish among the categories of late-type stars by means of observations of the longitudinal Zeeman effect using the photographic technique, and the results of Brown and Landstreet (1981) indicate that modern techniques of higher precision lead to the same conclusion.

REFERENCES

Boesgaard, A.M.: 1974, Astrophys. J. 188, p. 567.
Bopp, B.W. and Talcott, J.C.: 1978, Astron. J. 83, p. 1517.
Brown, D.N. and Landstreet, J.D.: 1981, Astrophys. J. 246, p. 899.
Brown, R.L. and Crane, P.C.: 1978, Astron. J. 83, p. 1504.
Burke, E.W. Jr., et al.: 1980, Astron. J. 85, p. 744.
Conti, P.S.: 1967, Astrophys. J. 149, p. 629.
Duncan, D.K.: 1981, Astrophys. J. 248, p. 651.
Eaton, J.A. and Hall, D.S.: 1979, Astrophys. J. 227, p. 907.
Feldman, P.A.: 1979, IAU Circular No. 3366.
Hall, D.S.: 1976, in "Multiple Periodic Phenomena in Variable Stars", Part I, ed. W.S. Fitch (Dordrecht, Reidel Publishing Co.) p. 287.
Hartmann, L., Dupree, A.K., and Raymond, J.C.: 1982, Astrophys. J. 252, p. 214.
Joy, A.H. and Wilson, R.E.: 1949, Astrophys. J. 109, p. 231.
Marcy, G.W.: 1981, Astrophys. J. 245, p. 624.
Owen, F.N. and Gibson, D.M.: 1978, Astron. J. 83, p. 1488.
Pallavicini, R., et al.: 1981, Astrophys. J. 248, p. 279.
Preston, G.W. and Pyper, D.M.: 1965, Astrophys. J. 142, p. 983.
Reimers, D.: 1982, Astron. Astrophys. 107, p. 292.
Robinson, R.D., Worden, S.P., and Harvey, J.W.: 1980, Astrophys. J. 236, p. L155.
Simon, T., Linsky, J.L., and Schiffer, F.H. III: 1980a, Astrophys. J. 239, p. 911.
Simon, T., Linsky, J.L., and Schiffer, F.H. III: 1980b, in "The Universe at Ultraviolet Wavelengths", NASA Conference Publication 2171, p. 435
Smith, M.A.: 1978, Astrophys. J. 224, p. 584.
Vaiana, G.S., et al.: 1981, Astrophys. J. 245, p. 163.
Walter, F., Charles, P., and Bowyer, S.: 1978, Astron. J. 83, p. 1539.
Wilson, O.C.: 1976, Astrophys. J. 205, p. 823.
Wolff, S.C. and Bonsack, W.K.: 1972, Astrophys. J. 176, p. 425.
Zirin, H.: 1976, Astrophys. J. 208, p. 414.

DISCUSSION

BASRI: I would like you to clarify what you mean by "organized magnetic fields". Do you mean that there should be more of one polarity on an entire hemisphere of the star?

BONSACK: By "organized magnetic fields" I mean a field appproaching the simple dipolar structure apparently present in early-type stars, rather than a solar-type configuration. In such a dipolar field, the observed hemisphere is usually dominated by a single polarity.

UCHIDA: Your sample did not include such active objects as UX Ari or V 711 Tau, which show coronae and flaring activities indicating the presence of strong magnetic fields. Is there any observational reason for this?

BONSACK: These stars were not available at the time of this single observing session. It clearly would be of interest to extend the observations. Note that Brown and Landstreet observed several additional RS CVn systems (including UX Ari and V 711 Tau) with a claimed sensitivity of about 10 G and found nothing.

WALTER: HR 5110 is not a particularly good choice to use to claim that all RS CVn systems fail to show "organized" magnetic structure. Most of the light in this system is from the F2 IV; the active star is considerably less luminous, and is the star likely to have the strong magnetic fields. Better choices to test the hypothesis may be II Peg, as used by S. Vogt, or HR 7275.

BONSACK: We make no claim about systems which have not been observed. The primary of HR 5110 is not the most active star, but strong magnetic fields may link both stars, as suggested by Uchida et al. (*PosterPaper*). Other stars have been observed by others with null results, including II Peg by Vogt, and several by Brown and Landstreet (see answer to Uchida's question).

THEORY OF PHOTOSPHERIC MAGNETIC FIELDS

H.C. Spruit
Max-Planck-Institut für Physik und Astrophysik
Institut für Astrophysik
Karl-Schwarzschild-Str. 1
D-8046 Garching, FRG.

In this review I discuss from a theoretical point of view the magnetic fields seen at the solar (stellar) surface. Since magnetic field lines have no ends, and the photospheric fields are mostly vertical, the discussion necessarily includes some of the properties of the fields above and below the photosphere. A more general discussion of the theory of solar magnetic fields can be found in Priest (1982).

1. GENERAL STRUCTURE

An important observed property is that at each length scale L the typical time scale τ on which the field changes is long compared with the Alfvén travel time L/v_A. This means that the field is close to magnetostatic equilibrium:

$$-\nabla p + \frac{1}{4\pi}(\nabla \wedge B) \wedge B + \rho g \approx 0 \qquad (1)$$

where p, B, ρ, g are the gas pressure, field strength, density and gravity. Except for occasional more dramatic events the field evolves slowly in time. Superimposed on it there may be wave-like disturbances. The solution of (1) for a field in a stellar atmosphere, in which the density decreases rapidly with height, is a nonlinear problem for which no analytic methods seem to be available. Numerical solutions were obtained by Gabriel (1977), and Schmidt and Wegmann (1982). Nevertheless, a qualitative picture of the field configuration above the photosphere can be drawn easily (Figure 1, Spruit 1981a). The photospheric field, divided into isolated tubes, fans out with height. The fields merge at a height h_m, given approximately by

$$h_m = 2H \ln B_p/\bar{B} , \qquad (2)$$

where H is the scale height of the atmosphere, B_p the field strength in the tube at the photospheric level, and $\bar{B}$ the horizontally averaged field strength. Above the merging level, the plasma is everywhere magnetized, this is the "coronal" regime. In general, the magnetic field dominates over the gas pressure in this regime ($P << B^2/8\pi$). Below it is the "flux

J. O. Stenflo (ed.), Solar and Magnetic Fields: Origins and Coronal Effects, 41–60.

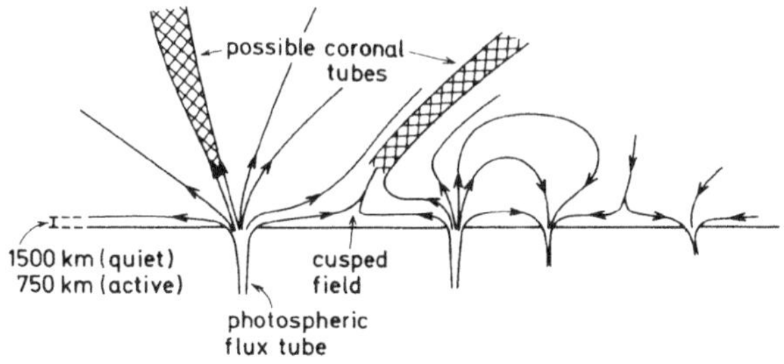

Fig. 1 Structure of the magnetic field near the photosphere (schematic)

tube" regime where the field is confined to flux tubes, separated by field free fluid. Here the gas pressure is of the same order as the magnetic energy density or larger. For the sun the merging height is expected to range from 700 to 1500 km. Observations (Giovanelli, 1980, Jones and Giovanelli, 1982, and this volume) indicate somewhat lower values. It is not quite clear what the cause of this discrepancy is (see however Anzer and Galloway, 1982).

2. FORMATION OF FLUX TUBES

The question naturally arises why there exists such a clear distinction, at least at the solar surface, between field free regions and concentrated patches of field. Secondly, why does the field in the flux tubes have the rather high value $B \simeq 1500$ G (at $z = 0$)? To the first of these questions we can answer, on the most heuristic level, that this is the way in which the field tries to minimize its interaction with the turbulent flow in the convection zone. In the absence of any backreaction of the field on the turbulent flow, and in the absence of ohmic diffusion, the field would very rapidly become tangled-up. Both the magnetic energy density $B^2/8\pi$ and the intermittency ("patchiness") would then increase exponentially in time with a time scale on the order of a few turnover times. This was shown, for example, by Kraichnan's (1976) numerical studies. Still assuming that the effects of magnetic diffusion are small (more about this below), it is clear that the Lorentz forces become important and start interfering with the tangling-up process within a short period of time. Thus they have to be included in any realistic treatment of MHD turbulence. Numerical simulations of MHD turbulence were done by Pouquet and Patterson (1978), Meneguzzi, Frisch and Pouquet (1981) and, in the context of a realistic model for granulation, by Nordlund (this volume). Such simulations confirm the tendency of the magnetic field to form intermittent structures ("spongy" distribution of field lines between turbulent elements). The most extensive numerical calculations on the formation of flux tubes are those by Weiss and coworkers (Weiss, 1981a,b, Peckover and Weiss, 1978, Galloway and Moore, 1979, and references therein, see also Proctor, this volume). In these studies, magnetoconvection in a Boussinesq fluid is calculated in a 2-dimensional (plane or cylindrical) geometry. The calculations follow the development of a single convective cell. For a given strength of the convective forces (measured by the Rayleigh number) the results depend mostly on the average field strength across the cell (measured by the Chandrasekhar number) and the magnetic Prandtl number $\sigma = \nu/\eta$ where ν and η are the viscosity and the magnetic diffusivity. Let us define the equipartition field strength B_e as the value at which the energy density in the field is the same as the energy density stored in the unstable thermal stratification. Equivalently, it is the value for which the Alfvén speed is equal to the typical convective flow speed in the absence of a field. If the initial field is stronger than about Be, the convective forces are not strong enough to form an overturning cell (Cowling, 1976); instead, convection with a reduced efficiency occurs in the form of an oscillating flow, provided that the thermal diffusivity κ is larger than η. This mode of energy transport may be important in the strong field of a sunspot umbra (Meyer et al 1974). The average field strength in an active region is well below the critical value B_e (which is about 600G for the top of the convection zone). Overturning convection occurs easily. The numerical experiments then show the formation, in about one turnover time, of a flux rope in which the fluid is stagnant, and a convective cell containing a tangled field which vanishes on a diffusive time scale $\tau_d = d^2/\eta$, where d is the size of the cell. The field strength B_m in the tube is approximately (Galloway et. al 1978)

$$B_m \sim B_e(\nu/\eta)^{1/2}(\ln Rm)^{-1/2} \qquad (3)$$

where $R_m = Ud/\eta$ is the magnetic Reynolds number, and U the typical convective flow speed. Clearly, the field strength obtained depends rather critically on the values of ν and η used in the calculations. These are usually interpreted as turbulent diffusivities that mimick the effects of motion on small scales that are not resolved numerically. In order to get a field strength of the order 1500G , (3) shows that one needs $\nu/\eta \sim 25$, for ln Rm = 2, and more for higher R_m. Estimates of ν and η are obtained using some statistical theory of turbulence. The simplest estimates (short-correlation time or first order smoothing approximations) yield $\nu = \eta$ (see e.g. Moffat, 1978). More elaborate theories yield widely diverging values and no convincing estimates seems to be available (see also section 7). Taking the simplistic value $\nu = \eta$ would $B_m \sim B_e$, too low to explain the observed fields. Thus, though the calculations demonstrate the process of flux tube formation and show that field strengths different from B_e can be obtained, they do not yield a value that can be compared directly with observations.

3. FLUX TUBES IN A STRATIFIED ATMOSPHERE

Accepting that flux tubes exist (from the above, or from observations) we can go a step further and study their properties under astrophysically more realistic conditions. This is possible in the so-called thin-tube-approximation (Defouw , 1976, Roberts and Webb, 1978, Spruit, 1981b). In this approximation one assumes that the component of the field perpendicular to the tube axis is small compared with the parallel component. The tube is thus approximated by a strand of field lines whose thickness varies slowly along the length of the strand. In an exponential atmosphere with scale height H a vertical flux tube expands with height h approximately as exp(h/4H). For the thin tube approximation to be valid, it is therefore necessary that at all heights considered the radius of the tube is less than 4H. For the small scale field of the sun this is satisfied roughly up to the temperature minimum or lower chromosphere. Details of the internal structure of the tube are ignored by characterizing the field strength by a single function of time t and length l along the tube:

$$B = B(l,t)\hat{l} , \tag{4}$$

where $\hat{l}$ is the unit vector along the tube. Since the tube is thin we may assume that during the subsequent evolution of the tube lateral pressure balance is maintained at all times:

$$P_i(l,t) + B^2(l,t)/8\pi = P_e(z) . \tag{5}$$

In this approximation the MHD equations become a set of equations in one spatial dimension (Spruit 1981b,c). Due to this simplification, compressibility and stratification can be taken into account realistically without invoking turbulent diffusivities.

3.1 Convective collapse

As an example, we first consider the stability of a tube, embedded vertically in a convective envelope. Suppose that in its equilibrium state the tube is vertical, and in temperature equilibrium with its surroundings:

$$T_{io}(z) = T_e(z), \tag{6}$$

then the pressure scale height in the tube is at each level equal to that of the surroundings (except for a small difference due to the effect of pressure on the molecular weight μ). In the equilibrium state the ratio of gas to magnetic pressure, $\beta_o = 8\pi P_{io}/B_o^2$ is therefore a constant independent of z. If the superadiabaticity of the stratification $T_e(z)$ is sufficiently strong, the tube is unstable to flows along the tube. During this flow the tube expands and contracts so that its field strength changes, in such a way that (5) remains satisfied. This instability was proposed by Nordlund (1976) and Parker (1978) as a mechanism to increase the field strength of the tube to the observed values. It has been studied in detail by Roberts and Webb (1978), Spruit and Zweibel (1979), Spruit (1979) and Unno and Ando (1979). For a stratification corresponding to the solar convection zone, the tube is unstable to adiabatic perturbations if $\beta_o > 1.8$ (Spruit and Zweibel 1979). This means that in a tube with a field strength less than 1200G (at $\tau = 1$ inside the tube) a downflow will occur that empties the tube and increases its field strength. Since at large field strengths tubes are stable, the process is self-limiting, and transforms the tube into a new equilibrium at a higher field strength (convective collapse). Starting with an initial β_o of 6.7 (corresponding to a field strength equal to the equipartition value) the final field strength (again at $\tau = 1$ inside the tube) is 1800G (Spruit 1979). These calculations show that the convective concentration of an initially weak field does not stop at the equipartition field strength. Though at this value the tube is strong enough to resist the external flow, and thus is expelled from the convective cell by the mechanism described in the previous section, it is still unstable to a convective downflow inside the tube. This effect is due to compressibility and the strong stratification of the atmosphere; it would not occur in a Boussinesq fluid.

The evolution of a magnetic field in the solar granulation was demonstrated in a spectacular numerical simulation by Nordlund (report in this volume). His results show the formation of flux tube - like structures between granules with field strengths close to the observed values.

Calculations of the collapse process for convection zones other than the sun's are not available yet. We may speculate however that the ratio of the final field strength Bf in the tube to the equipartition value B_e is a constant, since both values are determined by the degree of superadiabaticity in the convection zone. B_e, as estimated from mixing length models, is fairly constant along the main sequence. The field strength in the small scale fields at the surface of these stars is thus expected to be fairly constant as well, at about 1800G (at $\tau = 1$, corresponding to about 1200 - 1500G for observations in spectral lines). For stars with a different gravity, the field strength probably scales with the square root of the surface pressure.

Unfortunately, the above results cannot be extrapolated to large flux tubes like sunspots. The structure of spots may well be different from that of small tubes. In particular the reason why the field strength of spots has a value of about 3000G is essentially unknown. For details I refer to the proceedings of the recent sunspot workshop (Cram and Thomas, 1981).

3.2 Buoyancy

In a tube in temperature equilibrium with its surroundings the density is

$$\rho_i = \rho_e \, \beta/(1+\beta) \tag{7}$$

For the strong photospheric fields ($\beta \approx 1$) the density is thus much less than the surrounding density ρ_e, and the tubes are strongly buoyant. This implies that as long as they are rooted somewhere deeper in the convection zone, they have a strong tendency to remain vertical. The turbulent convective flow is strong enough to shuffle them around between granules or supergranules and to bend them slightly, but not strong enough to pull them back below the surface (see also section 8).

3.3 Energy balance of tubes

The strong field of the tube allows only the less efficient oscillatory type of convection. At the photosphere, the tube therefore cools until its radiation losses are balanced by the reduced convective flux, augmented by the radiation that leaks in from the sides. For larger tubes the lateral influx of radiation is small so that they will be dark, like sunspots. In small tubes however the lateral influx is strong, at least near the photosphere, and the internal temperature is only slightly less than the surrounding temperature. Since the internal opacity is less (due to the reduced pressure) one sees a higher than photospheric temperature when looking down on such a tube. Small tubes are therefore brighter than their surroundings even though at each geometrical level the internal temperature is lower than the surroundings. Calculations of this effect were made by Spruit (1976,1977). They predict that the tubes are bright structures when their radius is less than about 2 pressure scale heights (300 km). This agrees well with observations (Spruit and Zwaan, 1981). Small tubes thus act as net emitters of continuum radiation. One can say that due to their lower opacity they form leaks through which photons from the surrounding convection zone escape. The extra emission can be as high as 100% of the photospheric flux, averaged over the surface area occupied by the tube. The chromospheric emission in flux tubes has a different origin, probably the dissipation of mechanical energy from the convection zone carried along the tube in the form of MHD waves.

4. WAVES IN FLUX TUBES

The propagation of waves in an inhomogeneous field in the presence of gravity is in general an intractable problem. The wave modes of a cylindrical flux fube of finite diameter, in the absence of gravity, have been studied in detail (Wilson 1978,1979, Wentzel,1979b,Edwin and Roberts, 1981). The full spectrum of such a tube is rather complex consisting of surface type modes (concentrated at the tube boundary) and body waves, either of the slow or fast magnetoacoustic type. In addition there are simple Alfvén waves, consisting of torsional oscillations of the tube. If the boundary of the tube is not sharp, as assumed in these studies, but has a smoothly varying field strength, the spectrum is far more complicated. This is because the linearized MHD equations in general are singular, so that they have solutions which cannot be described in terms of eigenmodes of the system. For such a solution there are points in the plasma at which the local Alfvén speed matches the phase speed of the wave (the resonant points). Mathematically, the situation is similar to that of hydrodynamic waves in a shear flow (see, e.g. the book by Drazin and Reid, 1981). These singular wave problems have been studied in some detail in the tokamak context (see e.g. Sedlaček, 1971, Tataronis and Grossmann, 1976, Goedbloed, 1975, Hasegawa and Chen, 1976, Hameiri, 1981). Applications to astrophysics were made by Ionson (1978), Wentzel (1979a), Rae and Roberts (1981) and Rae (1982).

Most of these complications can be avoided elegantly by considering again the thin tube limit. In doing so we concentrate on the behaviour of the tube as a whole, skipping over the details of its internal dynamics. We now explore the consequences of this approach, bearing in mind that strictly speaking its validity should be checked in each particular application.

A thin tube has three wave modes, which are analogous to the waves in an elastic wire under tension (Spruit, 1982a). There is a torsional wave, a longitudinal tube wave and a transversal tube wave. The torsional wave is an Alfvén wave, satisfying the equation

$$\partial_{tt}\phi = v_A^2 (l)\, \partial_{ll}\, \phi \ , \tag{8}$$

where ϕ is the torsion angle and l the arclength along the tube. We have assumed that the tube is untwisted in its equilibrium state; this assumption is implicit also in the following discussion. In the absence of gravity the tube is straight and of constant cross section. The longitudinal wave consists of fluid motions along the tube, accompanied expansions and contractions ("sausage mode"). Its propagation speed is v_1:

$$v_1^2 = c^2 v_A^2/(c^2 + v_A^2) \ . \tag{9}$$

where c and v_A are the sound and Alfvén speeds inside the tube. The transversal wave consists of bends in the tube, without flow along the tube, resembling an Alfvén wave. Its propagation speed is, however, less than v_A:

$$v_t^2 = \rho_i v_A^2/(\rho_i + \rho_e) \ . \tag{10}$$

When gravity is present the fluid is stratified and the cross-section of the tube varies. The longitudinal and transversal motions of the tube are in general coupled. We consider the case of a vertical flux tube. In this special case the longitudinal and transversal motions are uncoupled (except when nonlinear effects are considered). The amplitude of the horizontal displacement ξ is governed by the equation (Spruit 1981b)

$$\partial_{tt}\xi = -g \frac{\rho_i - \rho_e}{\rho_i + \rho_e}\, \partial_z\xi + \frac{\rho_i}{\rho_i + \rho_e}\, v_A^2\, \partial_{zz}\xi \ , \tag{11}$$

where g is the acceleration of gravity and z the depth (positive into the sun). The second term on the RHS is the restoring force due to magnetic tension. The first term is the restoring force due to the buoyancy of the tube. It changes the character of the waves significantly: in contrast with Alfvén waves, transversal tube waves are dispersive. The transversal wave is incompressive, so that it is not affected by radiative damping.

The longitudinal tube wave is compressive, so that its propagation depends on the exchange of heat with the surroundings. If the motion is adiabatic, the wave equation for the vertical displacement ζ in a tube which is initially in temperature equilibrium with its surroundings, is (Defouw, 1976, Roberts and Webb, 1978, Spruit and Zweibel, 1979):

$$(2/\gamma + \beta)\, \partial_{tt}\zeta + 2gH\partial_{zz}\zeta + g\partial_z\zeta + \delta(1 + \beta)g/H\, \zeta = 0 \ . \tag{12}$$

where H(z) is again the pressure scale height and $\delta = d\ln T/d\ln P - (1-1/\gamma)$ is the superadiabaticity of the stratification. If δ is sufficiently positive, (12) has unstable solutions. These correspond to the convective collapse instability described above. The effects of heat exchange with the surroundings on the longitudinal tube wave are discussed in Roberts and Webb (1980a,b).

In the simple case of an isothermal atmosphere we have

$$\frac{v_A^2}{c^2} = \frac{\gamma}{2}\beta \tag{13}$$

The transversal and the (adiabatic) longitudinal wave are then of the form (Spruit 1981b)

$$\exp(i\omega t + ikz - z/4H) . \tag{14}$$

The dispersion relation is of the form

$$\omega^2 = v^2 k^2 + \omega_c^2 , \tag{15}$$

where, for the transversal wave $v^2 = v_t^2$ (eq.10), and the cutoff frequency is given by

$$\omega_{ct}^2 = \frac{g}{8H} \frac{1}{2\beta+1} \tag{16}$$

The speed of the longitudinal wave is $v=v_l$ (eq.9), and ω_{cl} is given by

$$\omega_{cl}^2 = \frac{g}{8H} [1 + 8(1-1/\gamma)(1+\beta)] /(2/\gamma+\beta) \tag{17}$$

The torsional tube wave, described by equation (8) has solutions of the form

$$\phi \sim \exp(i\omega t + ikz) , \tag{18}$$

with

$$\omega/k = v_A$$

Since the radius r of the tube varies as

$$r \sim \exp(-z/4H) , \tag{19}$$

the torsional velocity amplitude $v_\phi = r\partial_t\phi$ varies as

$$v_\phi \sim \exp(i\omega t + ikz - z/4H) \tag{20}$$

The dispersion relations are shown in Figure 2 for solar conditions. For comparison, the dispersion relation for a vertically propagating adiabatic sound wave is shown. Its solution is of the form $\exp(i\omega t+ikz-z/2H)$, with $v=c$, $\omega_c^2 = \gamma g/4H$.

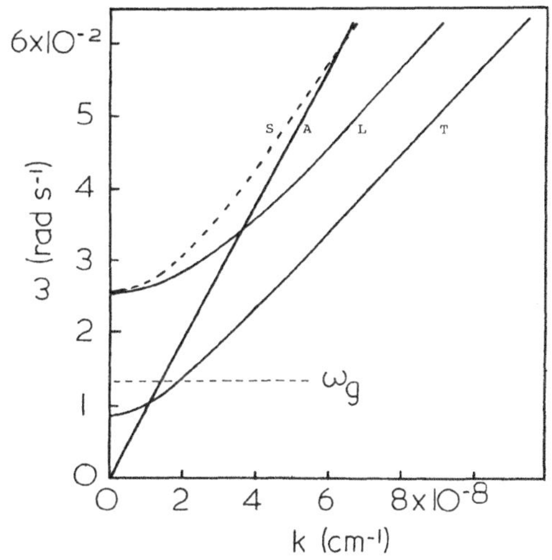

Fig.2. Diagnostic diagram for flux tubes in an exponential atmosphere with scale height 160 km, sound speed 8.3 km s^{-1} and Alfvén speed 9.4 km s^{-1}. A: torsional Alfvén wave, L: longitudinal tube wave, T: transversal tube wave. A vertically propagating sound wave (s) is shown for comparison. The characteristic frequency ω_g corresponding to granulation is shown.

4.1 Nonlinear tube waves

The amplitude of the motions observed in the chromosphere are of the same order as the sound speed, or larger. The waves are therefore nonlinear in these layers. In a nonlinear wave, transversal or torsional motions are always accompanied by longitudinal motions. This is because they produce "centrifugal" accelerations along the tube. Longitudinal motions can exist on their own at finite amplitude (though, at large amplitude such flows can be unstable to transversal motions, due to the "fire hose " instability). We consider finite amplitude longitudinal waves first. Roberts and Mangeney (1982) studied waves at small but finite amplitude in a flux sheet (rather than a tube) of small but finite thickness, in the absence of gravity and stratification. They show that such a sheet supports solitons. In this soliton the nonlinear effects, which tend to steepen the wave, just balance the dispersion of the wave introduced by the finite thickness of the sheet. As a result a soliton propagates, in the form of a single pulse of permanent shape. The importance of this result is that it shows that a nonlinear longitudinal wave need not form shocks. This contrasts with the case of a sound wave in one dimension, which forms shocks at arbitrarily low amplitude.

In a stratified atmosphere the nonlinear propagation of tube waves can only be studied numerically. Hollweg (1982) studied the longitudinal wave in a rigid tube of varying cross section, for solar conditions. In such a tube, the wave is equivalent to a sound wave. When driving occurs in the form of a sufficiently strong (1.4 km s^{-1}) and short pulse (90sec duration), a shock easily forms, followed at some distance by a "wake" which lifts the chromosphere upwards. This wake has many of the properties shown by solar spicules. In a related calculation, Hollweg et.al. (1981) followed the nonlinear development of a torsional wave, again in a tube with a given cross section (but varying with height). The results are a bit more complex in this case, but also show a strong upward flow resembling a spicule. The flow is initiated by the longitudinal acceleration in the strongly nonlinear torsional pulse.

In reality a flux tube can expand and contract in radius. In the usual thin tube approximation this is taken into account using Eq.(5). Solutions of the full set of thin tube equations for longitudinal flow, taking these changes into account, are given by Ulmschneider et.al. (1983) These solutions also show a tendency to form shocks. The concept of a shock in an expandable thin tube leads to certain philosophical problems. It seems likely however that some form of shock wave, and the associated heating, can indeed occur in such tubes (Ulmschneider et.al. 1983).

No calculations of nonlinear transversal tube waves are available at present.

5. ENERGY TRANSPORT FROM THE PHOTOSPHERE TO HIGHER LAYERS

It is generally realized that most of the mechanical energy that heats the chromosphere and corona is apparently transmitted from the convection zone to the atmosphere via the magnetic field. There are two conceptually different mechanisms for this transmission. The first is steady reconnection releasing in a quasicontinuous fashion the energy fed into the atmospheric magnetic field by motions of its photospheric footpoints. I refer to Chiuderi (this volume) Parker (1979,1981a,b) and Rosner et. al. (1978) for discussions of this mechanism. The detailed structure of the photospheric field is not of great importance in this mechanism, all that is needed is some knowledge about the motions of the foot points of the coronal field. Another mechanism is transport by MHD waves. Here knowledge of the field structure and the atmospheric stratification near the photospheric level is crucial since they determine the propagation characteristics in an essential way. Typically the photosphere is the level which is hardest to pass by a wave; an example is a pure acoustic wave which becomes evanescent at the photosphere for wave periods comparable with the time scale of convection. We now discuss the various kinds of waves from this perspective. For nonmagnetic atmospheric waves I refer to Stein and Leibacher(1981).

The efficiency with which magnetic waves are generated by turbulence in a homogeneous unstratified fluid was estimated by Kato (1968) and Stein (1981). The energy flux F according to these estimates is of the form $F = \rho u^3 f$; where u is the typical convective flow speed, and $f=(u/c)^5$ for acoustic waves, $f=(u/v_A)^5$ for fast mode waves, $f=u/c$ for slow modes and $f=u/v_A$ for Alfvén waves. Slow modes and Alfvén waves are therefore produced most effectively. Inserting typical numbers however, one finds that F is very large, of the order of a few percent of the total solar energy flux. This is evidently unrealistic, it is due to the neglect of stratification on the waves. Though it can be argued that the estimates still may give an impression of the variation of the heating rate with spectral type (Ulmschneider and Stein, 1982), this may equally well be a rather optimistic point of view.

Waves in a stratified atmosphere permeated by a magnetic field are called magneto-atmospheric waves. A case which has been studied in some detail (Thomas, 1982a,b, Schwartz and Leroy, 1982, Leroy and Schwartz, 1982) is a horizontally homogeneous atmosphere with either a completely uniform field or a horizontal field which varies with height in magnitude and/or direction. Thomas (1982) showed that in the case of a uniform horizontal field the magnetoatmospheric waves have a cutoff frequency $\omega_c = c/2H$, equal to the acoustic cutoff of the corresponding nonmagnetic atmosphere.

Homogeneous vertical fields have been considered by Leroy and Schwartz (1982) and Schwartz and Leroy (1982). They calculated the propagation of Alfvén waves and compressive waves in a stratification simulating solar conditions. Alfvén waves are strongly reflected by the strong density drop between the photosphere and the transition zone. Nevertheless, an energy flux of a magnitude which is potentially interesting for coronal heating can be obtained. The behavior of the compressive magnetoatmospheric waves depends strongly on their frequency. Below the acoustic cutoff frequency the waves tend to be evanescent so that little energy penetrates into the corona. If driving occurs above the cutoff large energy fluxes can be produced. Like in the case of heating by acoustic waves the main question is therefore which part of the spectrum of driving motions in the convection zone lies above the acoustic cutoff. The observational upper limits (Athay and White, 1978) on the level of acoustic waves in the atmosphere suggest that this fraction is small. This is consistent with the fact that the typical time scale of granulation is three times as long as that corresponding to the cutoff. Schwartz and Leroy (1982) conclude that the Alfvén mode, rather than one of the compressive waves is the most promising wave for heating the corona. This does not rule out the possibility that compressive waves, including sound waves, could be important for heating the lower atmosphere (Ulmschneider and Stein, 1981).

Because of the concentrated nature of the photospheric field, wave propagation calculated in a homogeneous field gives a misleading picture. In a photospheric flux tube for example the Alfvén speed is around 8 km s^{-1} and varies slowly with height. In a homogeneous field of 20 G on the other hand the Alfvén speed increases rapidly with height and is only 9 m s^{-1} at the photosphere. Let us consider the flux tube waves described above as a means to transport energy to the corona. Of course the flux tube picture breaks down at the merging level h_m (~1200 km). Above this level it is more appropriate to think in terms of a more or less uniform field.

The torsional tube wave propagates at all frequencies (Figure 2). In an isothermal atmosphere the Alfvén speed in the tube is constant so that no reflection occurs until the wave reaches the merging level. Once generated, the wave therefore easily reaches the corona. The generation of such waves by convective motions presents some conceptual difficulties. It would require, first, a substantial amount of vertical vorticity at a scale compared to that of the flux tube (100 km, say) and, second, a way to couple this motion to the tube. Neither of these conditions is easily satisfied. If torsional waves are present, they are more likely due to the gradual unwinding of twists stored in the tube before its eruption at the surface, possibly by a dynamo-related process (Parker, 1974a).

The longitudinal tube wave is excited by pressure fluctuations in the convection zone. Since its cutoff frequency is close to the acoustic cutoff (Figure 2), the amplitude of such waves depends critically on the amount of power present in the high frequency tail of the convective spectrum. Yet longitudinal waves have some advantage over acoustic waves because they are channeled along the field. Like ordinary slow mode waves this leads to a higher generating efficiency (Kato, 1968, Stein 1981).

Transversal tube waves are probably the easiest waves to excite. Their cutoff frequency lies just below the typical frequency of convective motions, so they respond to the bulk of the convective spectrum. The waves are excited by horizontal motions in the granulation (Spruit, 1982a). Observations show ample evidence for transversal tube waves. They correspond with the "swaying" motion that is typical for $H\alpha$ mottles (e.g. Bray and Loughhead, 1974). The wave motions seen in $H\alpha$ fibrils (Giovanelli, this volume) could represent the response of the chromospheric magnetic field to such waves (i.e. the field near the merging level). The expected periods and wave amplitudes agree well with these observations (Spruit, 1981b). Hollweg (1981) has calculated the propagation of these waves in open as well as closed field regions by assuming that they behave in the same way as torsional Alfvén waves. He finds large amplitudes in the corona, especially in closed field regions. A model for the propagation of transversal tube waves in the solar atmosphere is given in Spruit (1983). The tubes are matched to a homogeneous field above the merging level in this model. The tube waves continue as Alfvén waves in this homogeneous field. The wave amplitude at the base of the corona is of the order of 12 km s^{-1} for an average field strength of 20 G. At such amplitudes the waves are no longer linear. In particular, they will generate strong longitudinal flows, like in the case of the torsional waves considered by Hollweg et.al. (see previous section), that could be the cause of spicules.

The discussion in this section is not meant to suggest that coronal heating by MHD waves is the most viable mechanism. The mechanisms of the "DC" type are in the present stage of the theory at least equally attractive.

6. MISSING FLUX

Since sunspots are dark, they reduce the local heat flux at the surface. The question where the blocked heat flux goes is known as the "missing flux" problem (e.g. Sweet 1955, Wilson, 1971, Parker, 1974b, Foukal and Vernazza, 1979). The blocked flux is either redistributed over the remaining surface, or temporarily stored elsewhere inside the star. The answer to the question is of obvious importance when one tries to interpret the light variations of RS CVn and BY Dra stars in terms of star spot models (Hartmann and Rosner, 1979). A similar question arises concerning the small tubes in the small scale magnetic field since theory predicts that they are net emitters of radiation (section 3.3).

To calculate the effect which a spot or a thin tube has on the temperature field in its surroundings, one needs a model for the transport of heat in the convection zone. The only model used so far is a "turbulent diffusion" model but this is probably adequate for the main conclusions reached below.

To specify the problem more precisely, let us "switch on" a sunspot at $t=t_0$ and ask how the thermal structure of the convective envelope will develop in time. Detailed numerical calculations of this problem were done by Foukal et.al. (1983), Fowler et.al. (1983). If we wait long enough, the star returns to thermal equilibrium, at the same luminosity as before. The blocked heat flux then reappears elsewhere at the surface. Numerical calculations for this case by Spruit (1977, cf. also Parker, 1974b, Clark, 1979, Eschrich and Krause, 1977), showed that the blocked flux spreads out over a large area, and does not cause a bright ring around the spot. Since the timescale on which spots appear and disappear is short compared with the thermal timescale of the envelope, spots may cause deviations from thermal equilibrium, i.e. some of the blocked heat flux may be temporarily stored in the convective envelope. Let f be the fraction of the surface covered by spots, and express the change δL in the surface luminosity L of the star as

$$\delta L/L = -(1 - \alpha)f \ . \qquad (21)$$

The quantity α represents that fraction of the heat flux blocked by spots that reappears at the surface; $1-\alpha$ is the fraction that is stored in the envelope. At $t=t_0$ we have $\alpha=0$, and for $t \to \infty$ $\alpha \to 1$. It was shown by Spruit (1982b,c) that for changes in spot area that occur on a timescale which is short compared with the thermal (Kelvin-Helmholtz) timescale of the envelope, α is at most of the order

$$\alpha \sim [1 + d/(3\ H_o)]^{-2} \ , \qquad (22)$$

where d is the depth at which the blocking of the heat flux takes place and H_o the pressure scale height at the surface. Since d/H_o is large for any reasonable sunspot structure, $\alpha \ll 1$, and $\delta L/L \approx -f$. The blocked heat flux is therefore effectively stored in the convection zone. Note that this holds for the variable part of the sunspot area f; the average of f on timescales long compared with the thermal timescale causes a change in radius and surface temperature, but no change in luminosity. The effect of sunspots on the solar luminosity has been observed (Willson et. al. 1981; Willson, 1982) and shows a small value of α ($\alpha \lesssim 0.1$).

Since thin flux tubes are net emitters of heat, facular areas have an effect similar to that of spots but of opposite sign. The precise magnitude of this effect is much harder to estimate than in the case of spots, since it depends on the details of the convective energy transport process in the top of the convection zone, as well as on the energy balance of a thin tube in these layers. For estimates see Spruit (1977).

7. ERUPTION OF NEW FLUX

The rise of a flux tube through the supergranulation layer, under the influence of magnetic buoyancy, aerodynamic drag, and an external flow, was modeled numerically by Meyer et.al. (1979). The results show that after a section of the tube crosses the surface, its "ends" very quickly become vertical due to the strong buoyancy force. The ends move apart under the influence of magnetic tension; the rate of separation depends rather strongly on the size of the flux tube, but is in broad agreement with the rate of separation observed in emerging flux regions. More detailed calculations of the motion of flux tubes in the convection zone, taking into account also the flow along the tubes, were done by Moreno Insertis (1983).

The segment of the tube which has crossed the photosphere finds itself in a vacuum, and quickly expands. This expansion stops when the field above the surface has assumed a potential (or more generally, a force-free) configuration. Since the Alfvén speed in the atmosphere is high, the time needed to establish this equilibrium configuration is short. Observations indicate that the eruption of new flux is a gradual process. Instead of one big flux tube, a series small tubes surfaces in the same place (for a discussion see Zwaan, 1978). The eruption of the entire tube typically takes an hour or so; the process is seen in H_α as an "arch filament system". The tubes therefore do not expand in a vacuum but into the potential field set up by their predecessors.

Since the process is slow compared with the typical Alfvén travel time, it can be described elegantly in a model given by Shibata (1980). In this model, the field configuration in the atmosphere is a potential field defined by two point sources of opposite polarities at the surface. The strength of the field increases slowly as new flux emerges. Since the geometry of the field is known at all times, the hydrodynamics of the process can be calculated relatively easily. The results of this calculation agree with physical intuition as well as with observations. These observations (Tarbell and Title, 1983, Brands and Zwaan 1983, see also references in Shibata, 1980) show an upward motion of relatively dense and cool matter in the middle of the emerging loops, and a downward flow along their legs.

The process of flux eruption can therefore be described as slow and steady rather than as impulsive. The mass loss from the sun associated directly with the eruption process is likely to be negligible, the main reason being that the matter carried up through the photosphere finds itself contained in a strong quasi-static field ($P << B^2/8\pi$). Rather than expanding into interplanetary space, it drains back to the surface, along the field lines. Secondary effects, like the unwinding of twist stored in the emerging flux tubes (Parker, 1974a) and the interaction of the new flux with a preexisting field (Priest and Heyvaerts, 1974) are more likely to be responsible for the coronal activity seen in young active regions.

8. DISTRIBUTION OF FIELDS OVER THE SURFACE

The distribution of magnetic fields over the surface and its evolution in time is typically described in terms of a turbulent diffusion process (see for example Sheeley, this volume). A satisfactory theoretical description of the statistical behavior of a magnetic field in a turbulent environment like the solar convection zone is absent and not likely to be obtained in the near future. The three major obstacles are the high magnetic Reynolds number (R_m) of the flow, the stratification of the convection zone, which makes the turbulence extremely inhomogeneous in the vertical direction, and the strong magnetic forces (buoyancy and tension). For the kinematic problem (no magnetic forces) for homogeneous unstratified turbulence exact results exist only in the first order smoothing approximation, valid for small magnetic Reynolds numbers (Moffat, 1978, Krause and Rädler, 1981, Parker, 1979). Theories for arbitrary R_m have been developed by Kraichnan (1976), Knobloch (1977, 1978a,b, 1980), but the results of these theories are not consistent with each other. The possiblity exists (Kraichnan, 1976, Parker, 1979, Knobloch, 1977) that the turbulent diffusivity is negative in some cases. It has been argued that this would explain the observed concentration of the magnetic field into small tubes. Even if this is a correct interpretation, however, the turbulent diffusion can surely not be represented by a single value, since on larger scales the diffusion of fields appears to be positive (spreading of active regions).

Some progress is being reported with theories based on closure approximations. Homogeneous dynamic MHD turbulence (i.e. with magnetic forces but in the absence of gravity or stratification) has been treated numerically in the so-called EDQNM approximation (Leorat et.al., 1981, Grappin et.al., 1982, Frisch, this volume, see these references for a translation of the acronym). These calculations yield the evolution in time of the magnetic and kinetic fluctuation spectra. They cannot yet be applied to the convection zone since they do not include thermal and magnetic buoyancy effects.

Some insight might be obtained by combining observations with a judicious choice of simplified turbulence models. This approach was taken by Knobloch (1981) and Knobloch and Rosner (1981). These authors conclude that the observed spectrum is formed at a depth of at least 15000 km, that the fluid motions are three dimensional, and that the field diffuses with negative diffusivity. In an alternative approach, Knobloch (1982) used the steady convection cells of Weiss and coworkers as building blocks for a turbulent magnetic cascade, and calculated a theoretical spectrum of the magnetic field fluctuations.

A simple picture of the diffusion of fields results when we assume that the field is vertical down to a certain depth z_o. Observations, as well as the theoretical calculations using a flux tube approximation indicate that this is at least a good approximation near the surface (section 3.3,6). The evolution of the field is then governed by a) some vertical average (down to z_o) of the horizontal velocity field, and b) magnetic tension forces at $z=z_o$. Under the influence of a) alone the field evolves passively and two-dimensionally. It will quickly seek the vertices of granules and supergranules and tends to stay there. This behavior agrees well with observations (Tarbell, this symposium). There are some indications however (Zwaan, 1978) that the expansion of ageing active regions is faster than expected on the basis of process a) alone. It was shown by Van Ballegooijen (1982) that magnetic tension effects are indeed important, and could even be the main cause of the expansion of active regions. In his model, the magnetic field is stored at the base of the convection zone, and the fields seen in active regions are connected with this field by nearly vertical flux tubes extending through the convection zone. This model is related more closely to Leighton's (1969) classical model of the dynamo than to the models based on mean field electrodynamics. More on this subject is to be found in Schüßler's review (this volume).

REFERENCES

Anzer, U., Galloway, D.J.: 1982, Mon.Not.Royal Astr.Soc., in press.
Athay, R.G., White, O.R.: 1978, Astrophys.J. 226, p.1135.
Brands, J., Zwaan, C.: 1983, in preparation.
Bray, R.J., Loughhead, R.E.: 1974, The Solar Chromosphere, Chapman and Hall, London.
Clark, A.Jr.: 1979, Solar Phys. 62, p.305.
Cowling, T.G.: 1976, Magnetohydrodynamics, Adam Hilger, Bristol.
Cram, L.E., Thomas, J.H., eds.: 1981, The Physics of Sunspots, Sacramento Peak Observatory, Sunspot, N.M. 88349.
Defouw, R.J.: 1976, Astrophys.J. 209, p.266.
Drazin, P., Reid, W.H.: 1981, Hydrodynamic Stability, Cambridge University Press, Cambridge.
Edwin, P.M., Roberts, B.: 1982, Solar Phys. 76, p.239.
Eschrich, K.-O., Krause, F.: 1977, Astron.Nachr. 298, p.1.
Foukal, P., Livshits, M., Fowler, L.A.: 1983, Astrophys. J.,in press.
Foukal, P., Vernazza, J.: 1979, Astrophys.J. 234, p.707.
Fowler, L.A., Foukal, P., Duvall, T.Jr.: 1983, Astrophys.J. submitted.
Gabriel, H.A.: 1977, in Bonnet and Delache, eds., The Energy Balance and Hydrodynamics of the Solar Chromosphere and Corona, IAU Coll.36, Bussac, Clermont-Ferrand.
Galloway, D.J., and Moore, D.: 1979, Geoph.Ap.Fl.Dyn. 12, p.73.
Galloway, D.J., Proctor, M.R.E., Weiss, N.O.: 1977, Nature 266, p.686.
Galloway, D.J., Proctor, M.R.E., Weiss, N.O.: 1978, J.Fl.Mech. 87, p.243.
Giovanelli, R.G.: 1980, Solar Phys. 68, p.49.
Goedbloed, J.P.: 1975, Phys.Fluids 18, p.1258.
Grappin, R., Frisch, U., Léorat, J., Pouquet, A.: 1982, Astron.Astrophys., 105, p.6.
Hameiri, E.: 1981, Phys.Fluids 24, p.562.
Hartmann, L., Rosner, R.: 1979, Astrophys.J., 230, p.802.
Hasegawa, A., Chen, L.: 1976, Phys.Fluids 19, p.1924.
Hollweg, J.V.: 1981, Solar Phys. 70, p.25.
Hollweg, J.V.: 1982, Astrophys.J. 257, p.345.
Hollweg, J.V., Jackson, S., Galloway, D.J.: 1981, Solar Phys. 75, p. 35.
Ionson, J.A.: 1978, Astrophys.J. 226, p.650.
Jones, H.P., Giovanelli, R.G.: 1982, Astrophys.J., in press.
Kato, S.: 1968, Publ.Astr.Soc. Japan 20, p.59.

Knobloch, E.: 1977, J.Fluid Mech. 83, p.129.
Knobloch, E.: 1978a, Astrophys.J. 220, p.330.
Knobloch, E.: 1978b, Astrophys.J. 225, p.1050.
Knobloch, E.: 1980, Phys.Lett.A 78, p.307.
Knobloch, E.: 1981, Astrophys.J. 248, p.1126.
Knobloch, E.: 1982, Astrophys.J., in press.
Knobloch, E. and Rosner, R.: 1981, Astrophys.J. 247, p.300.
Kraichnan, R.H.: 1976, J.Fluid Mech. 77, p.753.
Krause, F., Rädler, K.H.: 1981, Mean-Field Magnetohydrodynamics and Dynamo Theory, Pergamon Press, Oxford.
Leighton, R.B.: 1969, Astrophys.J. 156, p.1.
Leorat, J., Pouquet, A., Frisch, U.: 1981, J.Fluid Mech. 104, p.419.
Leroy, B., Schwartz, S.J.: 1982, Astron.Astrophys., in press.
Meneguzzi, M., Frisch, U., Pouquet, A.: 1981, Phys.Rev.Lett. 47, p.1066.
Meyer, F., Schmidt, H.U., Weiss, N.O.: 1974, Mon.Not.Roy. Astron.Soc. 169, p.35.
Meyer, F., Schmidt, H.U., Simon, G.W., Weiss, N.O.: 1979, Astron.Astrophys. 76, p.35.
Moffat, H.K.: 1978, Magnetic Field Generation in Electrically Conducting Fluids, Cambridge University Press, Cambridge.
Moreno Insertis, F.: 1983, in preparation.
Nordlund, Å.: 1977, Highlights of Astronomy 4 part II, remark on p.272.
Parker, E.N.: 1974a, Astrophys.J. 191, p.245.
Parker, E.N.: 1974b, Solar Phys. 36, p.249.
Parker, E.N.: 1978, Astrophys.J. 221, p.368.
Parker, E.N.: 1979, Cosmic Magnetic Fields, their Origin and Activity, Clarendon Press, Oxford.
Parker, E.N.: 1981a, Astrophys.J. 244, p.631.
Parker, E.N.: 1981b, Astrophys.J. 244, p.644.
Peckover, R.S., Weiss, N.O.: 1978, Mon.Not.Roy.Astron.Soc. 181, p.189.
Pouquet, A., Patterson, G.S.: 1978, J.Fluid Mech. 85, p.305.
Priest, E.R.: 1982, Solar Magnetohydrodynamics, D. Reidel, Dordrecht.
Priest, E.R., Heyvaerts, J.: 1974, Solar Phys. 36, p.433.
Rae, I.: 1982, Plasma Physics 24, p.133.
Rae, I., Roberts, B.: 1981, Geophys.Astrophys.Fluid Dyn. 78, p.197.
Roberts, B., Mangeney, A.: 1982, Mon.Not.Roy.Astron.Soc. 198, p. 7P.
Roberts, B., Webb, A.R.: 1978, Solar Phys. 56, p.5.
Roberts, B., Webb, A.R.: 1979, Solar Phys. 59, p.249.
Roberts, B., Webb, A.R.: 1980a, Solar Phys. 68, p.71.
Roberts, B., Webb, A.R.: 1980b, Solar Phys. 68, p.87.
Rosner, R., Golub, L., Coppi, B., Vaiana, G.S.: 1978, Astrophys.J. 222, p.317.
Schmidt, H.U., Wegmann, R.; 1982, in Brosowski and Martensen, eds., Proceedings "Methoden und Verfahren der Mathematischen Physik" (Oberwolfach 1982).
Schwartz, G.J., Leroy, B.: 1982, Astron.Astrophys., in press.
Sedlaček, Z.: 1971, J.Plasma Phys. 6, p.187.
Shibata, K.: 1980, Solar Phy. 66, p.61.
Spruit, H.C.: 1976, Solar Phys. 50, p.269.
Spruit, H.C.: 1977, Solar Phys. 55, p.3.
Spruit, H.C.: 1979, Solar Phys. 61, p.363.

Spruit, H.C.: 1981a, in S.D. Jordan, ed., The Sun as a Star, NASA SP-450, p.385.
Spruit, H.C.: 1981b, Astron.Astrophys. 98, p.155.
Spruit, H.C.: 1981c, Astron.Astrophys. 102, p.129.
Spruit, H.C.: 1982a, Solar Phys. 75, p.3.
Spruit, H.C.: 1982b, Astron.Astrophys. 108, p.348.
Spruit, H.C.: 1982c, Astron.Astrophys. 108, p.356.
Spruit, H.C.: 1983, in preparation.
Spruit, H.C., Zweibel, E.G.: 1979, Solar Phys. 62, p.15.
Spruit, H.C., Zwaan, C.: 1981, Solar Phys. 70, p.207.
Stein, R.F.: 1981, Astrophys.J. 246, p.966.
Stein, R.F., Leibacher, J.W.: 1981, in S.D. Jordan, ed., The Sun as a Star, NASA SP-450, p.289.
Sweet, P.A.: 1955, Vistas Astron. 1, 675.
Tarbell, T.D., Title, A.M.: 1983, in preparation.
Tataronis, J., Grossmann, W.: 1976, Nucl. Fusion 16, p.667.
Thomas, J.H.: 1983a, Astrophys.J., in press.
Thomas, J.H.: 1983b, Ann.Rev.Fluid Dyn. 15.
Ulmschneider, P., Stein, R.F.: 1982, Astron.Astrophys. 106, p.9.
Ulmschneider, P., Herbold, G., Spruit, H.C.: 1983, in preparation.
Unno, W., Ando, H.: 1979, Geophys.Astrophys.Fluid Dyn. 12, p.107.
Van Ballegooijen, A.A.: 1982, Astron.Astrophys. 106, p.43.
Weiss, N.O.: 1981a, J.Fluid Mech. 108, p.247.
Weiss, N.O.: 1981b, J.Fluid Mech. 108, p.273.
Wentzel, D.G.: 1979a, Astrophys.J. 233, p.756.
Wentzel, D.G.: 1979b, Astron.Astrophys. 76, p.20.
Wilson, P.R.: 1971, Solar Phys. 21, p.101.
Wilson, P.R.: 1978, Astrophys.J. 221, p.672.
Wilson, P.R.: 1979, Astron.Astrophys. 71, p.9.
Willson, R.C., Gulkis, S., Janssen, M., Hudson, H.S., Chapman, G.A.: 1981, Science 211, p.700.
Willson, R.C.: 1982, J.Geophys.Res. 87, p.4319.
Zwaan, C.: 1978, Solar Phys. 60, p.213.

DISCUSSION

ROBERTS: I would like to make two comments on wave cut-off. Firstly, the behaviour of an impulsively generated disturbance is different from that of a fixed frequency disturbance. In an impulsively generated disturbance, all frequencies are effectively involved, and the final disturbance is a Fourier integrated sum. The result is the creation of a wake, which oscillates at the mode's cut-off, trailing behind a wavefront (which propagates at the tube speed of the mode). This behaviour, which has been described in greater detail in Rae and Roberts (*Astrophys.J.*, 1982), is also evident in Hollweg's spicule calculations. The second comment concerns the modifications in the concept of cut-off that are necessary if conditions are no longer adiabatic — which is the case in the sun's photosphere and low chromosphere. There is no longer a clear division between propagation and non-propagation: all frequencies have a propagating component. In deciding on the relative efficiency of a particular mode, both the above aspects have to be borne in mind.

FOING: Two questions about the thermal structure inside the fluxtubes: (1) Due to the net excess of emission from the walls, is the thermal structure inside the tubes affected by the radiation (heating by the walls). (2) What is the time scale of thermal relaxation compared to the dynamic time of downfall of material inside the tube to allow an isothermal equilibrium?

SPRUIT: (1) Yes. (2) The thermal relaxation time scale depends strongly on depth, since the opacity in the solar atmosphere increases so enormously with depth. Near the surface, tubes with say a width of one sec of arc can relax on a time scale of a minute or less. If you go deeper, although the fluxtube itself shrinks in size the opacity increases so rapidly that these time scales become of the order of days or months or longer. So you may assume that near the surface there may be something like an energy equilibrium state, while in the deeper layers there will in general be thermal disequilibrium.

ROXBURGH: When describing your tube and uniform field model you stated that the energy density is still an increasing function of the field strength. Can you give us some idea of the dependence of energy flux on field strength?

SPRUIT: For this particular model the energy flux is roughly proportional to the average field strength. But this result depends critically on one of its assumptions, namely that the base of the corona coincides in height with the merging level of fluxtubes. This is about right for the sun, but the height of the coronal base is itself a function of the coronal heating mechanism. One therefore needs to include a coronal heating mechanism in order to make a real prediction.

RIBES: (1) If the magnetic tubes are very thin, say 100 km or less below the photosphere, I would expect that convection would ignore them and not be much reduced. So, I do not see how you can get a large excess of flux from deep convective layers, in order to explain the center-to-limb variation of the continuum faculae. (2) If the downflow exists in the magnetic fluxtubes, which seems to be the case, then the hydrodynamical solutions will be quite different from the hydrostatic ones through the energy equation (the term of entropy is added). One characteristic of the hydromagnetic solutions is the decrease of the field strength at the bottom of the photosphere ($B_0 \lesssim 1$ kG) and the rapid fanning out of the lines of force.

SPRUIT: (1) The center-to-limb variation is determined by the geometry of the depression in the visible surface created by the tube. The net heat flux excess produced by a tube is a consequence of the thermal properties of a convective envelope. If you deal with the diffusion of heat in the convection zone as a turbulent diffusion process, you will find that the thermal conductivity is small near the surface and very large in the deeper layers. Near the surface one thus requires large temperature fluctuations to get significant fluctuations of the heat flux; in the deeper layers one needs much smaller fluctuations, about one degree or so, to get large flux variations. So in this model you can look at the surface layers as insulating, the deeper layers as superconducting for heat. A fraction of the heat flux entering the tube sideways comes in through the deeper layer and thereby from a very large part of the convection zone. How large this fraction is relative to the contribution near the surface depends critically on how deep the fluxtube is compared with the superadiabatic surface layer (which is about 500 km deep). (2) Of course the downdrafts will have a significant effect because they have large amplitudes and carry a lot of heat. Certainly for the thermal effects, static models would be somewhat inappropriate. For the lateral pressure balance it may not be too bad, because the flows are still subsonic. A problem in dealing with this question is that the nature of the flow is not known, though it is most likely not a steady flow. The reason for this statement is simply that you cannot get enough mass from the corona. If you take one single fluxtube, you will find that it drains the entire mass of the corona in a few minutes. This is just because the density at the photospheric levels is so high, and the velocities are also high. You have to get this mass from somewhere. You may assume that there is an anomalous diffusion process into the fluxtube, as has been done by Giovanelli. I personally tend to think that this is not a real steady flow at all, but an oscillating or stochastic flow. An apparent net downflow would then be the result of a correlation between brightness and velocity in the fluxtubes.

SOLITONS IN MAGNETIC FLUX TUBES

B. Roberts
Dept. Applied Mathematics
St. Andrews University
Scotland

ABSTRACT

The highly inhomogeneous nature of the solar atmosphere leads us to suggest that solitons may occur in magnetic structures such as coronal loops or photospheric flux tubes. The theory is outlined for the simple case of a magnetic slab in a field-free atmosphere and shown to lead to the Benjamin-Ono equation.

1. INTRODUCTION

A soliton is a nonlinear solitary wave of permanent form that behaves in many respects like a particle. Solitary waves were first observed by John Scott Russell, in 1834, who, in a famous horseback ride, pursued one such wave as it propagated along a canal near Edinburgh. This observational discovery conflicted with the theoretical understandings of the day, but eventually inspired, in 1895, the celebrated equation of Korteweg and de Vries. Numerical solutions of the KdV equation, exhibiting no permanent loss of identity when two solitary waves interact, led Zabusky and Kruskal (1965) to refer to such waves as solitons. For a recent review of the literature on water waves, see Miles (1980).

Here we consider the possibility that MHD solitons may occur on the sun. Recall that solitons can arise when both nonlinearity (wave breaking, shock forming) and dispersion occur in a physical system. The importance of nonlinearities on wave propagation in the solar atmosphere, where the gas density changes by many orders of magnitude, is well-known and needs no emphasis here. Dispersive effects, however, have tended to be ignored.

Dispersion typically arises whenever an atmosphere is structured, so that the system possesses a natural length-scale. In ideal MHD, with gravity neglected, dispersion of waves does not occur. But in a structured field the magnetoacoustic modes are dispersive, their phase-speeds depending upon the wavenumber. Thus, since the solar atmosphere abounds

J. O. Stenflo (ed.), Solar and Magnetic Fields: Origins and Coronal Effects, 61–66.

with magnetic structure — intense tubes, sunspots, coronal loops, magnetic canopies, for example — we may anticipate that solitons will also occur.

Indeed, we expect solitons to occur whenever magnetic structuring and nonlinearity are important. The coronal loop and the photospheric flux tube are two obvious examples. Here we will consider one simple example of a magnetically structured atmosphere, the magnetic flux tube, ignoring the effect of gravity. This is the example discussed by Roberts and Mangeney (1982).

2. LINEAR AND NONLINEAR WAVES

Consider a magnetic slab of field B, with width 2a, embedded in a field-free atmosphere. The undisturbed pressure and density of the gas within the magnetic field are p and ρ, their external values being p_e and ρ_e.

The linear free modes of vibration for such a configuration have been studied in detail in Roberts (1981) and Edwin and Roberts (1982). Figure 1 shows the types of wave modes allowed.

Note that geometrically we can have either symmetric (sausage) modes or antisymmetric (kink) modes of vibration. Additionally, a mode may be either a surface wave (for which the amplitude declines away from the slab boundaries) or a body wave (in which the motions penetrate into the slab interior). The dispersion diagrams for these modes are given in Roberts (1981) and Edwin and Roberts (1982).

Our interest here is in the behaviour of the surface sausage mode in the long wavelength, thin tube approximation (ka << 1). In this approximation, the sausage mode is a longitudinal sound wave guided along

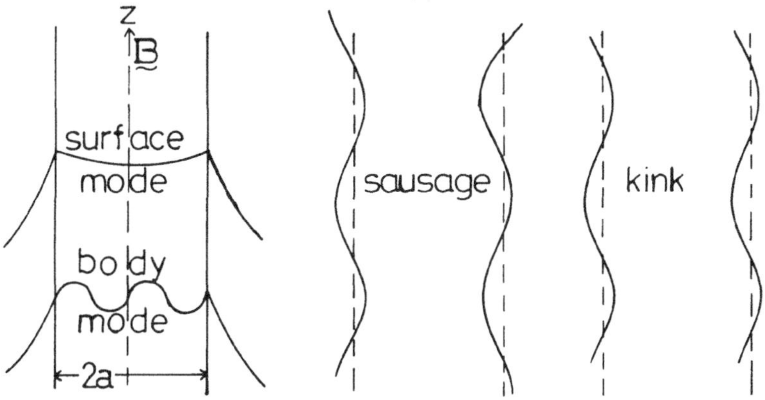

Figure 1. Magnetoacoustic modes of a slab.

the magnetic slab; for a mode travelling along the slab, its frequency ω and longitudinal wavenumber k are approximately related by (Roberts, 1981)

$$\omega = kc_T - \alpha k|k| \quad . \tag{1}$$

Here c_T is the tube speed, given in terms of the sound speed $c_s = (\gamma p/\rho)^{1/2}$ and Alfvén speed $v_A = B/(\mu\rho)^{1/2}$ by

$$c_T = c_s v_A/(c_s^2 + v_A^2)^{1/2} \quad , \tag{2}$$

and α, given by

$$\alpha = \frac{1}{2} (\rho_e/\rho) a \left(\frac{c_T}{v_A}\right)^3 c_T \quad , \tag{3}$$

is a measure of the inertial properties of the slab's environment. It is the presence of an environment that gives the tube (magnetoacoustic) mode its dispersive character and, ultimately, permits a soliton to occur.

The dispersion relation (1) is quite different in form from the well-known relation ($\omega = k - k^3$) associated with the KdV equation. This difference arises here as an immediate consequence of magnetic structuring: the fact that there is a length-scale (namely, 2a) for disturbances within the tube, but none for the disturbance outside the tube. The form of (1) arises in the propagation of water waves on an internal density inhomogeneity, and there leads to the Benjamin-Ono equation when nonlinearities are allowed for (see Benjamin, 1967; Ono, 1975).

Much the same kind of analysis may be made for the magnetic slab, with the result (Roberts and Mangeney, 1982):

$$\frac{\partial v}{\partial t} + c_T \frac{\partial v}{\partial z} + \beta v \frac{\partial v}{\partial z} + \frac{\alpha}{\pi} \frac{\partial^2}{\partial z^2} \int_{-\infty}^{\infty} \frac{v(z', t)}{z' - z}\, dz' = 0 \quad . \tag{4}$$

This is the Benjamin-Ono equation for the longitudinal velocity v(z,t) of gas within the magnetic slab. The nonlinearity coefficient β is given by

$$\beta = \frac{1}{2} v_A^2 \, [3\, c_s^2 + (\gamma + 1)\, v_A^2\,] \,/(c_s^2 + v_A^2)^2 \quad . \tag{5}$$

The derivation of (4) is lengthy and so will not be given here; suffice it to remark that it follows from a multiple-scale analysis of the thin flux tube equations (with gravity ignored) with the external pressure terms fully incorporated.

Equation (4) describes the behaviour of weakly nonlinear long waves in a magnetic slab. It is known to possess soliton solutions. In particular, the one-soliton solution is (Benjamin, 1967)

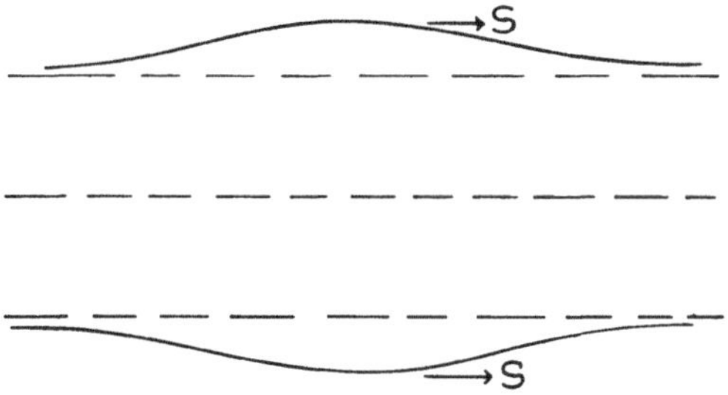

Figure 2. The soliton in a magnetic slab.

$$v(z,t) = \frac{a}{1 + \left(\frac{z - st}{\ell}\right)^2} , \tag{6}$$

where a is the velocity amplitude, and s and ℓ are the speed and characteristic scale of the soliton:

$$s = c_T + \frac{1}{4}\beta a , \qquad \ell = \frac{4\alpha}{a\beta} . \tag{7}$$

Figure 2 illustrates the propagation of the soliton. It resembles a symmetrical bulge, like a knee cap, propagating at a speed slightly above the speed c_T. Under photospheric conditions, this gives a soliton speed of about 7 km s^{-1}.

3. DISCUSSION

What conclusions can be drawn from the above analysis? Firstly, it is evident that solitons <u>are</u> likely, at least on theoretical grounds, to propagate in the solar atmosphere. Our example gives a general illustration of this possibility. Of course, we have made a number of simplifying assumptions in our analysis, and so it would be premature to claim that a soliton of precisely the form calculated can arise. Nonetheless, the indications that solitons can arise are clear enough. Such solitons may occur, for example, in coronal loops — though here the influence of an external magnetic field becomes important (Edwin and Roberts, 1983) — or in intense flux tubes. The difficulty in applying the present soliton theory to an intense tube is the neglect of gravity, and it is clear from the study of linear waves that gravitational effects are particularly important (e.g., Spruit, 1983; Roberts and Webb, 1978). However, ignoring such effects allows us to speculate that the tube soliton may become manifest in the chromosphere as a spicule. Indeed, many of the properties of spicules are reminiscent of those of a soliton.

In conclusion, we have demonstrated (following Roberts and Mangeney, 1982) that magnetoacoustic solitons may arise in ideal MHD if the effects of magnetic structuring are allowed for. This suggests, since the solar atmosphere is highly structured, that the sun may exhibit the soliton phenomenon. However, the theory is not yet sufficiently advanced as to permit direct applications to the solar atmosphere with any confidence.

ACKNOWLEDGEMENT

It is a pleasure to acknowledge the support of Professor J.O. Stenflo while the author was a visiting scientist at the Institut für Astronomie, ETH Zürich.

REFERENCES

Benjamin, T.B.: 1967, J. Fluid Mech. 29, p. 559.
Edwin, P.M. and Roberts, B.: 1982, Solar Phys. 76, p. 239.
Edwin, P.M. and Roberts, B.: 1983, in preparation.
Miles, J.W.: 1980, Ann. Rev. Fluid Mech. 12, p. 11.
Ono, H.: 1975, J. Phys. Soc. Japan 39, p. 1082.
Roberts, B.: 1981, Solar Phys. 69, p. 39.
Roberts, B. and Mangeney, A.: 1982, Monthly Not. Roy. Astron. Soc. 198, p. 7p.
Roberts, B. and Webb, A.R.: 1978, Solar Phys. 56, p. 5.
Spruit, H.C.: 1983, these proceedings.
Zabusky, N.J. and Kruskal, M.D.: 1965, Phys. Rev. Letters 15, p. 240.

DISCUSSION

GOKHALE: Since the slender fluxtube approximation is used I wonder how the result could be applied to magnetic fields in the solar atmosphere, for which the fluxtubes diverge very rapidly with height.

ROBERTS: As I stressed in my talk, stratification will have important consequences, and I suspect the soliton that I have described will be of limited relevance to that problem. That is an area currently under investigation. But my general aim is to demonstrate the theoretical existence of an MHD soliton in a structured field, and to do this I have taken the simplest example of a structured field I could think of. Of course, the neglect of gravity will be valid for coronal disturbances, so coronal solitons are entirely possible and may be similar to the type described here.

IONSON: You mentioned that thin fluxtubes can support MHD solitons. Are these solitary wave solutions of the KdV equation or nonlinear Schrödinger equation, and are they two or three dimensional?

ROBERTS: The solitons are given by the Benjamin-Ono equation, which is quite distinct from the KdV or nonlinear Schrödinger equations. The soliton I give is two dimensional. It may be in a slab (as described in my talk) or in a cylindrical fluxtube.

STIX: Does the occurrence of solitons on the sun require special conditions of excitation (i.e. special initial conditions)?

ROBERTS: I don't think so. If we look at the laboratory results for internal waves in a fluid (reported by Benjamin (1967), Davis and Acrivos (1967), and very recently by a group in California, all published in the *J. Fluid Mechanics*), we find that solitons are very readily generated. A wavemaker at one end of a tank generates the disturbance. As soon as the transients die out — which is very quickly — we see solitons formed and reflecting in a preserved form from the ends of the tank. In the sun I would guess that granular buffeting, for example, could generate waves in the fluxtube, which would then quickly result in solitons. But for photospheric tubes we must not forget the effect of stratification, which has yet to be included in our description of soliton dynamics.

FRISCH: Is there any evidence for the existence of *torsional* solitons as predicted for slender vortex tubes?

ROBERTS: None that I can think of, but it is interesting to note that vortex motions of a non-magnetic fluid in a cylindrical geometry may give rise to solitons of a type not very different from the magnetic soliton described here. There are some extentions of the theory that can be made, which may give rise to other soliton solutions, e.g. to consider the kink ($m = 1$) mode in a cylindrical fluxtube.

ROXBURGH: A colleague of mine, Dr. Chris Ovenden, has obtained soliton solutions for waves propagating in the solar wind and shown that scattering of these solitary waves can produce transverse temperatures greater than longitudinal temperatures within the solar wind plasma. This is hardly direct evidence that solitary waves are generated, but it is at least suggestive. Perhaps these solitons are actually produced in the solar surface layers.

ROBERTS: Presumably these are Alfvén solitons, depending upon plasma effects for dispersion rather than the ideal MHD case I have described here.

SELF-CONSISTENT MODELS FOR SMALL PHOTOSPHERIC FLUX TUBES

W. Deinzer, G. Hensler, D. Schmitt, M. Schüssler, E. Weisshaar
Universitäts-Sternwarte Göttingen
Federal Republic of Germany

We give a short summary of some results of a numerical study of magnetic field concentrations in the solar photosphere and upper convection zone. We have developed a 2D time dependent code for the full MHD equations (momentum equation, equation of continuity, induction equation for infinite conductivity and energy equation) in slab geometry for a compressible medium. A Finite-Element-technique is used. Convective energy transport is described by the mixing-length formalism while the diffusion approximation is employed for radiation. We parametrize the inhibition of convective heat flow by the magnetic field and calculate the material functions (opacity, adiabatic temperature gradient, specific heat) self-consistently. Here we present a nearly static flux tube model with a magnetic flux of $\sim 10^{18}$ mx, a depth of 1000 km and a photospheric diameter of $\sim$ 300 km as the result of a dynamical calculation. The influx of heat within the flux tube at the bottom of the layer is reduced to 0.2 of the normal value. The mass distribution is a linear function of the flux function A: $dm(A)/dA = \text{const}$. Fig. 1 shows the model: Isodensities (a), fieldlines (b), isotherms (c) and lines of constant continuum optical depth (d) are given. The "Wilson depression" (height difference between $\tau = 1$ within and outside the tube) is $\sim$ 150 km and the maximum horizontal temperature deficit is $\sim$ 3000 K. Field strengths as function of x for three different depths and as function of depth along the symmetry axis are shown in (e) and (f), respectively. Note the sharp edge of the tube.

We have used the 2D radiative transfer code of Mihalas, Auer and Mihalas (1978, Astrophys. J. 220, 1001) for LTE diagnostics of the radiation field. The emergent intensity is found by integrating along rays through the multicomponent source function. Fig. 2 shows the centre-to-limb-variation (CLV) of the continuum intensity emerging at different horizontal positions while Fig. 3 gives the x-dependence of the intensity for different aspect angles (see figure captions for more information). Fig. 4 displays line profiles for the magnetically unchanged ($g < 0.1$) line Ti I λ5222.7 for $\mu = 1$. and different horizontal positions. The thick curve shows the horizontally averaged

J. O. Stenflo (ed.), Solar and Magnetic Fields: Origins and Coronal Effects, 67–71.

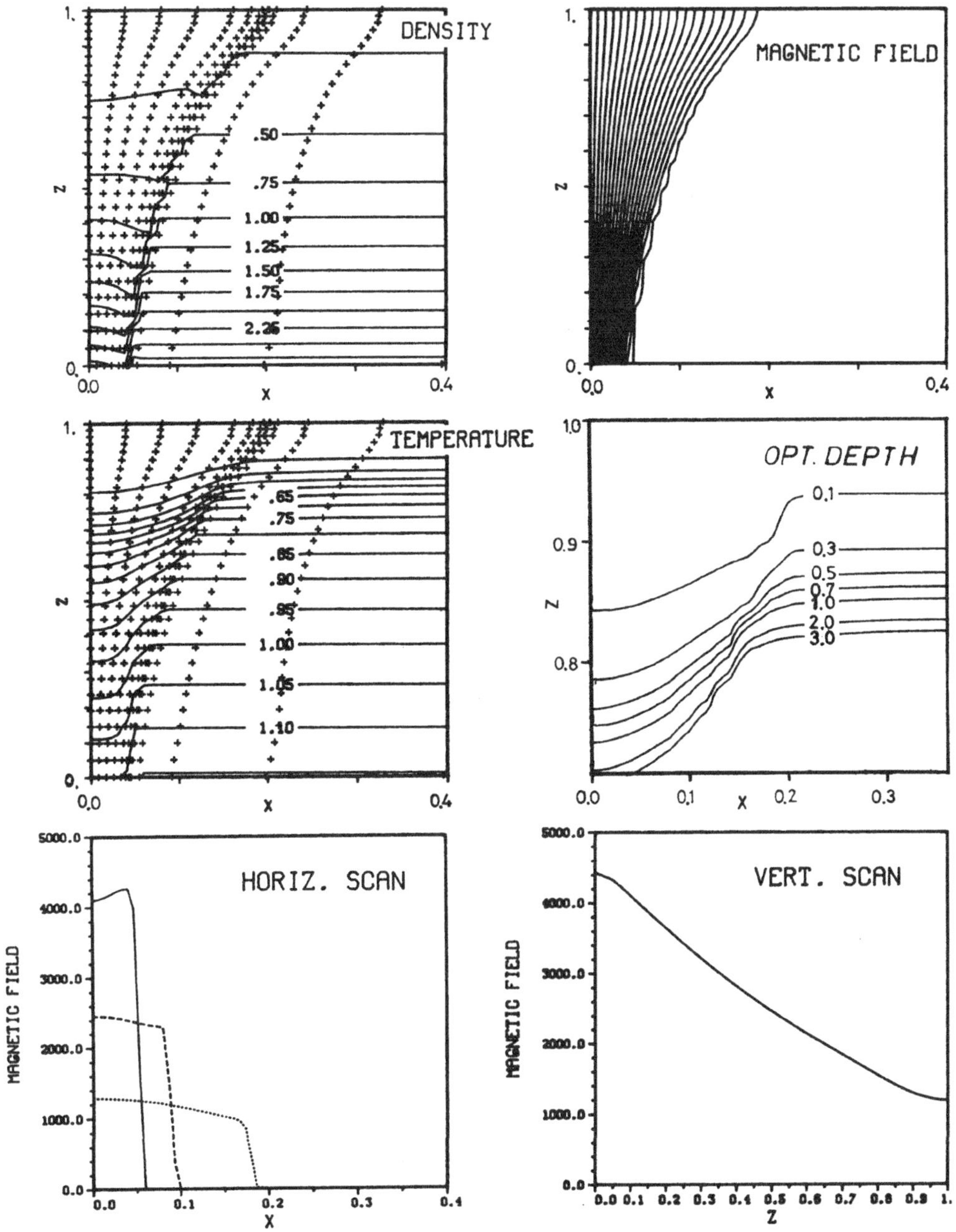

Fig. 1: Isodensities (a), fieldlines (b), isotherms (c) and lines of constant continuum optical depth (d) for the flux tube model. The horizontal scale has been expanded in (a) - (c). The length scale is 1000 km. (e) and (f) give the field strength as a function of x for different depths and as a function of z along the symmetry axis (x = 0).

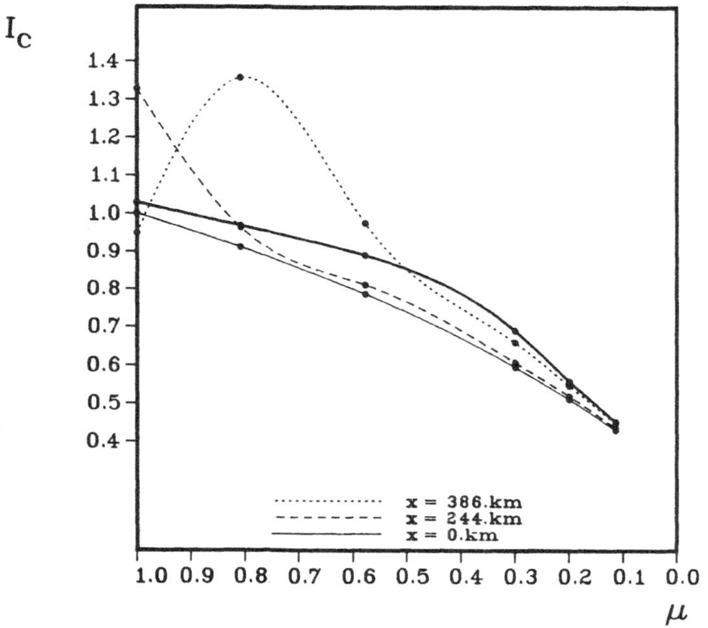

Fig. 2: CLV of the continuum intensity emerging at different horizontal positions. x = 0 km: undisturbed photosphere; x = 244 km: "hot ring" near the boundary of the tube; x = 386 km: centre of the tube. The thick line gives the CLV of the horizontally averaged intensity, i.e. the observed intensity at finite resolution. The tube appears slightly brighter (∿ 3 %) than the undisturbed photosphere at the centre of the disc ($\mu = 1$) and the intensity contrast increases towards the limb with a maximum of 17 % for $\mu \sim 0.4$ and decreases afterward.

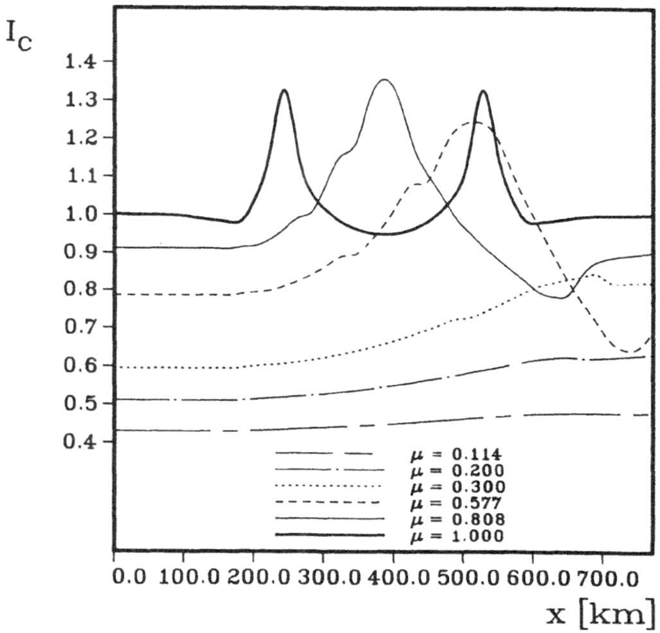

Fig. 3: x-dependence of the intensity for different aspect angles. The curve for $\mu = 1$ shows that the intensity enhancement at disc center comes from the "bending walls" which lead to a bright ring at the periphery of the tube. When viewed under an angle $\mu < 1$, only the opposite wall can be seen and we find only a single intensity maximum. Furthermore, the maximum decreases because the ray has to penetrate the edge of the structure and reaches $\tau = 1$ at a smaller temperature for decreasing μ.

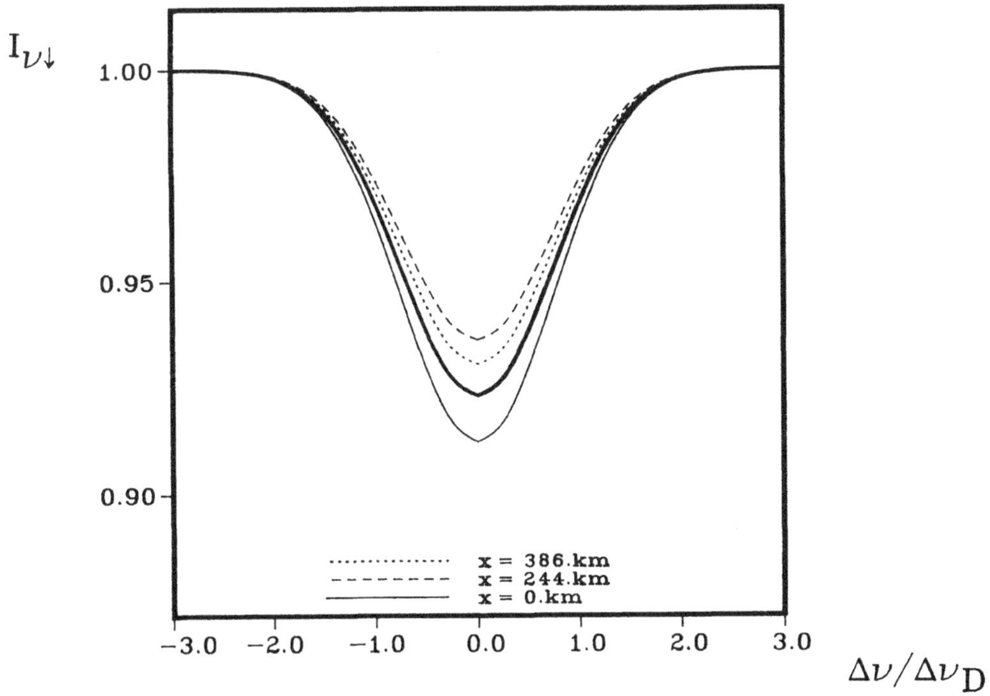

Fig. 4: Line profiles for Ti I λ5222.7 for $\mu = 1$ and different x positions. The thick line is the horizontally averaged profile

profile. By comparison with the profile for the undisturbed photosphere (x = 0. km) we notice that the line is weakened. This "line gap" is enhanced by the slightly larger continuum for the averaged case, which leads to a total intensity enhancement of 4 % in the line core. This is a rather low value and points to the necessity of non-thermal heating for the explanation of the observed line gaps.

ACKNOWLEDGEMENTS

We are grateful to Drs. F. Kneer and E. Wiehr for helpful discussions and to Dr. H. Schleicher for putting some of his routines at our disposal. This work has been supported by the Deutsche Forschungsgemeinschaft (DFG).

DISCUSSION

GARCIA DE LA ROSA: (1) Is there any observation of the bright ring that you mentioned? (2) Where do you think the size limit between large (dark) and thin (bright) fluxtubes is? Dr. Spruit suggested $\sim$ 0.8 arcsec, but we have observations at -1 Å from the Hα line center on the emergence of spot groups, which show bright features eventually converging to form dark pores, but these have diameters larger than 2 arcsec.

SCHÜSSLER: (1) Dr. Tarbell will comment on this. (2) Our results are compatible with the value given by Dr. Spruit.

TARBELL: My group at Lockheed has filtergrams in temperature-sensitive lines and magnetograms with a resolution of 0.5 arcsec or better. In pores which are clearly resolved (1–3 arcsec diameters), the entire structure is dark in continuum wavelengths, but at the line center we see a dark core with a bright ring around it. In smaller but still resolved fluxtubes ("knots" with 0.5–1.5 arcsec diameters), the contrast is neutral in the continuum and shows a neutral core with a bright ring at line center. In the smallest magnetic elements which are probably not fully resolved, we often see an offset between the bright dot at line center and the magnetic point, as your models indicate for an inclined line of sight. These offsets are sometimes seen at disk center but are more common at offset disk positions. Thus there is good qualitative agreement with your models, and we look forward to more detailed comparisons when your results have been published.

SPRUIT: Your solutions for a small tube have an almost constant field strength across the tube. To what extent is this a result of the lower boundary condition assumed?

SCHÜSSLER: It is a result of the initial mass distribution as a function of the vector potential, which does not change during the calculation due to infinite conductivity. We are currently performing calculations for other mass distributions.

SPRUIT: So it is an effect of the initial conditions?

SCHÜSSLER: Yes.

KOUTCHMY: You showed several nice pictures giving the detailed distribution of the magnetic lines of force and the distribution of the downward velocity field over the whole fluxtube. It seems to me that there are motions *across* the field lines. Is that true?

SCHÜSSLER: No. We have infinite conductivity and no flow across field lines. These are carried bodily with the fluid.

ROBERTS: Have you considered the timescale for the convective collapse process and how it compares with available linear thin tube results?

SCHÜSSLER: We have not yet considered the convective collapse realistically, because we have been mainly interested in static models which evolve. Preliminary dynamical calculations we have done show that the surface-layer structure of the tube (depths $\lesssim$ 300 km) settles to a steady state within a few sound-travel times, i.e., a few minutes.

TIME-DEPENDENT CONVECTIVE COLLAPSE OF FLUX TUBES

S.Sirajul Hasan
Indian Institute of Astrophysics
Bangalore-560034. India.

ABSTRACT

The time-dependent collapse of a slender flux tube extending vertically in the convection zone of the Sun is modelled. Starting from an initial state in which the flux tube is in hydrostatic equilibrium, the non-linear MHD equations are used to examine its temporal evolution. A detailed study of the flow variables and magnetic field within the tube is presented. It is seen that asymptotically in time a unique state of dynamic equilibrium is established, irrespective of the value of β_0 (the ratio of the thermal to magnetic pressure at the initial epoch).

Observations of the solar photosphere have revealed the existence of intense magnetic fields (1-2 kG) concentrated into isolated flux tubes (see review by Stenflo, 1976 and references therein). Parker (1978) suggested that these large fields are produced due to a convective instability. This paper aims to quantitatively study the temporal development of the collapse of a slender flux tube as a result of convective instability.

Consider a flux tube extending vertically into the convection zone. We use the ideal MHD equations for a slender flux tube as given in Roberts and Webb (1978). Noting that these equations are hyperbolic, we solve them by the method of characteristics (for details see Hasan and Venkatakrishnan, 1980).

We assume an initial state in hydrostatic equilibrium. Furthermore, the temperatures inside and outside the tube at each depth are taken to be the same. The convection zone model of Spruit (1977) along with the model atmosphere of Vernazza et al (1976)are used to specify the thermodynamic state of the fluid outside the tube. The pressure and density inside the tube are lower by a constant factor. It can be seen that the plasma parameter $\beta_0 = 8\pi p_0/B_0^2$, where p_0 and B_0 denote the pressure and magnetic field strength respectively at the initial time, is independent of depth.

J. O. Stenflo (ed.), Solar and Magnetic Fields: Origins and Coronal Effects, 73–77.

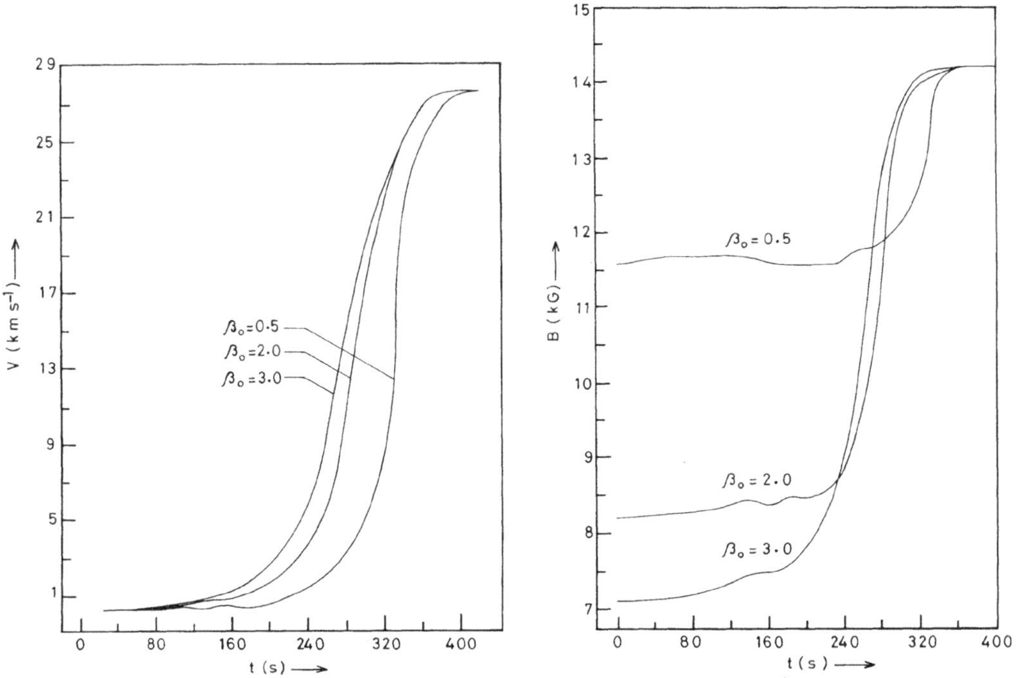

Fig.1. The variation with time of (a) velocity and (b) magnetic field for different β_0 at $z = 1400$ km.

The initial equilibrium state is perturbed by a velocity perturbation of the form $v_0 = V_\alpha \sin \{ \pi(z-z_i)/(z-z_f)\}$, where V_α denotes the amplitude of the downflow velocity and z_i and z_f refer to the depths of the boundaries. For purposes of computational economy, we chose $z_i = -200$km and $z_f = 3000$km (z is positive into the Sun). In the computations a fixed value of 200 m s^{-1} for V_α was used.

Figure 1a shows the variation with time t of the fluid velocity v near the central depth for various β_0. For $\beta_0 = 0.5$, v initially exhibits oscillatory behaviour, but subsequently increases monotonically with t. In contrast, for $\beta_0 = 3.0$, v increases monotonically with t from the initial instant. For the intermediate value $\beta_0 = 2.0$ there is an initial gradual increase followed by a plateau-like region and then a rapid increase of v. All the curves asymptotically approach a state of constant v. From a least squares analysis, it is possible to calculate a linear growth rate (for $\beta_0 = 3.0$) of about 200s which is in broad agreement with the value of about 220s one infers from fig.1 of Spruit and Zweibel (1979).

Figure 1b depicts the magnetic field strength B as a function of t (at the same depth). The temporal behaviour in this case again exhibits an initial gradual variation followed by a rapid transient

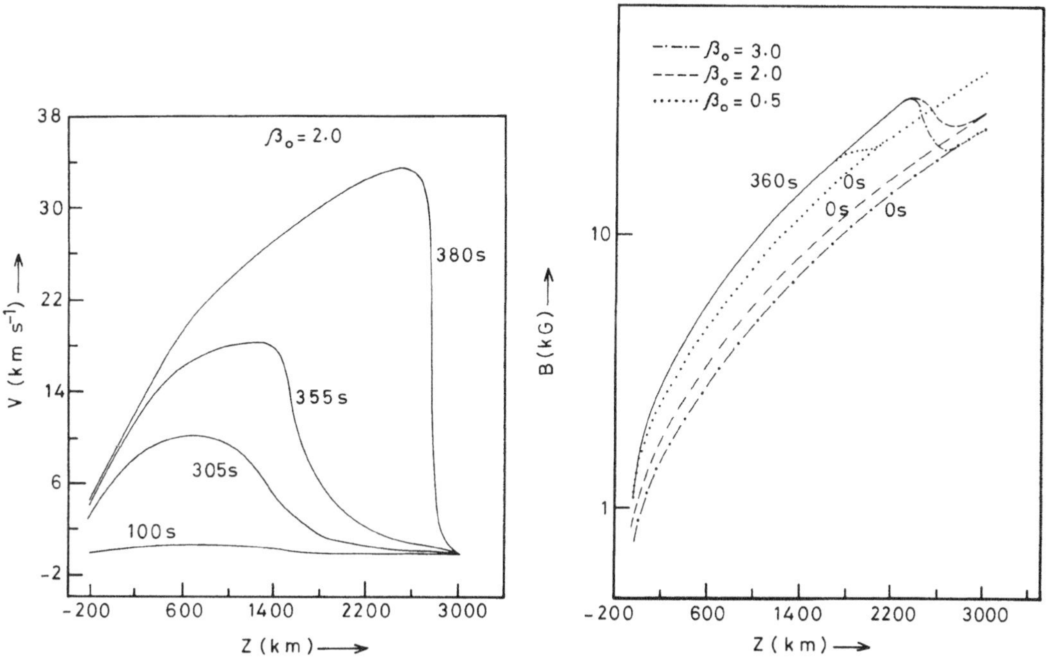

Fig.2 (a) v versus z at different times for β_o=2.0. (b) B versus z for various β_o at t = 0s and t = 360s. The solid line is used for the portion where the curves overlap.

phase and an eventual saturation to a constant value of B, which is independent of β_o.

In figures 2a and 2b, the spatial variation of v and B is shown. In figure 2a the steep gradient in v near the boundary at z = 3000 km is a consequence of assuming zero velocity there. From figure 2b, we see that curves of different β_o after sufficient time merge into one.

On the basis of our time-dependent study it is possible to discern several interesting features. Firstly, we see that a velocity perturbation in the form of a downflow can trigger a convective instability in a slender flux tube in hydrostatic equilibrium. This leads to a collapse of the tube and an intensification of the magnetic field. Secondly, there does not appear to exist a critical value of β_o (at least in the range 0.5 - 3.0) below which the instability is suppressed. This is contrary to the results of Spruit and Zweibel (1979) who find a critical value for β_o close to 2. Lastly, we notice that after large enough time the instability saturates leading to a halt of the collapse and to the establishment of a new state of dynamic equilibrium. The physical properties of this state are the same, irrespective of the choice of β_0.

Let us briefly consider the observational implications of this study. An important finding that emerges is that flux tubes formed as a result of convective collapse should be characterized by a unique value of v and B at each level (for some choice of boundary conditions). From figures 2a and 2b we find that at z=0, $B \simeq 1.5$ kG and $v \simeq 9$ km s^{-1}. In order to actually observe the collapse of a single flux tube, observations with a spatial resolution better than 0.5 arc sec and time resolution of about 50s would be desirable. Since measurements of magnetic fields are comparatively difficult, it might be easier to attempt to detect changes in downflow velocity in magnetic elements. Observations showing a rapid increase of downflow velocity could perhaps be considered as a signature of convective collapse.

REFERENCES

Hasan, S.S. and Venkatakrishnan, P.: 1980, Kodaikanal Obs.Bull.Ser. A 3, pp.6-16.
Parker, E.N.; 1978, Astrophys.J.221, pp.368-377.
Roberts, B. and Webb, A.R.: 1978, Solar Phys.56, pp. 5-35.
Spruit, H.C.: 1977, thesis, University of Utrecht, pp.26-34.
Spruit, H.C. and Zweibel, E.G.: 1979, Solar Phys.62,pp. 15-22.
Stenflo, J.O.: 1976, IAU Symp.No.71, pp. 69-99.
Vernazza, J.E., Avrett, E.H. and Loeser, R.: 1976, Astrophys.J.Supp. 30, pp. 1-60.

DISCUSSION

SPRUIT: You used a boundary condition allowing free inflow at the top at constant pressure. In my opinion this is the cause of the differences with other calculations. A nonlinear calculation of the possible equilibrium states of a fluxtube (Spruit: 1979, *Solar Phys.* **61**, p. 363) with a no-flow condition at the top shows the existence of equilibria with lower energy and higher field strength for tubes with $\beta > 2$.

HASAN: Initially, when I started this calculation, I used the no-flow condition, but this led to a practical computational problem because the pressure dropped very rapidly at the boundary. In your nonlinear calculation you assumed hydrostatic equilibrium, which may not be valid as observations show the existence of downflows in fluxtubes.

SPRUIT: I still think that a calculation with zero flow at the top would be very interesting, and I would urge you to remove these numerical difficulties.

CRAM: Are magnetic fields "concentrated" only after they emerge through the photosphere, or is it possible that the *subphotospheric* field is already in strong-field form — i.e. stronger than equipartition — before the field can be observed? *If* the field is intensified only after it emerges, then the discussed model is probably not particularly relevant, since it predicts collapse times of about 5 min, and this would not be short compared with the time scale for flux emergence itself.

HASAN: The purpose of my calculation was to study the final states that result due to the convective collapse of a fluxtube. The results show that intense fields $\sim$ 1.5 kG at the solar surface can be produced by this process. To answer your question whether the subphotospheric field is already in strong form when it emerges, one would need a separate calculation studying the dynamical evolution of a fluxtube as it rises up towards the photosphere.

WEISS: Your models imply fields of the order of 14 kG relatively close to the surface — much greater than anything that has ever been observed. These field strengths follow, I believe, from your initial assumption of a uniform β_0. Would the results be significantly affected by allowing a more plausible field distribution?

HASAN: The field strength of 14 kG that you allude to is at a depth of about 1500 km below the surface. At the surface, the field strength is only 1.5 kG, which is within the observed range. My calculations indicate that the final results depend more sensitively upon the boundary conditions, rather than on the initial conditions.

RIBES: I would like to stress the importance of dynamics on the thermal and magnetic structure of fluxtubes. Steady downdrafts in thin magnetic tubes may give various theoretical solutions: A strong field solution ($B_0 \approx 1.3$ kG) is possible with either a strong downdraft ($v_0 \approx 6$ km s^{-1}) and no excess temperature ($\Delta T_0 = 0$), or with a large excess temperature ($\Delta T_0 \approx 2400$ K) (cf. Webb's thesis) with subsonic flow. The latter case is difficult to accept because the radiative cooling time is much shorter than the dynamical time. The former case has to be ruled out if large velocities are not observed. So, the alternative is to reduce the field strength to moderate values (500 G$< B <$ 1 kG), or to give up the thin tube approximation.

HASAN: I entirely agree with your comment.

NUMERICAL 3-D SIMULATIONS OF THE COLLAPSE OF PHOTOSPHERIC FLUX TUBES

Åke Nordlund, Copenhagen University Observatory

The interaction of photospheric granular convection with a small scale magnetic field has been simulated numerically in a three-dimensional model, with an extension of techniques recently used to simulate field-free granulation. The evolution of an initially homogeneous magnetic field was followed numerically, both in a kinematic (weak-field limit) description, and in a dynamic description, where the back-reaction of the field on the motion through the Lorentz force is taken into account. The simulations illustrate the strong tendency for the field to be swept up and concentrated in the inter-granular lanes because of the topology of the granular flow. The convectively unstable stratification allows field concentration up to a kilogauss field because the temperature reduction in the magnetic plasma.

1. INTRODUCTION.

As pointed out by Parker many years ago (Parker, 1955), a simple explanation for the observed concentration of magnetic flux in the solar photosphere is provided by a reduced temperature of the magnetic plasma, with a resulting lower internal gas pressure following from (almost) hydrostatic equilibrium.

Similar explanations have been given by Spruit (1976, 1977, 1979) and others (Parker, 1978; Webb and Roberts, 1978) for the concentration of magnetic field to kilogauss strength on a small scale, as first evidenced by indirect observational methods (cf. Stenflo, 1977). Spruit (1979) showed that, for an initially homogeneous magnetic field of sufficient strength, a convectively driven instability may concentrate the field up to strengths corresponding to the external gas pressure, by "evacuation" of the interior of

J. O. Stenflo (ed.), Solar and Magnetic Fields: Origins and Coronal Effects, 79–83.

the magnetic flux concentrations. A large number of other explanations for the field concentration have also been proposed (cf. Parker, 1979 and referencs cited therein). The purpose of the work reported here is to demonstrate, by numerical simulation of the actual situation, that the convective instability is indeed responsible for the concentration of magnetic flux on a small scale in the solar photosphere. This is done by an extension of techniques recently used to simulate field-free convection in the solar photosphere (Nordlund, 1982).

2. NUMERICAL CONSIDERATIONS.

Relatively little extra numerical effort (some 15 %) is required to step the full set of MHD equations forward in time, given the effort of stepping the equations of ordinary hydrodynamics. The induction equation determines the evolution of the magnetic field density $\underline{B}$, in terms of the rotation of the electric field $\underline{E}$,

$$\partial \underline{B}/\partial t = -\mathrm{rot}\underline{E} \quad .$$

The induction equation ensures that an initially divergence free magnetic field stays divergence free, since the divergence of a rotation vanishes. Numerically, this holds true if and only if the space derivatives in the div and rot operators commute. For the horizontal variations, this is automatically the case, because of the Fourier representation of the horizontal variations. For the vertical variations, which are represented by cubic splines, commutation is ensured by defining second derivatives by a similar cubic spline representation of the first derivatives.

In the equation of motion, the Lorentz force term $\underline{i}\mathrm{x}\underline{B}$ has to be included. This requires one rot operation and one vector product operation. It should be noted that the common expansion of the Lorentz force in terms of a pressure and a tension term is unfavourable in terms of numerical effort.

3. THE KINEMATIC CASE.

In the kinematic case, the ordinary equations of hydrodynamics alone determine the evolution of the temperature, pressure, and velocity, and the induction equation just describes the evolution of the magnetic field with a given velocity field.

The initially homogeneous and vertical magnetic field is carried along by the horizontal flow from the granules towards the inter-granular lanes. Since typical horizontal velocities are 4-6 kms^{-1} and typical granular radii are 500-1000 km, field lines are carried into the intergranular lanes in a few minutes time. Below the surface, the flow is mostly vertical, and in the shape of "fingers" of descending gas. This topology favours the concentration of field lines in the downdrafts, in contrast to situations with little or no stratification, where up/down symmetric cells may form (Galloway and Moore, 1979)

In the kinematic case, there is nothing to stop the field from concentrating without limit. In a numerical description of the evolution, the field must not be allowed to concentrate to a smaller scale than the resolution element. Therefore, a diffusion term has to be included in the induction equation (resistive term in $\underline{E}$), of a magnitude sufficient to prevent excessive concentration. On the other hand, in order for the dynamic case to be realistic, diffusion must not prevent concentration before the Lorentz force becomes dominant in the equation of motion. Taken together, these two conditions imply a minimum total flux in the model determined by the numerical resolution; the higher the numerical resolution, the smaller total flux may be allowed, with small scale flux concentrations still reaching sufficient flux density. In the case considerd here, with a horizontal period of 3000x3000 km, and 256 Fourier components in the plane, an initial flux density of 35 mT (350 Gauss) was found suitable.

4. THE DYNAMICAL CASE.

When the Lorentz force is included in the equaion of motion, the concentration of magnetic flux in the intergranular lanes is retarded by the Lorentz force, directed away from the flux concentrations (magnetic pressure term in the pressure plus tension expansion). However, the retarded flow results in a smaller supply of entropy to the intergranular lanes by the flow. The layers where the radiative continuum is formed are losing entropy at a significant rate. The reduced horizontal inflow therefore results in a further reduced temperature of the descending gas. The reduced temperature corresponds to a reduced gas pressure scale height, and consequently the gas pressure in the intergranular lanes is reduced further, relative to the case of field-free convection. This offsets the retardation by the Lorentz force and allows further concentration of the magnetic flux. A (quasi-) stable situation is reached when the internal gas pressure is well below the external gas pressure, because then further reductions of the internal gas pressure have only marginal effects on the external-internal gas pressure difference. In this situation, the magnetic flux is concentrated in the intergranular lanes, with almost field-free plasma in the granular updrafts. The flow in these field-free areas evolves in a way typical of normal (field-free) granular convection; the granules grow and then split (or "explode"), because the central updraft eventually reverses, after radiative cooling overtakes the stagnating upflow. Thus, new downdrafts form in the field free areas. The magnetic flux expands horizontally into these newly formed downdrafts, because the external gas pressure in the downdrafts are insufficient to balance the internal magnetic pressure of the flux concentrations.

5. DISCUSSION, CONCLUSIONS.

The numerical simulations reported here illustrate the mechanisms responsible for the formation of small-scale magnetic flux concentrations in the solar photosphere: The topology of the granular flow is such that magnetic field is carried towards the intergranular lanes, and the heat cycle of the convective flow is such that it aids in the concentration of the field up to flux densities where the field is essentially confined by the external gas pressure of the field free fluid.

However, these flux concentrations are not stationary; the magnetic field continually readjusts in response to the evolution of granules in the field-free areas. In a certain sence, the lifetimes of the flux concentrations are thus of the order of the granular life times, even though no flux is destroyed. The stresses (shear, twist) caused by this continual readjustment propagate up into the chromosphere and the corona, along the fieldlines rooted in the photosphere. To the extent that Alfven transit times are short compared to characteristic times of evolution (short loops), the field in the corona will evolve through a series of (almost) force-free states. The magnetic field aligned currents associated with such twisted fields may cause resistive MHD instabilities similar to Tokamak instabilities (Waddell et al. 1979, Carreras et al. 1980), and may thus be of importance for the small scale chromospheric and coronal activity.

The video-tape sequence illustrating this contribution was produced with a MATROX RGB-256 frame buffer, mounted on a PDP 11/35 computer. The financial support of the Danish Space Board and the Danish Natural Science Research Council, and the access to the Niels Bohr Institute PDP 11/35 is gratefully acknowledged.

REFERENCES

Carreras, B., Hicks, H.R., Holmes, J.A., Waddell, B.V.: 1980, Phys. Fluids 23, 1811.
Galloway, D.J., Moore, D.R.: 1979, Geophysic. Astrophys. Fluid Dyn. 12, 73.
Nordlund Å. 1982, Astron. Astrophys. 107, 1.
Parker, E.N.: 1955, Astrophys. J. 121, 491.
Parker, E.N.: 1978, Astrophys. J. 221, 368.
Parker, E.N.: 1979, "Cosmical Magnetic Fields", Clarendon Press, Oxford
Spruit, H.: 1976, Solar Phys. 50, 269.
Spruit, H.: 1977, Solar Phys. 55, 3.
Spruit, H.: 1979, Solar Phys. 61, 363.
Stenflo, J. O.: 1977, in R.M. Bonnet and P. Delache (eds.), "The Energy Balance and Hydrodynamics of the Solar Chromosphere and Corona", IAU Coll. 36, 143.
Waddell, B.V., Carreras, B., Hicks, H.R., Holmes, J.A.: 1979, Phys. Fluids 22,896.
Webb, A.R., Roberts, B.: 1978, Solar Phys. 59, 249.

DISCUSSION

PROCTOR: Did you include diffusion in the model? Presumably it is essential for a steady state to be established. Does the value of the diffusion coefficient affect the final state?

NORDLUND: Yes, as explained in the text of my contribution, diffusion is necessary to prevent the flux from concentrating below the limit of numerical resolution. The value of the diffusion coefficient does not qualitatively affect the final state, since the initial value of the vertical field was chosen large enough to make the Lorentz force dominate before diffusion does. In other words, the flux density reaches values of several hundred mT in the kinematic case, whereas pressure balance limits the flux density to some 150 mT in the dynamic case. However, the value of the diffusion coefficient does influence the amount of downflow in the flux concentrations, because it determines the amount of cross-field flow allowed.

KÖMLE: Which numerical code did you use for your calculations?

NORDLUND: An extension of the pseudo-spectral code used for numerical simulation of field-free granular convection, as described in *Astron. Astrophys.* **107**, p. 1, 1982.

ROSNER: Is your code fully compressible? Furthermore, what is your grid spacing compared with the pressure or density scale height?

NORDLUND: The code is compressible, in the sense that the model extends over ten pressure scale heights. However, the $\partial\rho/\partial t$ in the continuity equation is neglected, which has the effect of filtering out sound waves (and fast-mode waves). The numerical resolution used was: Horizontally 3000 km/16 points $\approx$ 190 km, vertically $\approx$ 50 km.

CHROMOSPHERIC AND CORONAL INDICATORS OF STELLAR MAGNETIC STRUCTURE

Cornelis Zwaan
The Astronomical Institute, Zonnenburg 2,
3512 NL Utrecht, The Netherlands

1. INTRODUCTION: SOLAR AND STELLAR MAGNETIC STRUCTURE

This paper discusses attempts to derive information on magnetic structure in late-type stars, only from emission fluxes in the cores of the Ca II H and K lines, in UV emission lines and in soft X-rays - and from relations between these emission fluxes. However simpleminded and indirect this approach may seem, it brings out stimulating results. Studies of this kind are guided by solar research, so let us look first at the solar data.

The emission from the solar chromosphere and corona originates in an intricate magnetic structure. Chromospheric spectroheliograms obtained in the cores of the Ca II and Mg II resonance lines or in chromospheric UV lines show the active regions and the magnetic network - these spectroheliograms resemble magnetograms but with no indication of the polarity. Apparently chromospheric emission flux is a measure for the total magnetic flux in the chromosphere.

In soft X-rays the "coronal condensations" in active regions are very bright, yet much of the solar X-ray emission originates from fainter coronal structure outside active regions. Comparison between X-ray heliograms and calculated coronal magnetic field structure shows that the coronal emission is strongest from *closed* magnetic loops, whereas the emission is much weaker from magnetic structure that is open to the interstellar space, such as coronal holes (see Vaiana and Rosner (1978)). The X-ray flux is a measure for the magnetic flux in the solar atmosphere, but in a more complicated fashion than the chromospheric flux.

If indeed the structure of the chromospheres and coronae of late-type stars is similar to that in the Sun, then some positive albeit crude correlation is to be expected between the coronal emission flux and the chromospheric flux. In principle such a relation may depend on parameters characterizing the global stellar structure (T_{eff} and g).

The simplest measure for chromospheric and coronal emission is the energy flux per unit area at the stellar surface. This surface flux measures the average product of the brightness and the fraction

J. O. Stenflo (ed.), Solar and Magnetic Fields: Origins and Coronal Effects, 85–97.

of the stellar surfac covered by features emitting in the particular wavelength band.

The "filling factor" α, which is the fraction of the stellar surface covered with a particular feature, is another useful measure for magnetic activity. The magnetic filling factor α_B in the photosphere, which is the fraction of the area covered by a strong field of either polarity, is well defined. For the Sun α_B may be derived from magnetographic measurements, and for stars from analysis of the profiles of magnetically sensitive lines (see Marcy's invited paper during this symposium). In the photosphere the magnetic filling factor α_B approaches 1.0 only in sunspots; in dense plages within active regions it is estimated to be $0.1 \lesssim \alpha_B \lesssim 0.3$. Over the area of a mature active region we estimate $0.05 \lesssim \alpha_B \lesssim 0.10$. For the quiet network, and for the quiet sun as a whole, a rough estimate is $\alpha_B \lesssim 0.005$, and for the solar disk at sunspot maximum $\alpha_B \approx 0.01$.

Because of the fanning of the fluxtubes the filling factor α_B increases with height. Hence we expect that the filling factor for the structure that is bright in the Ca II H and K line cores, α_{H+K}, is a few times larger than the photospheric factor α_B. (Note that in routine measurements the Ca II plage areas are overestimated because of the lack of spatial resolution).

In the corona the magnetic pressure exceeds the gas pressure, hence the magnetic filling factor is unity. However, the coronal soft X-ray emission depends both on the strength and the geometry of the magnetic field, hence filling factors must be specified for each of the various features that show up in coronal emission.

2. CHROMOSPHERIC STRUCTURE; CA II H AND K EMISSION

Figs. 1a. and b. show the Ca II H and K line-core flux F_{H+K} against color (B-V) for samples of F-, G- and K-type stars. The surface flux F_{H+K} has been derived by Middelkoop (1982a,b,c) from the H and K line-core index S measured at Mt. Wilson Observatory by various observers. (The unit of F_{H+K} is approximately $7.6\ 10^5$ erg $cm^{-2}s^{-1}$). These diagrams show:

(i) Both for main-sequence stars (Fig. 1a) and for evolved stars (Fig. 1b) there are lower limits $F_{H+K,min}$(B-V) to the distributions of F_{H+K} against (B-V). (There are some points significantly below $F_{H+K,min}$ (B-V), but probably all of these are due to peculiar stars, such as subdwarfs or (sub-) giants with peculiar abundances).

(ii) There is a large spread in F_{H+K} upward from $F_{H+K,min}$. The spread in the fluxes F_{H+K} for stars of similar spectral type and luminosity class has been attributed to a spread in stellar rotation rate. Recently some quantitative relations between Ca II H and K flux and rotation rate have been derived for main-sequence stars (Vaughan et al. 1981, Middelkoop, 1981a). These matters will be discussed in session 2 of this symposium.

In this context let us try to relate the F_{H+K} values for main-sequence stars at (B-V) $\approx$ 0.66 (see Fig. 1a) with solar Ca II K line profile data collected by Oranje (1982a,b). Between summer 1979

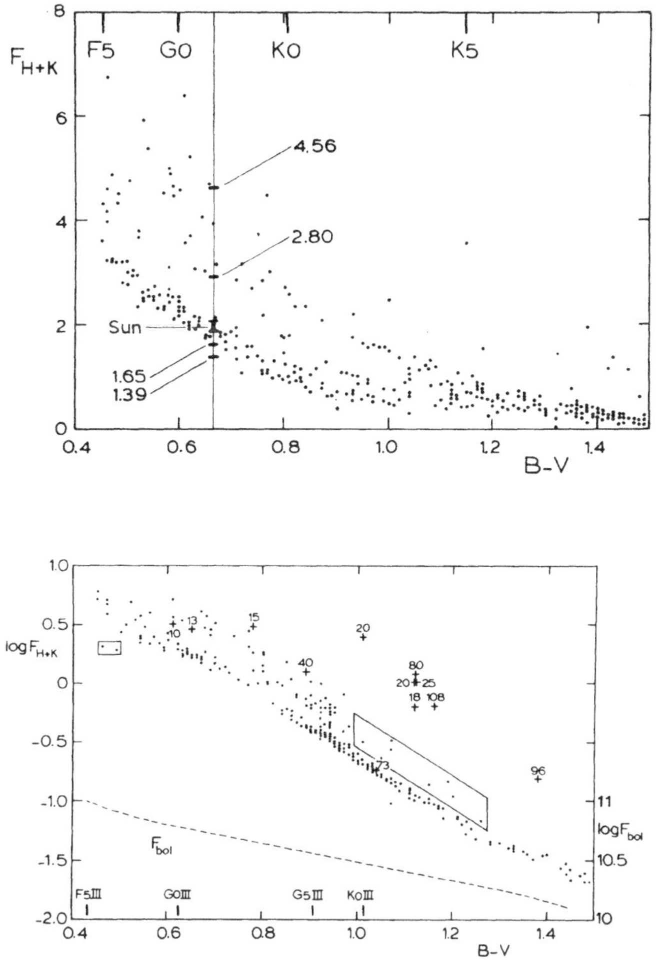

Figs. 1a and b: The Ca II H and K line-core flux F_{H+K} (in units of $7.6\ 10^5$ erg cm^{-2} s^{-1}) versus color (B-V):
a. Main-sequence stars (adapted from Middelkoop, 1982a,b). The flux levels indicated at (B-V) = 0.66 are discussed in the text.
b. Giants (Middelkoop 1982b,c). The crosses refer to spectroscopic binaries with periods P < 120 days, indicated in days. The data points enclosed in boxes refer to intrinsically bright giants.

and summer 1982 about two hundred Ca II K line profiles of the full solar disk have been recorded. Fig. 2 shows one mean profile for about ten particularly active days and one mean profile for relatively quiet days. In addition, Fig. 2 presents a profile for the part of the disk remaining when the active regions are blocked off in the

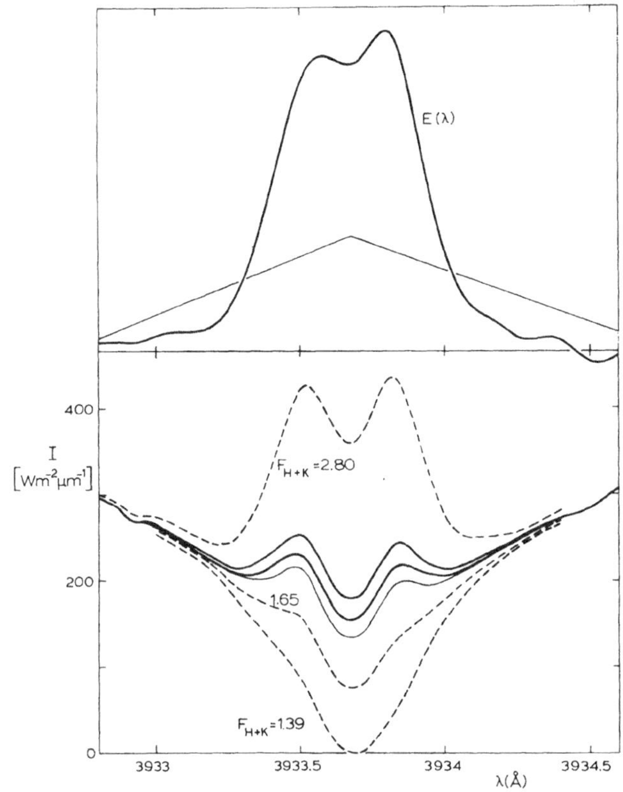

Fig. 2. Profiles of the Ca II K line core derived from full-disk solar spectra by Oranje (1982a,b). The two heavy full curves in the lower frame are average profiles, one for very active days, and one for relatively quiet days in the period summer 1979 - summer 1982. The thin line is the profile for the part of the disk remaining when the plages are blocked off. The top frame shows the emission profile E(λ) which is obtained by subtracting observed profiles. The dashed lines are profiles computed by extrapolating the solar data; the associated fluxes F_{H+K} are indicated (and also shown in Fig. 1a). The thin triangular curve is the transmission profile of the Mt. Wilson Ca II H and K photometer.

intermediate image. The resulting profile is within the error of measurement identical with the Ca II K profile in the atlas of the solar flux spectrum (Beckers et al. 1976), which has been recorded during a day without a trace of activity on the disk. This suggests that the flux from the chromospheric network does not vary significantly during the solar cycle.

The comparison of the Ca II K profiles shows that any of these profiles $I_i(\lambda)$ can be obtained from any other profile $I_j(\lambda)$ by adding (or subtracting) the *same emission profile* E(λ) multiplied with a wavelength-independent factor A_{ij}:

$$I_i(\lambda) = I_j(\lambda) + A_{ij}\, E(\lambda) \qquad (1)$$

Apparently the emission profile $E(\lambda)$, shown in Fig. 2, represents the emission added by chromospheric plages to the line profile of the quiet disk. Clearly, the emission profile is an average over the solar disk. Note that the familiar two emission peaks in the Ca II K profile can be reproduced by adding the emission profile $E(\lambda)$ to a V-shaped absorption profile. Note that in the emission profile itself the depression between the two maxima is barely visible.

The factor A_{ij} is proportional to the difference of the filling factors for the plages representative for the profiles $I_i(\lambda)$ and $I_j(\lambda)$.

Oranje (1982b) assumed that main-sequence stars with $(B-V) \approx 0.66$ differ from the Sun by the filling factor of the plages α_p, but that the emission profile is identical. If so, then the set of solar Ca II K profiles may be extrapolated to the case of stars that are more active than the Sun. Fig. 2 shows the profile for a star with $F_{H+K} = 2.80$, situated just above the Vaughan-Preston gap (see Fig. 1a). For a very active star, with $F_{H+K} = 4.56$ (see Fig. 1a), the Ca II K emission peaks reach a level between 800 and 900 W m^{-2} µm^{-1}.

In order to compute profiles that may represent stars less active than the Sun, Oranje introduced the additional assumption that the emission profile for the elements of the chromospheric network is identical to the emission profile for plages. Two profiles are shown in Fig. 2: the one labeled $F_{H+K} = 1.65$ represents a star with the lower-limit flux $F_{H+K,min}$ (see Fig. 1a). The profile $F_{H+K} = 1.39$ yields a black line core; the physical meaning of this slightly asymmetric profile depends on the soundness of the extrapolation procedure. In any case, the large corresponding flux $F_{H+K} = 1.39$ reminds us that the instrumental profile of the Mt. Wilson Ca II H and K photometer transmits quite some radiation in the line wings.

Estimates for the filling factors for plages on stellar surfaces may be obtained from the comparison between the stellar fluxes, the flux $F_{H+K} = 2.09$ associated with the upper solar profile in Fig. 2, and the flux $F_{H+K} = 1.92$ from the solar disk without active regions. We find that for the star with $F_{H+K} = 2.80$ the plage filling factor is about 5 times the plage filling factor for the active Sun. For an extremely active star with $F_{H+K} = 4.56$ (see Fig. 1a) the plage filling factor is 15 times larger than for the active Sun. Since the plage filling factor for the Sun during days of large magnetic activity is about 0.04, for an extremely active star we estimate $\alpha_p \approx 0.60$. It may seem difficult to reconcile this large filling factor with the rotational modulation observed for very active stars (Vaughan et al. 1981). Recall, however, that the magnetic filling factor of solar chromospheric plages is quite small, less than 0.1, hence the plages on active stars may be more compact.

An estimate for the filling factor of the chromospheric network may be obtained by assuming that the lower-limit $F_{H+K,min} = 1.65$ represents stars without a trace of magnetic structure left. Then comparison with the quiet Sun (F_{H+K} = 1.92) and the active Sun (F_{H+K} = 2.09) suggests that the network filling factor is about 1.5 times the filling factor of the plages during sunspot maximum.

The F_{H+K} (B-V) diagram for evolved stars (Fig. 1b) differs in several ways from the diagram for main-sequence stars (Fig. 1a). The lower limit $F_{H+K,min}$ (B-V) drops off very quickly with increasing (B-V). There is a considerable upward spread for G-type giants (B-V < 1.0). Most of the K-type giants are confined to a narrow band along the lower limit. There are two groups of evolved K-type stars that are more active: (i) many of the intrinsically bright (M_V < + 1.0) early K-type giants, and (ii) synchronized spectroscopic binaries (for details and interpretation see Middelkoop 1982b,c).

3. EMISSION FROM CORONAE, TRANSITION REGIONS AND CHROMOSPHERES

According to the expectation in Sect. 1, a crude positive correlation is found between the soft X-ray surface flux F_x and the Ca II H and K line-core flux F_{H+K} (Mewe and Zwaan, 1980). The correlation is greatly improved by plotting F_x against the Ca II H and K *excess* flux ΔF_{H+K}, defined as

$$\Delta F_{H+K} \equiv F_{H+K} - F_{H+K,min} \text{ (B-V)} \qquad (2)$$

where $F_{H+K,min}$(B-V) is the lower-limit flux introduced in Sect. 2 (Mewe et al. 1981). Fig. 3 includes recent results derived from observations obtained with the imaging proportional counter onboard the HEAO-2 Einstein Observatory. It is surprising that all stars in the available sample follow the same relation: single stars of different spectral type and luminosity class, and synchronized spectroscopic binaries. A statistical common-factor analysis yields the relation

$$F_x = 2.4\ 10^{-3}\ \Delta F_{H+K}^{1.4} \qquad (3)$$

(Schrijver et al. 1982); this relation does not depend on parameters determining the global stellar structure (e.g., radius or mass).

Note that some less active giants show a mean flux F_x that is smaller than the flux from a solar coronal hole.

Probably the spread in Fig. 3 is mainly due to the variability of the emission fluxes (the X-ray and the Ca II H and K measurements have not been obtained simultaneously). Moreover, for stars of a low activity level the excess flux ΔF_{H+K} is inaccurate.

These causes of scatter do not apply in the comparison of fluxes in various chromospheric and transition-region (TR) lines observed simultaneously with IUE in the 1200 - 2000 Å region. In order to reduce the noise we define the chromospheric flux F_{chrom} by the sum of the fluxes in lines of Si II (1808, 1817 Å) and of O I (1304 Å), and the transition-region flux by the sum of the fluxes in lines of Si IV (1394, 1403 Å), C IV (1550 Å) and N V (1204 Å) - see Oranje, Zwaan and

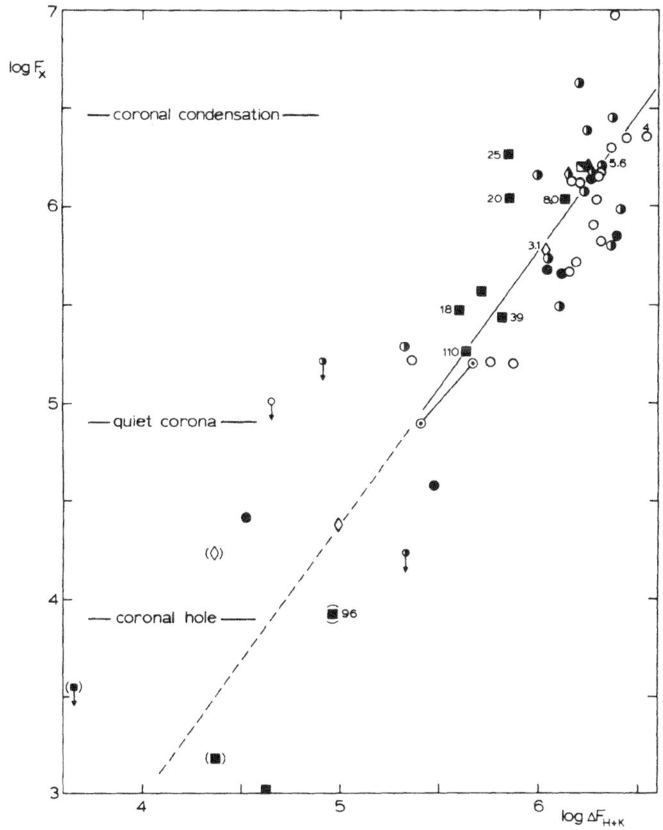

Fig. 3: Soft X-ray flux F_x versus excess Ca II line-core flux ΔF_{H+K} (surface fluxes in erg $cm^{-2}s^{-1}$). Symbols:

B-V LC	≤ 0.60	0.60...≤ 0.80	> 0.80
V	○ F6 - G2	◑ G0 - K0 ⊙ Sun	● K1 - K5
IV	◇ G0	◆ G4, G5	◆ G8, K2
III, II	-	◩ G5	■ G5 - K2

Arrows indicate upper limits, brackets mark uncertain values. For short-period spectroscopic binaries the periods are given in days. For the sources of F_x, ΔF_{H+K} data: see Schrijver et al. (1982). The solar data are from Vaiana and Rosner (1978). The line represents relation (3).

Figs. 4: Transition-region flux F_{TR} versus chromospheric flux F_{chrom} (surface fluxes in erg cm^{-2} s^{-1}): a. (below): dwarfs, yellow giants and supergiants, specific active stars; b. (opposite page): orange and red giants and supergiants, Ke and Me dwarfs, superimposed on the data points from Fig. 4a.
The line represents relation (4). Short-period binaries are marked by their period in days. Arrows indicate upper or lower limits.

Fig. 4a:

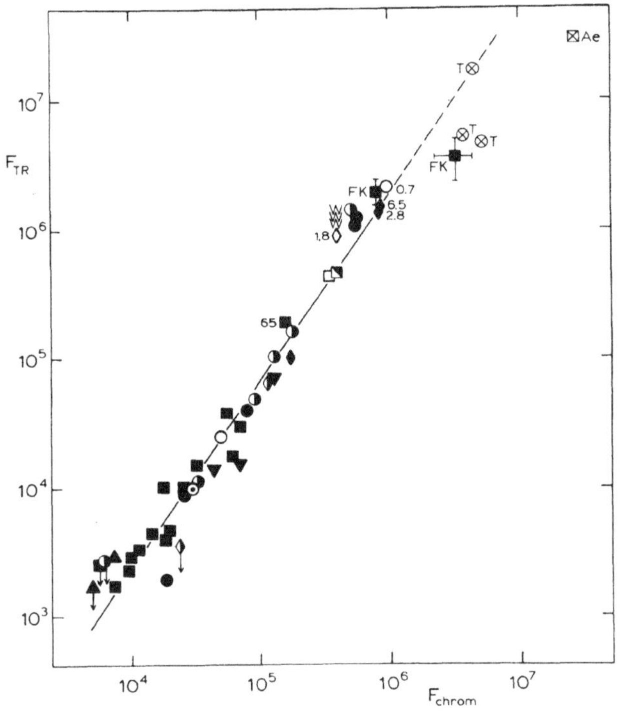

LC \ B−V	< 0.60	0.6...≤ 0.8	0.8...≤ 1.0	> 1.0	
V	○ G0	◑ G2,G8	● K1,K2	● K5	
IV	◇ F6	◈ G4,G5	◆ (G2,G5)	-	see
III	□ F9	◪ G0	■ G5 - K0	-	Fig. 4b
II	-	-	▲ G8,K2	-	
I	-	-	▼ G0,G2	-	

Sources: Ayres, Marstad and Linsky (1981); Ayres et al. (1982); de Castro et al. (1981); Hartmann, Dupree and Raymond (1982); Linsky et al. (1982); Mullan and Stencel (1982); Oranje (1982b); Oranje, Zwaan and Middelkoop (1982); Simon, Linsky and Stencel (1982).

W ◑ or W ● : W UMa stars: Rucinski and Vilhu (1982)
FK ■ : FK Com stars: Bopp and Stencel (1981)
T ⊗ : T Tau stars: Cram et al. (1980); Tjin A Die (1982)
Ae ⊠ : Herbig Ae star: Tjin A Die et al. (1982)

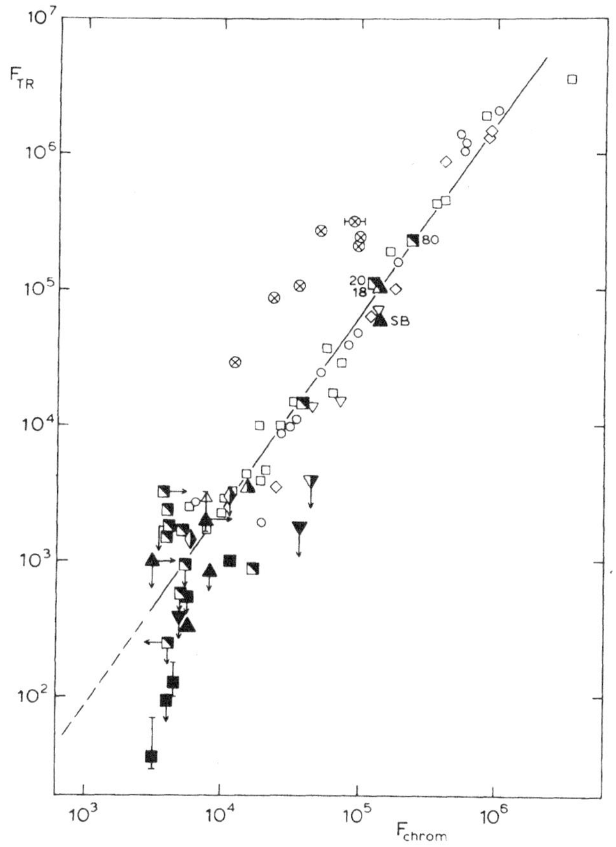

Fig. 4b: Small open symbols: data points from Fig. 4a.

LC \ B-V	1.0... ≤ 1.2	> 1.2	
IV	◆ G8 – K2	–	sources: see caption Fig. 4a
III	◪ K0 – K2	■ K3 – K5	
II	▲ K0, K1	▲ K0, K3	
I	▼ G5	▼ G8, K5	
V	⊗ dKe, dMe stars: Linsky et al. (1982)		

Middelkoop (1982). The results are shown in Figs. 4a and 4b.

Fig. 4a shows main-sequence stars (except the Ke and Me dwarfs) and evolved stars with (B-V) < 1.00. The fluxes F_{TR} and F_{chrom} are closely related. The data are fitted well by the relation suggested by Oranje et al. (1982):

$$F_{TR} = 4.3\ 10^{-3}\ F_{chrom}^{1.44} \qquad (4)$$

The W UMa contact binaries and the very active FK Com stars follow the same relation. Even the pre-main-sequence stars, the T Tau stars (in the quiescent stage) and the Herbig Ae star HR 5999 (A 7 IIIe), seem to fit to the top of the relation (Tjin A Die, 1982).

Fig. 4b shows the evolved stars with (B-V) > 1.00, superimposed on the data in Fig. 4a. The bin 1.00 < B-V ≤ 1.20 is chosen to enclose the region (V-R) ≈ 0.80. There a dividing line has been suggested (Linsky and Haisch, 1979), separating yellow giants, which may show evidence of 10^5 K plasma, from red giants and supergiants which show no evidence of plasma hotter than 10^4 K. The bin (B-V) > 1.20 is well beyond (V-R) ≈ 0.80.

Fig. 4b shows that some of the K-type giants follow nicely the trend indicated by Fig. 4a - all the synchronized spectroscopic binaries do. However, the K-type evolved stars of low activity clearly fall below the relation (4), which suggests a vertical asymptote near $F_{chrom} \approx 3\ 10^3$; the flux measurements from IUE spectra by Simon, Linsky and Stencel (1982) provide most of the data for this result. The present data suggest that there is a lower limit to the UV chromospheric emission, a ubiquitous emission which is not associated with a detectable TR emission at $T \approx 10^5$ K. We point out that this finding is similar to the lower-limit for the Ca II H and K flux.

The present sample supports the notion of a lower-limit chromospheric flux associated with disappearing transition-region flux only for red giants and supergiants. It is not yet known whether such a lower-limit F_{chrom} exists for extremely quiet yellow (sub-)giants or for main-sequence stars. The present sample is too limited to find a possible dependence of the lower-limit F_{chrom} on T_{eff} or g, although there is a hint in Fig. 4b that the lower-limit F_{chrom} is larger for supergiants than for giants.

There is one category of stars that markedly deviates from the relation in Figs. 4a and 4b: the Ke and Me dwarfs (Linsky et al. 1982). These active stars show transition-region lines which are enhanced with respect to the chromospheric lines (see Fig. 4b).

4. DISCUSSION

Stars with convective envelopes show a wide range of chromospheric and coronal activity: the data discussed in this paper cover 3 decades in chromospheric surface flux, 5 decades in transition-region flux and 4 decades in coronal soft X-ray flux (Figs. 3, 4a, 4b).

There are surprisingly close relations between the chromospheric, transition-region and coronal fluxes. These relations by themselves do not depend on the parameters characterising the global stellar structure; only the Ke and Me dwarfs depart clearly from the relations. So, if one of the fluxes ΔF_{H+K}, F_{chrom}, F_{TR} or F_x is given, the other fluxes may be estimated with reasonable accuracy.

Apparently there is one "activity parameter" which determines the structure of the outer stellar atmosphere, as far as the emission fluxes from the various atmospheric components are concerned. Probably this activity parameter is entirely determined by the magnetic structure in the atmosphere, for instance by the mean magnetic flux density. It is now generally accepted that the activity parameter depends strongly on the stellar rotation rate. The wide range of activity levels among the G-type main-sequence stars and giants is readily explained by the large range of rotation rates among these stars. However, it is to be expected that the classical parameters characterizing the global stellar structure, the effective temperature T_{eff} and surface gravity g, also enter into the activity parameter. For instance, the stellar dynamo probably depends on the (relative) depth of the convective envelope, hence on T_{eff} and g. Moreover, the part of the coronal magnetic structure that consists of closed loops, and the parameters characterizing these loops, are expected to depend on the surface gravity g (see contributed paper by Mewe et al. during session 2).

In order to obtain a close and general relation between the X-ray flux and the Ca II H and K line-core flux (Fig. 3) it is necessary to subtract a lower-limit flux which depends on (B-V) and luminosity class. The discussion in Sect. 2 shows that at least for solar-type main-sequence stars a substantial fraction of this lower-limit flux is due to photospheric radiation in the Ca II H and K line wings. However, there is a truly chromospheric component. In Fig. 4b we find an indication that there is a lower-limit chromospheric flux in the UV emission lines which is not associated with a detectable flux in transition-region lines, at least for red giants and supergiants. We suggest that these lower-limit fluxes come from *ubiquitous* chromospheric emission that may originate from the part of the magnetic network that is connected with coronal holes, and from the non-magnetic part of the chromosphere.

Both the X-ray flux and the transition-region flux increase with the chromospheric flux to a power $\alpha \approx 1.4 > 1$. Similar non-linearities, with somewhat different exponents, have been found by others, e.g. by Ayres, Marstad and Linsky (1981). Qualitatively this non-linearity may be understood: the number of closed coronal loops (origins of F_X and, indirectly, of F_{TR}) increases faster than linearly with the number of magnetic poles in the photosphere and chromosphere (origins of ΔF_{H+K} and F_{chrom}). However, quantitatively the closeness of the relations with $\alpha \approx 1.4$ over the wide range of activity levels is surprising and yet unexplained.

Finally, let us consider some observational studies that may set the interpretation of the chromospheric and coronal activity on a quantitative footing.

The first problem is the calibration of the "chromospheric magnetometers" F_{H+K} and F_{chrom}. This requires solar studies comparing magnetic flux densities measured with a magnetograph and simultaneously measured chromospheric fluxes F_{H+K} and F_{chrom} in young, mature and old active regions, and in the network, at various positions on the disk. Does the relation between chromospheric and magnetic flux depend on the type of the region? Results obtained in this way may be generalized to stars that are very similar to the Sun. Clearly, this job should be supplemented by a stellar program simultaneously measuring chromospheric fluxes (e.g. F_{H+K}) and magnetic fields in active stars, as described by Dr. Marcy during this symposium).

The interpretation of the close relations between F_x, F_{TR} and F_{chrom} or F_{H+K} for stars asks for a renewed study of coronal, transition-region and chromospheric emissions from various solar regions at various positions on the disk. These regions should be selected according to coronal criteria, such that a coherent coronal area is covered, for instance a complete coronal condensation, a large area in a coronal hole, etc. From these data "solar-type stars" may be synthesized for comparison with real stars. This program may be a considerable job of teamwork, but the results will be rewarding. I take it that the necessary observational data are in the ATM archives.

ACKNOWLEDGEMENTS

H.R.E. Tjin A Die communicated IUE UV line fluxes for pre-main-sequence stars. B.J. Oranje permitted to include results prior to publication; he and C.J. Schrijver were helpful in the preparation of this paper. A.G. Hearn commented on a draft. E. Landré has drawn the figures and Mrs. J. Odijk has prepared the typescript.

REFERENCES

Ayres, T.R. et al. 1982: Astrophys. J., in press.

Ayres, T.R., Marstad, N.C., and Linsky, J.L. 1981: Astrophys.J. *247*, 545.

Beckers, J.M., Bridges, C.A., Gilliam, L.B. 1976: A High Resolution Spectral Atlas of the Solar Irradiance from 380 to 700 Nanometers, AFCL-TR-76-0126(I), Environment Research Paper, No. 565.

Bopp, B.W., Stencel, R.E.: 1981, Astrophys. J. *247*, L 131.

Castro, E. de, et al. 1981: Astron. Astrophys. *102*, 207.

Cram, L.E., Giampapa, M.S., Imhoff, C.L. 1980: Astrophys. J. *238*, 905.

Hartmann, L., Dupree, A.K., Raymond, J.C. 1982: Astrophys. J. *252*, 214.

Linsky, J.L. et al. 1982: Astrophys. J., in press.

Linsky, J.L., Haisch, B.M. 1979: Astrophys. J. *229*, L 27.

Mewe, R., Zwaan, C. 1980: Proc. Workshop Cool Stars, Stellar Systems and the Sun, Smithsonian Astrophys. Obs. Special Report 389, p. 123.

Mewe, R., Schrijver, C.J., Zwaan, C. 1981: Space Sci. Rev. *30*, 191.

Middelkoop, F. 1982a: Astron. Astrophys. *107*, 31.

Middelkoop, F. 1982b: Thesis Utrecht.

Middelkoop, F. 1982c: Astron. Astrophys., in press.

Mullan, D.J., Stencel, R.E. 1982: Astrophys. J. *253*, 716.

Oranje, B.J. 1982a: Astron. Astrophys. *109*, 36.

Oranje, B.J. 1982b: in preparation.

Oranje, B.J., Zwaan, C., Middelkoop, F. 1982: Astron. Astrophys. *110*, 30.

Rucinski, S.M., Vilhu, O. 1982: preprint.

Schrijver, C.J., Mewe, R., Zwaan, C. 1982: Advances in Space Research, Proc. 24th COSPAR Conference, Ottawa, in press.

Simon, T., Linsky, J.L., Stencel, R.E. 1982: Astrophys. J., in press.

Tjin A Die, H.R.E. 1982: private communication.

Tjin A Die, H.R.E. et al. 1982: Astron. Astrophys. *106*, 98.

Vaiana, G.S., Rosner, R. 1978: Ann. Rev. Astron. Astrophys. *16*, 393.

Vaughan, A.H. et al. 1981: Astrophys. J. *250*, 276.

DISCUSSION

HAISCH: With regard to the determination of chromospheric fluxes from the strength of the O I triplet (1302–1305 Å), I would like to point out that, especially for giant stars, this line is pumped by hydrogen Lyman α and thus may not be a valid chromospheric diagnostic.

ZWAAN: The use of the O I triplet is questionable. However, there are very few strong, low-temperature lines in the $\lambda < 2000$ Å region, and we were happy to see from the work of Ayres, Marstad, and Linsky that the O I complex appears to behave as a chromospheric line.

BASRI: Can you comment on the fact that your results do not agree with the results I have presented here, or those of Ayres, Marstad, and Linsky. These show, for example, a much steeper dependence of X-rays on chromospheric flux.

ZWAAN: We have noticed the differences, but as yet we have no explanation. Oranje is analysing the relations between the fluxes in the Ca II H and K, Mg II h and k, and the chromospheric lines with $\lambda < 2000$ Å.

DUPREE: The search for an "activity parameter" is complicated by observations of the four Hyades giant stars ($\sim$ K0 III) by Baliunas, Hartmann, and Dupree, which show a factor > 6 in the range of C IV flux, and > 10 in soft X-ray flux for the three giants observed by Stern et al. Since these stars are thought to be similar in most physical parameters (T_{eff}, g, age, chemical composition, and rotation rate), such observations strongly suggest that a long-term activity cycle may be present that is similar to the one known in the sun.

II. ROTATION – ACTIVITY – CYCLE – AGE CONNECTION: OBSERVATIONS

THE OBSERVED RELATIONSHIPS BETWEEN SOME SOLAR ROTATION PARAMETERS AND THE ACTIVITY CYCLE

Robert Howard
Mount Wilson and Las Campanas Observatories,
Carnegie Institution of Washington
Pasadena, California 91101, U.S.A.

Barry J. LaBonte
Institute for Astronomy
University of Hawaii
Honolulu, Hawaii 96822, U.S.A.

ABSTRACT

Several parameters of the solar rotation show variations which appear to relate to the phase of the solar activity cycle. The latitude gradient of the differential rotation, as seen in the coefficients of the $\sin^2$ and $\sin^4$ terms in the latitude expansion, shows marked variations with the cycle. One of these variations may be described as a one-cycle-per-hemisphere torsional oscillation with a period of 11 years, where the high latitudes rotate faster at solar activity maximum and slower at minimum, and the low latitudes rotate faster at solar activity minimum and slower at maximum. Another variation is a periodic oscillation of the fractional difference in the low-latitude rotation between north and south hemispheres. The possibility of a variation in the absolute rotational velocity of the sun in phase with the solar cycle remains an open question. The two-cycle-per-hemisphere torsional waves in the solar rotation also represent an aspect of the rotation which varies with the cycle. We show that the amplitude of the fast flowing zone rises a year before the rise to activity maximum. The fast zone seems to be physically the more significant of the two zones.

1. INTRODUCTION

Variations of solar rotation in phase with the 11-year spot cycle have been sought by several investigators (Howard, 1979; Livingston and Duvall, 1979; Belvedere et al., 1979). The discovery of large-scale plasma flows in step with the cycle would provide important clues to the mechanism which generates the magnetic activity of the sun.

Measurements using sunspots as rotation tracers suffer from potential systematic effects, as the nature of the spots and their connection

J. O. Stenflo (ed.), Solar and Magnetic Fields: Origins and Coronal Effects, 101–111.

to the solar plasma may vary through the cycle. Spots are also limited in their latitude range, precluding studies of polar rotation. Doppler shift measures cover all latitudes, but may also have systematic errors in their absolute calibration (Scherrer et al., 1980). Because of such errors we expect that variations of the absolute rotation rate may be difficult to verify, but variations of the shape of the differential rotation, a relative measurement, may be easier to detect. A recent paper (Howard et al., 1983) describes and presents the Mount Wilson velocity data since 1967, and we use that data here to measure variations of solar rotation correlated with the activity cycle.

The torsional waves on the sun, discovered from the Mount Wilson velocity data (Howard and LaBonte, 1980; LaBonte and Howard, 1982a) are in a sense a parameter of the latitude distribution of the solar rotation. These 11-year oscillations are clearly associated with the solar cycle in both period and latitude motion, with the latitudes of sunspot appearance and torsional velocity shear coincident.

In this paper we will discuss correlation of the differential rotation and the torsional oscillations with the activity cycle. We will exclude from our study the global velocity fields that appear symmetric about the central meridian (e.g., limbshift and "ears"). Those velocity fields and their relation to magnetic activity are considered in detail elsewhere (LaBonte and Howard, 1982b). All the observations discussed here were made with the 5250.2 Å line of Fe I, with a square aperture of 17.5 arcseconds until 1975, and 12.5 thereafter. Details of the instrument, observation procedure, and reduction procedures are given in earlier papers.

The dataset of Howard et al. (1983) contains 3164 full-disk velocity maps in the interval January 1, 1967 to March 31, 1982. The data are fitted by least squares to determine the angular rotation coefficients $\underline{A}$, $\underline{B}$, $\underline{C}$, in the polynomial equation

$$\omega(\phi) = \underline{A} + \underline{B} \sin^2\phi + \underline{C} \sin^4\phi,$$

where ϕ is the solar latitude. Plots of the coefficients are given in that paper. In practice, the fit is done for the full disk and for the north and south hemispheres separately. In Section 2 we consider those variations of solar rotational velocities which are measured by the coefficients $\underline{A}$, $\underline{B}$, and $\underline{C}$ in this simple representation. In Section 3 we consider those velocity variations that are not measured by the coefficients, because of their higher latitudinal wavenumber.

2. DIFFERENTIAL ROTATION VARIATIONS

A. Absolute rotation rate

The absolute value of the equatorial rotation rate, $\underline{A}$, is most sensitive to systematic error. By examination of a plot of the time

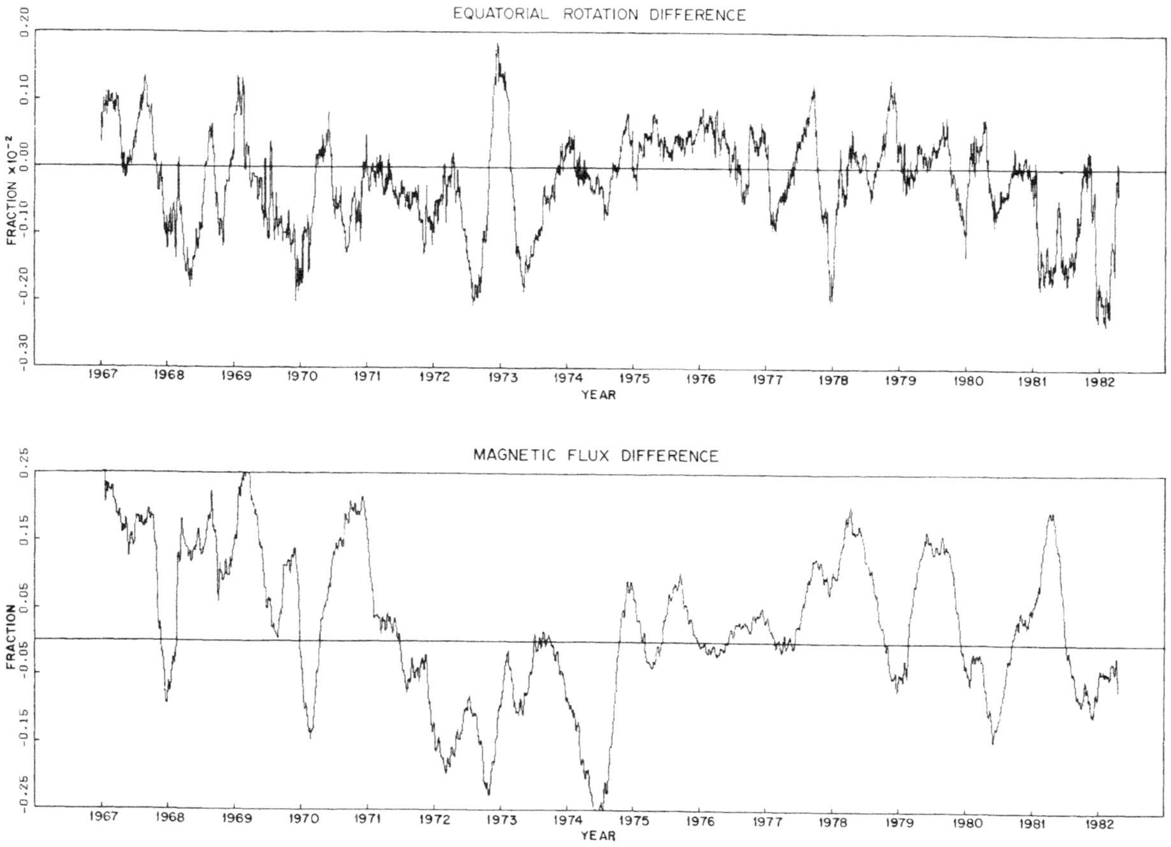

Fig. 1. (Top) Fractional difference of the rotation coefficients A determined from data in the north and south hemispheres separately (N-S/N+S). (Bottom) Fractional difference of the total magnetic flux in the two hemispheres (N-S/N+S). Daily values are averaged over intervals of 4 Carrington rotations (109.1012d). Although A is strictly the equatorial rotation rate, it is influenced by data over a range of low latitudes. The roughly sinusoidal variation of the rotation difference is not seen in the magnetic flux difference.

series of *A*, we can set a limit on the 11-year variation of ∿2% (0.06 μ rad s^{-1}) sinusoidal amplitude. Better limits from the Mount Wilson data will require an improved understanding of the remaining sources of systematic error. Livingston and Duvall (1979) reported an increase of the equatorial rotation rate of 3.7% over the interval 1966 to 1978, with no apparent 11-year periodic variation. Scherrer *et al.* (1980) found no variation >0.5% in the equatorial rotation rate over the interval 1975 to 1980, from activity minimum to maximum.

B. Hemispheric rotation difference

The coefficient *A* in the solution for the angular rotation is affected by the data over a range of latitudes near the equator, and in a sense, represents a mean rotation rate over that range. In Figure 1 is shown the fractional difference of the equatorial rotation rates derived from the data in the north and south hemispheres separately. There is a variation of this parameter, which may be periodic at a period of 11 ±3 years. The sense of the difference is that the north hemisphere rotated faster in the years 1974 to 1980. The peak-to-peak amplitude of the fractional difference is 1.5±0.3 x 10^{-3} m $\sec^{-1}$. This corresponds to a peak velocity difference between hemispheres of 1.5 m s^{-1}. Thus each hemisphere rotates faster and slower by ∿1 m s^{-1}, with roughly an 11-year period. This can be considered a torsional oscillation of wavenumber 0.5 hemisphere^{-1}.

This rotation difference between the hemispheres may be a deep-seated phenomenon, or it could be a simple response to the unequal distribution of magnetic flux in the two hemispheres. Figure 1 also shows the fractional difference of the total magnetic fluxes in the north and south hemispheres. There is no clear relation between the fractional flux difference and the fractional rotation difference on long time intervals. The relative motion of the north and south hemispheres is not simply related to the distribution of surface magnetic flux.

C. Latitude gradient of rotation

The coefficients *B* and *C* measure the latitude gradient of the angular rotation rate. The ratios *B/A* and *C/A* should be free of systematic error, and LaBonte and Howard (1982a) showed that those ratios vary in phase with the 11-year cycle. The behavior of *C/A* is shown in Figure 2; *B/A* is anticorrelated with *C/A* and is shown in Howard *et al.* (1983).

This variation of latitude gradient can be considered a torsional oscillation of wavenumber 1 hemisphere^{-1}. The ratio *C/A* varies about its average value (-0.171) by 0.020 ±0.003 sinusoidal amplitude; the ratio *B/A* varies about its average value (-0.122) by 0.009 ±0.002 sinusoidal amplitude. The velocity amplitude is ∿10 m s^{-1} peak to peak, and may be concentrated at high latitudes (Livingston and Duvall, 1979). Our measures of the true latitude distribution may be colored by the form of the differential rotation curve adopted.

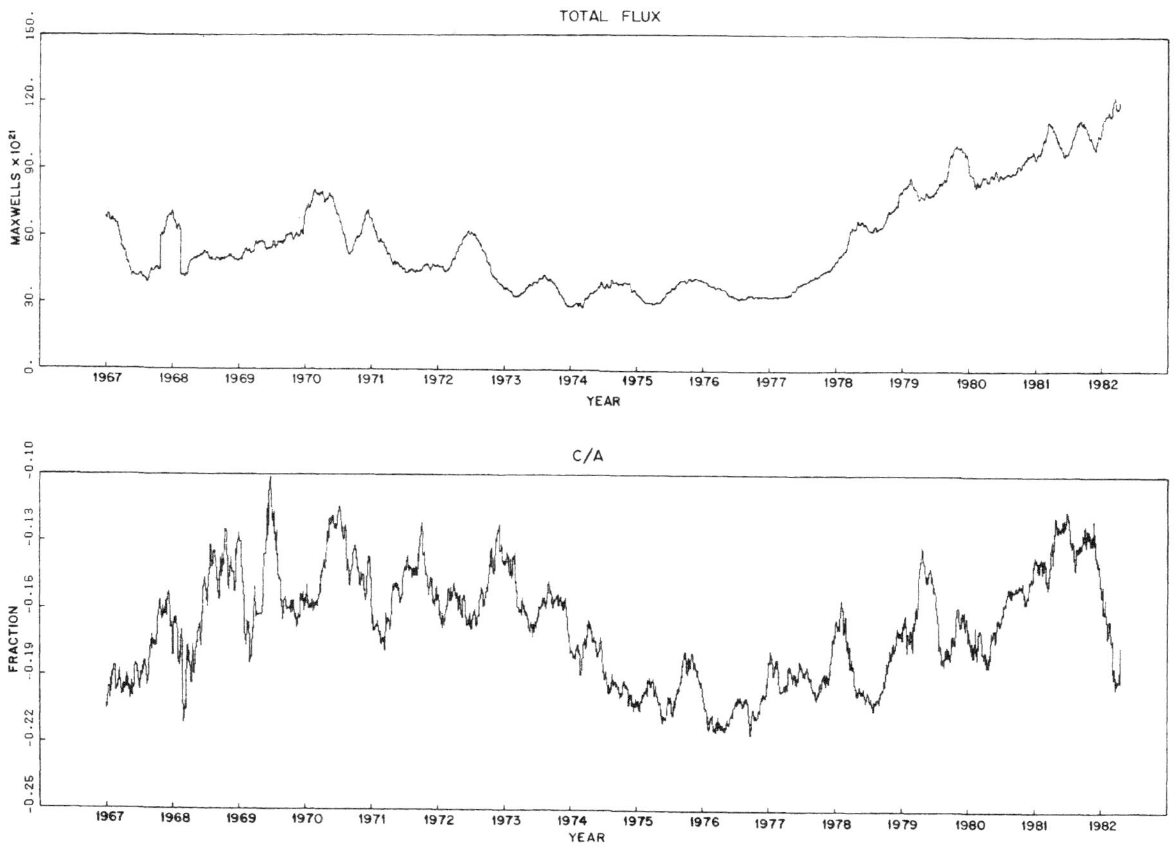

Fig. 2. (Top) Disk total magnetic flux. (Bottom) Ratio of the full-disk rotation coefficients C/A. Daily values are averaged over intervals of 4 Carrington rotations. The ratio C/A shows a clear cycle-related variation. Note that the total magnetic flux has continued to increase after the time of maximum sunspot number.

For comparison, the disk total magnetic flux is also shown in Figure 2. The sense of the variation of $\underline{B}/\underline{A}$ and $\underline{C}/\underline{A}$ is that high latitudes rotate faster at spot maximum and slower at spot minimum.

3. TORSIONAL OSCILLATIONS

The difference of the rotation rates measured independently in each of 34 zones of equal interval in sine latitude from the rotation rates given by the polynomial fit is a residual east-west velocity field. In previous papers we have shown that this residual velocity field takes the form of a travelling torsional wave of wavenumber 2 hemisphere^{-1}. The data representation used here is improved from that used in the earlier papers. The dataset now extends to April, 1982, well past the sunspot maximum of Cycle 21. In each time interval, the average of the residual velocities in the central 32 latitude zones was removed; this corrected the residuals for a slight bias ($\sim$1 m s^{-1}) that was a part of the average solar rotation rate. The time averaged values of the two equatormost latitude zones were also removed, because there is a constant rotation deficit $\sim$5 m s^{-1} at the equator which is too small in latitude extent for the polynomial representation of the differential rotation to remove. The resulting residual velocity map is shown in Figure 3. All systematic bias has been reduced to <1 m s^{-1}. For comparison with the magnetic activity, a map of the latitude distribution of magnetic flux (butterfly diagram) is also shown in Figure 3.

The improved representation of the torsional velocity more clearly shows the motion of the wave to the equator. The continuity of the waves as they travel from high to low latitudes is also more easily seen. The westward zones of the waves which will be cospatial with the Cycle 22 activity zones later in this decade are now distinctly seen to have moved from the poles to latitudes $\sim$45 degrees north and south.

We noted previously (LaBonte and Howard, 1982a) that the westward flowing zones of the waves appeared to have increased amplitude in the years before spot maximum. The velocity amplitude has now returned to its former value. In Figure 4, the average velocity amplitude of the Cycle 21 westward flowing zones is plotted, along with the total magnetic flux in the Cycle 21 activity zones. The velocity amplitude is seen to precede the magnetic activity, increasing $\sim$1 year before the total magnetic flux, as judged from the times of the steep increases in their relative amplitudes. An equal lead of the torsional velocity relative to the sunspot number has also been found, but is not plotted here. The data are not good enough at this time to measure the time difference between the torsional wave and magnetic flux in the north and south hemispheres separately. Only the westward flowing zone of the wave shows this amplitude increase; the eastward flowing zone has unchanged amplitude throughout this time interval.

Note also that although the maximum sunspot number (1 year smoothed value) was reached in 1980.0, the total magnetic flux in the activity

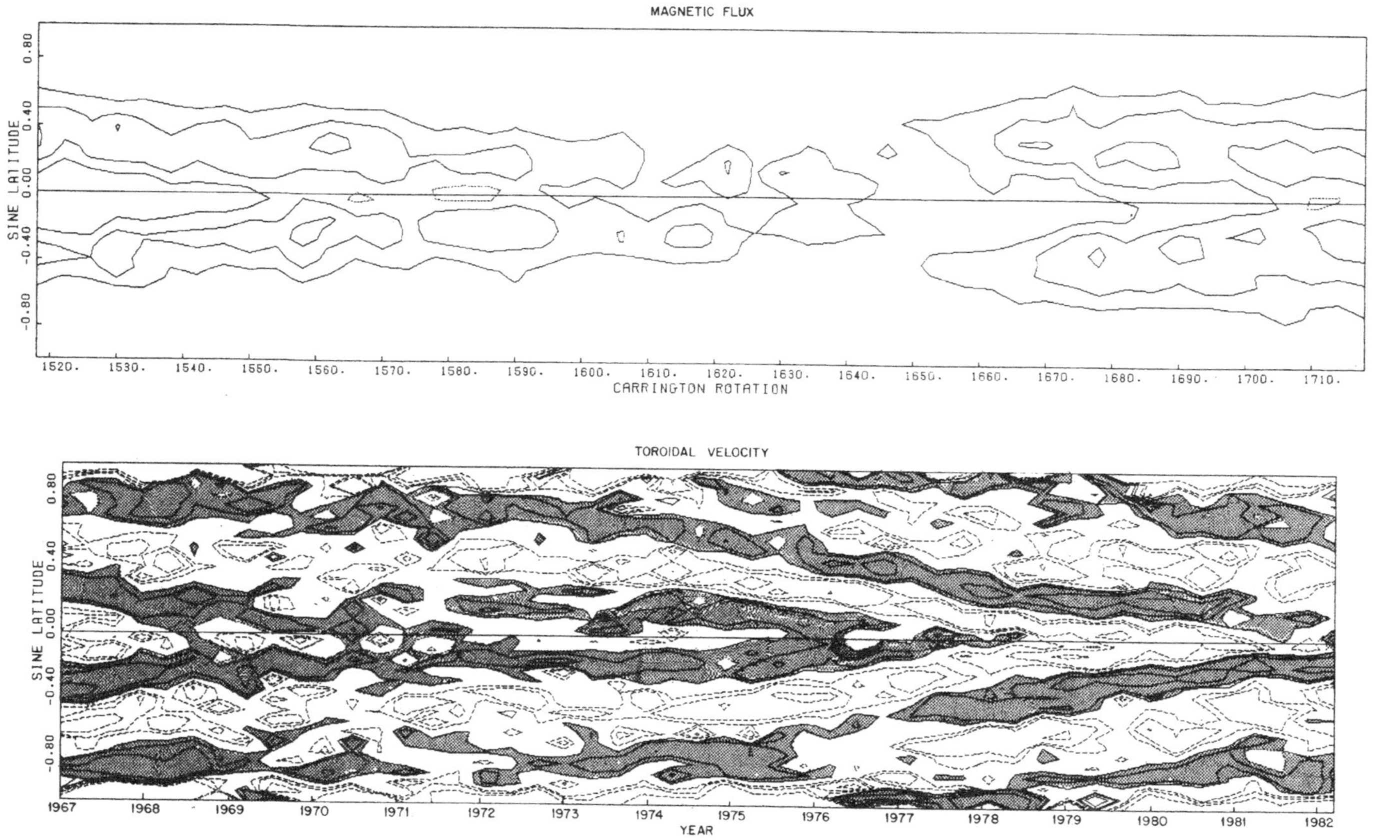

Fig. 3 (Top) Total magnetic flux. There are 34 zones of equal intervals in sine latitude. (Bottom) Zonal velocity plot from the Mount Wilson data since 1967. Daily values are averaged over intervals of 4 Carrington rotations. The zonal velocity is the residual from the smooth polynomial differential rotation curve. A constant rotation deficit at the equator has also been removed, which cannot be fit by the polynomial form. The average velocity in each time average has been set to zero. Velocity contours are ±1.5, 3, 6 m s^{-1}. Flux contours are 1.5, 3, 6 x 10^{21}Mx. Positive velocities (solid contours) are westward flows.

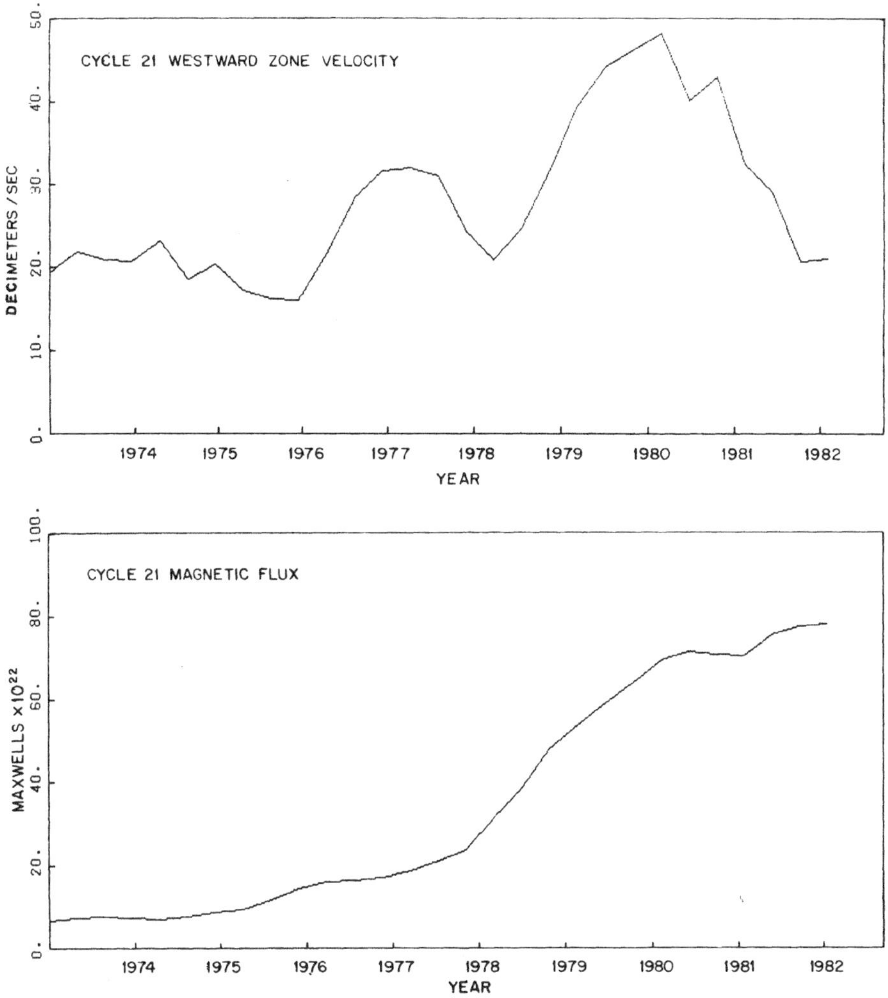

Fig. 4 (Top) Average velocity of the westward flowing zone of the torsional wave associated with Activity Cycle 21. (Bottom) Total magnetic flux in Cycle 21. Data is averaged over intervals of 12 Carrington rotations. Data from north and south hemispheres is combined. Magnetic fluxes were summed over a latitude band of width 0.47 in sine latitude, centered on the shear zone of the torsional velocity wave. The velocity increases ∿1 year before the rapid rise in magnetic flux.

zones has continued to rise since that time. This continuing increase in magnetic flux is presumably due to the presence of fewer but larger active regions at the current time, compared to the time of spot maximum. A similar effect was seen in Cycle 20, with the spot maximum in 1968.8 while the magnetic flux continued to increase until 1970.3.

4. INTERPRETATION

The most interesting result of this study is the possibility of predicting the onset of solar activity with a lead time of ∿1 year. The amplitude increase of the westward flowing zones of the torsional wave can serve as a marker of the impending rapid rise of sunspot and magnetic activity on the surface. Close observations of the torsional wave during the next spot minimum will test whether this is a reliable harbinger of activity.

The correlation of only the westward flowing zone amplitudes with the later activity increase also indicates that it is those zones which are the most physically meaningful in the operation of the wave. It appears that we must understand a mechanism which accelerates the westward flow (faster rotation), rather than a mechanism which decelerates the eastward flow (slower rotation). Some explanations of the torsional wave, for example, invoke a deceleration of the eastward zones by magnetic tension forces; such a model seems inconsistent with the constancy of the eastward flow amplitudes and the activity correlation of the westward flow.

The lag of the magnetic activity relative to the torsional velocity increase provides further evidence that the wave is not an artifact of small-scale activity-related velocity fields (e.g., Evershed flow). Previous papers have shown that the wave and magnetic activity do not have the perfect correlation one expects of an artifact; the time difference reinforces that conclusion.

The interpretation of the time variation of the rotation difference between the north and south hemispheres (Section 2B) is unclear. The small amplitude does not allow a clear determination of an 11-year period, in contrast to the variations in $\underline{B}/\underline{A}$ and $\underline{C}/\underline{A}$. The phase relative to the activity cycle is not obviously important. Further, the general symmetry of the magnetic activity makes a hemispheric rotation difference difficult to explain in terms of a dynamo mechanism, unlike the oscillations of wavenumber 1 hemisphere^{-1} and higher. A simple explanation in terms of the surface imbalance of magnetic flux does not work.

ACKNOWLEDGEMENTS

Grant support for this research has come from the Office of Naval Research (Contract N00014-18-C-0065), the National Aeronautics and Space Administration (Grant NGR-09-140-015), and the National Science Foundation (Grant AST-8020445). The solar observers have provided valuable assistance over the years: J. M. Adkins, T. A. Cragg, T. S. Gregory, S. P. Padilla, and L. Webster. John E. Boyden has provided valuable programming help, and Maynard K. Clark has provided important electronics engineering help.

REFERENCES

Belvedere, G., Zappala, R. D., D'Arrigo, C., Motta, S., Pirronello, V., Godoli, G., and Paterno, L.: 1979, Workshop on Solar Rotation, Catania, p. 189.

Howard, R.: 1979, Astrophys. J., 228, L45.

Howard, R. and LaBonte, B. J.: 1980, Astrophys. J., 239, L33.

Howard, R., Adkins, J. M., Boyden, J. E., Cragg, T. A., Gregory, T. S., LaBonte, B. J., Padilla, S. P., and Webster, L.: 1983, Solar Phys., submitted.

LaBonte, B. J., and Howard, R.: 1982a, Solar Phys., 75, 161.

LaBonte, B. J., and Howard, R.: 1982b, Solar Phys., 80, 15, 361, 375.

Livingston, W. and Duvall, T. L.: 1979, Solar Phys., 61, 219.

Scherrer, P. H., Wilcox, J. M., and Svalgaard, L.: 1980, Astrophys. J., 241, 811.

DISCUSSION

GOKHALE: Are there any basic difficulties in determining the north-south asymmetries by including *odd* powers of $\sin\theta$ in the expansion of the rotation rate?

HOWARD: I suppose the effect might show up in a term of odd power in $\sin\theta$, but we have not tried it. In general we find that it is best to solve for as few terms as possible.

LIVINGSTON: You have ascribed the temporal increase of total flux past sunspot maximum to the subsequent emergence of large regions containing a lot of flux. Could an alternative mechanism be the distribution (by random walk) of flux of umbral origin over the entire disk, thus making this flux more available to sensing by the magnetograph?

HOWARD: This mechanism may contribute to the phase lag of the flux, but I think the flux disappears rather quickly from the surface of the sun, so the size of the regions must be an important factor.

STIX: If the upper limit of cellular velocities is 10 m s^{-1} *per longitudinal wave number*, then what is a possible velocity of actual cells, which may consist of a whole spectrum of Fourier modes?

HOWARD: The actual limit is quite small for some configurations, of course, but the lifetime of the cells also enters in a complicated way. Long-lived cells have a much lower limit. Remember that the results apply to cells, which are extended in the latitude direction.

GIOVANELLI: I think that the torsional oscillation is the most important discovery concerning the solar cycle that has been made in recent times. It contains the clue to the whole cycle mechanism. Our thanks should be due to the whole Mt. Wilson team for this magnificent new phenomenon.

ROXBURGH: At the levels you are talking about, you should be able to pick up, or should have had to subtract, the gravitational redshift from the surface of the sun. Do you find this systematic effect?

HOWARD: No, because we do not have an absolute velocity scale. Until this summer, all we have known is the position of a line in mm at the focus of our spectrograph. We have not known what the wavelength is at all. So, to find the residual velocities, we simply average all the velocities over the sun and call that zero.

KUKLIN: The westward zonal streams or local accelerations of the sun's rotation are connected with an increased pressure difference in direction from the pole to the equator. May we conclude from your results that in average the pressure inside the sunspot zone is larger than in quiet regions of the sun?

HOWARD: We have not thought about this effect in these terms. We believe that it is a deep-seated phenomenon, not a surface effect driven by pressure differences.

BONNET: If you could increase the accuracy of your velocity measurements, would you expect some additional velocity patterns to show up between the main patterns which are presently evident?

HOWARD: The rms velocity-signal error in our spectrograph in about 15 m s^{-1}. Improving that greatly or reducing it to zero would have little or no effect on the results I have just shown. Systematic instrumental effects are much more important, and we know less about them. But effects from solar velocity fields (i.e. supergranular motions) also set an important limit, and it is not so easy to avoid this.

SHEELEY: Have you made Dopplergrams in Zeeman-insensitive lines to check for possible cross-talk between the Doppler and Zeeman signals?

HOWARD: Yes, and there seems to be no influence from the Zeeman effect.

KOTOV: I believe that you have found a very interesting and important phenomenon in the Sun, which is quite similar in some sense, as I said yesterday, to the jet-streams clearly seen in the giant planets, for example in the Jovian atmosphere. But I wonder whether you have checked the possibility of a significant influence of the spectral line asymmetry potentially appearing in active regions, which are drifting towards the equator during the solar cycle. Thus they can, in principle, produce certain lanes in the Doppler maps. Have you estimated the amplitude and time behaviour of the effect of active regions?

HOWARD: We have considered this possibility. But for several reasons we have concluded that it cannot explain the observed torsional oscillations. A curious configuration of line asymmetries would be required, antisymmetric about the north-south axis of active regions on average and about the central meridian of the sun. Furthermore, we see active regions only at latitudes lower than about 30°, whereas the torsional motions appear to start near the poles.

THE MOUNT WILSON PROGRAM FOR STELLAR ACTIVITY CYCLES

Arthur H. Vaughan
Perkin-Elmer
Mount Wilson & Las Campanas Observatories

ABSTRACT

The present review will focus upon the incidence, form, and characteristic timescale of long-term chromospheric variations that, from the work of O.C. Wilson and his successors, can be descerned in records of CaII H and K emission now extending over 16 years, and the relation, if any, between properties of activity cycles and stellar mass, age, and rate of rotation, in the light of current evidence.

INTRODUCTION

In 1966, O.C. Wilson began what became the first successful attempt to uncover and study, in other main sequence stars, chromospheric variations analogous to those observed in the course of the solar cycle.

This work has been continued and considerably expanded in scope by Wilson's successors since the completion of the first phase by Wilson himself (Wilson 1978). In fact, July 1982 marks the beginning of the third year in which the Mount Wilson 1.5m telescope has been dedicated almost exclusively and continuously to various aspects of this work, a level of effort which has been sustained in the form of a team project involving individuals associated with Mount Wilson, Sacramento Peak Observatory, The Harvard-Smithsonian Center for Astrophysics, and the Astronomical Institute at Utrecht.

The effort has paid off. Even in the past year, there have emerged a number of extremely interesting results (and perhaps an equal number of new questions) bearing upon the rotation rates of stars on the lower main sequence, and the rotation - activity connection. The results also bear to some extent upon the detailed surface properties of stars exhibiting chromospheric activity.

J. O. Stenflo (ed.), Solar and Magnetic Fields: Origins and Coronal Effects, 113–132.

At the risk of proceeding not entirely in a logical order (since the critical discussion of many of the things I just mentioned will follow in the review by Bob Noyes), I would like in the present discussion to focus upon the observed properties of activity cycles as they are found in stars, in the light of what has been accomplished in this field since the publication of Wilson's monumental original work.

OBSERVATIONAL APPROACH

The method of observation introduced by Wilson is based upon the fact that the chromospheric CaII H and K line emission flux varies in synchronism with other indices of the state of magnetic activity in the Sun.

The photoelectric spectrometer (Vaughan, Wilson and Preston 1978) currently in use to measure the calcium flux index in stars is different in detail but identical in principle to that used by Wilson. It has thus been possible to preserve continuity of measurement for over 15 years.

The CaII H and K line-core index is defined by the expression

$$S = k\ (N_H + N_K)\ /\ (N_V + N_R),$$

where $\underline{k}$ is a normalizing factor derived from observations of standard stars to remove instrumental effects, and N_H, N_K, N_V, and N_R refer to the number of photon counts in two band 1Å wide centered at H and K and in two photospheric reference bands 20Å wide centered at 4001Å and 3901Å to the red and violet sides of the H-K region. In the present instrument the fluxes in these four bands are measured sequentially at a switching frequency of about 30 Hz with a single detector. The errors of measurement of S are close to those introduced by photon noise. The error is usually held to about 2 percent in the observations of Wilson's stars. To allow for wavelength shifts corresponding to a star's geocentric velocity a precomputed offset is introduced at the time of observation, centering the passbands on the H-K lines to within about 0.05Å. All of the investigations being pursued in the Mount Wilson stellar activity program are made in exactly this way.

THE STARS UNDER INVESTIGATION

Since we will be discussing Wilson's collection of stars in which chromospheric variations have been followed up to this point, it is appropriate to recall how these stars were selected.

In the first 1.5 years of Wilson's study, observation was concentrated on a rather large number of stars from the (never published) Strömgren-Perry catalog for which 10 Å/mm spectrograms had earlier been obtained. The results of this study were reported by Wilson (1968). Beginning in the fall of 1967 many of the S-P stars were deleted from the program; retained were those in which H-K emission has been seen in the spectrograms, as well as some in which emission might be weakly present and others with minimal H-K emission to serve as standards. Also in 1967 a number of late type (G2 - M0) main sequence stars were added. The ultimate collection thus consisted of about 90 stars ranging from early F to M0, selected on the basis of CaII emission to the representative (characteristically but not statistically) of the observed range in H-K flux along the lower main sequence.

One of the global attributes of this sample is that it shares the property found by Vaughan and Preston (1980) in a survey of a much larger and presumably unbiased sample of nearby stars, that F and G dwarfs in the Solar neighborhood are either roughly as active as the Sun, or about twice as active, as measured by the calcium emission index. As yet there is not satisfactory explanation for this dichotomy, which is illustrated in Figure 1.

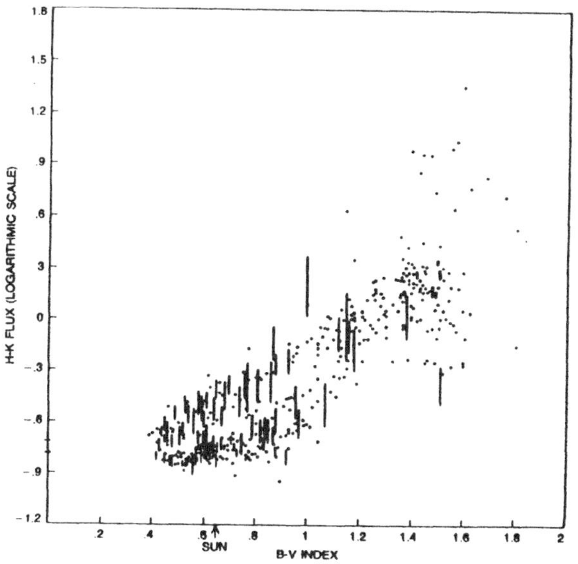

Fig.1: The log S, B-V diagram for stars in the solar neighborhood survey of Vaughan and Preston (1981) [points], and stars in the Wilson study of long-term chromospheric variations [bars indicate the range of their variations]. The plot reveals an apparent dichotomy in which H-K emission is either comparable to the Sun's, or about twice as great, in late F and G stars.

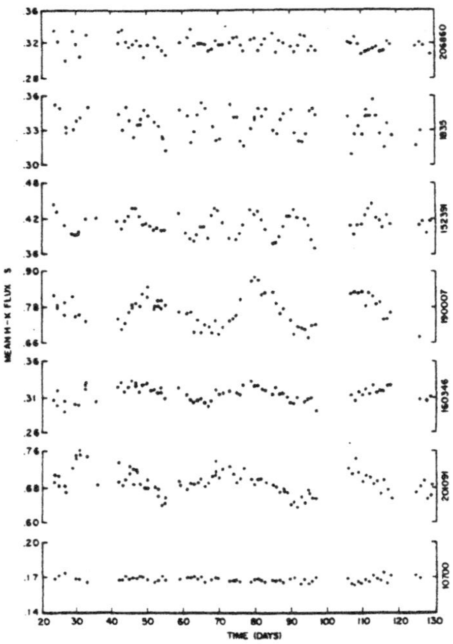

Fig. 2: Mean H-K fluxes vs. time for selected main sequence stars, illustrating the rotational modulation effect. From Vaughan, Baliunas, Middelkoop, Hartmann. Mihalas, Noyes and Preston (1981).

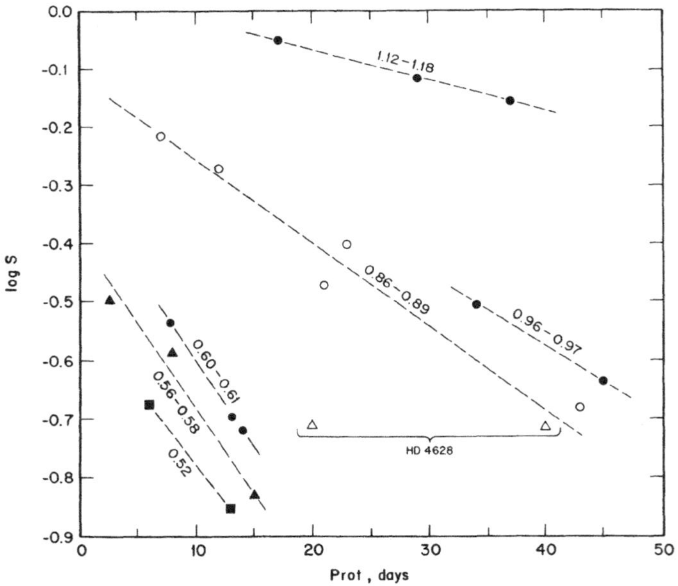

Fig. 3: Log S vs. period of rotation for stars in selected narrow intervals of B-V. For a given color, the points lie along well-defined sequences indicative of the rotation-activity relation. Recent observations have shown that HD 4628 rotates with a period of 36 days. From Vaughan, Baliunas, *et. al.*, 1981.

Furthermore, as was also noticed in 1979 (Vaughan 1980), cyclic variations closely resembling the solar cycle seem to be encountered most commonly (if not exclusively) among the less active stars.

CALCIUM EMISSION AS A TRACER AND PREDICTOR OF ROTATION

Perhaps the most recent major advance in the study of chromospheres has come about through the use of chromospheric active regions as rotation tracers, a possibility made evident by Wilson's earliest observations. In fact, Wilson's observations have been shown to contain significant rotational information (Stimets and Giles 1980; Noyes 1981).

The rotational modulation effect, illustrated in Figure 2 (taken from Vaughan, et.al 1981; see also Baliunas, et.al. 1983) actually leads to two methods for inferring the rate of stellar rotation, as will be discussed more fully by Noyes.

In the first method the rotational period is inferred by autocorrelation analysis (in many cases by inspection) of a time series of observations showing the rotational signature. LaBonte (1982) has shown in the case of the Sun that when the fundamental rotational modulation is present in a time series of disk-integrated surface magnetic flux measurements (an index proportional to Wilson's CAII index), the rotation period inferred corresponds to the solar rotation at the latitude of maximum magnetic activity.

The second method, calibrated by the first method but otherwise entirely independent of it, makes use of the relation illustrated in Figure 3 (from Vaughan, et.al. 1981), between the mean level of calcium emission and the rotation rate.

In HD 4628, one of the stars identified in Figure 3, autocorrelation analysis at first indicated a rotation period of about 20 days. However, the apparent flux-rotation relation would have been better satisfed if the period of rotation of this star were 40 days, and we suggested that the period found by autocorrelation analysis might be a harmonic of the true period of rotation. This has now been found to have been a correct prediction, on the basis of more recent observation (see Baliunas, et.al. 1983) in which a clear modulation emerged with, in fact, a period close to 36 days.

CURRENT OBSERVATIONAL RESULTS

Having indicated the direction in which the discussion will go, I would like now to focus attention upon the present state of affairs in the observational study of activity cycles, in the light of what has been learned up to this point.

We can proceed by first examining a few up-to-date examples of the chromospheric variations thus far observed in main sequence stars. I have selected 9 examples (Fig. 4a-i) illustrating the variety of forms encountered; they are not meant to be statistically representative of the reppertoire found in nature.

HD 81809 exhibts a cyclic variation that would be hard to distinguish from the solar cycle in form, amplitude, noisiness, and the relative timescales of rise and decay of activity in the course of the cycle. Such a pronounced cycle is rarely seen except in main sequence stars that are of considerably later spectral type (cooler) than the Sun.

HD 160346 and HD 201091 (61 Cyg A) are examples of later-type stars that exhibit highly pronounced cycles. Parts of three successive cycles have been recorded for these stars. Note that certain differences can be seen between one maximum and the next, reminiscent of the irregularity of the solar cycle.

In the stars just discussed, there is high confidence in the reality of the observed cycles, and in the periods (or characteristic timescales) that might be assigned to them.

HD 120136 is a case in which I would say there is only fair confidence that a cycle is present, having a timescale of about 10 years, as judged in part from the variation on the amplitude of the seasonal scatter which in previous examples (and in the Sun) is largest at the peak of the cycle.

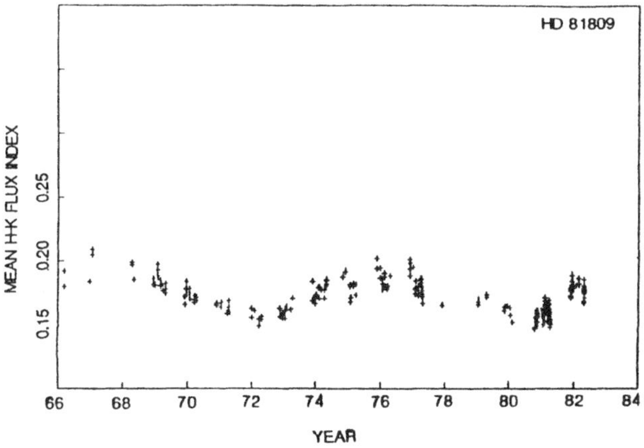

Fig. 4(a): H-K Flux records from Mt. Wilson project, 1966 to 1982.

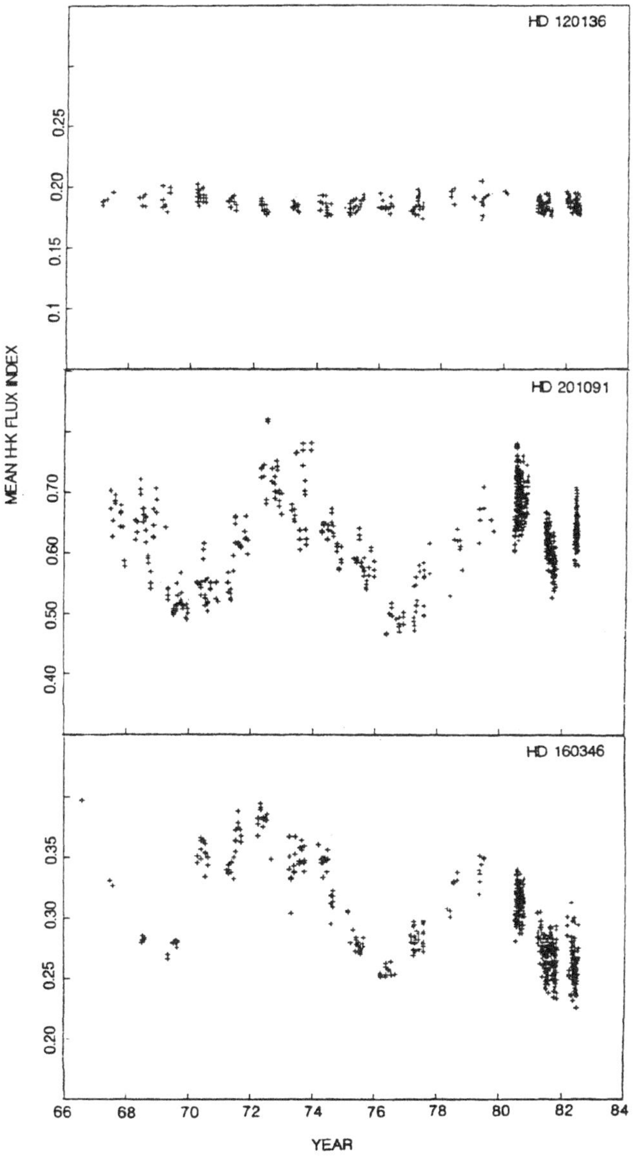

Fig. 4(b)-(d), continued.

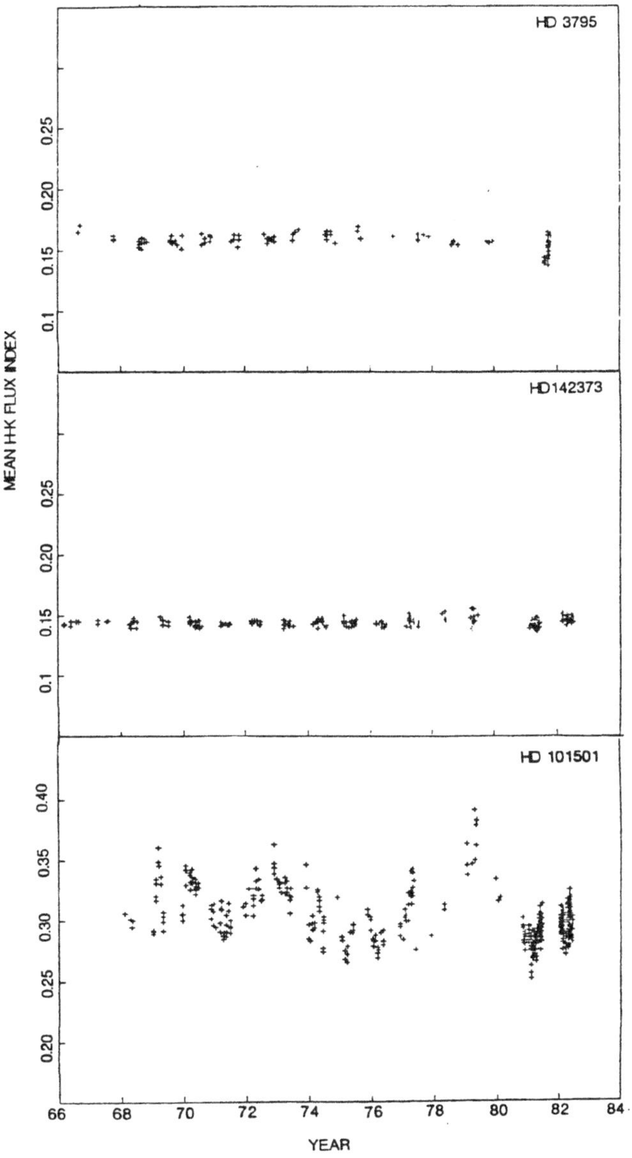

Fig. 4(e)-(g), continued.

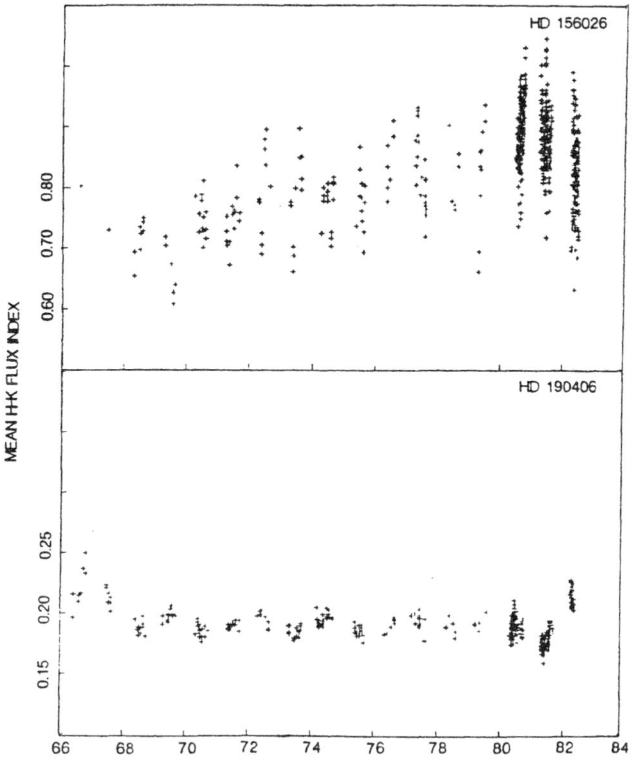

Fig. 4(h)-(i), continued.

HD 3795, with only 4.5% standard deviation in H-K index, is a case in which confidence in the existence of a cycle is low, but not zero. In comparison with HD 142373, the quietest standard with s.d. =2%, HD 3795 seems to be not quite flat; it could be varying on a timescale of about 12 years, or longer. It is likely that all stars, even the standards, show some real change.

There are variations in some stars that suggest limited resemblance, if any, to the solar cycle. HD 10151 is one of several stars in which large changes occur in the course of time, but the variation appears to be entirely chaotic. The solar cycle as we know it is irregular, but not random.

HD 156026 has increased steadily in mean flux during the entire interval of observation and its nearly uniform seasonal scatter is the largest of any star in Wilson's collection. The gradual rise in mean flux could be part of a cycle whose timescale exceeds 16 years. It may be that HD 190406 is a case of a very short cycle of perhaps 2.6 years. In what follows I have not taken it to be such. At a later time perhaps this will have to be reconsidered.

WHICH STARS ARE CYCLIC?

From examination of the entire collection of observations I have listed in Table 1 those stars for which I think a characteristic timetable of solar-cycle-like chromospheric variation might plausibly be assigned for comparison with other stellar parameters. In most respects, this list is in accord with lists independently drawn up by Wilson and by Noyes from the same set of material, although there are some differences arising out of the subjectiveness of the criteria one uses to decide whether to include weakly or perhaps irregularly varying stars.

In the resulting collection of 27 cyclic or possibly cyclic stars, there is high confidence ("H") in the reality of the cycle in 15 of them, fair confidence ("F") in eight, and low confidence ("L") in five.

TABLE 1

27 STARS THAT SEEM TO HAVE ACTIVITY CYCLES SIMILAR TO THE SOLAR CYCLE*

HD	B-V	S_{min}	S_{max}	P_{rot} (d)	P'_{rot} (d)	T_{cyc} (y)	H	F	L	COMMENTS
SUN	.66	.165	.21	25.5	25.4	8-12	✓			
3651	.85	.16	.24	-	39.9	10	✓			
3795	.73	.14	.17	-	29.8	≥12		✓		Weak
4628	.88	.18	.30	38	36.0	8-9	✓			
10476	.84	.16	.26	-	34.6	10	✓			
12235	.62	.15	.18	-	21.4	10		✓		
16160	.97	.20	.28	45	44.8	11-12	✓			
26965	.82	.18	.25	30:	33.4	10	✓			61 Cyg B Type
32147	1.07	.24	.39	-	48.1	9-10	✓			
81809	.64	.15	.21	-	22.3	9	✓			
100180	.57	.15	.18	-	18.3	>12		✓		
103095	.75	.16	.21	-	30.6	6-7	✓			
160346	.96	.24	.39	33.5	36.4	7-8	✓			
161239	.65	.13	.16	-	29.0	>12	✓			Weak
166620	.87	.18	.25	42	38.1	14?	✓			
182572	.78	.14	.18	-	36.3	≥12		✓		
201091	1.19	.46	.80	38	41.0	7	✓			
201092	1.38	.70	1.30	48	48.0	11-12	✓			t(rise)> t(decay)
219834	.79	.14	.18	30:	28.7	10		✓		Weak
120136	.48	.175	.21	-	8.3	10		✓		Weak
115404	.93	.45	.63	17-20	17.7	≥12			✓	61 Cyg B Type?
131156	.76	.38	.55	6.5	7.2	13?			✓	61 Cyg B Type?
149661	.81	.30	.49	21.3	17.6	>11?			✓	
155886	.86	.30	.51	20.3		>13?		✓		
152391	.76	.33	.48	11.1	10.3	≥12	✓			
165341	.88	.28	.55	20	19.7	≥12		✓		t(rise)$_{<<}$ t(decay)
190007	1.12	.58	.98	29	28.0	11?			✓	

* Compiled July 1982

For the stars in Table 1, the distribution of H-K flux as a function of B-V is plotted in Figure 5, in which the amplitude of flux variation of each star is shown by a vertical bar. The bar is solid for "H-rated" stars, and dashed otherwise. All but one of the stars rated "H" lie in the lower portion of the diagram which includes the Sun. Conversely all but one of the stars rated "L" lie in the upper branch of the same diagram in which the level of activity tends to be twice the Sun's. This is the dichotomy mentioned earlier. It is convenient to preserve the distinction in listing separately, in the upper and lower sections of the Table, the stars belonging to the two categories.

As can be seen in Table 1, cycle timescales estimated for stars in the upper branch are for the most part lower limits of around 12 years, possibly of significance, but uncertain at best.

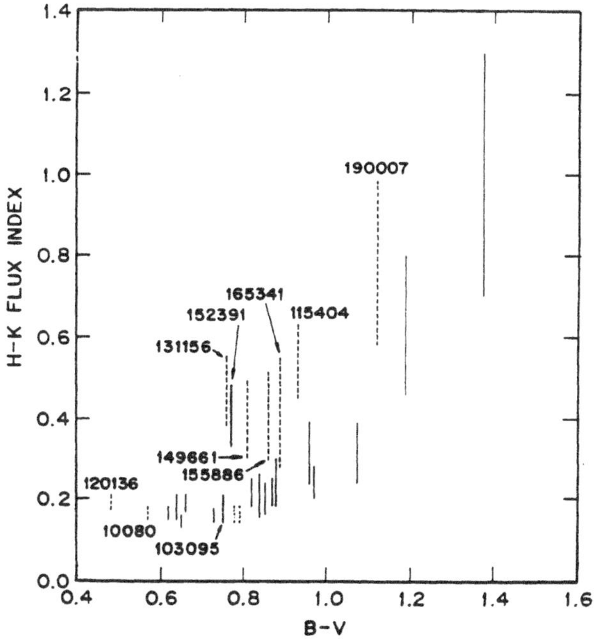

Fig. 5: The S, B-V diagram for stars listed in Table 1. Solid bars indicate the range of variation in stars considered with high confidence to have a cycle. Dashed bars indicate the range of variation in stars whose variations might, with fair or low confidence, be considered as cyclic. Compare Fig. 1.

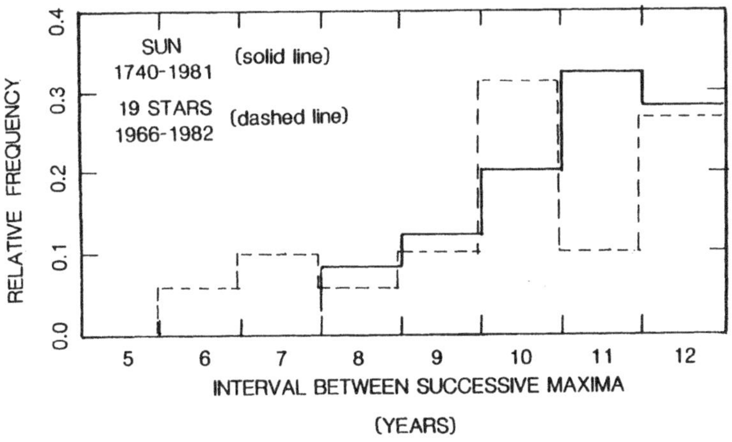

Fig. 6: Histogram of the intervals in years between successive sunspot maxima after 1740 [solid line, after Waldmeier], compared with histogram of the same parameter for 19 stars in which analogous variations are seen.

For stars in the lower branch, the estimated cycle timescales have a mean value of 10 y ± 1.9 y (s.d.). As shown in Figure 6, the cycle periods of these solar-branch stars exhibit a frequency distribution somewhat resembling the distribution of the intervals of time between successive solar mixima as given, from records long maintained at the Swiss Federal Observatory here in Zurich, by Waldmeier(1961; 1976). Stellar activity cycles are found, however, that are definitely shorter in duration than any so far seen in the Sun. It is of interst, therefore, to explore the question of whether individual deviations from the mean cycle period are in any way correlated with other parameters that might be considered, such as stellar mass or rate of rotation, or both.

As can be seen in Figure 7, if there is any correlation between cycle timescale and stellar color index (mass), it is not particularly striking.

THE ROTATION PERIODS OF CYCLIC AND POSSIBLY CYCLIC STARS

As mentioned earlier, there are two methods whereby the rotation period of a star can be inferred from observations of its calcium emission index. Thus, by means of autocorrelation analysis, values of P_{rot} have been determined for 16 of the stars listed in Table 1. In 10 others, rotation periods could not be inferred by this method, presumably because sufficient modulation is not present in the data so far collected.

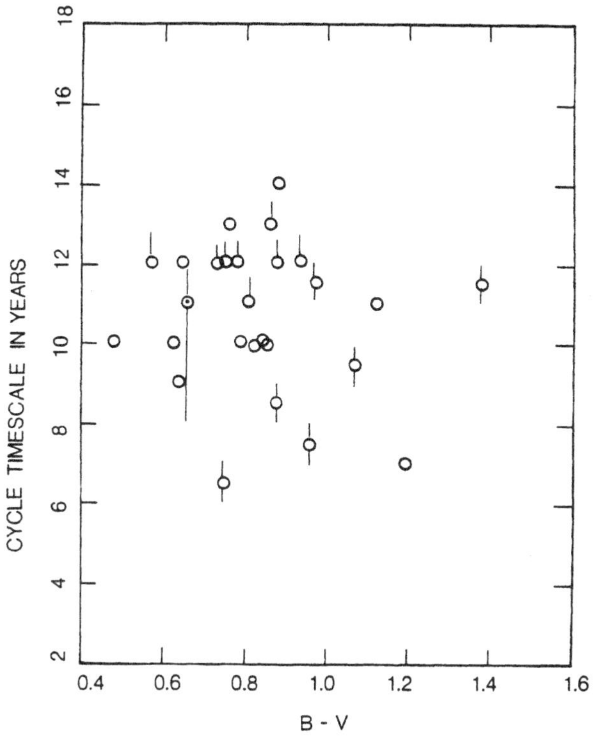

Fig. 7: The cycle timescale in years vs. B-V for the 27 stars in Table 1. The solar cycle timescale (sun symbol) varies from about 8-12 years. Some stars are found to have a shorter timescale, but there is no clear correlation of this parameter with B-V.

By appealing to the rotation-activity relation, however, it is possible to supply predicted values of the missing rotational periods for those cases in which these are not yet known by a more direct method.

Vaughan, et.al. (1981) gave a tentative calibration of the activity-rotation relation, of the form

$$S = 0.78 \exp(-P_{rot}/30), \quad 0.86 \leq B\text{-}V \leq 0.89$$

but this applies only to a narrow interval of color index. To gain wider applicability it is necessary to calibrate also the color dependence of the relation. Recently Noyes and I, working partly together and partly independently, have followed two different approaches in attempting to arrive at such a more general calibration. One approach, which I pursued, but will mention only briefly, was simply to fit to the observations an empirical expression of the form

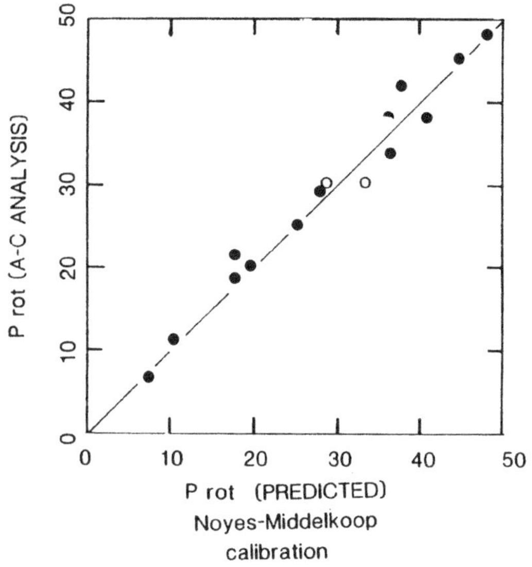

Fig. 8: Periods of stellar rotation inferred from autocorrelation analysis of H-K flux variations compared with periods of rotation as predicted from the activity-rotation relation.

$$\log S = \underline{a} + \underline{b}\, P_{rot}$$

in which a and b are assumed to be linear functions of B-V.

The result, solved for P_{rot}, is the expression

$$P_{rot} = \frac{\log S + 0.63 - 0.59 \quad (B-V)}{0.0.48 \quad (B-V) \quad - 0.056}$$

This expression actually reproduces the observed value of P_{rot} fairly well, but only in the range $0.54 \lesssim B-V \lesssim 0.89$

The second and much better approach followed recently by Noyes is based upon the use of Middelkoop's (1982) calibration of $\underline{S}$ in absolute flux units. Leaving further discussion of the approach to Noyes, therefore, I have adopted, in Column (6) of the Table, his predicted rotation periods, given to me, in fact, over the telephone. I believe Noyes regards these as preliminary, but most likely not far from the final values based on present data.

The likely validity of this procedure is evident when one compares those cases in which P_{rot} has both been inferred from A-C analysis and predicted from the rotation-activity relation. As shown in Figure 8, the resulting periods are in remarkably good agreement over the full range in period of rotation and spectral type exhibited by the stars under discussion.

COMPARISON OF CYCLE TIMESCALE, ROTATION RATE, AND COLOR

For the first time, it becomes possible to examine, more or less directly from observation, the possible connection between cyclic behaviour and the rate of stellar rotation. This we shall do by means of a series of four diagrams. The discussion should be considered preliminary, as mentioned earlier, being the result of work still in progress.

To proceed, it is expedient first to examine the variation of P_{rot} with B-V, as is done in Figure 9. Filled symbols identify stars whose periods of rotation were found by A-C analysis; open symbols identify stars whose rate of rotation was predicted by the use of the Noyes-Middelkoop calibration. Also circles identify stars whose periods of rotation were found by A-C analysis; open symbols identify stars whose rate of rotation was predicted by the use of the Noyes-Middelkoop calibration. Also circles identify stars that lie in the lower (solar) branch of the flux-color-diagram, and triangles identify stars in the upper (Hyades-like) branch. The dichotomy in H-K emission is not apparent in stars as red as 61 Cyg A and B.

As can be seen in Figure 9, there appears to be two entirely distinct sequences, one containing all but one of the circles, the other all of the triangles and in addition 61 Cyg A and B. Evidently

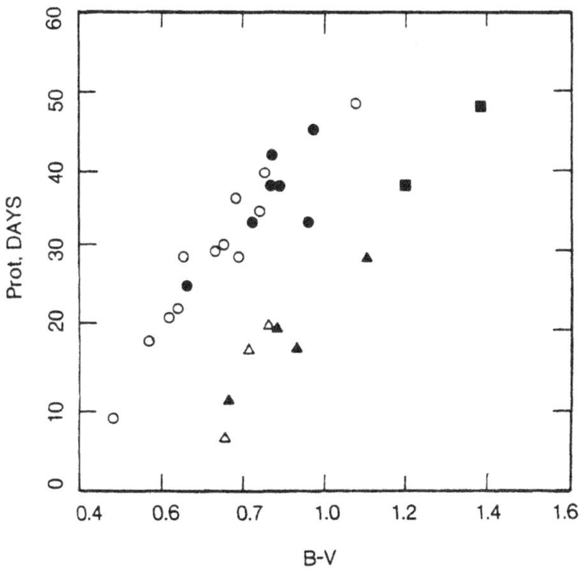

Fig. 9: Period of rotation vs B-V for the stars listed in Table 1. Filled symbols denote stars whose P_{rot} is known from A-C analysis. Circles represent stars belonging to the solar sequence in Fig. 1.

this diagram reflects the dichotomy encountered previously. However, here it appears also in the distribution of rates of rotation in the limited sample of stars collected for the study of cyclic behavior. This seems to me to be a surprising result that should be investigated on the basis of an unbiased and larger sample.

One way to look for a relation between cycle period, rotation rate, and color is to plot the ratio of two of the variables against the third. In this case the dimensionless ratio T_{cyc}/P_{rot} (the number of rotations per cycle) would seem plausibly to be a physically significant choice. A plot of this parameter against B-V is shown in Figure 10. Circles and triangles have the same meaning as earlier. As a result of the small dispersion in the cycle timescales and the tight correlation (existing in the present possibly biased sample) between rotation rate and B-V, the points (especially the circles) lie along sequences in this diagram. The variation in the parameter T_{cyc}/P_{rot} over a factor of about five is contributed primarily (indeed, as will be seen, entirely) by the color-dependence of the period of rotation. Is there, then, any causal relationship between T_{cyc} and the color (or mass) of a star exhibiting a cycle ?

To find out, we may plot, as a function of color, the difference between T_{cyc}/P_{rot} (using observed values of the cycle timescale) and (using the mean value T_{cyc} = 10 y) the ratio T_{cyc}/P_{rot}, and examine

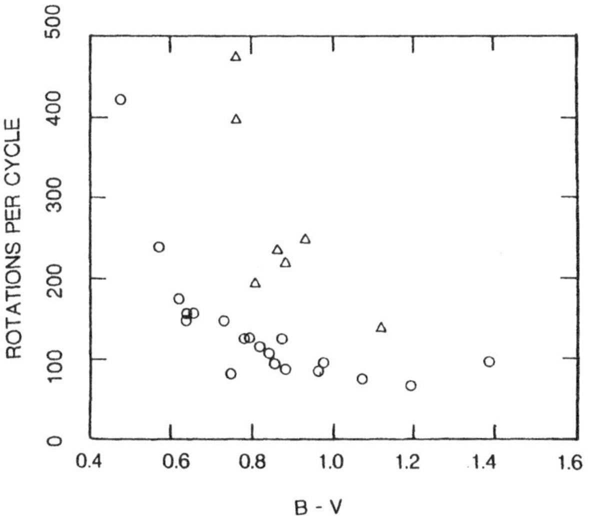

Fig. 10: The ratio of cycle timescale to rotation period (rotations per cycle) vs. B-V for stars in Table 1. Circles depict stars in the solar sequence of H-K flux vs. B-V. Triangles depict stars in the active (Hyades-like) sequence.

the resulting diagram for systematic trends. Such a diagram is given in Figure 11. It seems that no systematic variation of the deviation from the mean cycle timescale as a function of B-V is evident.

Finally, in Figure 12, a similar procedure serves to test whether there is systematic variation of the deviation of T_{cyc} from the mean as a function of rate of rotation. The plot is again restricted to solar-branch stars. Except for the fact already noted by Vaughan et.al (1981) that cycles seem to occur only when P_{rot} exceeds about 20 days (or equivalently B-V exceeds about 0.6 as seen in Figure 9.) there is no evidence that deviations of individual cycle timescales from the mean value depends systematically upon the rate of stellar rotation.

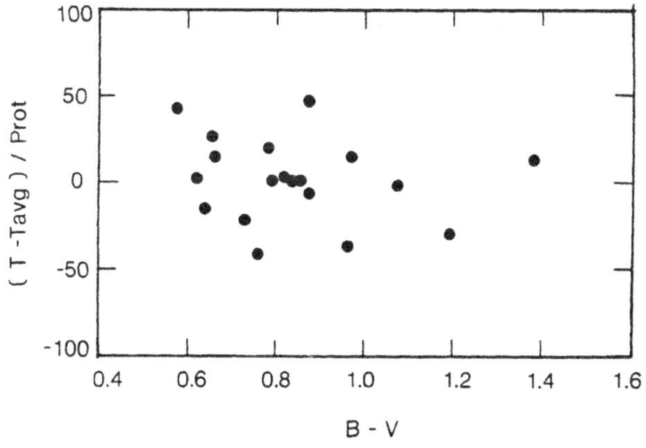

Fig. 11: Deviation from the mean cycle timescale (T_{avg} = 10 y) divided by P_{rot}, plotted as a function of B-V.

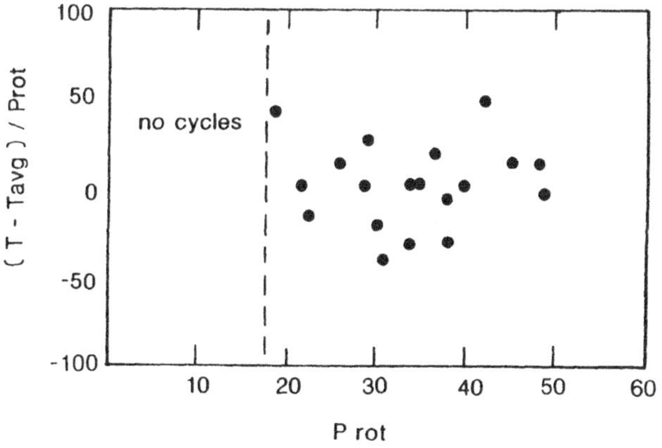

Fig. 12: Deviation of the mean cycle timescale (T_{avg} = 10 y) divided by P_{rot}, plotted as a function of P_{rot}.

CONCLUDING REMARKS

As mentioned earlier, these are preliminary results, but they confirm what previously has been suspected from comparison of the distribution of activity cycle periods in stars with irregularities in the solar cycle timescale.

Whether or not a solar-type cycle exists seems to depend upon the period of rotation, but if such a cycle exists, its period does not seem to depend upon the parameters we are currently considering.

Can it be that the stellar-mass dependence of T_{cyc}/P_{rot} (the "windup ratio" of an activity cycle), so beautifully illustrated on Figure 10, is entirely coincidental and of no causal significance? The observations do not by any means serve to disprove that the cycle is casually related to stellar mass or rotation; neither, it seems, do they suggest any direct or simple causal connection of these parameters with the details of the process once the cycle becomes established.

In any case, it would appear that if a theoretical model is constructed in which T_{cyc} is affected by the value of the rotation rate as a parameter, this effect must be compensated as a function of some other parameter related to stellar mass.

It seems likely that significant further progress might come about if it should become possible in some way to measure the magnitude of differential rotation in a star, upon which any dynamo process depends perhaps more directly than upon rotation itself. The observational difficulties involved seem very substantial at this point, but it must be remembered that the whole effort has only just begun.

In the meantime, important questions remain to be explored such as the question of the significance of the dichotomy in F-G dwarfs and the related question of the age dependence of calcium emission and rotation both in dwarfs and (as investigators from the Utrecht group have begun to explore) in evolved stars. Thus, it is worth mentioning that in preparation for studies that will take place beginning this Fall, the H-K spectrometer is about to be modified with installation of a new chopper wheel to allow measurement of H&K simultaneously, thus gaining factor of two improvements in throughput for observations of faint Hyades and Pleiades cluster stars. At the same time a new exit mask will provide both 1 Å and 2 Å bandwidths at H and K, the latter being designed for the study of chromospheric emission in giants.

ACKNOWLEDGEMENTS

The Mount Wilson program is supported in part by a grant from the National Science Foundation to the Carnegie Institution of Washington, with additional support coming from NASA (1975) and currently from the Smithsonian Astrophysical Observatory. My participation in this Symposium was sponsored by Perkin-Elmer. Most of the observations in the course of the past year were made by Dr. Douglas K. Duncan (Research Associate, MWO), James Frazer, Howard H. Lanning, and Laura A Woodard (MWO), Tony Misch (SAO-MWO), and Dr. Sallie L. Baliunas (CFA). I am continually grateful to these individuals and the many others whose enthusiastic cooperation has made a large undertaking possible and fruitful.

REFERENCES

Baliunas, S.L., Vaughan, A.H., Hartmann, L., Middelkoop, F., Mihalas, D., Noyes, R.W., Preston, G.W., Frazer, J., and Lanning, H.: 1983 (in preparation).
LaBonte, B.J.: 1982, Astrophys. J. (in press).
Middelkoop, F.: 1982, Astron. Astrophys. 107, p. 31.
Noyes, R.W.: 1981, in R. Bonnet and A.K. Dupree (eds.), "Solar Phenomena in Stars and Stellar Systems". Dordrecht, D. Reidel Publ. Co., pp. 1 - 31.
Stimets, R.W. and Giles, R.: 1980, Astrophys. J. Letters 242, p. L37.
Vaughan, A.H., Preston, G.W., and Wilson, O.C.: 1978, Publ. Astron. Soc. Pacific 90, p. 267.
Vaughan, A.H., Baliunas, S.L., Middelkoop, F., Hartmann, L., Mihalas, D., Noyes, R.W., and Preston, G.W.: 1981, Astrophys. J. 250, p.276.
Waldmeier, M.: 1961, "The Sunspot-Activity in the Years 1610 - 1960", Swiss Federal Observatory Zürich.
Waldmeier, M.: 1976, "The Sunspot-Activity in the Years 1961 - 1975", Astron. Mitt. Eidgenössische Sternwarte Zürich, Nr. 346.
Wilson, O.C.: 1968, Astrophys. J. 153, p. 221.
Wilson, O.C.: 1978, Astrophys. J. 226, p. 379.

DISCUSSION

SODERBLOM: Of the stars you have shown us, I would pick HD 142373 as the one star most like the sun in terms of mass, age, rotational velocity, etc., yet this star shows no apparent variations. How would the sun appear if observed in the same way as your stars? Would its 11-yr cycle be obvious?

VAUGHAN: Of all 91 stars in Wilson's collection, I think that HD 81809 most resembles the sun in its variation of H-K emission and mean H-K flux. Yes, the sun's cycle would be obvious.

SODERBLOM: Do you feel that the lack of apparent cyclic variations in stars above the "gap" is due to the greater noise in the observations? In other words, if you observe these stars long enough, might you see the cycles?

VAUGHAN: I mentioned in a *Publ.Astron.Soc.Pacific* article (1980) that another decade of observations will be required to decide whether these stars are periodic (or have an underlying periodic component). That may be an optimistic statement.

SEVERNY: The presented paper represents an example of a unique, long-term investigation, and we should congratulate the Mt. Wilson observers for such a large amount of efficient work.

I think it would be interesting to complement this investigation with simultaneous estimates of the brightness of the stars, which might be obtained from other sources, e.g. from surveys. Do you have some simultaneous data on the fluctuations of brightness (m_{ph} or m_V)? Have you been monitoring the brightness?

VAUGHAN: Thank you. Our instrument (the H-K photometer) is not equipped for magnitude measurements. In the case of λ And, Baliunas has shown that H-K flux is anti-correlated with photospheric brightness, using data from the Mt. Wilson survey and from Ed Guinan.

WALTER: You mentioned an anti-correlation between stellar brightness and Ca II index in λ And. Is the chromosphere getting brighter, or is it constant while the photosphere is getting fainter?

VAUGHAN: Ed Guinan's photometry showed 20–30% change in the V and U bands, while the H-K emission varied in the opposite direction by a much larger factor. This must mean that the chromosphere actually brightened.

GOLUB: Both in the solar case and for the stars in your survey there is interest in learning about the polar fields. I would like to know whether there are any stars for which you have Ca data, and which you think are being observed pole on.

VAUGHAN: One suspect is HD 10700, which is remarkably constant in H-K emission, but whose emission level ($S \approx 0.17$) is well above that of Wilson's standards. More generally, I suppose stars for which we fail to see rotational modulation might be candidates, especially if the activity level is not minimal, and lack of modulation has characterized the star for a long period of time. As observations continue, cases may turn up in which one can believe computed values of $\sin i$ from spectroscopic and Ca II observations.

FOING: What stellar parameters distinguish the cyclic from the non-cyclic stars?

VAUGHAN: The stars showing cycles lie (with some possible exceptions) in the lower half of the H-K flux vs. $B - V$ diagram, and are slow rotators ($P_{rot} > 20$ days).

THE RELATION BETWEEN ROTATION AND MAGNETIC ACTIVITY ON LOWER MAIN SEQUENCE STARS AS DERIVED FROM CHROMOSPHERIC INDICATORS

R.W. Noyes
Harvard-Smithsonian Center for Astrophysics
Cambridge, Massachusetts 02138 USA

It is now firmly established that lower main sequence (LMS) stars show a qualitative correlation between rotation rate and chromospheric and coronal emission. By analogy with the Sun, the emission is believed to be intimately associated with surface magnetic fields. This association is especially close on the Sun for the Ca II H and K lines, for which the spatial correlation between chromospheric emission and photospheric fields is essentially one-to-one down to scales at least as fine as a few arcseconds and for which the emission flux from an area on the Sun is approximately proportional to the total magnetic flux passing through the same area in the underlying photosphere (Leighton 1959; Skumanich, Smythe, and Frazier 1975; Frazier 1971). The extension of the association to other LMS stars, while based on appeal to analogy, has been strengthened by recent detections of strong magnetic fields covering large fraction of the surface area of chromospherically active stars (see review by Marcy in this symposium).

The association between stellar coronal X-ray emission and rotation is discussed by Vaiana and by Walter elsewhere in this symposium; here we concentrate on the details of the association between rotation and chromospheric emission, and in particular emission in the Ca II H and K lines. We concentrate on the Ca II H and K lines mainly because those emissions are by far the most thoroughly-observed indicator of stellar chromospheric activity. However, as is apparent from the review of Zwaan in this symposium, the emission of other chromospheric or transition zone emission lines is also correlated with magnetic activity, and should also show a dependence on rotation or other stellar parameters qualitatively similar to (even if quantitatively different from) the relations discussed below.

A close relation between rotation rate and magnetic field-induced chromospheric emission for Sun-like stars is to be expected given the widely-accepted picture that magnetic activity in the Sun and other LMS stars is caused by the coupling of rotation and convection to produce a magnetic dynamo. By the same token, we may expect the variation of convective properties from star to star to affect chromospheric

J. O. Stenflo (ed.), Solar and Magnetic Fields: Origins and Coronal Effects, 133–147.

activity, and inasmuch as these properties vary significantly with spectral type (see for example, Durney and Latour 1978, or Gilman 1980) we may anticipate a strong dependence on spectral type. To isolate this dependence, we shall compare below the chromospheric activity level of stars with similar rotation rates, but different spectral types. Of course other stellar properties, such as surface temperature, mass, radius, etc., vary along the lower main sequence, so we must be cautious in attributing any single physical cause to activity variations with spectral type which may be revealed by the data.

A third stellar parameter, in addition to rotation rate and spectral type, which has long been associated with varying levels of chromospheric activity, is stellar age. However, as has been clearly shown in recent years, age seems to affect stellar activity mainly insofar as it affects a star's rotation rate through the phenomenon of spindown. The RS CVn stars, whose rotation is maintained by tidal coupling in spite of their aging, and whose activity is commensurately high, provide strong support for this conjecture, (see for example, Hall 1981; Ayres and Linsky 1980; or Walter 1981).

A long-standing view has been that stellar chromospheric emission decays smoothly with age as a star spins down (e.g. Schatzmann 1962; Kraft 1967): Both emission level and rotation rate decrease with age approximately as $(\text{age})^{-1/2}$ (Skumanich 1972; see also Skumanich and Eddy 1981). This picture is compatible with the hypothesis of steady spindown through angular momentum shedding by stellar winds, assuming mean surface magnetic field strength is proportional to angular velocity (see Durney 1972; Durney and Stenflo 1972). This scenario has been complicated by the recent discovery of the Vaughan-Preston "gap" in chromospheric emission (Vaughan and Preston 1980). The gap, which was pointed out in data from a survey of lower main sequence stars in the solar neighborhood, is also quite visible in the stars of Wilson's (1978) long-term activity survey, as may be seen in Figure 1 (see also Vaughan 1980). There are at least four possible explanations for the gap: (a) It may be simply a chance distribution and not a statistically significant description of properties of stars in the solar neighborhood. (b) It may be a real characterization of the solar neighborhood, but one produced simply by a paucity of stars of a given age (and hence emission level), rather than by an intrinsic stellar mechanism. (c) The gap may be real and represent a physical change in the character of magnetic activity generation at a given (perhaps spectral type-dependent) age and activity level. As example of this possibility, the gap might be produced by an abrupt decrease of dynamo efficiency at a given rotation rate. Several suggestions to this effect have been made (e.g. Duncan 1981; Middelkoop 1982) and specific physical mechanisms have been suggested by Durney, Mihalas, and Robinson (1981) and Knobloch, Rosner, and Weiss (1981). On the other hand (d), it is conceptually possible that it is not the relation between rotation and activity which changes, but that the rotation rate itself suddenly slows when a critical rotation rate is reached. (For example, at a critical

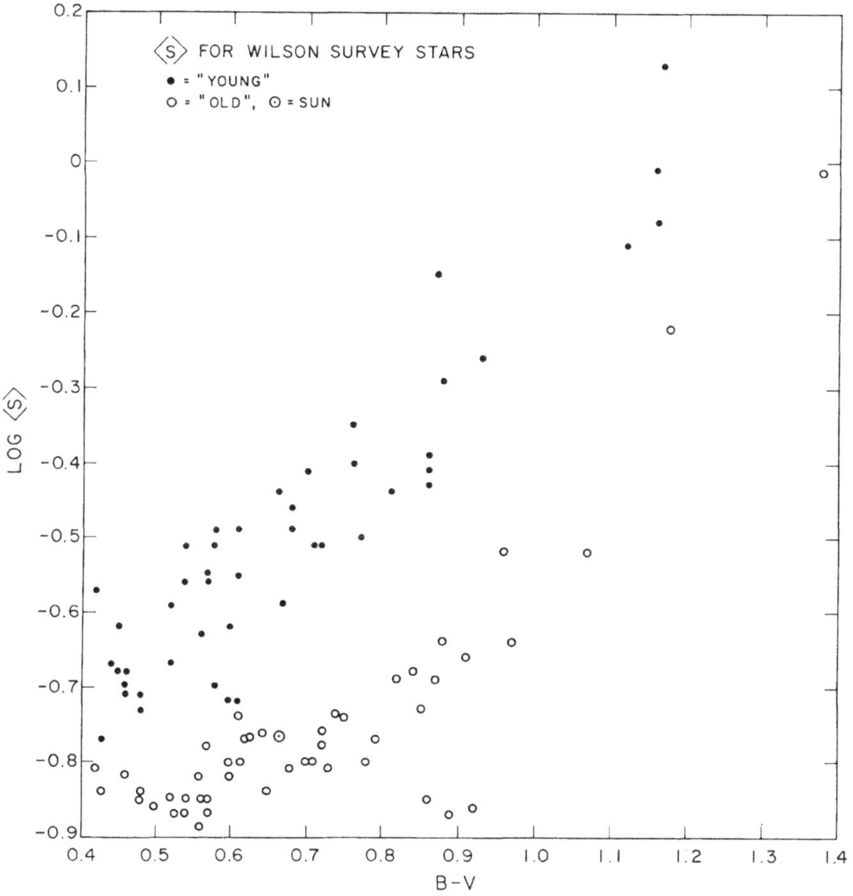

Figure 1. Mean values S of the Mt. Wilson H-K flux index, as determined from the long-term survey data of Wilson (1978). Closed and open circles represent "young" and "old" stars respectively (Vaughan 1980).

rotation rate, previously closed field lines might open up and increase the stellar wind angular momentum loss).

Clearly detailed empirical relations between rotation rate, spectral type, age, and chromospheric emission level would go far toward understanding the nature of the gap. For example, empirical evidence for a drop in activity at a particular rotation rate would argue for explanation (c) above. Evidence of this sort has been presented by Middelkoop (1982), but it is based on measured values of v sin i near the threshold of sensitivity of such measurements, and is scarcely conclusive. Conversely, evidence for continuity of the relation between rotation, activity, and spectral type across the gap would argue against explanation (c).

Even the "zero-order" spin-down scenario mentioned earlier has still to be quantitatively defined and tested by detailed measurements of many stars of different rotation rates, ages, and activity levels. However, because of the increasingly clear importance of such a test, there have been many recent studies aimed at this goal. The largest body of systematic data so far available on chromospheric losses as a function of rotation, age, and spectral type are measurements of emission in the Ca II H and K lines, and we shall now focus our discussion on these lines. Vaughan et al. (1981) found that rotation periods generally increase toward later spectral types, and that for stars of a given spectral type slower rotators had less chromospheric activity. Their quantitative relation between rotation and activity was spectral type-dependent, primarily because the chromospheric emission indicator which they used, the H-K flux index, is a quantity which depends on spectral type as well as on chromospheric activity. Middelkoop (1982), after converting the H-K flux index into absolute surface fluxes F_{HK} ($erg/cm^2/sec$) within the Mt. Wilson H-K photometer bandpass, found much less dependence on spectral type, with $F_{HK} \propto \log v_{eq}$. Catalano and Marilli (1982) find the surface luminosity L_K (erg/sec) to vary as exp($-P_{rot}$/27 days), independently of spectral type. Also, as discussed elsewhere in this symposium, several investigators have derived relations between coronal activity indicators (i.e. X-ray luminosity, surface flux, or ratio of surface flux to bolometric luminosity) and rotation as a function of spectral type (see for example, Walter 1982; Pallavicini et al. 1981).

The above discussion points up a certain arbitrariness in describing the dependence of emission on spectral type, for the dependence will vary significantly as one expresses emission in terms of absolute luminosity L_{HK}, surface flux F_{HK}, or ratio R_{HK} of luminosity to the star's bolometric luminosity. In the discussion below, we will use the last of these, which may be equivalently described as the flux ratio $R_{HK} = F_{HK}/\sigma T_{eff}^4$. We note in passing that because virtually all of a LMS star's luminosity is carried upward through the convection zone as a mechanical energy flux, the ratio R_{HK} may be thought of as approximately porportional to that fraction of the mechanical energy flux in the convection zone which is converted into magnetic field-related chromospheric heating.*

We shall not attempt here to derive a relation between H-K emission,

*This statement is strictly true of course only when R_{HK} is corrected for the non-magnetic contribution to the emission, and even then the proportionality will depend on spectral type, insofar as the ratio of chromospheric H-K losses to all chromospheric losses (including Lyman lines, Mg II h and k, the Ca II infrared triplet, etc.) varies with spectral type.

rotation, activity, and age sufficiently definitive to lead to conclusions about physical mechanisms. Our rather more modest goal is two-fold: (a) to address the question of the physical meaning of the Vaughan-Preston gap by studying how the empirical rotation-activity-spectral type relation varies across the gap, and (b) to derive an empirical relation between activity, spectral type, and rotation sufficient to allow estimation of rotation periods from the measured activity level of a star of a given spectral type.

The most extensive homogeneous set of data on chromospheric activity for LMS stars is the Ca II H and K flux measurements of Wilson (1978). For many of these same stars, Vaughan and colleagues (Vaughan et al. 1981, Baliunas et al. 1983) have determined rotation periods from the rotational modulation of Ca II flux. The Ca II flux data of Wilson have another very important attractive feature: Because the observations of these stars extend from 1966 up to the present (when combined with the more recent data obtained by Vaughan and his collaborators), it is possible to make allowance for the variations in Ca II flux level associated with long-term activity variations, as well as short-term rotational modulation. Figure 1 shows the mean value S of the Ca II H-K flux index $\langle S \rangle$, as taken from Wilson's (1978) data, updated to the present. The flux index S is proportional to the equivalent width of the emission in 1A bands centered on H and K, and referenced to the mean quasi-continuum at 3922 and 4001A (Vaughan et al. 1978). The mean value $\langle S \rangle$ is a value approximately halfway between the minimum and maximum levels of S attained for each star between 1966 and the present. Because of the greater reliability of $\langle S \rangle$ as an indicator of overall chromospheric emission level for a particular star, we shall confine the remainder of this discussion to the relation between the data in Figure 1 and the rotation period and spectral type determined for these stars.

Before proceeding further, we convert the observational parameter S into the chromospheric flux ratio $R'_{HK} = F'_{HK}/\sigma T^4_{eff}$, where F'_{HK} is the chromospheric flux per unit area in the H and K lines (the prime denotes subtraction of the photospheric contribution, as we describe below). This requires a two-step process:

1) The flux index S is the ratio of the total flux in instrumental bandpasses centered on the H and K line cores, to that in the nearby "continuum." The "continuum" flux however decreases steeply toward later spectral types; this effect is principally responsible for the general increase of $\langle S \rangle$ toward later spectral types in Figure 1. Middelkoop (1982) has determined an empirical conversion function C_{cf} (B-V), such that the $C_{cf}S$ yields the ratio of total flux F_{HK} within one Angstrom bandpasses centered on H and K to the stellar bolometric flux σT^4_{eff}. We denote this ratio R_{HK}. The function C_{cf} was determined from Mt. Wilson observations of stars in the solar neighborhood by comparison of the fluxes in the two "continuum" channels of the H-K photometer to the bolometric flux inferred from the measured color of

each star and its visual magnitude.

2) The flux F_{HK} is made up of radiation from both the chromosphere and the photosphere. The photospheric contribution is due primarily to the fact that the instrumental profile of the Mt. Wilson H-K photometer is triangular with a halfwidth of 1.09Å (Figure 2) and is sensitive well beyond the K_1 and H_1 minima of intensity in the line profiles. Hartmann (personal communication) in analyzing the Vernazza et al. (1981) model C for the quiet Sun, notes that there is close agreement between the amount of radiative losses in the K line from the chromosphere and the totality of emission inside the K_1 minima. Thus the regions outside the K_1 and H_1 minima, shown shaded in Figure 2, are essentially photospheric in origin, whereas the unshaded regions inside the K_1 and H_1 minima are mainly completely chromospheric in origin.

The upper curve of Figure 2, illustrating the spectrum of a solar plage, may be considered as comparable to that of an active chromosphere G2 star (i.e., in the upper branch of Figure 1). We note that in the plage the part of the K line profile outside the K_1 minimum is enhanced much less than that inside the K_1 minimum, which implies that the photospheric response to magnetic fields is much less than that in the chromosphere. Thus, to a first approximation the fraction of the emission detected outside the K_1 minima (shaded in Figure 2) does not respond to the magnetic flux through the underlying photosphere. We therefore correct the observed flux emission F_{HK} by subtracting off the photospheric component F°_{HK}. We write the true chromospheric flux as

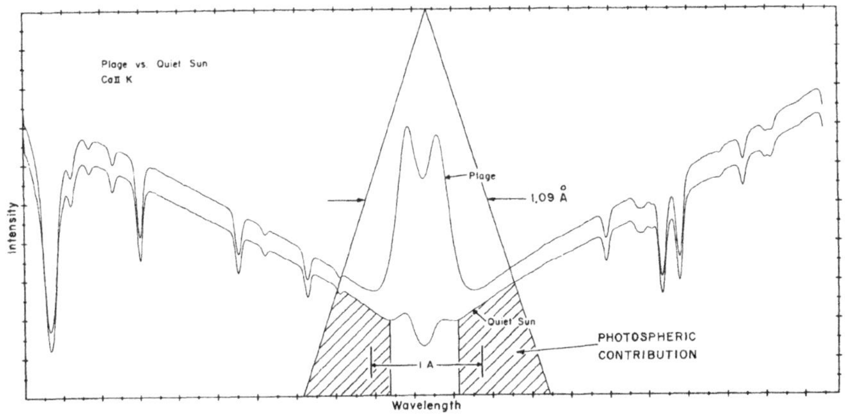

Figure 2. The triangular H-K photometer profile (Vaughan et al. 1978) is superimposed on profiles of the solar Ca II K line for the quiet Sun and for a solar plage, as observed by White and Livingston (1981). The emission outside the K_1 minima which is recorded by the instrument (schematized by the shaded area) is attributed to photospheric radiation.

$$F'_{HK} = F_{HK} - F^{\odot}_{HK},$$

and the chromospheric flux ratio as

$$R'_{HK} = F'_{HK}/\sigma T_{eff}^{4}.$$

For the quiet Sun, the photospheric correction is about 0.5 (i.e., $R^{\odot}_{HK} \sim 0.5\ R_{HK}$).

Hartmann et al. (1983) have determined the ratio $R^{\odot}_{HK}/R_{HK}$ from high-resolution line profiles of four representative stars with weak chromospheric activity and from the observed values of R_{HK} for these stars have derived an empirical relation* for $R^{\odot}_{HK}$ (B-V). Values of the residual chromospheric flux, obtained for the stars in Figure 1 by using this correction, are plotted in Figure 3.

Zwaan (this symposium) has discussed the sharp lower boundary seen in F_{HK} (or, equivalently in the value of $\langle S \rangle$ plotted in Figure 1), and suggests that it could be produced by residual H-K emission in stars whose magnetic activity is so low as to contribute negligibly to the total observed H-K flux. However, the emission $F^{\odot}_{HK}$ we have subtracted is less than that corresponding to the lower boundary of points in Figure 1. In other words, even the most inactive stars still have some residual chromospheric emission. In terms of the search for a relation between rotation and magnetic activity, the question is whether this residual chromospheric emission is associated with surface fields or with purely acoustic heating. An important test is to determine the rotation rate of the most inactive stars in Figure 1 (a difficult task since these stars show little rotational modulation). If these stars are indeed rotating very slowly, one may infer a low level of surface magnetism and hence that the residual chromospheric emission is nonmagnetic. On the other hand, rotation rates not grossly slower than somewhat more active stars such as the Sun would suggest surface magnetic fields on even the least active stars, which would create the observed chromospheric emission by magnetic-associated heating. (In this case, the sharp lower boundary of Figure 1 could be produced by a decreased rate of spindown for very old stars.) In the absence of definitive data on the above point, we shall proceed in this discussion simply to subtract off the photospheric part $R^{\odot}_{HK}$, and investigate how $\langle R'_{HK} \rangle = \langle R_{HK} \rangle - R^{\odot}_{HK}$ varies with spectral type and rotation period, recognizing the uncertainties implied by the above discussion.

---

*The stars used were HD 114378, B-V = .45; HD 115383, B-V = .58; the Sun, B-V = .66; and HD 23249, B-V = .92. The resulting fit is $\log R^{\odot}_{HK} = -4.90 + 1.92\ (B-V)^2 - 2.89(B-V)^3$, valid for $0.45 < B-V < 1$; for B-V > 1 the correction is negligible for any of the stars considered here.

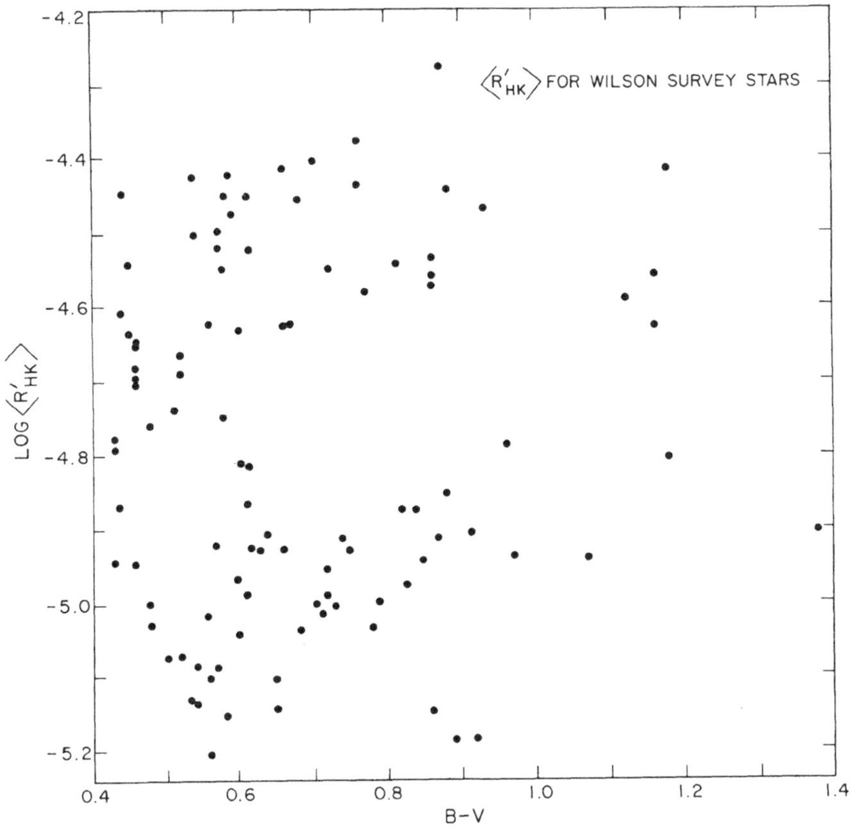

Figure 3. The mean chromospheric H-K flux ratio $\langle R'_{HK} \rangle$, as obtained from the mean Wilson (1978) survey data plotted in Figure 1.

Figure 4 displays the $\langle R'_{HK} \rangle$ values of Figure 3, for those stars with measured rotation periods, plotted versus rotation period. There is a general tendency for a decrease in $\langle R'_{HK} \rangle$ with increasing rotation period, but with considerable scatter. However, examination of the (B-V) colors associated with each point shows that the scatter is clearly color-dependent: for a given period later-type stars show a larger value of $\langle R'_{HK} \rangle$ than do earlier-type stars.

We note that the general trend of chromospheric emission with rotation period for stars of a given color goes approximately as 1/P, and for the purposes of this discussion we have therefore divided the data by the function $f(P) = 7.94 \times 10^{-5}/P$ (dashed line in Figure 4). The ratio of the observed data to f(P) is plotted in Figure 5a. It may be seen that

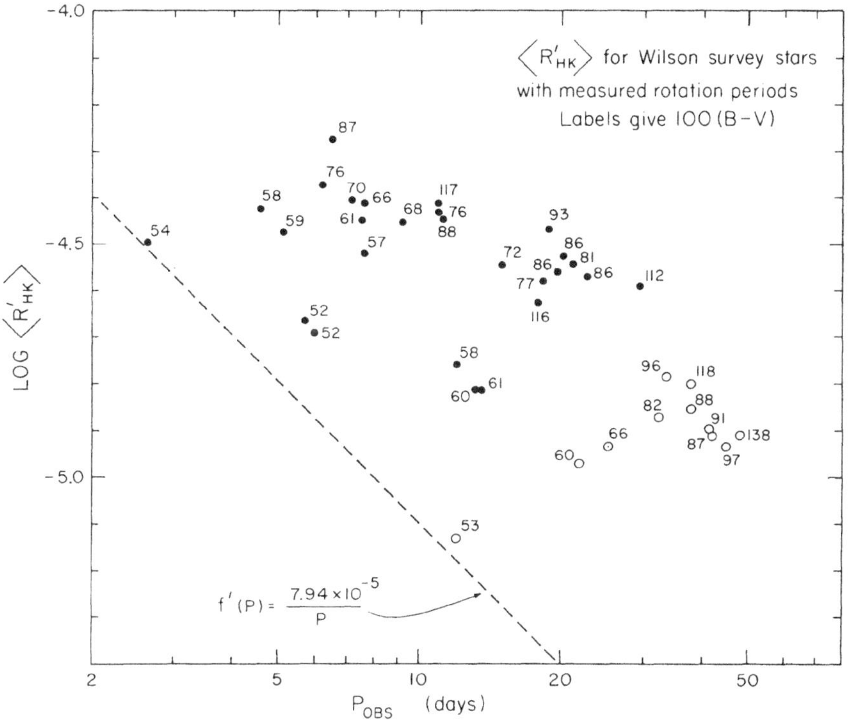

Figure 4. Variation of the mean chromospheric H-K flux ratio $<R'_{HK}>$ with observed rotation P_{obs}, for stars whose rotation period has been measured by rotational modulation of Ca II K (Vaughan et al. 1981; Baliunas et al. 1983; Noyes, in preparation). Closed and open circles represent "young" and "old" stars respectively.

the residual shows a rise of nearly one order of magnitude from earlier to later spectral types.

We have represented the residual in Figure 5a by a function g(B-V), as shown by the solid line. Dividing the data points by g(B-V), we get the plot shown in Figure 5b, illustrating the decreased scatter in the data after the spectral type dependence is factored out.

The factorization $<R'_{HK}> = f(P)\, g\,(B-V)$, where $f(P) \sim 1/P$, has not been chosen by least squares or other optimization technique, and indeed there is not strong a priori reason for picking a power law or a different relation for fitting. The purpose of this factoring is simply to point out that chromospheric activity level in LMS stars appears to first order to be a separable function of rotation rate and spectral type (through the variation of convection zone properties, mass,

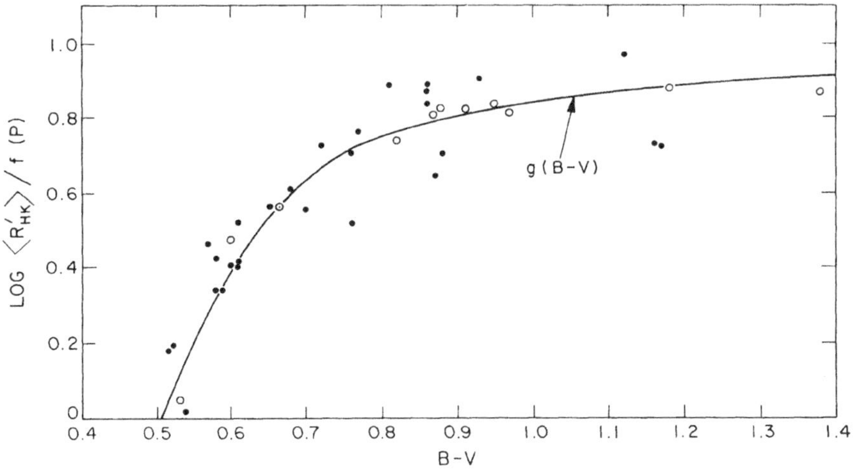

a

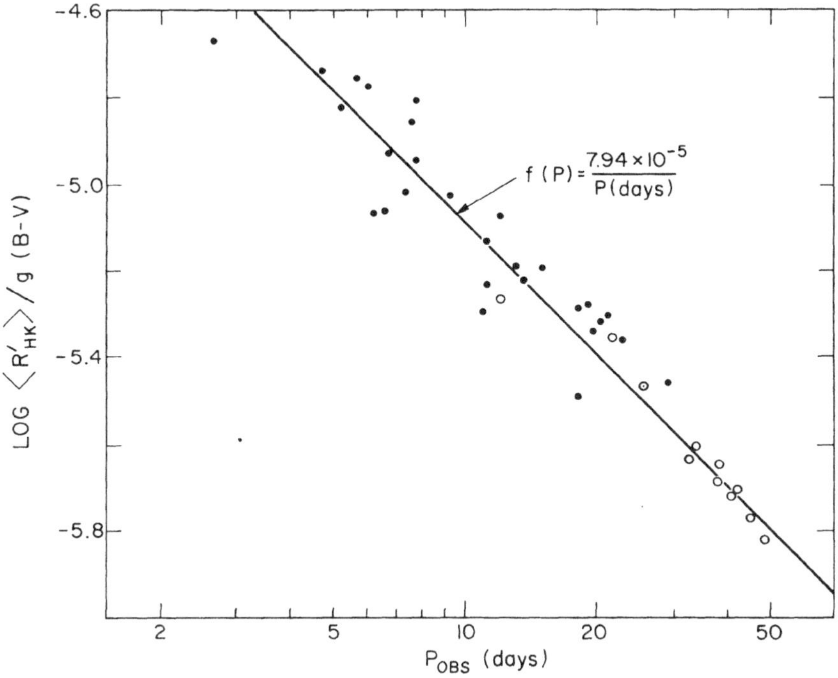

b

Figure 5. (a) Data of Figure 4, divided by the function f(P) = 7.94 x 10^{-5}/P, and replotted versus (B-V). The function g(B-V) is a polynomial fit through the plotted points. (b) Data of Figure 4, divided by the function g(B-V). Closed and open circles represent "young" and "old" stars respectively.

effective temperature, etc.) While an increase of activity with advancing spectral type has been noted before for stars of the same rotation rate (e.g. Walter 1981), the relatively clean separability of the functional dependence is somewhat surprising.

The significance of the shape of the function g(B-V) is not clear. Many potentially significant parameters increase with advancing spectral type, including for example the depth of the convection zone, the convective turn over time, (e.g. Durney and Latour 1978), differential rotation (Belvedere et al. 1980) or the fractional surface area of a star covered by magnetic field (Durney and Robinson 1982). For example, the curve g(B-V) has essentially the same dependence on B-V as does the convective turnover time τ_c (for mixing length to scale height ratio of 1.0), calculated by Gilman (1980). This would imply that chromospheric activity ratio R'_{HK} is approximately proportional to the inverse Rossby number

$$R^{-1} = \tau_c/P_{rot}.$$

The possible importance of the Rossby number in determining overall magnetic activity level is discussed by Schussler in this symposium. However, as Schussler points out, many other physical effects related to the above list of parameters which vary along the main sequence, could lead to a growth of magnetic activity with advancing spectral type (see also Rosner and Vaiana 1980; Durney and Robinson 1981). It is to be hoped that, as the quantity and quality of the data base describing the dependence of chromospheric (and coronal) activity on rotation rate and on spectral type improves, it will be possible to distinguish definitively between different physical mechanisms by the shape of the empirical dependences found.

In Figure 1, 4, and 5, open and closed circles represent stars labeled "old" and "young" respectively by Vaughan (1980), i.e., stars below and above the "gap."* There does not appear to be a significant difference in the dependence of chromospheric emission on rotation rate or color between the two classes of star. This may be taken as suggestive that the cause of the gap is not simply a change in the characteristics of convection or the dynamo at a critical value of the rotation rate (e.g. Durney et al. 1981, Knobloch et al. 1981).

The fact that chromospheric emission ratios R'_{HK} can be empirically represented functionally as f(P) g(B-V) with only moderate scatter permits us to predict rotation rates of stars whose H-K flux index and color are known. This predictive capability has already been used by Vaughan (this symposium) to investigate the rotation period dependence of activity cycles for stars whose period has not yet been measured by

*The stars 61 Cyg A and B, labeled by Vaughan as "young," are here reclassified as "old" based on their clear activity cycles and their low absolute emission level; see also Skumanich and Eddy (1981).

rotational modulation observations.

In another application of this method of rotation period estimation, we have used observed H-K fluxes of Hyades main sequence stars (Duncan, personal communication)) to predict rotation periods for Hyades stars (Figure 6). The variation of H-K flux with (B-V) suggests that for this presumably coeval group of stars (age $\sim$ 600 million years), rotation rates vary smoothly with mass, decreasing by a factor of about 6 from Log $M/M_\odot$ = .09 (B-V = .45) to Log $M/M_\odot$ = -.10 (B-V = .9). This prediction, which needs to be tested by rotational modulation observations of Hyades stars, is of potential importance for two reasons. First, the smoothness of the relation of H-K flux to mass (Figure 6a) suggests that late-type stars either are born on the main sequence with rotation varying smoothly with mass, or that any

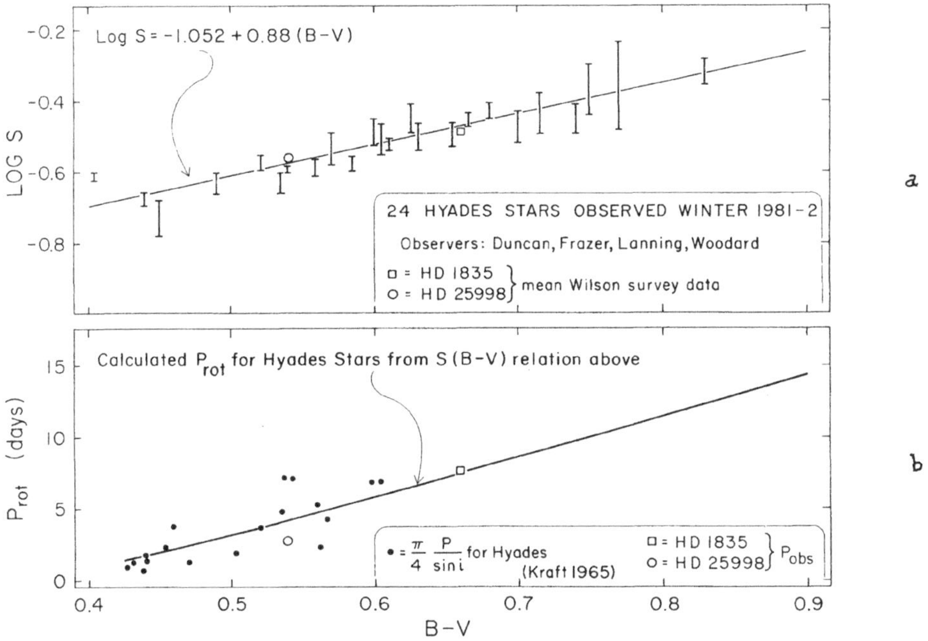

Figure 6. (a) H-K flux index values for 24 Hyades stars (data kindly supplied by D. Duncan), plus the co-moving solar neighborhood stars HD 1835, HD 25998. Solid line is a hand-drawn linear fit to the data. (b) Solid line is the calculated rotation for stars with flux index defined by Log S = -1.052 + .88 (B-V), if the stars obey the relation R_{HK} = f(P) g (B-V) (Figure 5). Also shown are observed rotation periods for HD 1835 and HD 25998, plus periods from Kraft's (1965) values of v sin i of Hyades stars.

idiosyncrasies of rotation rate of individual stars at birth are smoothed out in the process of their spindown as they age to 600 million years. Second, the detailed shape of the curve of rotation versus spectral type (Figure 6b), when compared with similar curves which should be derivable for other clusters of different ages, will allow one to determine the explicit mass-dependence of spindown as a function of age in lower main sequence stars.

The author is grateful to L.W. Hartmann and D. Soderblom for numerous discussions, to A.H. Vaughan and colleagues for permission to use some rotation periods which have not been published in the preparation of Figure 4 and 5, and to D. Duncan for permission to use previously unpublished H-K fluxes in Figure 6.

REFERENCES

Ayres, T. and Linsky, J.: 1980, Astrophys. J. 241, p.279.
Baliunas, S.L., Vaughan, A.H., Hartmann, L.W., Middelkoop, F., Mihalas, D., Noyes, R.W., Preston, G.W., Frazer, J., and Lanning, H.: 1983, in preparation.
Belvedere, A., Paternò, L., and Stix, M.: 1980, Astron. Astrophys. 88, p.240.
Catalano, S. and Marilli, E.: 1982, Astron. Astrophys., submitted.
Duncan, D.K.: 1981, Astrophys. J. 248, p.651.
Durney, B.R.: 1972, in "Solar Wind", eds. C. Sonett, P. Coleman, J. Wilcox, NASA Sp-308, p.282.
Durney, B.R. and Latour, J.: 1978, Geophys. Astrophys. Fluid Dyn. 9, p.241.
Durney, B.R., Mihalas, D., and Robinson, R.D.: 1981, Proc. Astron. Soc. Pacific 93, p.537.
Durney, B.R. and Robinson, R.D.: 1982, Astrophys. J. 253, p.290.
Durney, B.R. and Stenflo, J.O.: 1972, Astrophys. Space Sci. 15, p.307.
Frazier, E.N.: 1971, Solar Phys. 21, p.42.
Gilman, P.A.: 1980, in "Stellar Turbulence", eds. D. Gray and J. Linsky, Springer-Verlag, p.19.
Hall, D.S.: 1981, in "Solar Phenomena in Stars and Stellar Systems", eds. R. Bonnet and A.K. Dupree, D. Reidel Publ.Co., Dordrecht, p.487.
Hartmann, L.W., Burnham, N., Soderblom, D., and Noyes, R.W.: 1983, in preparation.
Knobloch, E., Rosner, R., and Weiss, N.O.: 1981, Monthly Not. Roy. Astron. Soc. 197, p.45P.
Kraft, R.P.: 1965, Astrophys. J. 142, p.681.
Kraft, R.P.: 1967, Astrophys. J. 150, p.551.
Leighton, R.B.: 1959, Astrophys. J. 130, p.366.
Mewe, R., Schrijver, C.J., and Zwaan, C.: 1981, in "Proc. 15th ESLAB Symposium", ed. R.D. Andresen, Space Sci. Rev., 30, p. 191.
Middelkoop, F.: 1982, Astron. Astrophys. 107, p.31.

Pallavicini, R., Golub, L., Rosner, R., Vaiana, G.S., Ayres, T., and Linsky, J.L.: 1981, Astrophys. J. 248, p.279.
Rosner, R. and Vaiana, G.S.: 1980, in "X-Ray Astronomy", eds. R. Giacconi and G. Setti, D. Reidel Publ.Co., Dordrecht, p.129.
Schatzmann, E.: 1962, Ann. Astrophys. 25, p.18.
Skumanich, A.: 1972, Astrophys. J. 171, p.565.
Skumanich, A. and Eddy, J.: 1981, in "Solar Phenomena in Stars and Stellar Systems", eds. R. Bonnet and A.K. Dupree, D. Reidel Publ.Co., Dordrecht, p.349.
Skumanich, A., Smythe, C., and Frazier, E.N.: 1975, Astrophys. J. 200, p.747.
Vaughan, A.H.: 1980, Proc. Astron. Soc. Pacific 92, p.392.
Vaughan, A.H., Baliunas, S.L., Middelkoop, F., Hartmann, L.W., Mihalas, D., Noyes, R.W., and Preston, G.W.: 1981, Astrophys. J. 250 p.276.
Vaughan, A.H. and Preston, G.W.: 1980, Proc. Astron. Soc. Pacific 92, p.385.
Vaughan, A.H., Preston, G.W., and Wilson, O.C.: 1978, Proc. Astron. Soc. Pacific 90, p.267.
Vernazza, J.E., Avrett, E.H., and Loeser, R.: 1981, Astrophys. J. Suppl. 45, p.635.
Walter, F.M.: 1981, Astrophys. J. 245, p.677.
Walter, F.M.: 1982, Astrophys. J. 253, p.745.
White, O.R. and Livingston, W.C.: 1981, Astrophys. J. 249, p.798.
Wilson, O.C.: 1978, Astrophys. J. 226, 379.

DISCUSSION

CRAM: I was surprised to see such a tight correlation between period and $B-V$ for the Hyades. Do you have any comment on how those stars get the fact that they are presumably born with different angular momentum?

NOYES: Presumably their angular momenta are not at all that different from what they were when the stars arrived at the main sequence. David Soderblom made the suggestion to me that the rate at which the stars spun down to arrive at that particular rotation period is simply related to the amount of time it took them to reach the main sequence.

WEISS: Could you clarify whether there is any clear evidence for a gap if the rotation rate is plotted against spectral type?

VAUGHAN: In a plot of P_{rot} against $B-V$ for the main sequence stars showing cycles or possible cycles, the dichotomy between active and inactive stars is very striking, in the sense that two sequences appear, one resembling the plot of P_{rot} vs. $B-V$ for the Hyades, the other a sequence containing the sun. Obviously the plot should be made for the full sample of solar neighborhood stars.

MOUSCHOVIAS: Quite justifiably, yesterday and to-day there has been a considerable amount of discussion on what the physical parameters are on which stellar and solar activity depends. Naive as the answer may seem, to somebody who "comes from interstellar space" the answer is evident. There are very few dimensionless parameters (ratios of such quantities as the gravitational, magnetic, thermal, and rotational kinetic energies) available at the protostar formation epoch. These parameters (and, more precisely, their distribution within a protostar) determine the subsequent evolution of a star, including its surface activity. It would therefore be useful and appropriate to plot some index of stellar activity against the energy ratios (magnetic/gravitational) and (rotational/gravitational); it is not necessary to plot activity against the ratio (thermal/gravitational), because stars are supported by thermal pressure gradients. Certainly the distribution of these parameters within a star is significant, but is itself determined from the protostar stage.

Quantities such as effective temperature, surface period of rotation, etc., against which activity is plotted, are important but not fundamental, in that they are derivable from the parameters mentioned above. We may miss the underlying fundamental physics if we focus only on the latter quantities as possible indicators of stellar activity.

MAGNETIC FIELDS AND CONVECTION: NEW OBSERVATIONS

W. C. Livingston
Kitt Peak National Observatory

ABSTRACT

Fourier Transform Spectrometer observations of Fraunhofer line displacement and asymmetry suggest that granular convection is inhibited in regions of magnetic activity. We discuss the observations and a method for removing the effects of solar rotation. In integrated sunlight ("the sun-as-a-star") records of line asymmetries indicate that a significant reduction in the amplitude of the sun's convective signature took place between 1980 and 1982. To the extent that full disk line asymmetry arises strictly from convective motions, the results constitute strong evidence that magnetic activity influences (inhibits) convective motion on a global scale.

1. INTRODUCTION

A number of theoretical papers have considered the possibility that changes in surface, or immediately subsurface, magnetic fields can affect the efficiency of convection in the granulation layer and thus modulate the sun's luminosity (Thomas, 1979; Spiegel and Weiss, 1980; Dearborn and Blake, 1982). The analytic tools for diagnosing photospheric convective motion, viz. by Fraunhofer line asymmetry, has recently advanced (Dravins, et al., 1981; Dravins, 1982). The brightness-velocity correlation arising, from granular motions, causes line bisectors to have a 'C' shape; the middle of the C being blue-shifted relative to the laboratory value. For strong lines, such as Fe 5302, the core of the line is formed sufficiently high to be largely above the influence of granules and so is less displaced. In the far wings of the line, increased density in the line formation region coupled with the condition of flow continuity also results in reduced displacement. Realistic modeling by Nordlund (1982) provides a quantitative basis for the interpretation of the 'C' shape.

Although line bisectors can be obtained from conventional spectrograph data, the Fourier Transform Spectrometer (FTS) offers several advantages. First the FTS instrument profile is inherently symmetric so that no corrections are necessary. Second, many lines are obtained

J. O. Stenflo (ed.), Solar and Magnetic Fields: Origins and Coronal Effects, 149–153.

simultaneously. Third, the wavelengths are absolute after a zero point correction.

Inspired then by the predictions of theory, armed with the diagnostic tools of the line bisector, and taking advantage of the Brault 1-m FTS, we have made observations of line asymmetry in and out of magnetic regions. We have also examined the convection signature of the sun as a star. Details on methology and results are now presented.

2. OBSERVATIONS AND REDUCTIONS

While high spatial resolution information on granule dynamics could be valuable (see e.g., Muller, 1977), the slow scanning speed of the FTS eliminates that possibility. Rather than work at intermediate resolution we have chosen to observe at the very low resolution of 1 arc-min to better average over a large number of granules. Our observing procedure is as follows.

Referring to the day's full disk magnetogram as a guide, we center the FTS entrance aperture on a magnetic region, avoiding sunspots and pores. A scan is then made which takes about 90^s, covers $\lambda 0.500$ to $\lambda 0.600$ μm. This scan is designated "Plage 1". Next we move to a nearby position that is relatively clear of fields and at about the same limb distance. Another scan is made and this becomes "Quiet 1". We continue taking such pairs of active-quiet scans until the magnetic concentrations on the solar disk have been fairly sampled. Typically a dozen regions are observed on a given day and to date about 50 pairs of scans in all have been acquired.

After transforming the data, for each line on the Dravins, et al. (1981) list, we insert the laboratory wavelength of Crosswhite (1958). We then compute the profile bisector on an absolute scale taking account of the gravitational redshift and the usual Earth-Sun Ephemeris corrections.

There remains a component of velocity between Plage 1 and Quiet 1 due to solar rotation. To remove this rotation part we note the difference in wavelength between the cores of Mg 5172 in Plage 1 and Quiet 1, and apply this difference as a correction to the other pairs of Fe line data sets. The assumption here is that the granulation signal is practically absent in the core of Mg 5172 and that the only other velocity difference is due to rotation. No doubt the 5-minute oscillation introduces some disturbing error, but the effect is reduced by the large sample population.

3. RESULTS

Figure 1 illustrates the behavior of 95% of the 50 regions sampled. The non-magnetic (or quiet) bisector almost always has greater displacement to the blue and a greater amplitude of curvature than the magnetic region (or plage). Only for regions very near the limb does this rule fail. We may mention that g = 0 lines are indistinguishable from Zeeman

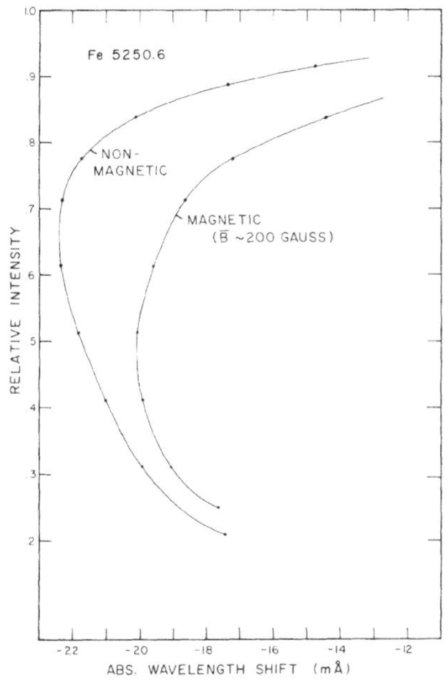

Figure 1. Line bisectors of Fe 5250.6 in and out of magnetic regions. In this example solar rotation has not been removed.

sensitive lines; from which we conclude that polarization effects play little or no role.

We have not yet been successful in determining the FTS zero point correction and thus placing the bisectors properly on an absolute velocity frame. The problem has been the wavelengths of the A-band of atmospheric oxygen which were observed for this purpose. The Babcock and Herzberg (1948) values show an internal dispersion of more than 1 mÅ which is totally inadequate. Brault and Lerner (1982) are preparing a set of revised O_2 wavelengths.

In addition to observing disk features it is easy to feed unfocused sunlight to the FTS and thus observe the "sun as a star". The resulting bisectors are remarkably stable. If we assign a velocity of zero to the core of a strong line such as Fe 5302.3 then the standard deviation of the blue shifted component is less than 0.1 mÅ for 6 observations on a given day. We have obtained a number of full disk observations since May 1980. Within the λ0.500 -λ0.600 μm window there are several relatively unblended lines similar in strength to Fe 5302.3. The bisectors of these lines have been averaged to yield Figure 2.

4. CONCLUSIONS

In agreement with the theoretical predictions it appears that granular convection is inhibited in regions of high magnetic flux. As this

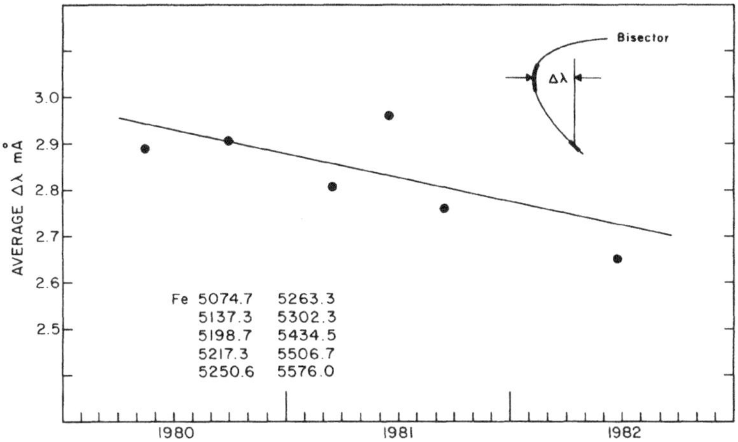

Figure 2. Full disk line bisector curvature, Δλ, showing decrease with time for the average of 10 similar Fe lines.

inference is based on low spatial resolution data, we still need high resolution observations to determine the role of facular points and decide whether or not the inhibition is global or localized.

The fact that full disk bisectors have diminished ~10% over the past two years suggests to us that there has been a global reduction in the vigor of granular convection (see also Livingston, 1982). Although the total surface magnetic flux has not increased lately, it has been redistributed. Initially confined to newly born sunspot umbrae, flux is now being distributed over the entire globe, presumably by the random walk process. Without additional new flux, a growing fraction of the sun's surface is threaded by magnetic fields with a consequent reduction in granular convection.

REFERENCES

Babcock, H. D. and Herzberg, L.: 1948, Astrophys. J. 108, pp. 167-190.
Brault, J. and Lerner, R.: 1982, private communication.
Crosswhite, H. M.: 1958, Johns Hopkins Spectroscopic Report No. 13.
Dearborn, D. S. P. and Blake, J. B.: 1982, Astrophys. J. 257, pp. 896-900.
Dravins, D., Lindegren, L. and Nordlund, Å.: 1981, Astron. Astrophys. 96, pp. 345-364.
Dravins, D.: 1982, Ann. Rev. Astron. Astrophys. 20 (in press).
Livingston, W. C.: 1982, Nature 297, pp. 208-209.
Muller, R.: 1977, Solar Phys. 52, pp. 249-262.
Nordlund, Å.: 1982, Astron. Astrophys. 107, pp. 1-10.
Nordlund, Å.: 1982, Astron. Astrophys., (in press).
Spiegel, E. A. and Weiss, N. O.: 1980, Nature 287, pp. 616-617.
Thomas, J. H.: 1977, Nature 280, pp. 662-663.

DISCUSSION

ZWAAN: Since you have measures of line asymmetries in plages, and you may estimate filling factors for plages, you may estimate the full-disk effect, and compare with your full-disk measurements — did you try?

LIVINGSTON: Yes, but our resulting plage filling factors were too large — about 20% . We plan more work on this.

SCHÜSSLER: I have a comment and a question. Comment: Macris has observed that granules are smaller than average in the vicinity of sunspots. This may have something to do with your work. Question: Would you not expect strong variations in emergent intensity for a strong magnetic influence on convection?

LIVINGSTON: Yes, and it was this expectation (with H. Holweger), looking for solar luminosity variations, that led us to this work in the first place.

SHEELEY: What have your integrated K-emission measurements done during this same 2–3 year interval? If the K-line flux were related to magnetic flux, then it should have gone up (according to Bob Howard's magnetic-flux measurements) rather than down as the sunspot number fell during 1980 and 1981.

LIVINGSTON: The K index has simply been fluctuating since 1979. We see no systematic trend up or down yet. As a matter of fact, in my opinion we are now at a new phase of the solar cycle with respect to the correlation between plage and the K intensity. We are at a phase when I think the sun is covered with so much network that the individual plages are beginning to affect the integrated signal from the sun less and less. This phase starts at sunspot maximum and continues on for some time, and it is a little different from the first leg of the cycle, when there was such a good correlation.

KEIL: I have two comments. (1) Pete Worden and I have noted that the K index and K intensity have not yet started to decrease with the declining phase of this solar cycle although the amount of plages has decreased considerably. There appears to exist a lag between the disappearance of plages and a decrease in chromospheric emission. (2) Peter Foukal and I have observed the difference between bisectors of quiet sun line profiles and profiles formed in the solar network. I have computed how much convection would have to be supressed to give the observed bisector shift between the quiet sun and the solar network. The shift can be modelled by either decreasing the amplitude of the convection by about 10% throughout the network, or by assuming that about 10% of the area in the network exhibits no convection, while the rest of the area has normal convection.

GRAY: When I received a preprint fom Bill Livingston showing the remarkable relation in his second figure, I did not chew on it the way the slide projector did, but I immediately went to my stellar bisector data to see if such an effect existed there. I first looked at several exposures of Arcturus and found no changes there. The errors, however, were found to be $\sim \pm 25$ m s^{-1}. Then I selected several pairs of stars by spectral type. In most of the pairs, there are no significant bisector differences at this error level, but in one K0 V set (70 Oph A vs. σ Dra) there is a distinct systematic difference shown by all lines. I do not know if there is any connection here with the solar results of Livingston, but at least we are able to see bisector differences, and are potentially in a position to find in stars what he has seen in the sun.

POSSIBILITIES AND ATTEMPTS TO DETERMINE THE DIFFERENTIAL ROTATION ON F-TYPE MAIN-SEQUENCE STARS

Hubertus W ö h l
Kiepenheuer-Institut für Sonnenphysik
Schöneckstraße 6, D-7800 Freiburg
Federal Republic of Germany

Abstract : A preliminary analysis of spectral lines by Fourier transformation was used to determine upper limits of differential rotation for 11 F-type stars, mainly from the main-sequence. No evidence for a much steeper differential rotation on these stars than on the Sun could be found.

1. INTRODUCTION

Differential rotation on stars may be determined by

(1) - a time analysis of changes of the stellar radiation due to the rotation of 'activity regions' and their drifts in latitude during a stellar activity cycle.

(2) - an analysis of differences of rotationally broadened spectral line profiles as compared to spectral lines from rigid body rotation.

While the first method is not yet proven for the Sun (but see Vogt,1981), the second is discussed since the pioneering work of Gray(1977):
If the differential effect is written in terms of angular velocity as a function of colatitude ψ

$$\omega(\psi) = \omega_o + \omega_2 \sin^2 \psi \qquad [1]$$

then the parameter $\alpha = \omega_2/(\omega_o + \omega_2)$ [2]

is a measure of the differential effect.
The Fourier transforms of the line profiles show characteristics of the differential effect: The strength, both height and width, of the first sidelobe decreases with increasing α, while the second sidelobe is nearly constant. For $\alpha \approx 0.5$ the first and second sidelobes are about the same.
Bruning(1981) and independently Garcia-Alegre et al.(1982) showed that this effect of the differential rotation can even be found in artificial integrated spectral lines

J. O. Stenflo (ed.), Solar and Magnetic Fields: Origins and Coronal Effects, 155–159.

contaminated by CLV, blends, line asymmetries, line shifts etc. For non-equator-on cases the effect produces even larger changes. They also confirmed that spectral line profiles with a high signal-to-noise ratio are needed to determine the sidelobes.
Gray(1982) reported the analysis of 7 F-type stars and found observational evidence against differential rotation in F stars.
The aim of this work was to investigate stars with higher rotation velocities and also compare the results with those of Gray(1982) and predictions of Belvedere et al.(1980).

2. OBSERVATIONS

The spectra were obtained with the linear RETICON array in the Coudé Echelle Spectrometer (CES) at the 1.4 m Coudé Auxiliary Telescope (CAT) of the European Southern Observatory (ESO) at La Silla,Chile, June 9 until 17,1982. The red light path was used and the exposure times were in general 0.5 h but sometimes less due to clouds. The spectral resolution was in general 100 000, in some cases 150 000. The wavelength region was 5534.4 to 5597.6 Å centered on the non-split Fe line at 5576.1 Å. In some cases also the wavelength region used by Gray(1982) from 6221.9 to 6274.2 Å was used. The pixel spacing was 23.1 mÅ in the first and 28 mÅ in the second case. The signal-to-noise ratio obtained in the continua of the spectra was in general 200 to 300. The census of data is given in Table 1, where the stars are arranged according to increasing rotation velocities given by Uesugi(1976).

3. REDUCTIONS AND RESULTS

The spectral line analysis was performed July 19-24, 1982 using the Image Handling Package (IHAP) at the ESO headquarter at Garching. The dark current and flat field corrections had to be performed separately for each spectrum. No smoothing of the original data was applied.
In a first analysis of the 36 useful spectra only the Fe 5576.1 Å or the Fe 6252.6 Å line were analysed. Most of the Fourier transformed spectral lines showed no indications for values of $\alpha > 0.4$ (see Table 1). Typical Fourier transformed spectral lines are given in double log scale in Fig.1. In those cases, where an indication for a steeper differentia rotation was found, in addition the spectral lines Fe 5569.6, Ca 5582.0 and Fe 5586.8 Å were analysed in the shorter wavelength region and the spectral lines Fe/Si 6254.2 Fe 6256.4, Fe 6265.1, Fe 6230.7 and Fe 6232.6 Å for the longer wavelength region.

Table 1 : Census of data of the analysed F-type stars

star name	spectral type	m_{vis} [mag]	v_{rot} [km/s] Uesugi	this work	first results, remarks	
ξ OPH	F 2 V	4.39	0	24	1S, 7N,	G1
γ SER	F 6 IV-V	3.89	8	11	1S,	G2
γ PAV	F 8 V	4.22	10	4-11	1S, 4N	
τ BOO	F 7 V	4.50	14	15-22	2S, 3N,	G3
α CRV	F 2 V	4.03	25	34-38	3N	
ψ CAP	F 5 V	4.14	37	48-61	1S, 5N	
ω PSC	F 3 V	4.01	38	48	2N	
μ VIR	F 3 IV	3.89	60	77	1N	
ζ SER	F 3 V	4.62	80	97	2N	
δ AQL	F 0 IV	3.36	85	97-108	2S	
η SCO	F 2 V	3.33	150	153	0.5S,0.5N, R	

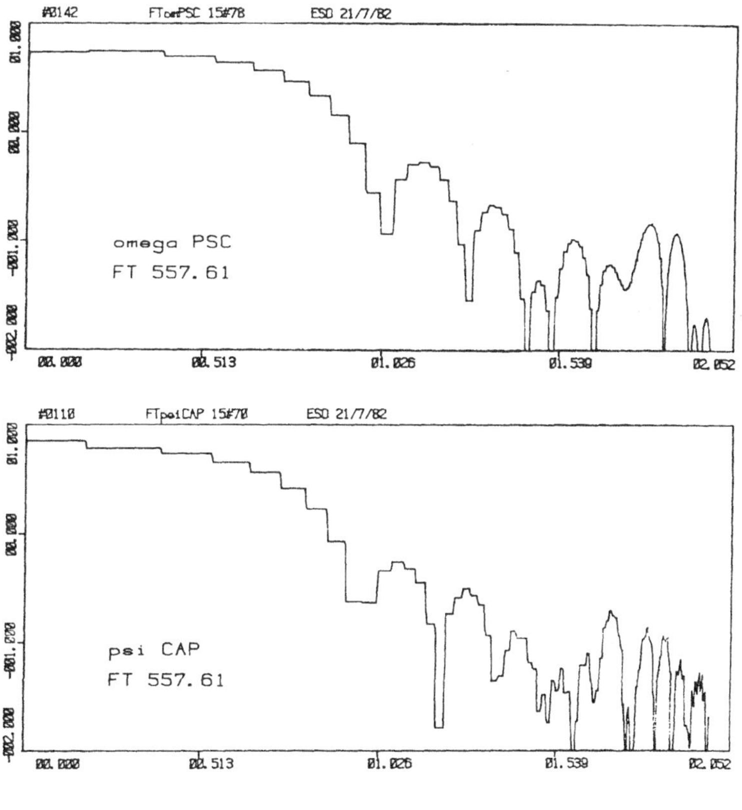

Fig. 1 : Typical examples of Fourier transforms of the line Fe 5576.1 Å. Given are normalized Fourier amplitudes against the first 200 points of the transform in a double log scale. The Fourier frequency spacing is about 0.0021 s/km.

In no case an indication for a value of $\alpha > 0.4$ remained for the majority of the spectral lines analysed from any star. In addition the rotation velocities for the stars investigated were determined from the zeros of the Fourier transforms; they are also given in Table 1.

4. CONCLUSIONS

No evidence could be found for a steep differential rotation on F-type stars as claimed by Belvedere et al.(1980) with $\alpha > 0.5$. This result is in agreement with that of Gray(1982), although in general other spectral lines and mainly different stars were used.
A more detailed analysis - comparable to the 'modeling' done by Gray(1982) - is needed to reduce the upper limit of $\alpha \simeq 0.4$ and to determine the parameters with more precision.

Acknowledgements : The cooperation with M.C.Garcia-Alegre and M.Vázquez in an earlier phase of this project is acknowledged. The observations and most of the data reductions were performed with equipment and financial support of the European Southern Observatory (ESO), which is gratefully acknowledged.

REFERENCES

Belvedere, G., Paternò, L., and Stix, M.: 1980, Astron. Astrophys. 88, p. 240.
Bruning, D.H.: 1981, Astrophys. J. 248, p. 274.
Garcia-Alegre, M.C., Vázquez, M., and Wöhl, H.: 1982, Astron. Astrophys. 106, p. 261.
Gray, D.F.: 1977, Astrophys. J. 211, p. 198.
Gray, D.F.: 1982, Astrophys. J. 258, p. 201.
Uesugi, A.: 1976, Revised Catalogue of Stellar Rotational Velocities (Kyoto: University of Kyoto Press).
Vogt, S.S.: 1981, Astrophys. J. 250, p. 327.

REMARKS TO TABLE 1 (previous page)

S : indication for steep differential rotation
N : no indication for steep differential rotation
R : spectral lines Na D 5890 and 5896 Å analysed, one with and the other without indication for steep differential rotation
other spectral classifications of this star : F 0 IVn and F 2 III

from Gray(1982): G1 : v_{rot} = 20.6 km/s, α = -0.15
G2 : v_{rot} = 10.8 km/s
G3 : v_{rot} = 14.8 km/s, α = 0.27

DISCUSSION

SODERBLOM: Your data have a very high signal-to-noise ratio and very high spectral resolution, yet your values of $v \sin i$ seem very rough; for example, for one star a range of 4 to 11 km s^{-1} is given. Why cannot the $v \sin i$ be specified more precisely? If the $v \sin i$ is poorly determined, how can differential rotation be detected?

WÖHL: In the preliminary analysis I did not try to determine the rotational velocities very accurately, but just from the double-log plots. In the case of slowly rotating stars, influences from motions in the atmosphere of the star have to be taken into account. They depend on the individual spectral line. In addition, more careful corrections (averaging of several dark currents, flat fields, etc.) will hopefully improve the internal accuracy within a detailed analysis. Nevertheless the statement about the upper limit of the differential rotation parameter α can be made, because then only the first sidelobe is compared with the second in height and width, not depending on the position, which gives the rotational velocity.

GILMAN: Do any of the stars you observed have magnetic activity, either cyclic or non-cyclic?

WÖHL: I did not yet check this, because I selected the stars according to the spectral type and rotational velocity. In addition I was guided by theories, which claim that differential rotation is independent of magnetic activity.

SODERBLOM: Response to Gilman: At least one of the stars, τ Boo = HD 120136, was illustrated by Dr. Vaughan earlier. It had at best only a slight cyclic variation.

GILMAN: The reason why I asked is that it is my opinion that stars having cyclic activity are the ones that have large differential rotations, and the ones which do not are likely to have small differential rotation, because differential rotation plays such a prominent role in producing a cycle.

GRAY: We should look carefully at the list of program stars used here. They are mostly early F stars, and for them the convection zones are thought to be very thin. So we might not expect differential rotation in most of these stars, and the calculations of Belvedere, Paterno, and Stix may not predict any.

WÖHL: As I understood the paper of Belvedere, Paterno, and Stix the steep differential rotation is mainly caused by the high rotational velocity. Therefore I selected these early F-type stars, which still should have convection zones.

STIX: I must admit that our model indeed produced such large differential rotation, if only a (even thin) convection zone was present. So maybe we have to modify this theory a bit.

IUE OBSERVATIONS OF THE CHROMOSPHERIC ACTIVITY-AGE RELATION IN YOUNG SOLAR-TYPE STARS

Theodore Simon and Ann Merchant Boesgaard
Institute for Astronomy
2680 Woodlawn Drive
Honolulu, Hawaii 96822 USA

INTRODUCTION

The difficulties of measuring magnetic fields in late-type stars other than the sun are well known, as one is reminded by other contributions to these Proceedings. This Symposium nevertheless comes at a very opportune time, as we are now at the point where we can begin to explore the relationship of stellar magnetism to flare activity and quiescent cool star chromospheres, transition regions (TRs), and coronae.

Except for the synoptic observations of the chromospheric Ca II H-K lines by Wilson (1978), in which he sought evidence for magnetic activity cycles, there is still scant data on stellar activity, especially at UV and X-ray wavelengths where 10^5 K TRs and 10^6 - 10^7 K coronae are expected to radiate. This paper presents new UV data, obtained with the IUE spacecraft, for a dozen solar-type stars in the field. The stars are of spectral type F6 V - G1 V; on the basis of their high Li content, they range in age from 0.1 to 2.8 Gyr (Duncan 1981). Our purpose is to study the evolution of TR and chromospheric emission with stellar age, and also the surface distribution of magnetically active regions as revealed by rotational modulation of UV emission line fluxes.

OBSERVATIONS

Low resolution (~6 Å) spectra from 1200-1950 Å were obtained with the SWP camera of IUE. We have measured the strengths of chromospheric and TR emission features (mostly unresolved blends) of O I, C I, C II, C IV, Si IV, N V, and He II. The bright continuum at $\lambda > 1750$ Å is generally overexposed in our deep exposures, saturating the Si II, Si III, and C III features. Echelle spectra of the Mg II lines at 2800 Å (resolution 220 mÅ) were obtained for all but one star with the LWR camera. A dozen exposures of χ^1 Ori, covering three rotational cycles, were made with both cameras in order to assess the effects of surface activity (Boesgaard and Simon 1982).

J. O. Stenflo (ed.), Solar and Magnetic Fields: Origins and Coronal Effects, 161–164.

CORRELATIONS: ACTIVITY AND AGE

We find that the hottest TR lines are more strongly enhanced above quiet sun fluxes in the young stars than are the lines formed in the low chromosphere, and thus the high temperature lines follow a steeper relation between emission strength and age than do the chromospheric lines. As representative of 10^5 K plasma, the observed C IV 1549 Å strengths are shown in Figure 1. The line fluxes are normalized to the apparent stellar bolometric luminosity, and the symbols are keyed to stellar rotational velocities from Soderblom (1980) or Kraft (1967). Normalized fluxes are also given for T Tauri variables (Imhoff and Giampapa 1982), corrected for UV extinction, for stars in the Hyades cluster (Zolcinski et al. 1982), and for the active and quiet sun. A power law, flux $\propto$ Age$^{-0.8}$, fits the C IV data reasonably well, despite some significant deviations. The range of variation observed for χ^1 Ori (at 0.6 Gyr), denoted by a shaded box in Figure 1, presumably gives a fair indication of the scatter in C IV brightness expected as stellar active regions or bright supergranulation network rotate on and off the visible stellar disk.

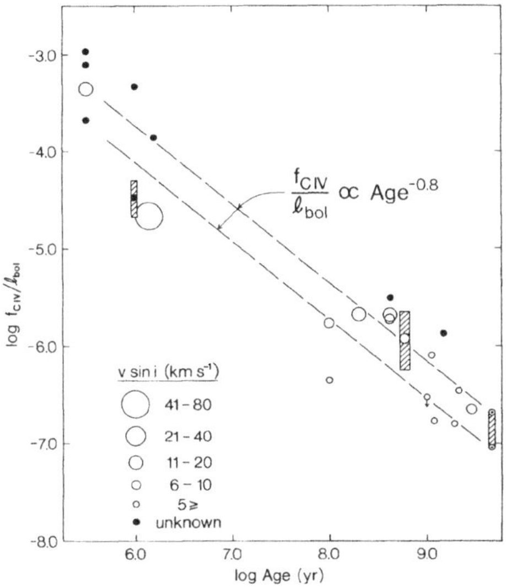

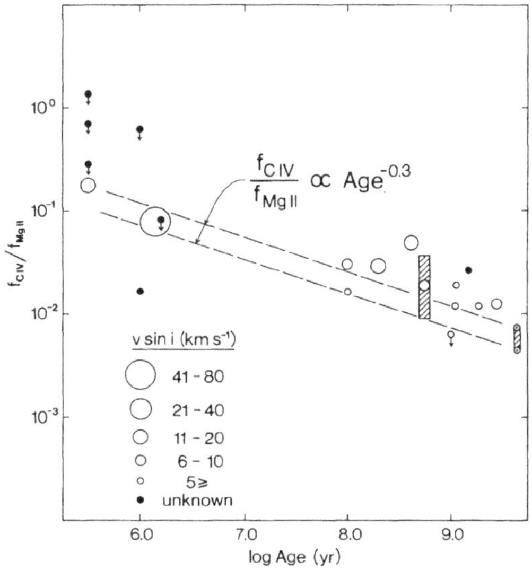

Figure 1 (left). Normalized flux of C IV 1549 Å. The range in line flux observed for RW Aur, χ^1 Ori, and the sun are shown.

Figure 2 (right). Ratio of C IV 1549 Å flux to the Mg II 2800 Å emission line flux.

UV emission lines of lower excitation (e.g., C II 1335 Å) follow similar power-law relations with shallower slopes. The chromospheric Mg II lines decline in strength with the square root of stellar age, a result given previously by Skumanich (1972) for the Ca II H-K reversals and axial rotation. The oldest stars in our sample tend to have both the weakest C IV and weakest Mg II emission. The $f_{C\ IV}/f_{Mg\ II}$ ratio, which measures the relative amounts of 10^5 K and chromospheric material, is correlated with age as shown in Figure 2: the younger the star, the larger the differential emission measure at high temperature, although the observational scatter is considerable. The scatter may be related to the fact that, in contrast to the large variation in the TR emission of χ^1 Ori, we found little variation in simultaneous observations of the chromospheric Mg II lines, with no more than a 13% range in average Mg flux over a year.

DISCUSSION

Our IUE observations indicate that emission fluxes of high temperature lines in the UV, and consequently the outer atmosphere heating and cooling rates, are correlated with the ages of solar-type stars. Stars one-tenth the age of the sun have emission measures in 10^5 K plasma an order of magnitude greater than the solar TR. This trend extends to the very young T Tau stars, although Hartmann et al. (1982) believe that their UV emission may arise from strong stellar winds. Among the young stars we find a greater enhancement of the emission strengths of high temperature lines (C IV, Si IV, N V) than those of low temperature lines (Si II, C I, Mg II).

Because a similar trend is present in UV spectra of solar active regions, and because high dispersion IUE spectra of active dwarf stars (including χ^1 Ori) show no opacity broadening by comparison with old solar-type stars like α Cen A, Ayres et al. (1982) conclude that magnetic structures on quiet and active dwarfs must be similar. They attribute differences in activity levels among dwarf stars to changes in the fraction of surface occupied by magnetic regions, e.g., plages. There are several difficulties with this model. First, in UV observations of the sun there is no difference in TR line widths between supergranulation cell interiors and boundaries (Feldman et al. 1976), coronal holes and the quiet sun (Doschek et al. 1981), or active regions and the quiet sun (Feldman and Doschek 1978). Furthermore, the Mg emission reversal, although more intense in solar active regions, is not perceptibly broader than quiet sun profiles from the chromospheric network or supergranulation cell interiors (Lemaire and Skumanich 1973). Second, the C IV surface flux of young field stars is ~20 times brighter than the quiet sun C IV emission. Since the contrast between active and quiet regions of the sun is a factor of 5-10 (Cook et al. 1980), the greater emission of young stars is difficult to explain by a larger coverage of solar plage, as suggested by Ayres et al. (1982), except possibly with filling factors very near to unity. But high values are precluded by the large rotational variations seen among the youngest stars. The surface fluxes of the T Tau stars, three orders of magnitude larger than the sun's, are impossible to understand with this kind of model. A more compact supergranulation network can't account for the enhanced C IV among young stars, especially the T Tauris, Since the solar TR lines are only 3-5 times brighter in the network than in cell interiors and the area coverage of the quiet sun network is ~40% (Reeves 1976). Therefore, a decided change in the temperature-pressure structure of young star atmospheres seems inescapable, and a change in the dimensions of their supergranulation patterns seems likely as well.

Although an enhanced network, varying in cell size across the stellar disk, is our preferred model for the UV activity-age correlations found here, intense active regions with lifetimes of a year or more may exist. We note however that brightness variations at the Mg II k_1 feature and the nearby continuum were not observed for χ^1 Ori, nor was any variation observed in the continuum below 2000 Å, as is found in the brightest solar plages (Cook et al. 1980).

REFERENCES

Ayres, T.R., Linsky, J.L., Brown, A., Jordan, C., and Simon, T.: 1982, in Advances in Ultraviolet Astron.: Four Years of IUE Research, in press.

Boesgaard, A.M. and Simon, T.: 1982, Poster paper, IAU Symp. 102.

Cook, J.W., Brueckner, G.E., and VanHoosier, M.E.: 1980, J. Geophys. Res. 85, 2257.

Doschek, G.A., Mariska, J.T., and Feldman, U.: 1981, Mon.Not.Roy. Astron.Soc. 195, 107.

Duncan, D.: 1981, Astrophys. J. 248, 651.

Feldman, U. and Doschek, G.A.: 1978, Astrophys. J. Supp. 37, 443.

Feldman, U., Doschek, G.A., and Patterson, N.P. 1976, Astrophys. J. 209, 270.

Hartmann, L., Edwards, S., and Avrett, E.: 1982, Astrophys. J., in press.

Imhoff, C.L. and Giampapa, M.S.: 1982, in Advances in Ultraviolet Astron.: Four Years of IUE Research, in press.

Kraft, R.P.: 1967, Astrophys. J. 150, 551.

Lemaire, P. and Skumanich, A.: 1973, Astron. Astrophys. 22, 61.

Reeves, E.M.: 1976, Solar Phys. 46, 53.

Skumanich, A.: 1972, Astrophys. J. 171, 565.

Soderblom, D.: 1980, Ph.D. thesis, Univ. of Calif. at Santa Cruz.

Wilson, O.C.: 1978, Astrophys. J. 226, 379.

Zolcinski, M.-C., Antiochos, S.K., Stern, R.A., and Walker, A.B.C. 1982, Astrophys. J. 258, 177.

STELLAR X-RAY EMISSION AS AN INDICATOR OF STELLAR MAGNETIC ACTIVITY

G. S. Vaiana
Osservatorio Astronomico di Palermo (Italy) and Harvard-Smithsonian Center for Astrophysics, Cambridge, MA (USA)

ABSTRACT

An overview of recent *Einstein* Observatory observations and theoretical modeling of stellar x-ray emission is presented, with particular emphasis upon the role such studies can play furthering our understanding of stellar magnetic activity. We argue that solar observations can be used to show that coronal emission is morphologically related to surface magnetic field activity; to establish a quantitative link between the observed soft x-ray flux and the mean surface (photospheric) magnetic flux; and, in sum, to demonstrate that soft x-ray emission is a sensitive diagnostic for the presence of surface magnetic fields, uncomplicated by radiative transfer effects and the resulting coupling to the underlying atmosphere (which may introduce unwanted correlations with spectral type and luminosity class). Recent analyses of stellar x-ray observations from the *Einstein* Observatory have focused on the interpretation of stellar coronal emission and possibly-related surface magnetic activity within the framework of our understanding of solar activity; including "loop" models of stellar coronae, observations of coronal emission variability on a wide range of time scales (which provide information on the morphology of the emitting plasma), and observations and modeling of low-resolution spectroscopic x-ray observations (which test the applicability of solar modeling). These recent results, which I review, all reenforce the argument that stellar coronal emission constitutes an excellent probe for studying stellar magnetic activity over a wide dynamic range, one which may prove to be uniquely suited to studying such activity in distant (and hence faint) sources.

1. INTRODUCTION

The problem of stellar surface activity is sufficiently broad and complex in and of itself that it is often easy to overlook its role in larger astrophysical problems. From the stellar point of view, surface activity directly affects the stellar spectral signature (for example, the optical light from dMe stars may be enhanced substantially during flaring; see Byrne 1983 and references therein); is directly responsible for mass and angular momentum loss during main sequence life, and thus affects the course of stellar evolution (Kraft 1967); and may play a significant role in despinning during the formative T Tauri phase of stellar evolution (cf. Imhoff 1978).

J. O. Stenflo (ed.), Solar and Magnetic Fields: Origins and Coronal Effects, 165–186.

From the galactic point of view, stellar surface activity is responsible for mass input to the interstellar medium via stellar winds; contributes to the galactic component of the soft X-ray background by virtue of X-ray emission from stellar coronae (cf. Rosner *et al.* 1981); and allows us to probe the low mass end of the stellar mass function through the relatively vigorous levels of X-ray emission from very late-type stars, and hence aids in establishing the late-type stellar space density and its contribution to the galactic mass. Finally, stellar surface activity provides us with a virtually unexcelled laboratory for plasma astrophysics: a wide range of plasma phenomena is now known to be taking place in the surface layers of stars as commonplace as the Sun; and the fact that we can now study the range of this behavior over the entire breadth of conditions encountered in the H-R diagram by means of stellar UV and X-ray emission has opened up new opportunities for studying these plasma processes (cf. Vaiana 1981a).

In the following, I will focus on an aspect of stellar coronal emission particularly relevant to this symposium, namely its possible use as a diagnostic tool for probing stellar magnetic activity. This point of view is one which is relatively well-established in the solar context: thus, studies of solar activity – as measured by the level of chromospheric or coronal emission – have long suggested a strong correlation between the vigor of emission and the level of photospheric magnetic field activity, showing in particular that *chromospheric* emission could be quantitatively correlated with the *photospheric* magnetic flux (Skumanich, Smythe, and Frazier 1975).

Our own perspective has been of course strongly influenced by our observational program in solar soft x-ray imaging; from our observational work, we have been led to the conclusion that the structured corona seen in solar soft X-ray images is best understood as the result of plasma heating by magnetic field-related processes. We therefore asked ourselves whether a quantitative connection between the level of soft X-ray emission in a given region of activity could be tied to the local level of photospheric magnetic activity. Using the extensive data base provided by the *Skylab* and earlier rocket flight observations (see, for example, Vaiana and Tucker 1974), Golub *et al.* (1980, 1982) showed that a good correlation exists between the total photospheric magnetic flux in an active region and its total soft x-ray emission (or, alternatively, the total coronal thermal energy content in the active region; Figure 1). If we define the average magnetic field strength in an active region as the ratio of the total flux to the active region area, a similar correlation can be shown to exist for intensive (i.e., non-volume-dependent) parameters characterizing the photospheric magnetic field and the corona, e.g., for the average longitudinal photospheric magnetic field strength and average coronal pressure (Fig. 1). Again, we find fairly good positive correlation, with the coronal gas pressure and photospheric mean magnetic field strength $\langle B \rangle$ related via a power law in $\langle B \rangle$ (Golub *et al.* 1980, 1982). This empirical result can be combined with specific theoretical models for magnetically-coupled heating in order to construct a theory which, given the surface magnetic flux, will predict the luminosity and temperature of the coronal plasma (viz., Rosner, Golub and Vaiana 1982; and Golub 1982 and in this volume) or, conversely, to predict the photospheric magnetic flux if one is given the coronal parameters.

These various morphological and quantitative arguments show that (1) coronal soft x-ray emission can be both qualitatively and quantitatively related to surface magnetic field activity; and (2) soft x-ray emission, as a sensitive diagnostic for the presence of surface magnetic fields, is uncomplicated by radiative transfer effects and the resulting coupling to the character of the underlying atmosphere (which may, as in the case of Ca II, introduce unwanted correlations with spectral type and luminosity class). With these arguments in mind, recent analyses of stellar x-ray observations from the *Einstein* Observatory have focused on the interpretation of stellar coronal emission and possibly-related surface magnetic activity within the framework of solar activity models. These studies have included correlation analyses of x-ray emission properties with other stellar parameters (in particular, with stellar rotation); modeling of stellar coronae with the aid of "loop" atmospheres; determination of coronal x-ray emission variability on a wide range of time scales (which, by comparison with analogous solar data, provides information on the morphology of the emitting plasma); and observations and modeling of low-resolution spectroscopic x-ray observations, which allow us to verify the applicability of the solar analogy. These recent results, which I shall review, are all consistent with the view that solar coronal modeling is an appropriate prototype for studying activity in late-type stars; and hence combine to argue forcefully that stellar coronal x-ray emission is an excellent probe for studying stellar magnetic activity over a wide dynamic range of activity levels, presently constituting one of our best tools for investigating magnetic dynamo behavior on stars other than the Sun.

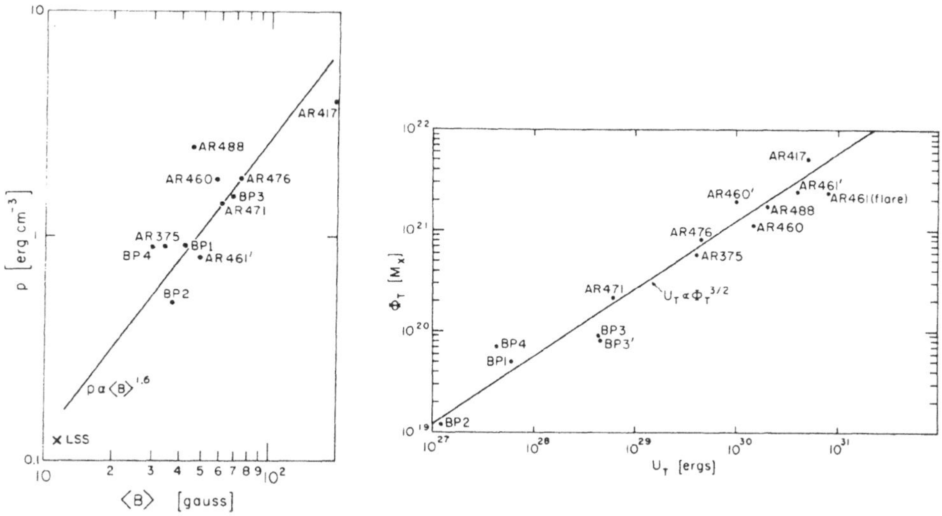

Figure 1: (left) Correlation between the average longitudinal photospheric magnetic field and the average coronal pressure. (right) Correlation between the total energy content in the coronal plasma and the magnetic flux in solar confined regions (from Golub *et al.* 1980).

2. OVERVIEW OF OBSERVATIONS AND MODELING

I begin by first summarizing the status of the observational results as reflected in the published literature, which dominantly derive from observations carried out by the *Einstein* Observatory; and give an overview of the modeling which we have done relevant to the study of stellar magnetic activity. This will set the stage for the presentation of our latest observational results on stellar coronal variability and coronal temperature/luminosity correlations, which will be discussed in §§ 3 and 4.

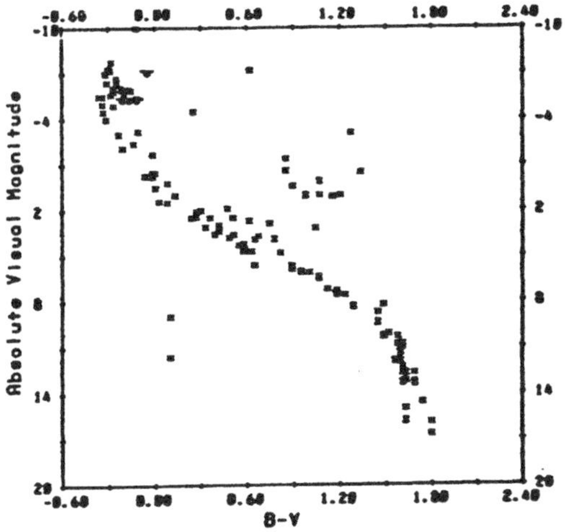

Figure 2: H-R diagram of optically well-characterized stars seen as soft x-ray sources by the Einstein Observatory. The stars shown were observed as part of the CfA stellar x-ray survey (cf. Vaiana et al. 1981), and include stars observed by collaborating guest observers (from Vaiana 1982).

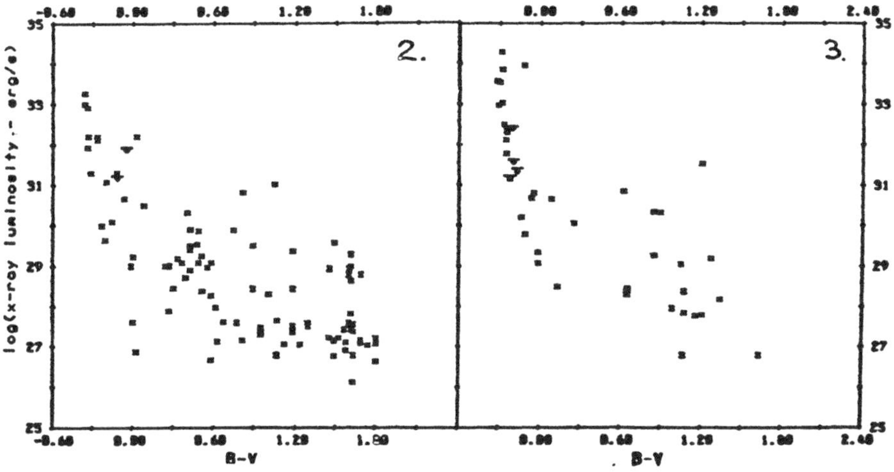

Figure 3: X-ray luminosity (0.2 - 4.0 KeV) versus spectral type for the main sequence stars shown in Figure 2 (from Vaiana 1982).

Figure 4: X-ray luminosity (0.2 - 4.0 KeV) versus spectral type for giants and supergiants plotted in Figure 2 (from Vaiana 1982).

(i) <u>*Status of observations.*</u> The overall behavior of x-ray emission throughout the H-R diagram is characterized (Figures 2-4) by the fact that essentially all stars emit in the range of 10^{26} to 10^{33} erg s^{-1} (Vaiana *et al.* 1981). Differences in luminosities appear to be tied to spectral type only in a gross sense; that is, there is a distinct difference in behavior between early-type and late-type stars, but relatively little differences within these two very general categories. For early-type stars, the behavior of luminosity seems to be dominated by a monotonic decrease from O to A characterized by a virtually constant ratio of X-ray to bolometric luminosity (see, for example, Pallavicini *et al.* 1981). For late spectral types (G to M), the behavior is instead dominated by a broad range of x-ray luminosity for each spectral type, and a lack of dependence on effective temperature.

What general conclusions can we draw from these data, particularly for the late-type (cool) stars? It is evident that x-ray emission is a common stellar attribute, so that nonthermal processes must be operative in the outer atmosphere of essentially all stars. Let us focus for the moment on the late spectral type stars. In order to elucidate the nature of the atmospheric heating processes for these stars, it is useful to ask whether the levels of coronal x-ray emission, as well as the spectral characteristics of this emission, are in any way correlated with stellar attributes other than those which place stars in the H-R diagram. Specifically, we would like to know if there exists a stellar attribute whose variation for fixed

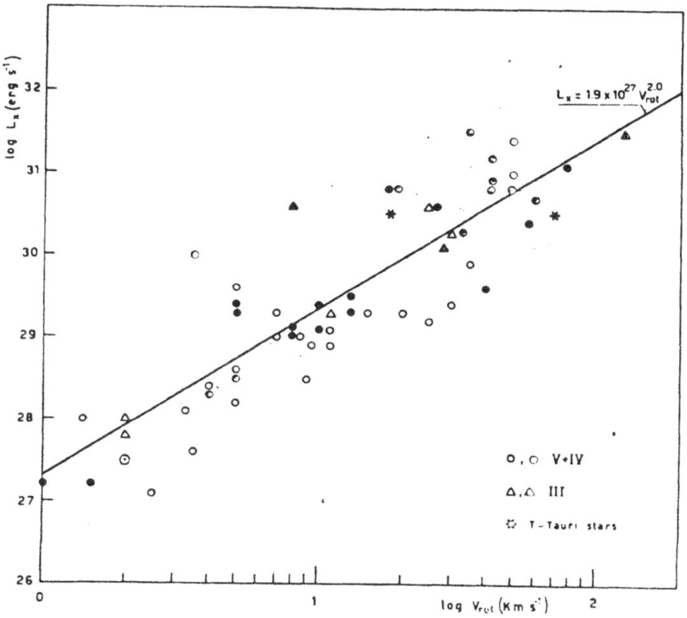

Figure 5: Plot of x-ray luminosity L_x vs. rotational velocity V_{rot} for a sample of late spectral type dwarf and giant stars (from Pallavicini *et al.* 1982). The x-ray observations are from the *Einstein*/CfA stellar survey; rotational data have been obtained with a variety of different methods and are discussed in the reference. The straight line represents a best fit relationship for the entire data sample.

spectral type and bolometric luminosity could account for the observed spread in soft x-ray luminosity. For present purposes, the most obvious such candidate stellar attribute is, of course, the stellar rotation rate. In fact, preliminary studies by Ayres and Linsky (1979), and more detailed work by by Pallavicini *et al.* (1981, 1982) and Walter (1981) and collaborators, have shown that for late-type stars there is correlation between x-ray emission and stellar rotation rate; and recent further studies have shown a correlation between stellar x-ray emission and stellar age (Vaiana 1981a, c; Stern *et al.* 1982). These results are of course reminiscent of the connection between Ca II h and k emission (Wilson 1966), activity, and rotation rate found almost a decade ago by Skumanich (1972). For example, the most recent study (by Pallavicini *et al.* 1982), has extensively investigated the correlation between observed stellar x-ray luminosities, total bolometric luminosities, and projected rotational velocities *v sin i* for a large group of late spectral type stars of various luminosity classes drawn from the *Einstein*/CfA stellar survey; the results of this study are shown in Figure 5. The apparently strong rotational dependence of stellar x-ray emission for late-type stars (whose precise functional form remains to be established by studies of statistically complete samples drawn from our survey) argues for a coupling mechanism between rotation and coronal heating which is naturally provided by coronal heating processes dependent on stellar magnetic fields, as already suggested by solar observations (Rosner and Vaiana 1980).

In light of the correlation of x-ray emission with rotation for late-type stars, one would expect a similar correlation between Ca II emission strength and x-ray luminosity as well. This is found, and has been reported by, among others, Mewe and Zwaan (1980) and collaborators, and by us (cf. Vaiana 1981b; Pallavicini *et al.* 1982). Figure 6 shows the corresponding result taken from the CfA data sample; note the steep dependence of x-ray emission on Ca II flux (Pallavicini *et al.* 1982). The observed correlation of coronal x-ray emission with calcium emission strengthens the hypothesis that the plasma heating processes leading to soft x-ray emission in late-type stars are largely governed by stellar magnetic fields coupling to surface turbulence.

(ii) *Status of modeling.* The ubiquity of x-ray emission throughout the H-R diagram raises the general problem of accounting for the presence of the hot plasma which must give rise to the observed emission. Several distinct experimental and theoretical questions are involved:

1. What are the basic processes that produce hot plasma under the very wide-ranging conditions found on stars' surfaces throughout the H-R diagram?
2. What are the appropriate plasma diagnostics for testing the various hypothesized energy supply and loss mechanisms?
3. What insights can we glean for understanding related plasma processes in other astrophysical systems?
4. How do these stellar x-ray observations impact our notions of stellar structure and evolution?

Some of these questions have been addressed in a number of recent studies of late-type stellar coronae (Walter and Bowyer 1981; Belvedere *et al.* 1981; Landini and Monsignori-Fossi 1981; Stein 1981; and Golub *et al.* 1982). Most of these models are distinguished by the fact that the coronal plasma is assumed to be con-

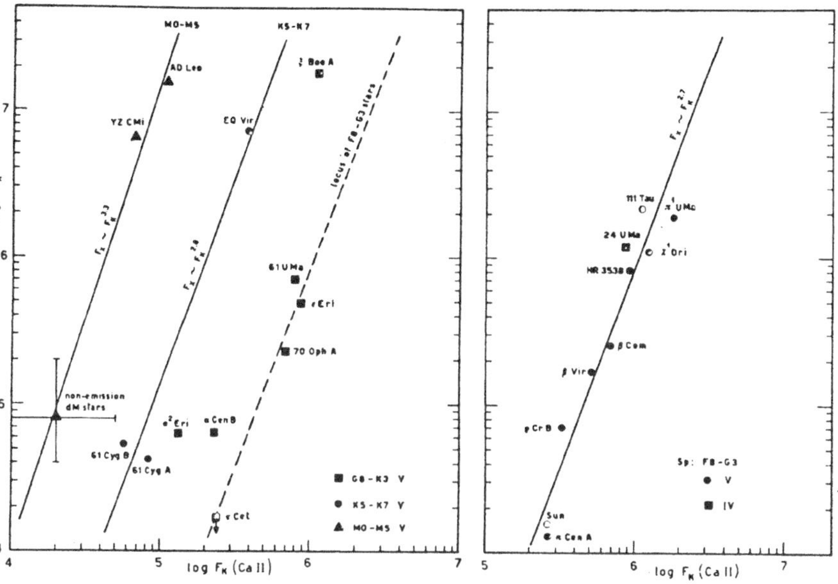

Figure 6: Plots of x-ray surface fluxes F_X (ergs/cm^2 sec) vs and Ca II K fluxes(ergs/cm^2 sec) for main sequence stars. The right-hand panel refers to stars of spectral types F8-G3; the straight line is a least-square fit. In the left panel, which refers to stars later than G8, different symbols are used to indicate subgroups of stars of similar spectral type. The two solid lines are best fit relationships to K5-K7 and M0-M5 stars, respetively. For reference, the locus of F8-G3 stars is plotted as a dashed line (from Pallavicini *et al.* 1982).

fined by surface magnetic fields. Because of this confinement, the coronal temperature is no longer constrained to be less than the stellar "escape" temperature, as defined by equating the thermal and surface escape velocities. In that case, one can take advantage of scaling laws connecting the temperature, pressure, and size scale of the confined plasma structures to "assemble" a prototype corona; and, by adopting a dynamo model for magnetic field generation, use the kind of magnetic field-coronal emission scaling discussed above to predict, for example, the variation of coronal emission along the main sequence (as has been done recently by, for example, Belvedere et al. (1981).

This sort of modeling does, however, have the disadvantage of applying dynamo calculations for magnetic flux production which are relatively poorly constrained by solar observations in and of themselves. A possible alternative procedure is to work backwards, and use the stellar coronal data, together with coronal models, to infer the rate of stellar magnetic flux production, and then to use this result – in the absence of sufficiently numerous direct observations of stellar magnetic fields (see however Marcy 1983 and Worden 1983) – as a constraint for stellar dynamo theory. We have recently carried out such an analysis (Rosner, Golub & Vaiana 1982; see also Golub *et al.* 1982, and Golub 1982, 1983), and I would like to briefly summarize the principal results.

We begin by assuming that the bulk of observed stellar coronal emission comes from "closed" coronal structures, in analogy with the Sun. If one adopts as a typical coronal scale length the emission scale height,

$$\ell = k_B T / 2\mu m_H g_o \,, \tag{1}$$

where T is the coronal temperature, μ is the mean atomic weight, and g_o is the effective gravity, and if the atmosphere is in hydrostatic equilibrium, one can solve

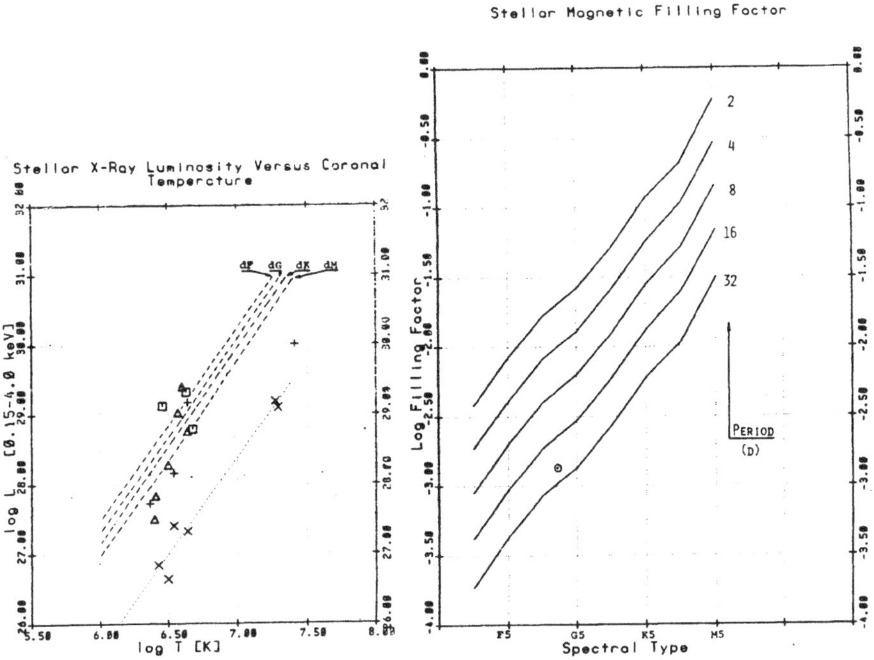

Figure 7: (left) Relationship between coronal x-ray luminosity and temperature, derived on the basis of a "loop" model for stellar coronae. (right) Variation of surface magnetic filling factor versus spectral type (with stellar rotation period as parameter). (from Rosner, Golub and Vaiana 1982).

the energy balance equation in the bounded coronal volume, and obtain a scaling law connecting the coronal soft x-ray luminosity with its temperature,

$$L_x = L(T\,;\ M, R)\,, \tag{2}$$

with the stellar mass M and radius R as parameters. This result is independent of any assumption regarding the coronal heating process. An example of the results of such a calculation is shown in Figure 7 for late-type main-sequence stars, in which we have also used the Pallavicini *et al.* (1981) rotation-emission analysis to calibrate the curves. These results can be directly tested against observations (as will be discussed in § 4 below); the agreement appears to be rather good.

If one, now, applies the Golub *et al.* (1980, 1982) scaling connecting x-ray emission with the average photospheric field strength, then, depending on the assumed coronal heating mechanism, one can derive the coronal temperature and luminosity as functions of photospheric stellar parameters; for the class of so-called non-resonant heating processes (Kuperus, Ionson & Spicer 1981), coronal temperature and luminosity depend on the typical surface convective velocity v, the surface equipartition magnetic field strength B, the stellar mass M and radius R, and the surface magnetic filling factor f_B (defined as the ratio of the mean surface field strength to the surface equipartition field strength)

$$T = 3.9 \times 10^7 \, K \cdot \delta(v, B, M, R)^{1/3} \, . \, f_B^{2/3} \qquad (3)$$

$$L_x = 5.3 \times 10^{31} \cdot \delta(v, B, M, R)^{5/6} \, . \, (M/M_o) \, . \, f_B^{5/3} \, . \, q(\ell/R) \qquad (4)$$

where q is a slowly-varying function of ℓ/R, and $\delta = (v/1 \text{ km sec}^{-1})^2 \, (B/10^3 \text{ G})^2 \, (M/M_o)^{-1} \, (R/R_o)^2$. If one adopts standard stellar models, the only unknown parameter is then the magnetic field filling factor f_B. Now the data have, as I've just mentioned, allowed us to correlate the level of coronal emission with the stellar rotation rate

$$L_x = L_x(\Omega); \qquad (5)$$

this additional, purely empirical relation thus allows us, together with our model, to connect the stellar rotation period with the surface filling factor, or equivalently, the total surface magnetic flux

$$f_B = \eta(v, B, M, R, \ell) \, . \, (\Omega/10^{-6})^{6/5}, \qquad (6)$$

where η is determined by Eqs. (3)-(5). This is a fundamental constraint for stellar dynamo models because, if B is defined (as above) as the equipartition magnetic field strength, then the relation

$$\Phi = f_B \cdot B \cdot 4\pi R^2 \qquad (7)$$

gives the total magnetic flux Φ predicted by the coronal model; this flux ought then to be compared with that predicted by models for stellar dynamos.

As an example, Figure 7 shows the result of such an analysis of coronal x-ray emission applied to late-type stars. Shown is the variation in filling factor as a function of spectral type and rotation period. The qualitative behavior corresponds to expectations (more rapid rotation leading to larger filling factors); and in quantitative terms, our results agree fairly well with observations in the very few cases to date in which comparison with data can be usefully made. For example, the model predicts filling factors of 1/10 or larger for rapidly rotating dM stars, which, of course, tend to be the well-known "spot stars", whose surfaces are in fact thought to be covered to a significant extent by surface magnetic field features.

We expect that a similar analysis will be carried out using other diagnostics, such as in the UV; it will then be of great interest to see if a consistent picture can be assembled.

3. NEW RESULTS: VARIABILITY

It has long been recognized that variability in coronal emission is a direct reflection of highly localized coronal plasma heating; the most obvious (and extreme) example is of course the solar flare. Let us however exclude such impulsive phenomena for the moment, and ask on what time scales one would expect to observe solar coronal variability if the Sun were regarded as a star. Figure 8 (adopted from Vaiana and Rosner 1978) gives a hint of some of the expected variability characteristics:

(i) *Very long-term fluctuations* related to the variation in amplitude of the rotationally-modulated integrated coronal flux occurs on a time scale of years; such variations are presumably related to a modulation in surface activity during the course of the solar (or stellar) magnetic dynamo cycle, and can be clearly seen in, for example, Wilson's (1978) classic Ca II monitoring data.

(ii) *Long-term fluctuations* related to the rotation of active regions and active region complexes onto the visible hemisphere occur on a time scale of several days to a month; this variability is especially noticeable in the solar record because it carries the (periodic) signature of the Sun's rotation (Fig. 8).

(iii) *Medium-term fluctuations* related to the emergence of active regions or active region complexes take place on the time scale of hours to days; such variations become accessible to *Einstein* observations on the orbital time scale.

(iv) *Short-term fluctuations* related to the impulsive brightening of individual "loop" structures (in the most extreme case, referred to as a flare) may take place on a time scale of minutes; these can be seen within a single orbit's coverage by the *Einstein* Observatory.

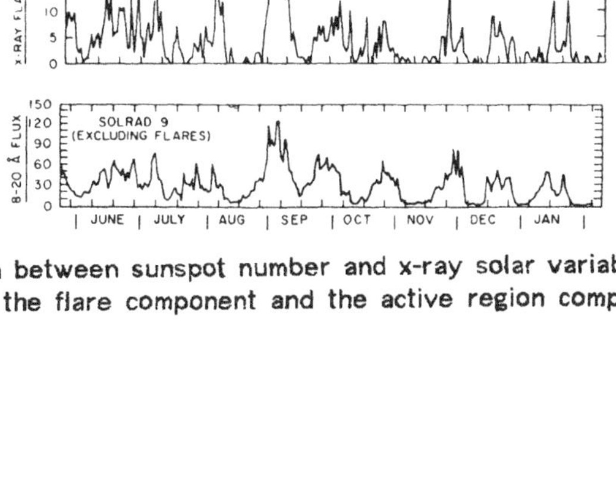

Figure 8: Correlation between sunspot number and x-ray solar variability. Solar rotation can be seen both in the flare component and the active region component (from Vaiana and Rosner 1978).

The *Einstein* data provide the most exacting constraints on stellar surface (coronal) variability to date; typically, one is able to test the hypothesis that a given star's luminosity is constant in time on some specified time scale to the 10-20% fluctuation level (3σ). Aside from the trivial limitation imposed by photon statistics (which constrains the ability to pick up fluctuations on very short time scale), the only other major constraints were the very tight scheduling demands imposed by the large variety of scientifically-distinct observation programs (which limited the ability to track, for example, rotational modulation) and the limited duration of the mission itself (which placed an absolute upper bound on the possible time scale of observable fluctuations of roughly two years). Within these constraints, we have been able to investigate variability on time scales ranging from the short-term to the long-term (though in a very few selected cases, data separated by up to roughly two years were obtained); what then was seen?

The basic algorithm underlying our variability analyses involves sorting the data into successive time bins of width Δt, and then to test the hypothesis that within some confidence level determined by a χ-square test, the count rate per bin is constant as a function of time. Typical bin widths range from Δt = 20 sec to 1,000 sec, and the available total time intervals over which variablity can be tested range up to 2 years. For a typical source, the actual count rate determines a sensitivity level for variability detection, which corresponds to the 3σ fluctuation level (expressed as a percentage of the mean count rate) about the mean bin count rate. The results presented here (some examples of which are shown in Figures 9-11) are preliminary, and reflect only a small portion of the available information; here I would only like to summarize some of the key results which are now emerging from our work.

(a) Long-term variability. Figure 9 illustrates some of the classic examples of long-term variability in coronal emission we have found for late-type dwarf stars, ranging from changes seen on a weekly (*EQ Peg*), monthly (*CN Leo*, *UV Ceti*, *YZ CMi*), and yearly (π^1 *UMa*, *EQ Vir*) time scales. The basic result is that most such sources which were observed by the *Einstein* Observatory on well-separated occasions appear to be variable at the level of 20-100%; however, notice that π^1 *UMa* appears to have had a constant luminosity for the three observations stretching over a period of 2 years. Some of the variability shown here is actually attributable to the occurance of a flare during one (or more) of the observing periods; but upon detailed investigation, we now know that this is not generally the case: late-type dwarfs are indeed substantially variable on time scales of weeks to years.

(b) Medium-term variability. A very convenient method for testing for medium-term variability (as defined above) is to compare the total source count rate on a orbit-by-orbit basis. Some examples from our survey which have been treated in this way are shown in Figure 10; from this figure, it is immediately clear that the targets available for such analysis can be segregated into two categories: those stars which have been specifically targeted for variability analysis in the original observing programs (typically, flare stars, such as *UV Ceti*), and those stars (such as *Capella*, *HR 1099*, and *Wolf 630*) chosen for detailed spectroscopic analysis by the Objective Grating Spectrometer (OGS) because they were x-ray luminous (so that extremely long exposures in the 0th order – which can be treated as a standard image – were available). The actually observed behavior is remarkably varied. Thus, *UV Ceti* appears to vary very smoothly over the course of six orbit (which can be interpreted either as the reflection of the emergence and subsequent dispersal of an active region or, more likely, as the reflection of rotational modula-

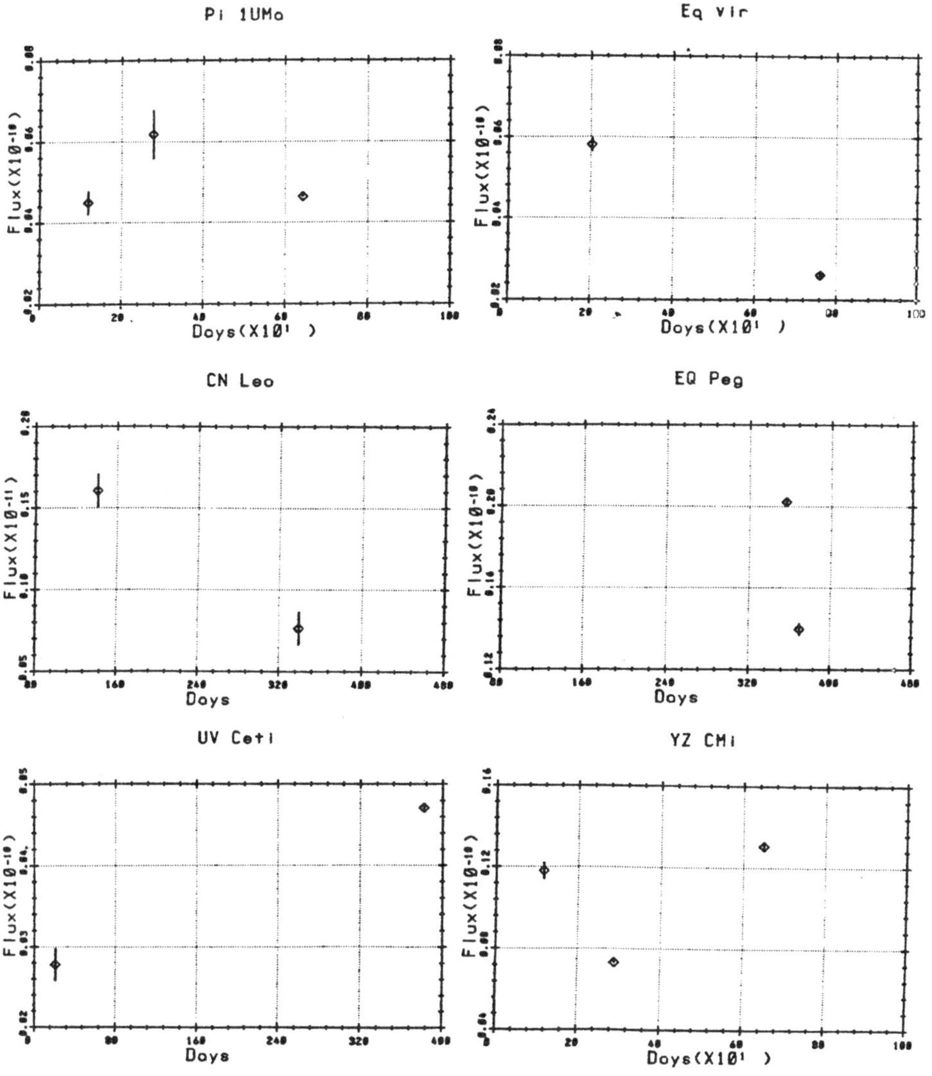

Figure 9: Example of long term variability in x-ray fluxes (erg cm^{-2} sec^{-1}) obtained from the Einstein Observatory for a number of late type stars sampled more than once at intervals ranging from a few weeks (EQ Peg), to months (π^1 UMa, CN Leo, YZ CMi) or years (π^1 UMa, EQ Vir, UV Ceti, YZ CMi). Most of the sources reobserved show substantial change in flux level well in excess of 3σs up to factors of two (π^1 UMa change is $< 3\sigma$). Some of the change is observed to be due to flaring behaviour, but base level changes of 10% to 50% are common.

tion of inhomogeneous surface x-ray emission). In contrast, the remaining three sources displayed in Figure 10 show no systematic behavior: both *HR 1099* and *Wolf 630* show large-amplitude variations on time scales of hours to a day; but *Capella* shows substantially smaller, if any, variability, and no long-term trends, in this section of the data.

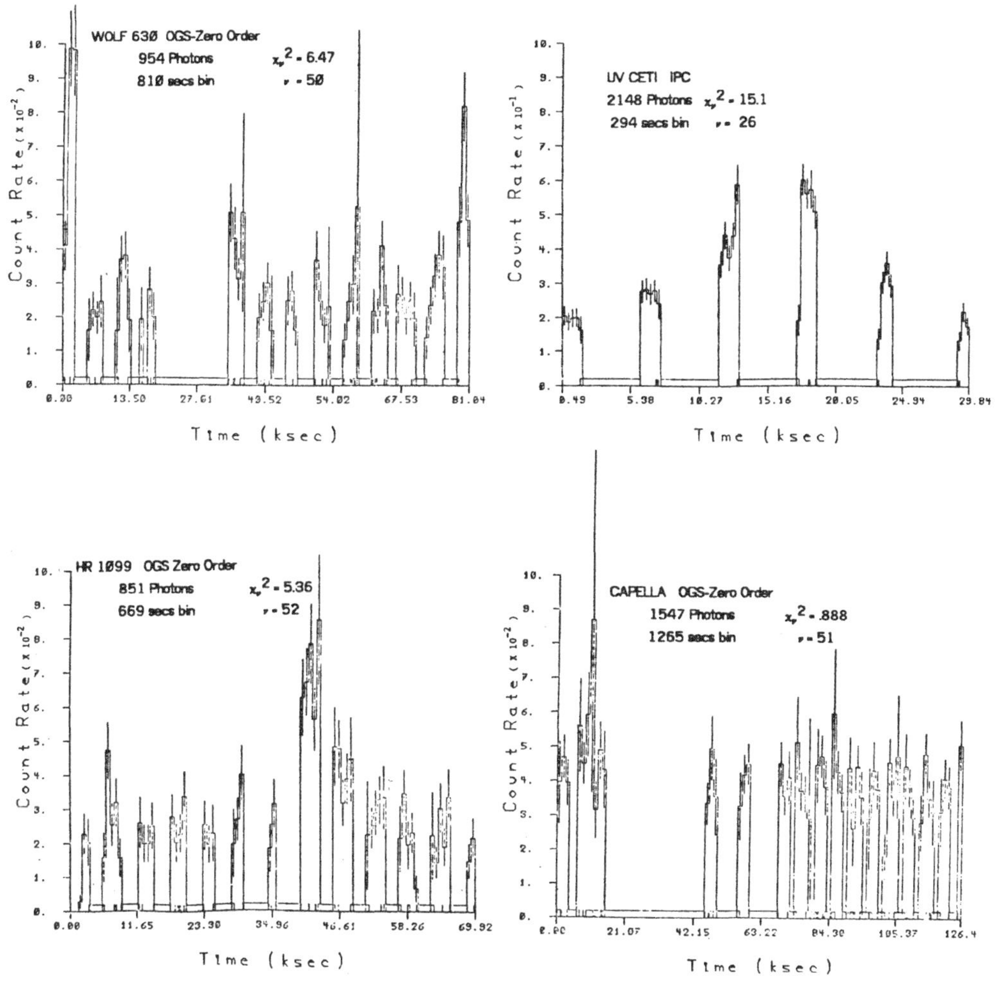

Figure 10: These four sources were observed more or less continously by the Einstein Observatory for periods ranging from 30 Ksec (6 orbits for UV Ceti with the Imaging Proportional Counter) to some 120 Ksec (16 orbits for Capella in the zero order of the Objective Grating Spectometer). The count rates shown here indicate orbit-to-orbit variability. Both UV Ceti and HR 1099 have increases of factors of 2 to 3 changes (in less than an orbit) lasting several orbits (3 to 4). A χ^2 test for flatness on several hundred seconds time-binning also indicate variability on this time scale for Wolf 630, UV Ceti and HR 1099 well above the sensitivity (at a level > 23%, > 24% and > 12% respectively), Capella does not vary on this time scale at fluctuation sensitivity level of 30%.

(c) <u>Short-term variability.</u> Finally, consider the variability of stellar soft x-ray emission on short time scales, as shown in Figure 11a, b. Several interesting and instructive results emerge: *first*, it is apparent that although a source may appear to be constant on long time scales, it may well be extremely variable on short time scale, as illustrated by π^1 *UMa*; this star shows variability at roughly the 20% level

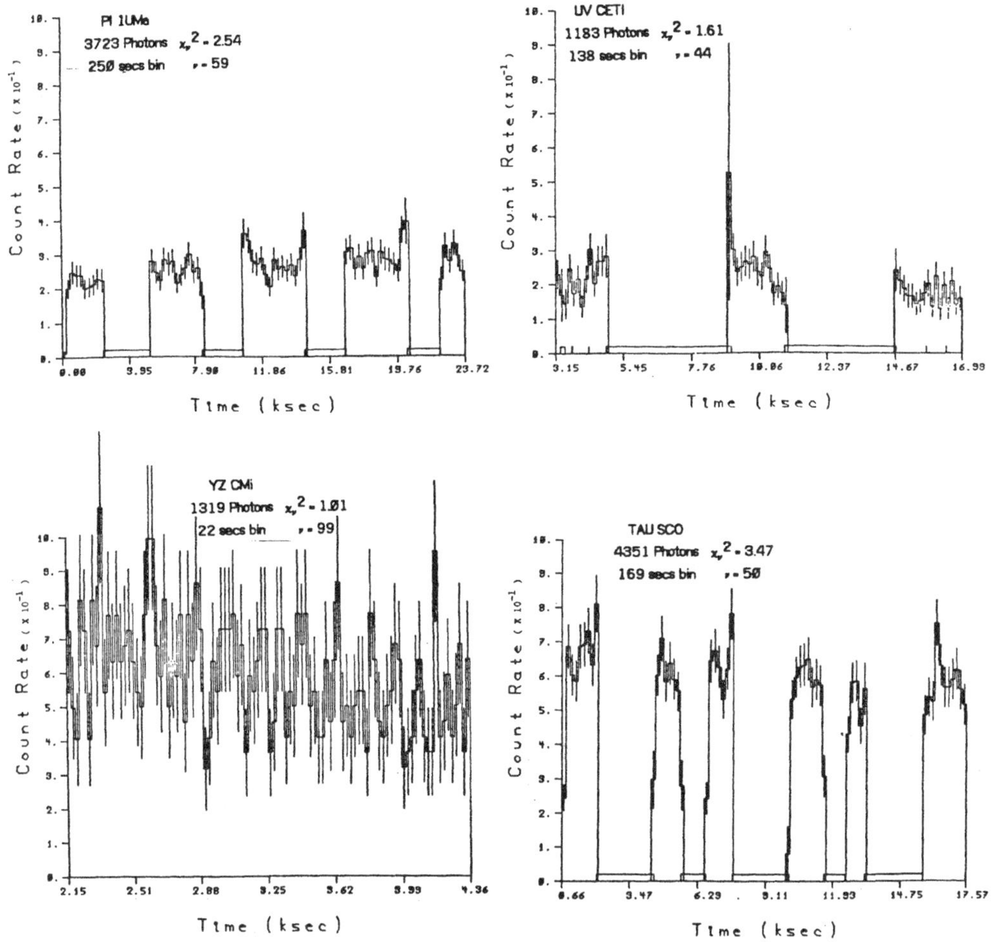

Figure 11a: Test for flatness with χ^2 statistics on a number of sources conducted with short time scale (binning of 25 to 250 seconds) show that variability on these time scales is also common in most sources at significant levels (> 10% - 20% levels). Variability in several orbits of both UV Ceti and π^1 UMa is here shown to be in excess of 15%. YZ CMi instead does not show sufficient excess χ^2 at the 30% level with a 22 second binning. This figure shows that variability is also present in early type stars: the hyphotesy that τ Sco's count rate is constant, for data binned on a 170 sec time scale, fails the χ^2 test severely.

when studied on short time scales. *Second*, the amplitude of long-term variability is little guide to the amplitude at short time scales, as shown by the virtually identical fluctuation levels at short time scale of π^1 *UMa* and *UV Ceti* (compare with Figures 9 and 10). *Third*, absence of variability at any given time is no guarantee that subsequent observations will not reveal substantial levels of variability; *EQ Peg* is a nice example of such change in variability level. Finally, *fourth*, I thought it instructive to show that one's (theoretical) notions of what ought to be are not always the most reliable guide: although x-ray emission from OB (early-type) stars is commonly thought to be due to rather different processes than are associated with x-ray emission from late-type stars, processes with which one would hardly associate fluctuations in x-ray emission on the minute time scale (viz. Cassinelli *et al.* 1981), it is nevertheless the case that *OB stars do vary on the minute time scale*, as shown by observations of τ *Sco*.

On a more detailed level, one can ask for the possible correlation between the fluctuation level ΔL and the stellar x-ray luminosity L; our results are still somewhat preliminary, but do suggest that the fluctuation level ΔL is correlated with the stellar x-ray luminosity, with $\Delta L/L$ in the range 0.2-0.6 for the luminosity range $L_x \sim 10^{26} - 10^{30}$ erg sec^{-1}. Furthermore, one can ask whether the fractional variability $\Delta L/L$ is correlated with other stellar attributes; the most obvious is of course

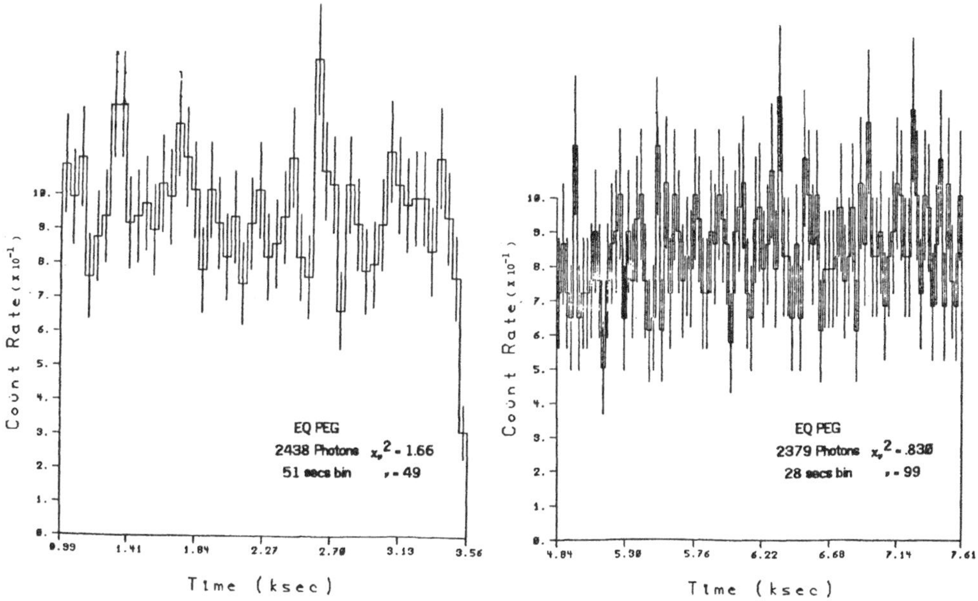

Figure 11b: Similar flatness tests are shown here for EQ Peg on two different orbits (weeks apart). In the first (left) a significant level of variability is shown at > 20% level, while in the second (right) no such variation is seen. This source behaviour is not uncommon in other sources.

spectral type. Our results are here still very preliminary; there is a suggestion that the "basement" level of variability seems to be fixed at roughly the solar level (and so is independent of spectral type), but that the maximum variability does show a trend with spectral type: the active low-mass stars seem to have exaggerated variability when compared with corresponding solar-mass stars.

What do we then conclude regarding stellar x-ray variability? Our basic conclusion is that *essentially all stars show variability on virtually all time scales*, given sufficient sensitivity. This conclusion encompasses not only the cool, solar-like dwarf stars, but also evolved late-type stars and hot, early-type (OB) stars of all luminosity classes. In retrospect, this result ought not be surprising -- at least for the late-type dwarfs -- if we recall our previous discussion of the underlying cause for stellar activity of such stars as revealed by the activity-rotation correlation; indeed, we here simply receive yet further confirmation of the aptness of the solar model.

4. NEW RESULTS: STELLAR CORONAL LUMINOSITIES AND TEMPERATURES

The *Einstein* Observatory carried four spectrometers onboard: the Imaging Proportional Counter (IPC), with $\Delta E/E \sim 1$; the Solid State Spectrometer (SSS), with $\Delta E/E \sim 50$-100; the Objective Grating Spectrometer (OGS), with $\lambda/\Delta\lambda \sim 50$-$100$; and the Bragg Crystal Spectrometer (BCS), with $\Delta E/E \sim 1000$. Of these four instruments, only the IPC carried out a sufficiently large number of observations so that systematic diagnostic studies and correlation analyses can be presently pursued for well-defined data samples. For this reason alone, I will focus in the following on the IPC results; it should, however, be kept in mind that high-resolution observations of a few selected (very active) stars were carried out by the other spectrometers (see, for example, Holt *et al.* 1979).

Typically, given sufficient counts, an *Einstein* stellar x-ray image could be analyzed by comparing the observed IPC spectrum with model spectra which assume a thermal source spectrum, take into account interstellar absorption, and are folded through the calibrated instrument response. The particular model spectra used include continuum and line contributions for an optically-thin plasma in ionization equiblibrium, provided by Raymond (1981). In the simplest case, we assumed that the stellar x-ray emission could be attributed to a plasma characterized by a single temperature; for such single temperature component fits, we typically constrained the interstellar hydrogen column density $N_H < 10^{18}$ cm^{-2} for nearby sources, so that the only remaining free parameters are the (emission measure-weighted) mean coronal temperature and the corresponding total emission measure. What are our results?

Let me begin by considering several specific cases which illustrate both the procedures of analysis and the range of results we have obtained. Figure 12 shows the IPC spectral data for four stars: twice for *UV Ceti* and *EQ Peg* (showing the difference in spectral characteristics for quiescent and active periods), and once for π^1 *UMa* and τ *Sco*. In each case, we show both the actual binned IPC spectrum and the best-fit (in the χ-square sense) single temperature model. Three major results are apparent:

(1) *There is a perceptible change in spectrum as the stellar activity level varies*: as shown by both the *UV Ceti* and *EQ Peg* observations, increasing activity level is accompanied by a hardening of the spectrum.

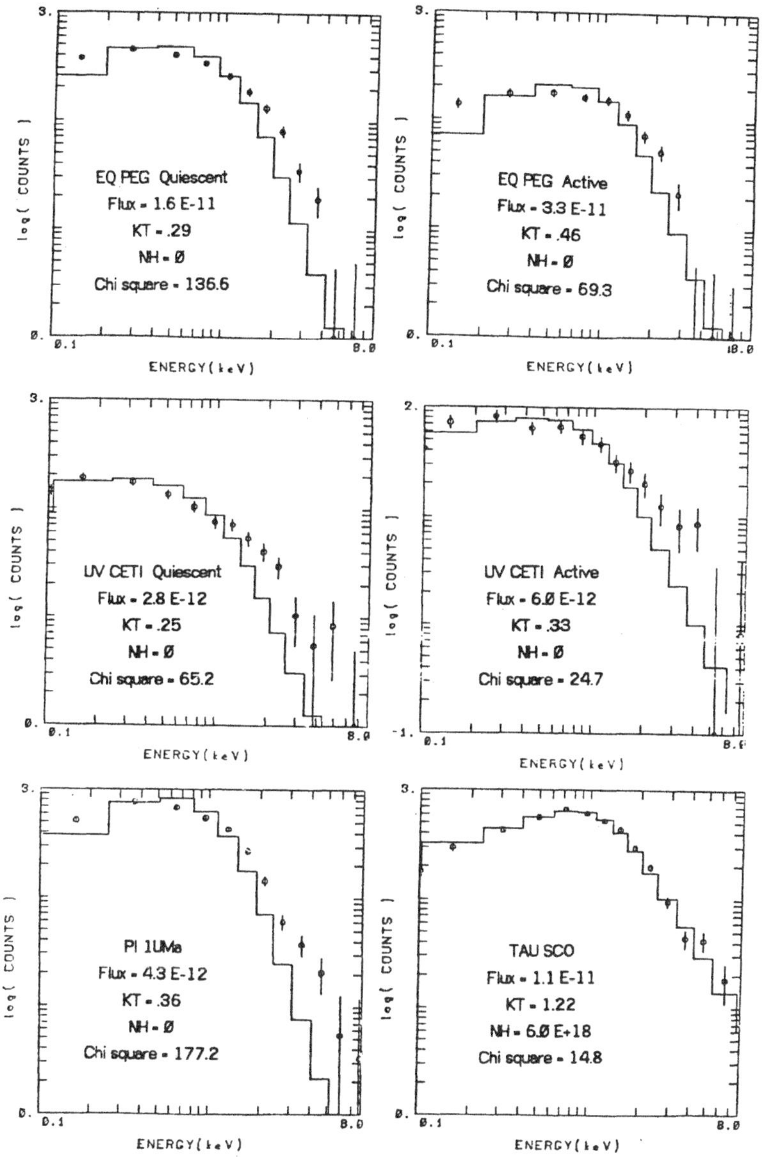

Figure 12: Single temperature best fit to IPC spectra for a number of sources are shown here. Indicated are the resulting flux in the 0.2 - 4.0 KeV energy band in ergs cm^{-2} sec^{-1}, the KT in KeV, the hygrogen column density NH, and the total χ^2. The poorness of the single temperature fits for these and many other sources suggests two temperature components. This is a remininiscent of the results obtained with the Solid State Spectometer for the RS CVns by Swank et al. 1981. It is also interesting to note that the active phases of UV Ceti and EQ Peg shown here correspond to those observations (Fig. 11) showing higher variability and fitted by harder spectra. Also shown for comparison is a best fit spectrum to the early-type star τ Sco.

(2) *In the overwhelming majority of cases, simple single temperature component models fail to give an adequate fit to the data*; in the cases shown, the fitting spectrum gives a good account of the soft portion of the observed spectrum, but seriously underestimates the flux at high (> 1.5 keV) photon energies. This inadequacy of the single-component models is apparent in all of the spectra for late-type dwarf stars shown in Figure 12. A hint of what might be going on is of course contained in the SSS observations of *RS CVn* stars reported by J. Swank and collaborators (cf. Swank and White 1980; Swank *et al.* 1981): these higher-resolution data show fairly clear evidence for the simultaneous presence of at least two distinct temperature components in the source spectrum. It is therefore tempting to argue that this effect is the cause of the discrepancy under discussion. We have in fact attempted 2-component fits to the data for those cases in which single-component analyses clearly failed; and it does appear that the multi-component fit gives substantially better account of the data (in particular, the improvement in the fit is substantially larger than would be expected on the basis of simply increasing the number of available degrees of freedom). How are we to interpret these results? In general, we must consider two distinct possibilities:

(a) Every one of the "two-component" x-ray sources is a binary, in which the low-mass secondary has not as yet been identified (or recognized); this situation might occur especially in those cases in which the primary is of relatively early spectral type (typically, earlier than G0), and relatively faint (typically, fainter that $m_V \sim 6$), so that detailed spectroscopic studies of the system in the optical are not as yet available. Because the later-spectral type secondary will tend to be active (it will be relatively young for its spectral type because of its presence in a binary dominated by a much earlier spectral type main-sequence star), and because of the above-mentioned connection between x-ray luminosity and temperature, it is likely that the secondary will be a harder x-ray source than the primary; hence the presence of two distinct temperature components.

(b) In the case of the Sun, it is well-known that there are two distinct classes of coronal structures: the active component, associated with compact regions of recently emerged magnetic flux, which has a relatively-high surface brightness and temperature; and the inactive ("Quiet") component, which consists of larger, evolved structures whose surface brightness and temperature are relatively lower. If we are to extend the solar analogy, it would not be farfetched to argue that a similar evolutionary segregation of coronal structural types occurs on stars other than the Sun. Hence, one is tempted to relate difficulties of fitting a single temperature models to the presence of this kind of "multi-component" corona, particularly for stars in which the above binary hypothesis does not seem appropriate. A good example is the apparently single active G0 star π^1 *UMa*, whose IPC spectrum shows precisely the kind of difficulty discussed here (Figure 12).

Which one of the two above cases actually holds in a given, specific instance cannot of course be determined on general grounds.

(3) *In spite of the supposed distinction between mechanisms accounting for x-ray emission from early and late spectral type stars, simple single-temperature thermal fits to IPC spectra of OB stars such as τ Sco* (Figure 12) can give an adequate description of the data. In fact, as is apparent from the figure, such a model gives better account of the spectrum of τ *Sco* than of the

spectra of the other (late spectral type) stars! I am of course not arguing that OB star x-ray emission ought to be modeled by using theories for late-type stellar coronae, but rather simply cautioning that both the variability and spectral data we have in hand are not to be easily accounted for by current models for x-ray emission from early-type stars.

Let me now turn from the details of the spectral fitting to an overview of these detailed results. Figure 13 shows our principal result to date, the correlation of stellar x-ray luminosity with mean coronal temperature. Comparison with the theoretically-predicted result, shown above in Figure 7a, suggests that what we are seeing is indeed appropriately described (at least to the level of detail allowed by the present data) by the solar analogy; that is, the data show a positive correlation between luminosity and temperature, with temperature varying over a rather narrow range, while the luminosity spans over three orders of magnitude (rather reminiscent of the brightness-temperature correlation observed for individual solar structures).

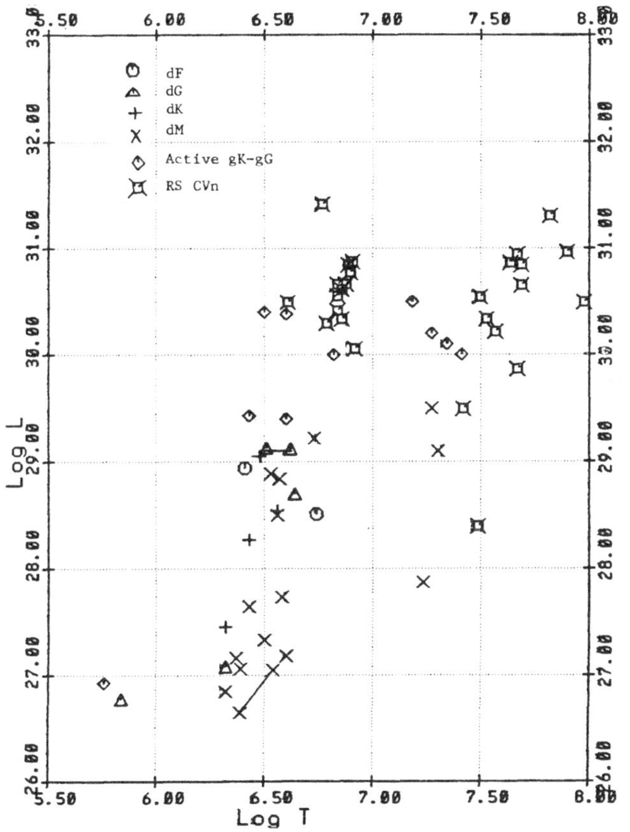

Figure 13: Luminosity vs. temperature results for a number of single temperature fit IPC spectra. RS CVn data points are from the two temperature fits of Swank et al. 1981. Present uncertanty for the IPC gain result in an uncertanty in the temperature determination of 0.5 in Log T for soft spectra. The general trend of L_x vs. T is quite suggestive of the model results of Fig. 7.

On a somewhat more detailed level, can one see the correlation between temperature and spectral type (for fixed luminosity) predicted by the model (Figure 7a)? Recall our original model assumption that the filling factor for coronal emission is unity, or at least fixed (neither assumption is likely to be universally applicable); and note that even in the ideal case that the filling factor is indeed fixed, the spread in temperature for fixed luminosity resulting from spectral type dependence is smaller than the uncertainty in a typical temperature determination. It is therefore evident that it is premature to use the present data to test for this particular correlation.

5. SUMMARY

It seems that, by using solar observations to establish the quantitative correlation between coronal activity (as reflected in the level of soft x-ray emission) and photospheric magnetic activity, observations of soft x-ray emission from late spectral type stars are able to shed light on the extent of stellar surface magnetic activity. The extensive data provided by the *Einstein* Observatory have been used to define a sequence of independent, redundant tests of the "solar hypothesis", using both the observed variability in emission, and the detailed luminosity-temperature dependence.

ACKNOWLEDGEMENTS

I would like to thank L. Golub, R. Rosner, and S. Sciortino for extensive discussions and collaboration in the course of this work. These studies were supported by the CNR (Italy) and by the Smithsonian Astrophysical Observatory.

BIBLIOGRAPHY

Ayres, T. R., and Linsky, J. L.: 1980, *Ap. J.*, **241**, 279.

Belvedere, G., Chiuderi, C., and Paterno, L.: 1981 *Astron. Ap.*, **96**, 369.

Byrne, P. B.: 1983, in *Proc. IAU No. 71 (Activity in Red Dwarf Stars)*, ed. M. Rodono and P. B. Byrne.

Cassinelli, J. P., Waldron, W. L., Sanders, W. T., Harnden, F. R., Jr., Rosner, R., and Vaiana, G. S.: 1981, *Ap. J.*, **250**, 686.

Golub, L.: 1982, in **Cool Stars, Stellar Systems, and the Sun, Vol. I**, ed. M. S. Giampapa and L. Golub (SAO SR-392), p. 39.

Golub, L.: 1983, this volume.

Golub, L., Maxson, C. W., Rosner, R., Serio, S., and Vaiana, G. S.: 1980, *Ap. J.*, **238**, 343.

Golub, L., Noci, G., Poletto, G., and Vaiana, G. S.: 1982, *Ap. J.*, **259**, 359.

Holt, S. S., White, N. E., Becker, R. H., Boldt, E. A., Mushotzsy, R. F., Serlemitsos, P. J., and Smith, B. W.: 1979, *Ap. J. (Letters)*, **234**, L65.

Imhoff, C. L.: in **Protostars and Planets**, ed. T. Gehrels (Tucson: Univ. Arizona Press), 699.

Kraft, R. P.: 1967, *Ap. J.*, **150**, 551.

Landini, M., and Monsignori-Fossi, B. C.: 1981, *Astron. Ap.*, **102**, 391.

Lucy, L. B., and White, R. L.: 1980, *Ap. J.*, **240**, 300.

Marcy, G. W.: 1983, this volume.
Mewe, R., and Zwaan, C.: 1980, in **Cool Stars, Stellar Systems and the Sun**, ed. A. K. Dupree (SAO SR-389), p. 123.
Pallavicini, R., Golub, L., Rosner, R., Vaiana, G. S., Ayres, T., and Linsky, J. L.: 1981, *Ap. J.*, **248**, 279.
Pallavicini, R., Golub, L., Rosner, R., and Vaiana, G. S.: 1982, in **Cool Stars, Stellar Systems, and the Sun, Vol. I**, ed. M. S. Giampapa and L. Golub (SAO SR-392), p. 77.
Rosner, R.: in **Cool Stars, Stellar Systems, and the Sun**, ed. A. K. Dupree (SAO SR-389), p. 79.
Rosner, R., Golub, L., and Vaiana, G. S.: 1982, *Ap. J.*, in press.
Rosner, R., and Vaiana, G. S.: 1980, in **X-Ray Astronomy**, ed. R. Giacconi and R. Setti (Dordrecht: Reidel), p. 129.
Rosner, R. *et al.*: 1981, Ap. J. (Letters), **249**, L5.
Skumanich, A.: 1972, *Ap. J.*, **171**, 565.
Skumanich, A., Smythe, C., and Frazier, E. N.: *Ap. J.*, **200**, 747.
Stein, R.: 1981, *Ap. J.*, **246**, 966.
Stern, R., Zolcinski, M.-C., Antiochos, S. K., and Underwood, J. H.: 1982, *Ap. J.*, **249**, 647.
Swank, J. H., and White, N. E.: 1980, in **Cool Stars, Stellar Systems, and the Sun**, ed. A. K. Dupree (SAO SR-389), p. 47.
Swank, J. H., White, N. E., Holt, S. S., and Becker, R. H.: 1981, *Ap. J.*, **246**, 208.
Vaiana, G. S.: 1980, *Highlights of Astronomy*, **5**, 419.
Vaiana, G. S.: 1981a, *Inst. Space Astronaut. Sci. (Tokyo)*, **597**, 1.
Vaiana, G. S.: 1981b, in **X-ray Astronomy with the EINSTEIN Satellite**, ed. R. Giacconi (Dordrecht: Reidel), pp. 1-18.
Vaiana, G. S.: 1981c, *Space Sci. Rev.*, **30**, 151.
Vaiana, G. S. *et al.*: 1981, *Ap. J.*, **245**, 163.
Vaiana, G. S., and Rosner, R.: 1978, *Ann. Rev. Astron. Ap.*, **16**, 393.
Vaiana, G. S., and Tucker, W. H.: in **X-ray Astronomy**, ed. R. Giacconi and H. Gursky (Dordrecht: Reidel), p. 169.
Vaughan, A. H. *et al.*: 1981, *Ap. J.*, **250**, 276.
Walter, F. M.: 1981, *Ap. J.*, **245**, 677.
Walter, F. M., and Boywer, S.: 1981, *Ap. J..*, **245**, 671.
Wilson, O. C.: 1966, *Science*, **151**, 1487.
Wilson, O. C.: 1978, *Ap. J.*, **226**, 379.
Worden, S. P.: 1983, in *Proc. IAU No. 71 (Activity in Red Dwarf Stars)*, ed. M. Rodono and P. B. Byrne.

DISCUSSION

MULLAN: Can you comment on the coronal structure in cool giants, where X-rays have not been detected? Do static loop models apply there?

VAIANA: You and others have advanced suggestions for these stars to have open coronae dominated by winds. Static loop coronae would not apply at all in this case.

MOUSCHOVIAS: Have I missed a point? In your talk you first mentioned that there is a wide range of X-ray luminosity at a given spectral type. Later on, if I understood you correctly, you said that spectral type very much determines the mean X-ray luminosity, as well as its variability. Could you, please, restate what the case is?

VAIANA: Experimentally, spectral type in late-type stars does not in a first approximation determine the X-ray luminosity; indeed, in this approximation, the population mean luminosity is independent of spectral type. Instead, it appears that rotation, or age, are far more important parameters in determining the luminosity. However, it may be that there is a residual dependence on other parameters, such as mass (or spectral type). Our model includes such a weak dependence. At present, however, this is not testable. With respect to variability, the preliminary data seem to indicate that M stars vary more than earlier solar-type stars.

MALTBY: Could you tell us how the stellar scaling laws you presented compare with the observations?

VAIANA: Within the limits of the present errors, the agreement is good. The present systematic errors are due to uncertainty in the IPC gain. This uncertainty will soon be removed, and we will be able to make a more detailed comparison. A two-temperature analysis will also help to subject the model to meaningful tests.

LINSKY: Could you comment on whether there are indications that coronal temperature depends on stellar gravity, and whether an active star is a member of a close binary system or is single?

VAIANA: We do not have many late-type giants that can be fit. Since they are not particularly strong sources as compared with main sequence stars, and because they are much more distant, we do not have many photons to consider. From the few giants I have seen, my impression is that their spectra are not harder: for instance, β Lep seems to be softer than π^1 UMa. In the early-type stars, κ Ori seems softer than τ Sco. Of course, the interesting comparison will have to be done along evolutionary sequences: β Lep may turn out to be softer than π^1 UMa, but harder than the Sun. With respect to single vs. binary stars, again π^1 UMa is a good example of a single star having a hard component as well. The difference between such single stars and active binaries such as RS CVn's may then be quantitative (rather than qualitative), but this remains to be looked at.

MEGESSIER: The first detections of X-rays in the stars indicated a gap in the H–R diagram near the A type. Is that always the case?

VAIANA: There will be a forthcoming paper by Cash and Snow and our group on A stars. Stars like Vega and Altair are at the bottom of the luminosity scale for early-type stars. Jurgen Schmitt and Leon Golub are investigating where the onset of the high luminosity regime associated with F dwarf stars is.

FRACTIONAL STELLAR CONVECTION ZONE DEPTH AND THE GENERATION OF MAGNETIC FLUX: AN EMPIRICAL APPROACH

Mark S. Giampapa
Harvard-Smithsonian Center for Astrophysics
60 Garden Street
Cambridge, MA 02138, U.S.A.

ABSTRACT

I present preliminary results from an observational investigation of very late M dwarf stars utilizing the Multiple Mirror Telescope facility. I find that dwarf stars later than spectral type M5 do <u>not</u> necessarily exhibit Hα line emission, contrary to the assertion by Joy and Abt (1974). The preliminary results I discuss herein tentatively suggest, but do not prove, that the generation of significant magnetic fields and magnetic flux is severely inhibited in fully convective stars.

INTRODUCTION

The chromospheres and coronae of the M dwarf stars are especially interesting since it is this region of the H-R diagram where the discrepancy between acoustic wave heating theories and the observations is most vivid. Also, to the extent that the chromospheric physical conditions in these cool dwarfs differ from those existing in the solar chromosphere, the M dwarfs can provide a unique laboratory to test theories of nonradiative heating of stellar atmospheres. For example, the importance of convection to the origin and structure of stellar chromospheres is not well understood, either empirically or theoretically. Hence, it is crucial to determine the degree of magnetic field-related nonradiative heating present in the atmospheres of a sample of stars which represent a range of fractional convection zone depths (i.e., convection zone depth / stellar radius). The very late M dwarf stars may be entirely convective. Thus these objects provide a sensitive test of the importance of the fractional convection zone depth to the generation of magnetic fields and the associated nonradiative heating in the atmospheres of late-type main sequence stars.

The main sequence M dwarf stars are subdivided into two classes, namely, the dMe stars and the non-dMe or, simply, the dM stars. The dMe stars exhibit Hα emission in their spectra while the spectra of dM stars

J. O. Stenflo (ed.), Solar and Magnetic Fields: Origins and Coronal Effects, 187–192.

reveal either Hα in absorption or no apparent Hα feature. Both Hα emission and Hα absorption are indicative of the presence of a chromosphere in a M dwarf star (Fosbury, 1974; Cram and Mullan, 1979). A variety of empirical evidence suggests that magnetic heating processes control the energy balance in stellar chromospheres. The compelling circumstantial evidence has been extensively reviewed by Giampapa (1980; see also Giampapa, Worden, and Linsky, 1982) in the case of the M dwarf stars. In basic summary, the strength of Hα line emission (or absorption) essentially constitutes a measurement of the degree of magnetic field-related, nonradiative heating present in an M dwarf chromosphere.

According to Joy and Abt (1974), all dwarf stars later than M5.5 are dMe stars. However, recent investigations by Liebert et al. (1979) disclosed the possible existence of very late (later than Mt. Wilson type dM5) M dwarf stars with no indication of Balmer line emission. Unfortunately, the observations by Liebert et al. (1979) were at low spectral resolutions (8 - 10 Å). I have therefore undertaken a program to obtain significantly higher resolution spectra in order to confirm the existence (or non-existence) of Hα line emission in the spectra of very late M dwarf stars.

OBSERVATIONS AND RESULTS

The observations were obtained with the Multiple Mirror Telescope (MMT) in conjunction with the Agassiz echelle spectrograph, a 100 μm slit and a photon counting Reticon detector. The spectral resolution of the configuration, as deduced from the FWHMs of unblended comparison lines is approximately 0.20 Å. The resulting spectra have been smoothed by successive applications of a triangular smoothing function. The final spectral resolution is typically 0.25 Å. The dark current has been subtracted from each spectrum. Finally, I found that the sky background with 1 arc-sec entrance apertures in the image stacker was negligible compared to the dark counts.

I display in Figure 1 a histogram of the results thus far obtained in this investigation. The program objects are segregated according to those M dwarfs that exhibit Hα emission and those that do not show an Hα emission feature. The abscissa is photographic $(m_{pg} - m_R)$ color deduced from data listed by Luyten (1979). The value of m_{pg} can be considered the "blue" apparent magnitude while m_R is the stellar apparent magnitude in the neighborhood of Hα. Many of the program stars have not been assigned exact M spectral types. I have therefore inferred the range of spectral types represented in Figure 1 in the following manner: I randomly selected 10 - 15 stars of each spectral type listed in the proper motion catalog of Luyten (1979) and computed their $(m_{pg} - m_R)$ colors. I then adopted the mean value of the photographic color as corresponding to the particular spectral type. This procedure yielded the calibration of spectral type versus photographic color shown in Figure 1. Although the calibration is

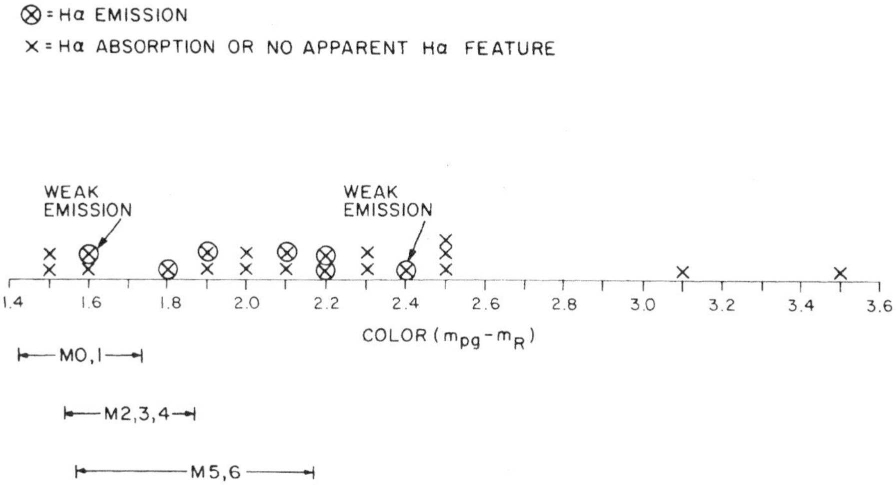

Figure 1. Histogram of the program M dwarf stars. The appearance of Hα is indicated. Note that nearly all the M dwarfs later than M6 are non-dMe stars.

crude, there are several program objects that are clearly later than M6. Furthermore, I demonstrate via Figure 1 that there exist dM stars later than M6 that are not dMe stars. I can therefore refute the claim by Joy and Abt (1974) that all M dwarfs later than dM5.5 are dMe stars.

Unfortunately the sample of very late-type dM stellar spectra thus far obtained is not statistically significant. I therefore cannot claim that all M dwarf stars later than $(m_{pg} - m_R) = 2.4$ are non-dMe stars. The test of such a hypothesis will require a larger sample of stars. Moreover, multiple observations should be obtained in order to investigate the possibility of variability. Nevertheless, the preliminary results displayed in Figure 1 tentatively suggest that there is a decline in chromospheric activity, and by implication, a decline in the generation of significant magnetic flux, toward the very late M dwarf stars.

DISCUSSION

The results given in Figure 1 invite a preliminary empirical exploration of the consequences for the importance of the fractional stellar convection zone depth to the generation of magnetic flux in late-type stars. I show in Table 1 the fraction of M dwarf stars that are dMe stars within a given spectral type. These data are taken from Joy and Abt (1974, their Table 2). I also list the fractional convection zone depth, $1 - r_c/R_*$, where r_c is the radius to the base of the convection zone and R_* is the stellar radius, following the stellar interior computations of Copeland, Jensen, and Jorgensen (1970).

Table 1. M Dwarf Emission and Convection Zone Depths

Sp. Type	% of dMe	$1 - r_c/R_*$
M0	3	0.343
M0.5	3	
M1	7	0.376
M1.5	8	
M2	9	0.451
M2.5	16	
M3	41	1.0
M3.5	35	
M4	33	1.0
M4.5	50	
M5	80	
M5.5	100	

The precise location in spectral type for the onset of complete convection is uncertain because the mass-luminosity law for the very late main sequence is ill-defined. Nevertheless, we can confidently presume that all stars later than dM6 must be wholly convective. The results of Table 1 appear to indicate an increase in the frequency of occurrence of chromospheric activity with increasing depth of the stellar convection zone. However, the histogram in Figure 1 tentatively suggests that those M dwarf stars which are surely wholly convective (even in view of the uncertainties in current model computations) suffer a decline in chromospheric activity. Thus, the generation of significant magnetic fields and magnetic flux may be severely inhibited in completely convective stars.

The results I discuss in this investigation may constitute corroborative evidence for the presence of $\alpha - \omega$ "shell" dynamos, as opposed to the so-called "distributed" dynamos, as the principal mechanism for the generation of significant magnetic flux in partially convective stars (e.g., see Parker, 1975; Rosner, 1980; Spiegel and Weiss, 1980; Golub et al., 1981). For those stars that are completely convective, the shell dynamo is no longer meaningful since there is, of course, no shear layer between a radiative core and the base of the stellar convection zone. In these cases, the rise time of toroidal buoyant magnetic flux through the convective interior may be much less than the amplification time (the ω-effect) to produce significant magnetic flux (Parker, 1975; Rosner, 1980; Spiegel and Weiss, 1980; Schmitt and Rosner, 1982; Durney and Robinson, 1982). Consequently, a low level of chromospheric and coronal activity is observed in the very late, fully convective M dwarf stars.

Interestingly, the T Tauri stars could constitute a contradiction to the view I present in this investigation. In particular, the T Tauri stars are pre-main sequence objects that are fully convective and exhibit intense chromospheric line emission (Giampapa et al., 1981, and references therein). If the speculative scenario I discuss in this paper is eventually verified, then the implication would be that the chromospheric (and coronal) activity observed in the T Tauri stars is associated with remnant primordial fields rather than dynamo generated magnetic fields, or that some T Tauri stars have radiative cores.

Finally, I recognize that the current sample of stars later than dM6 must be enlarged and carefully selected so as to avoid possible biases toward very old stars that may be introduced by samples selected from the Luyten proper motion survey.

I acknowledge interesting discussions with R. Rosner, K. Tsinganos, N. Weiss, and L. Golub during the course of this study. The results presented in this investigation are based upon data obtained at the Multiple Mirror Telescope Observatory (MMTO). The MMTO is operated jointly by the University of Arizona and the Smithsonian Institution.

REFERENCES

Copeland, H., Jensen, J.O., and Jorgensen, H.E.: 1970, Astron. Astrophys. 35, p. 12.
Cram, L.E. and Mullan, D.J.: 1979, Astrophys. J. 234, p. 579.
Durney, B.R. and Robinson, R.D.: 1982, Astrophys. J., in press.
Fosbury, R.A.E.: 1974, Monthly Not. Roy. Astron. Soc. 169, p. 147.
Giampapa, M.S.: 1980, Ph.D. dissertation, University of Arizona, Tucson.
Giampapa, M.S., Calvet, N., Imhoff, C.L., and Kuhi, L.V.: 1981, Astrophys. J. 25, p. 113.
Giampapa, M.S., Worden, S.P., and Linsky, J.L.: 1982, Astrophys. J. 258, p. 740.
Golub, L., Rosner, R., Vaiana, G.S., and Weiss, N.O.: 1981, Astrophys. J. 243, p. 309.
Joy, A.H. and Abt. H.A.: 1974, Astrophys. J. Suppl. 28, p. 1.
Liebert, J., Dahn, C.C., Gresham, M., and Strittmatter, P.A.: 1979, Astrophys. J. 233, p. 226.
Luyten, W.J.: 1979, in "LHS Catalog and Atlas", University of Minnesota: Minneapolis.
Parker, E.N.: 1975, Astrophys. J. 198, p. 205.
Rosner, R.: 1980, in "Cool Stars, Stellar Systems and the Sun", ed. A.K. Dupree, SAO Spec. Rep. 389, p. 79.
Schmitt, J.H.M.M. and Rosner, R.: 1982, Astrophys. J., in press.
Spiegel, E.A. and Weiss, N.O.: 1980, Nature 287, p. 616.
Worden, S.P., Schneeberger, T.J., and Giampapa, M.S.: 1981, Astrophys. J. Suppl. 46, p. 159.

DISCUSSION

SODERBLOM: In regard to the very cool stars that show no $H\alpha$ emission: Do you know the space motions for those stars? Are they kinematically old? If they turn out to be, how would that affect your conclusions about magnetic fields and chromospheric emission for these very low masses? I suspect that these stars may be kinematically old, because they are intrinsically so faint that the only way to find them is because of their large proper motions.

GIAMPAPA: I will soon measure the radial velocities for these stars and obtain their space motions. However, even if the stars are kinematically old, the conclusion that magnetic flux generation is inhibited in fully convective stars would remain. For example, current dynamo models predict very efficient field generation in the M dwarfs at solar rotation periods or longer resulting in high filling factors ($> 50\%$) of kG fields. Perhaps the spin-down mechanism is unusually efficient, but it is difficult to understand why there should be an apparently sharp drop in the frequency of dMe stars just after dM6. I remind you that dMe stars are found among the kinematically old disk as well as young disk populations.

GOLUB: The X-ray survey of M dwarf emission may provide an indication of the point at which M-S stars become fully convective, perhaps with greater sensitivity than the chromospheric indicators. The results show a very abrupt drop in the luminosity function of M dwarfs at a B-V of about + 1.6, corresponding roughly to dM5. Of course, we need some good theoretical work to tell us if this is really a workable hypothesis.

GIAMPAPA: I am happy to see corroborative X-ray evidence for my hypothesis that magnetic flux generation is inefficient in the fully convective dM stars. Additional theoretical work should include more sophisticated M dwarf interior calculations that utilize an improved equation of state and better opacity tables in order to accurately identify the critical mass for the onset of complete convection. Supporting observational work must include the determination of a considerably better mass-luminosity law than exists at present combined with a more consistent spectral classification scheme for the M dwarf stars.

STIX: The magnetic time scale of a fully convective star is the time of turbulent mixing which is very short even in comparison with the age of a T Tauri star. It is difficult therefore, for these stars to keep their primordial fields.

GIAMPAPA: I am aware of this. However, I do not see how to reconcile the apparent contradiction that there is a low level of chromospheric activity in fully convective M dwarf stars while there is a greatly enhanced level of activity in the T Tauri stars. Perhaps radiative cores are present in T Tauri stars or the relatively more rapidly rotating T Tauri stars operate efficiently as "distributed" dynamos.

WALTER: T Tauri stars which are X-ray sources with "normal" K star coronae (Walter and Kuhi: 1981, ***Astrophys.J.*** **250**, p. 245), do not necessarily have high masses ($\gtrsim 2M_\odot$) If put on Cohen and Kuhi's (1979) H-R diagram, many are $\sim 1M_\odot$ or less. In fact, 5 "post-T Tauri" stars (Mundt et al.: 1982, ***Astrophys. J.***, submitted) have $M \approx 0.6M_\odot$ ages of $\sim 2 \times 10^6$ yr (like the T Tauris), and apparently normal coronae, including surface spots.

ROSNER: In response to the previous question regarding activity of dF stars, Vaiana et al (1981) and Topka et al. (1982) have shown that the X-ray luminosity function of dF stars is "shifted" to higher luminosities when compared to the luminosity functions of later-type dwarf stars. Hence the "activity" of dF stars is, on average, somewhat ***higher*** than that of later-type dwarfs.

THE CORONAL ROTATION-ACTIVITY RELATIONS IN LATE-TYPE STARS

Frederick M. Walter
Joint Institute for Laboratory Astrophysics, University of Colorado and National Bureau of Standards, Boulder, Colorado 80309 USA

ABSTRACT

I discuss EINSTEIN X-ray observations of late-type (G and K) stars. For the RS CVn systems and the evolved G stars, there is a linear relation between X-ray surface flux and stellar angular velocity. The situation is more complex among the G-K dwarfs: rapid rotators, with rotational periods less than $\sim 10^d$, also show a linear relation between X-ray surface flux and angular velocity, but the slow rotators fall some two orders of magnitude below that relation.

I. INTRODUCTION

It has been evident for some time that rapid rotation in late-type stars influences the heating of their chromospheres and coronae. Kraft (1967) showed that F-G dwarfs with strong chromospheric Ca II H and K emission lines had larger <V sin i> than did those stars without detectable emission. Skumanich (1972) proposed, based on data for the Pleiades, Hyades, and the Sun, that the Ca II emission flux decreased as the stellar age $T^{-1/2}$, as did <V sin i>. Hence, the chromospheric emission flux scaled linearly with stellar rotational velocity.

By observing the relationship between velocity (either rotational or angular) and an appropriate measure of the activity (a measure of the non-radiative heating rate), one might hope to gain insight into the processes which drive the activity in these convective stars. These processes likely involve the stellar dynamo, given the correspondence of regions of high activity with regions of emerging magnetic flux on the Sun (Vaiana and Rosner 1978). Such studies may improve our understanding of convective dynamos and how they interact with the stellar atmosphere. In this work I have obtained EINSTEIN X-ray observations of a number of late-type stars with known rotation periods, in order to investigate the coronal rotation-activity relation.

II. THE OBSERVED ROTATION-ACTIVITY RELATIONS

I use the ratio of the X-ray to the bolometric flux L_x/L_{bol} as the measure of stellar coronal activity. The justification is that L_x/L_{bol} is a measure of the fraction of the stellar energy balance radiated from the corona and should scale with the energy in the closed magnetic flux

J. O. Stenflo (ed.), Solar and Magnetic Fields: Origins and Coronal Effects, 193–198.

tubes. In addition, L_x/L_{bol} is convenient in that it is proportional to the X-ray surface flux for stars of the same T_{eff}. As a rotational parameter I prefer the stellar angular velocity Ω, which is the inverse of the stellar rotation period P, and is not subject to uncertainties due to the stellar inclination.

A. K Stars

The first and largest sample of stars is the RS CVn systems (Hall 1976), which are close binaries showing extreme activity levels (Walter and Bowyer 1981). Rotation periods are the photometric periods, or in their absence synchronism is assumed. The bolometric flux is that of the active star in the system. In Figure 1, I plot L_x/L_{bol} vs. $P(=1/\Omega)$. There is a clear trend in the data, with activity increasing with more rapid rotation. The best fit power law relation is $L_x/L_{bol} \propto \Omega^{1.17\pm0.12}$, which is a linear relation within the errors.

There is no significant correlation between L_x/L_{bol} and V since the rotational velocities of the slow rotators, which are giants, are not significantly different from those of the shorter period systems. Among these stars, there is no dependence of L_x/L_{bol} upon the surface gravity (Walter and Bowyer 1981).

This linear relation is not unique to the RS CVn systems: active K dwarfs, T Tauri stars, and weak emission line PMS stars (Walter and Kuhi 1981; Mundt et al. 1982) all fall on this relation. Caillaut (1982) has found the same for BY Dra stars. In fact, among rapidly rotating G8-K5 stars, only the Algol systems (White 1982) fall significantly below the relation (although the secondary of Algol does fit). The data indicate that this linear relation is universal for K stars, both single and binary, as long as significant mass transfer is not occurring. This may not be surprising, if the dynamo is sensitive primarily to the depth of the convective zone and the stellar angular velocity.

B. G Stars

The second sample of stars I studied includes 16 F8-G5 dwarfs. The best least-squares power law fit is significantly steeper than the linear relation found for the K stars, with a slope of 2.5 (Walter 1982), and is consistent with the steep relation obtained by Pallavicini et al. (1981). However, the fit is poor and the residuals show a systematic trend: the data seem to require a shallower slope among the rapid rotators, and a steeper slope at long periods. Figure 2 shows the fit to only those stars with $P < 12^d$. This fit has a slope of 1.25 ± 0.21, which is identical to the linear K star fit.

The justification of cutting the data at this point lies in the observations of Vaughan (1980) and Vaughan and Preston (1980), who found among G dwarfs a bifurcation in chromospheric activity levels, with the active stars showing stochastic long-term chromospheric variability and the inactive stars showing cyclic variability. At G0, the location of the gap corresponds to an age of $\sim10^9$ yrs. By Skumanich's (1972) relation between V_{rot} and age, this also corresponds to a rotation period of $\sim12^d$. Hence the active, rapid rotators which show a linear relation between L_x/L_{bol} and Ω correspond to the chromospherically active, young stars. Slow rotators, like the Sun, fall some two orders of magnitude below the extrapolation of the linear relation. It is not clear if the slope of the relation changes, or if the linear relation is scaled down by two orders of magnitude.

I find these X-ray data, taken together with the chromospheric observations, to be convincing evidence of a change in the nature of the dynamo at a given rotation period in dwarfs. This may be due to a juxtaposition of many dynamo modes in the rapid rotators, with only the fundamental mode excited at low Ω (Durney et al. 1981), or to a change in the pattern of convection at a critical Ω. I note that there is marginal evidence, based on five K dwarfs, that there is a similar break in the L_x/L_{bol} vs. Ω relation for K dwarfs (Walter 1982).

Finally, a small sample of five G giants is consistent with the linear portion of the relation found for the G dwarfs, in that the scaling, within the errors, is identical for $P < 12^d$. However, for the giants as for the long-period RS CVns, the linear relation extends to periods of ~100 days without evidence of any break. While this is clear evidence of a difference between giants and dwarfs, it is not clear what the physical basis for the difference is.

C. F Stars

The last sample of stars includes 12 F3-F5 dwarfs with measured V sin i. These stars are interesting in that they have small convective zones, and may not have the same type of dynamo mechanism as later type dwarfs. The analysis is still in a preliminary state, but there is no

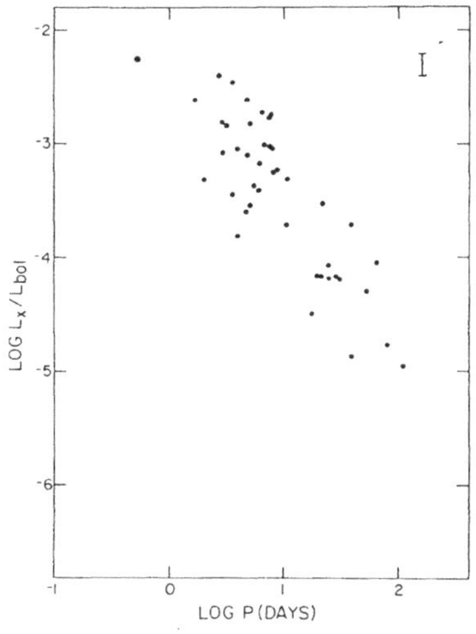

Fig. 1. The coronal rotation-activity relation for G8-K2 stars. Most of the stars are RS CVn systems, but some single K dwarfs are included. The relation is nearly linear, with $L_x/L_{bol} \propto \Omega^{1.2\pm0.2}$ ($\Omega = P^{-1}$).

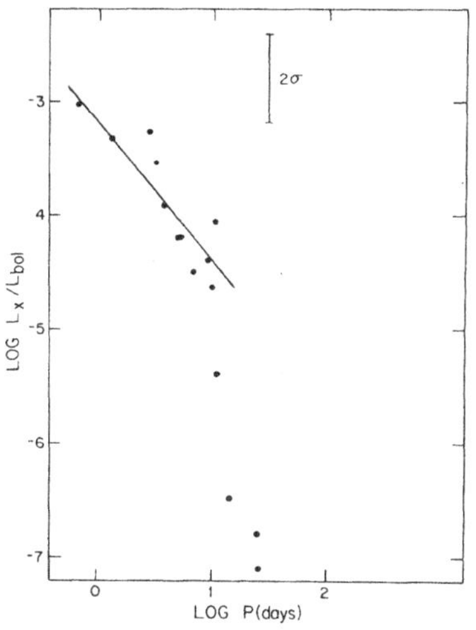

Fig. 2. The coronal rotation-activity relation for F8-G5 dwarfs. The line is the best fit relation for $P \lesssim 12^d$; it has a slope identical to that found for the K stars. The slow rotators including the Sun fall well below this relation.

indication for any correlation of L_x/L_{bol} with V sin i; all stars show similar L_x/L_{bol}, although V sin i varies from 50 km s^{-1} to <6 km s^{-1}. L_x/L_{bol} for these stars is $\sim 10^{-5}$, which is two orders of magnitude brighter than Pallavicini et al. (1981) found for the O, B, and A stars. These early F stars seem to behave neither like the A or G stars.

III. SUMMARY

There is no single coronal rotation-activity relation. G-K dwarfs with rotation periods between $\sim 12^h$ and 12^d, and G-K giants and subgiants exhibit a linear relation between L_x/L_{bol} and Ω. Slowly rotating G-K dwarfs have L_x/L_{bol} some two orders of magnitude lower than an extrapolation of the linear relation. The break in the relation occurs at the same rotation period as the chromospheric break observed by Vaughan (1980).

The W UMa systems appear to show a turnover in the relation, with $L_x/L_{bol} \propto \Omega^{-3}$ (Rucinski and Vilhu 1982). These stars are not only more rapidly rotating than the detached systems, but may have disturbed circulation patterns in the convective zones, because the stars are in contact.

Converting L_x/L_{bol} to surface fluxes by removing the T^4 term in L_{bol} shows that at a given P, the K stars show about twice the X-ray surface flux as the G stars. Swank et al. (1981) have shown that K RS CVn systems have a hot ($T \gtrsim 3 \times 10^7$ K) coronal component contributing half the X-ray flux, whereas earlier type RS CVn systems show a weak, if any, hot component (cf. Walter et al. 1982). It may be that the deeper convective zone of the K stars is fundamental to the existence of this hot coronal component.

The data show that the two stellar parameters which strongly affect the activity levels are Ω and the spectral type. Surface gravity is not important. Stellar duplicity appears important only in enforcing rapid rotation. Among the rapid rotators, the relation between Ω and coronal activity is linear.

REFERENCES

Caillaut, J.-P.: 1981, Astron.J. 87, p.558.

Durney, B.R., Mihalas, D., and Robinson, R.D.: 1981, Publ.Astron.Soc. Pacific 93, p.537.

Hall, D.S.: 1976, in IAU Colloq. No.29 (Dordrecht: Reidel).

Kraft, R.P.: 1967, Astrophys.J. 150, p.551.

Mundt, R., et al.: 1982, Astrophys.J. (submitted).

Pallavicini, R., et al.: 1981, Astrophys.J. 248, p.279.

Rucinski, S.M. and Vilhu, O.: 1982, Monthly Not.Roy.Astron.Soc. (submitted).

Skumanich, A.: 1972, Astrophys.J. 171, p.565.

Swank, J., White, N., Holt, S., and Becker, R.: 1981, Astrophys.J. 246, p.208.

Vaiana, G.S. and Rosner, R.: 1978, Ann.Rev.Astron.Astrophys. 16, p.393.

Vaughan, A.H.: 1980, Publ.Astron.Soc.Pacific 92, p.392.

Vaughan, A.H. and Preston, G.H.: 1980, Publ.Astron.Soc.Pacific 92, p.385.

Walter, F.M.: 1982, Astrophys.J. 253, p.753.

Walter, F.M. and Bowyer, S.: 1981, Astrophys.J. 245, p.671.
Walter, F.M., Gibson, D.M., and Basri, G.S.: 1982, Astrophys.J. (submitted).
Walter, F.M. and Kuhi, L.V.: 1981, Astrophys.J. 250, p.245.
White, N.E.: 1982, preprint.

DISCUSSION

ZWAAN: Let me take issue with your statement that stellar chromospheres may be heated by acoustic waves. If you mean to say *purely* acoustic, without any role of magnetic fields, then your statement is wrong, since you could not explain the large range in chromospheric surface fluxes from stars of similar T_{eff} and g.

WALTER: I agree. There is clear evidence as you, among others, have shown, that the chromospheric Ca II flux is magnetically influenced. However, as Basri et al. show (following paper), at least in the RS CVn systems the chromospheric flux does not vary with Ω. The situation is complex, but may be less so in the corona, where the X-ray emitting coronal gas must be magnetically confined, and does show strong scaling with Ω.

SODERBLOM: I would like you to comment on the spread in the L_x/L_{bol} vs. P (rotation period) diagram for RS CVn's. There is nearly an order-of-magnitude spread in L_x/L_{bol}. Given this, couldn't one line fit the G-dwarf data adequately (although I grant the two-line fit is clearly better)?

WALTER: The scatter about the best fit relation for the RS CVn stars is about a factor of 3 (Walter and Bowyer: 1981, *Astrophys. J.* **245**, p. 671). This is similar to the range of solar coronal variability, and may well indicate that the scatter is due to intrinsic stellar variability. For lack of a better estimate, I adopted this variance for the G dwarfs. Even with such large error bars, the single power law fit is not good ($\chi^2 \approx 1.3$; cf. Walter: 1982, *Astrophys. J.* **253**, p. 753), and a second power law significantly improves the fit. The quadratic trend in the residuals to the single power law fit is suspicious, and the similarity of the unit slope in the rapidly rotating G dwarfs to that of the RS CVn systems is unlikely to be coincidental.

GIAMPAPA: (1) What is the physical justification for using L_x/L_{bol} as opposed to L_x ? (2) How would your correlations change if you used L_x ? (3) If you use the L_x/L_{bol} parameter, you might find an increase in this ratio as you go towards the late M dwarf stars. However, I find a decline in chromospheric activity towards the very late M dwarf stars. Thus the L_x/L_{bol} parameter and observed Hα emission, or lack thereof, would contradict each other. Would you like to comment on this?

WALTER: (1) L_x/L_{bol} is a measure of the fraction of the total stellar energy balance which gets into the corona. It may be related to the energy field up in the stellar magnetic field. Additionally, for stars of a given spectral type, L_x/L_{bol} is proportional to a surface flux. This facilitates comparing stars of different radii. L_x is OK if the sample is, for example, the G dwarfs, but I hesitate to compare giants and dwarfs using L_x. (2) For the RS CVn systems, L_x is essentially independent of P (period), since L_{bol} scales with P (see answer to Rosner's question below). For the dwarfs there will be absolutely no changes. (3) I do not know what L_x does in late M dwarfs. It may well decline, in which case L_x/L_{bol} would decline, and there would be no contradiction.

ROSNER: If, as you stated, comparison of L_x vs. P (period) gives no correlation, but comparison of L_x/L_{bol} vs. P does, doesn't that mean that L_{bol} is correlated with P ?

WALTER: Yes. RS CVn stars are evolved, detached binaries. As such, there is a rough correlation between the radius of the active, evolved star and the semi-major axis of the system. All are of roughly solar mass, so L_{bol} does correlate with P. This was first shown by Young and Koniges (1977, *Astrophys. J.* **211**, p. 836), who found that M_V increased as $P^{10/3}$ in long-period RS CVn systems.Because of these variations of stellar radius, it is very important to normalize the X-ray flux. L_x/L_{bol} is a surface flux for stars of the same spectral type, since $L_{bol} \sim R^2T^4$. Hence in the RS CVn stars, L_x/L_{bol} ($\sim F_x$) $\sim \Omega \sim P^{-1}$.

VAIANA: Your points for RS CVn's are basically scattered around two regions, the rapid rotators and the slow rotators. You have assumed that the slow rotators are corotating. If you exclude them, can you see any dependence at all?

WALTER: Yes. Of course the slope is much more poorly determined because the range in P is only an order of magnitude, but it is consistent with unity (cf. Walter and Bowyer: 1981, *Astrophys. J.* **245**, p. 671). There is no physical justification for subdividing the RS CVn systems into subgroups, except for taxonomic purposes (i.e., differentiating the RS CVn *systems* as defined by Hall, with $1^d < P < 14^d$, from those stars exhibiting the RS CVn *phenomenon*; cf. comment by D. Popper in *IAU Symp.* **88**). As concerns synchronous rotation in the slow rotators, many are observed to rotate synchronously. λ And does not, but 7 of the remaining 14 have observed photometric waves, and are in synchronous rotation. The assumption of corotation in the other seven should be quite valid.

PERIOD-ACTIVITY RELATIONS IN CLOSE BINARY SYSTEMS

Gibor Basri[1,3], Robert Laurent[1] and Fredrick Walter[2]
[1]Univ. of California, Berkeley and [2]Univ. of Colorado, Boulder
[3]Guest Observer, International Ultraviolet Satellite

Since the advent of extensive ultraviolet observations of cool stars, it has been clear that the stellar activity observed is not directly correlated with the star's position on the HR diagram (Basri and Linsky 1979, Stencel et al. 1980). Observations of an important chromospheric diagnostic, the MgII resonance lines, led to the conclusion that stellar chromospheric activity had only a weak dependence on spectral type, and exhibited large variations within a given spectral type. Because of the strong observed correlation of solar activity with magnetic fields, the field is thought to be a natural candidate for the extra parameter which predicts the level of activity. Unfortunately, it is quite difficult to measure magnetic fields directly in most cool stars. Another method with which to examine correlations between magnetic field and stellar activity indirectly is the hypothesis that magnetic fluxes are directly related to a combination of the convective and rotational parameters of a star through its generation in a magnetic dynamo. The α-ω dynamo theory (Parker, 1979) predicts a direct correlation between differential rotational velocities and field generated. Durney and Robinson (1982) predict basically a linear dependence of the emergent flux on the angular velocity of the star. One might therefore expect that in stars with the same fundamental stellar parameters, the amount of activity observed would depend on the rotational velocities. This is difficult to test because most cool stars are slow rotators and only a few rotational velocities are known.

We have undertaken to study this question by looking at close binary systems which present a number of advantages over the single stars: (1) well known rotation periods covering two orders of magnitude, (2) generally strong (easily detected) activity, (3) well known evolutionary status and stellar parameters, (4) an extensive existing survey of their coronal activity. We have used orbital rather than rotational period; these are known to be equivalent for most of the stars (where a different rotation period is known we have used it) but the assumption probably breaks down for the longest period systems. The X-ray survey of RS CVn stars by Walter and Bowyer (1981) had already

J. O. Stenflo (ed.), Solar and Magnetic Fields: Origins and Coronal Effects, 199–204.

shown a linear period-activity relation. We have observed both RS CVn and less evolved close binary systems; we separate the K subgiant RS CVn stars for special consideration as a very homogeneous group of active stars. We have chosen to present the results to reflect surface fluxes vs. angular velocity. The alternative method of Pallavicini et al. (1982) (total flux vs. v sin i) was less appealing both because it retains errors in the stellar distance and radius (and the ambiguity of sin i) and because we feel our parameters are more fundamental to the question posed. We therefore plot ratios of fluxes against period. The ratio of a diagnostic flux to ℓ_{bol} (at the earth) gives us a measure of surface activity to total emergent flux; we have also examined the diagnostics relative to the emergent chromospheric flux (as measured in MgII). Our errors in ℓ_{bol} are limited to the bolometric corrections to m_V (small) and corrections for inactive components in some of the systems.

Our aim was to survey as many of the stars observed in X-rays as we could with the IUE satellite. As there was a known period-activity correlation in their coronae, we could learn how the relation behaves as one moves down through the transition region into the chromosphere. We have used the same diagnostics, measurement techniques and calibrations as in other IUE surveys (Ayres, Marstad, and Linsky, 1981; Hartmann, Dupree, and Raymond, 1981). We estimated the calibration and measurement errors each to be 15%. Of more concern was the intrinsic stellar variability, which could easily cause the stars to change their mean activity levels by a factor of two, and leads to at least the same level of scatter in any relations found. A linear least-squares fit to the points in the log (flux ratio) vs. log (P) plane provides power law relations between activity and period which we give as the slopes of the fitted lines. We excluded a few stars from the analysis: ER Vul because its period is too short for this class of stars and 33 Psc and o Dra because their coronal fluxes were anomalously low.

We first discuss the set of X-ray/ℓ_{bol} vs. P relations. For the full sample we find a slope of -0.9 which is basically the same as the Walter and Bowyer (1981) result; the K subgiant subsample yields almost the same slope (-0.8). All of the slopes and χ^2 are listed in Table 1. Note that χ^2 for the full sample is not particularly good; it is almost always larger than for the restricted sample because the spectral inhomogeneity of the full sample shows up as a real disparity between classes in the relation. In particular, the main sequence G star binaries lie systematically lower than the K subgiants for the same rotational periods. Thus relations which include a variety of spectral classes cannot be used directly to deduce correlations between magnetic field and activity without correction for this effect. We observe this same effect for almost all the diagnostics. The MgII observations give a very different period-activity relation than X-rays: a very small slope with almost no correlation between chromospheric activity and period. There is a division by decreasing activity levels into K subgiants, G subgiants, and G dwarfs--each exhibiting the same flat slope and covering similar ranges in period. The scatter for each class is rather small but the entire sample again shows larger scatter and slope. We note that ℓ_{bol} itself exhibits a slope with period of about 1.1 for the K subsample and .5 (but with a great scatter) for the full sample (this has been explained by Young and Koniges 1977, whose work originally inspired this study). This fact does not reflect on our relations, however, which are between surface flux and period. In looking at the full sample for the other diagnostics, we see three basic groupings. Those showing flat (0 to -0.2) slopes are MgII, OI, NV, HeII and SiIV. Intermediate slopes (-0.35 to -0.6) are seen in CI, SiII, CII, CIV, and X-rays show a large slope. A number of these relations have rather larger χ^2, but we believe there is a better way to look at the data.

Because stellar activity is not determined primarily by spectral type (except on average, as found above) and because MgII seems to exhibit little correlation with period, we have re-examined the relations using MgII flux rather than ℓ_{bol} as our normalizer. This has the effect of looking at magnetic field effects which are above the general level of chromospheric activity, rather than the total energy flux (which could be poorly related to activity). The effect of doing this is rather striking; all of the relations show less scatter and smaller χ^2, and the distinction between the various subclasses (and even the excluded stars) is largely eliminated. This effect cannot be due to errors in ℓ_{bol} (which are rather small and randomly distributed). It reflects the fact that stellar activity at all temperatures is correlated. One likely explanation is that the energy flux which heats the chromosphere has the same basic source as for higher layers, although the heating mechanisms could be different. This source is not directly correlated with photospheric flux. Our relations, especially for the K subgiants now have formal errors in slope of ±.05 at the 95% confidence level. The diagnostics are grouped in the following fashion: MgII, OI and NV are basically flat (0. to -0.2), the CI at -0.2, SiII, CII at -0.3 (SiIV shows anomalously large scatter), HeII and CIV at -0.4, and X-rays at -0.7. AR Lac always lies anomalously low. The behavior of NV is quite surprising as it is closest in temperature to X-rays, but we have no reason to distrust the data, despite the fact that it is a fairly weak line sometimes disturbed by the wing of Lyman-α at IUE resolution. The scatter is quite reasonable, and a separate analysis of the correlations among the diagnostics turns up nothing unusual.

The relations are of sufficient quality to warrant a number of conclusions. First, the activity in the RS CVn systems is not elevated purely because they are rapid rotators. The behavior of our sample suggests that activity increases with later spectral type and with evolution off the main sequence. These both could be explained by the increasing role of convection. Second, the chromosphere exhibits at best a small reaction to the increase in rotation velocity. The transition region diagnostics are somewhat more sensitive, but coronal fluxes are the most sensitive measure of magnetic activity. We note that this increasing sensitivity is rather consistent with the correlation between the diagnostics in these stars (Basri and Laurent, 1982); if MgII increases with a slope of .15 and X-rays increase relative to MgII with a slope of 4.5 (as they found) then we would get our observed results. This is true for most of the diagnostics. What isn't clear is whether those relations are largely due to the increasing effect of the magnetic field, or to a small increase of activity with field in the chromosphere which is amplified in the higher layers by some non-magnetic agency (like stability criteria). Adopting the former viewpoint, we find that the directly magnetic contribution to chromospheric activity (assuming $B \propto P$) is << 10% in the transition regions it is ~ 10%, and in the corona ~ 70-90%. There are several lines of evidence that there is a major component of chromospheric heating which is much less sensitive to the magnetic flux directly.

This work was supported by NASA grant NAG 5-69 to the University of California.

Table 1: Period-Activity Relations

			Slope	χ^2	[<σ>~.5(.2)]			Slope	χ^2
MgII	14[1]	K	-.1[2]	.1	CII	13	K	-.4(-.3)	.5(.3)
	33	All	-.25	.4		27	All	-.6(-.4)	.7(.2)
OI	13	K	-.0[2] (-.0)[3]	.3(.2)	CIV	14	K	-.5(-.4)	.4(.2)
	15	All	-.2(-.0)	.6(.2)		24	All	-.4(-.35)	1.0(.25)
CI	11	K	-.35(-.2)	.4(.2)	SiIV	11	K	-.3(-.3)	.9(.6)
	23	All	-.55(-.3)	1.3(.3)		19	All	-.15(-2.)	.8(.5)
SiII	12	K	-.4(-.3)	.4(.3)	NV	12	K	-.0(-.1)	.6(.2)
	15	All	-.35(-.3)	.4(.3)		21	All	-.0(-.0)	.7(.25)
HeII	12	K	-.4(-.4)	.4(.2)	X-ray	18	K	-.8(-.7)	.6(.35)
	21	All	-.2(-.2)	1.2(.4)		32	All	-.9(-.65)	1.6(.6)

[1] number of stars used
[2] slope of the relation log (line/ℓ_{bol}) vs. log Period (days)
[3] slope of the relation log (line/MgII) vs. log Period

REFERENCES

Ayres, T., Marstad, N., and Linsky, J.: 1981, Astrophys.J. 247, pp.545-55
Basri, G., and Laurent, R.: 1983, IAU Coll. No. 71, D. Reidel, p. 439.
Basri, G., and Linsky, J.: 1979, Astrophys.J. 234, pp.1023-1035.
Durney, B.R., and Robinson, R.D.: 1982, Astrophys.J., in press.
Hartman, L., Dupree, A.K., and Raymond, J.C.: 1982, Astrophys.J., pp.214-229.
Pallavicini, R., Golub, L., Rosner, R., Vaiana, G., Ayres, T., and Linsky, J.: 1982, Astrophys.J., in press.
Parker, E.N.: 1979, "Cosmical Magnetic Fields," Clarendon Press, Oxford.
Stencel, R., Mullan, D., Linsky, J., Basri, G., and Worden, S.: 1980, Astrophys.J. Suppl. 44, pp.383-402.
Walter, F., and Bowyer, S.: 1981, Astrophys.J. 245, pp.671-676.
Young, A., and Koniges, A.: 1977, Astrophys.J. 211, pp.836-843.

DISCUSSION

VILHU: Do you have any idea on the basis of your observational material how single stars may be different from binary stars at the same spectral type and rotational period, i.e., what are the effects of binarity?

BASRI: We see some increase in activity for our binaries compared to single stars rotating at the same rate. As we report elsewhere, we also see steeper correlations between the diagnostics compared with the single stars of Ayres, Marstad, and Linsky. It may well be that binarity is an added complication to understanding the amount of activity observed.

WALTER: I disagree with your statement that binaries and single stars show different activity levels. In my X-ray samples, there are *no* obvious differences between close binaries and single stars with the same rotation periods.

BASRI: My statement was meant to refer to the ultraviolet diagnostics, for which the binaries seem to be slightly more active.

HAISCH: It seems very curious that the ratios of transition-region and chromospheric lines should show such random behaviour, when you would expect a temperature-stratified correlation of some sort or other. Could you comment on the lack of height dependence of your calculations, especially for N V, the hottest of your diagnostics, which is apparently uncorrelated with rotation?

BASRI: The height dependence of the slopes shows a general increase with temperature of formation, with the glaring exception of N V. I do not think that this is an effect of the data, and totally agree that this result is very surprising. I can offer no explanation for it at this time.

LINSKY: It is very important that we use diagnostic spectral features that clearly indicate the physical quantities that we want. I have been concerned for some time that the Ca II H and K lines may give uncertain estimates of chromospheric radiative loss rates because the correction for the photospheric light is large. Since the photospheric correction for the Mg II lines is far smaller, I suggest that the Mg II lines be used rather than the Ca II lines for intercomparison of radiative loss rates from different levels of a stellar atmosphere. Another concern is the N V flux. At the 6 Å spectral resolution of IUE, the 1240 Å spectral feature includes the N V doublet, some chromospheric lines like C I and S I, and the wings of the Lyman α geocoronal feature. Thus the N V "data" may not be an accurate measure of the N V stellar flux.

BASRI: I certainly agree that we should use these diagnostics very carefully. In particular, the behaviour of N V in our data certainly lends weight to your caution, and perhaps you have given the reason for its anomalies. Concerning Ca II and Mg II, I fully concur that photospheric corrections are both important and easier to handle for Mg II. What concerns me, however, is a growing number of indications in various work that the Mg II lines show less variability and sensitivity to activity changes in other diagnostics (like C IV and X-rays) than does Ca II. In this conference, for example, we have seen several correlations of Ca II with both X-rays and rotation, which are stronger than in our Mg II data. Studies of rotational modulation of Mg II have also often been less striking than for Ca II. I believe the time is ripe for a close examination of these two diagnostics. I am myself looking at Ca II for the same stars reported here to see if it really shows a period – activity connection contrary to Mg II.

GIAMPAPA: Are your Mg II flux measurements high or low resolution observations? Morossi and Ramella at Trieste found serious background correction problems due to halation effects, which increase with stronger Mg II emission.

BASRI: We are measuring only the flux above any underlying stellar continua or background signal. The corrections due to this subtraction are in any case small compared to the emission flux and should not have much effect on our log – log slopes.

CRAM: I am almost reduced to tears! We have beautiful correlations between Ca-line indices and periods (Noyes). We have bad correlations between Mg-line indices and periods. We think that Mg and Ca are formed in similar circumstances, and Linsky suggests that Ca, rather than Mg, is the line whose diagnostics are to be suspected. We have to be very worried about this problem, since so many of our inferences in this area depend on *empirical* correlations, for which we have no (or poor) theoretical models. It is very important to study the relation between Mg and Ca emission.

BASRI: I quite agree that this is worrisome and should be dealt with. I do not necessarily agree that Ca is the suspect diagnostic, but certainly a resolution is required. I am currently studying the Ca period – activity relation for these same stars, which may provide the answer. It is possible that either Ca and Mg are not as similar diagnostics as we think, or that the RS CVn chromospheres really do not share in the relations seen for the less active single star samples which are usually discussed.

GOLUB: I believe that we can put together results from several of the talks we have just heard in order to clarify at least a small part of this puzzle. Vaiana showed a few minutes ago that the X-ray flux goes with a high power (close to 3) of the Mg II flux. Fred Walter showed that, for the sample of stars you have been using, the X-ray flux varies inversely with the rotation period with a power of -1. Putting these results together, we expect a very shallow dependence of Mg II flux on rotation period, just as you are finding.

BASRI: My own data show, of course, the same steep dependence of X-rays on Mg II that you mention. The data are consistent with each other; the mystery is why Mg II is so shallow in light of the other reports that Ca II is well correlated with X-rays and magnetic field.

CORONAL ACTIVITY IN F-, G-, AND K-TYPE STARS;
EMPIRICAL RELATIONS BETWEEN STELLAR PARAMETERS

R. Mewe, C.J. Schrijver, E.H.B.M. Gronenschild, C. Zwaan
The Astronomical Institute, Space Research Laboratory
Beneluxlaan 21, 3527 HS Utrecht, The Netherlands.

ABSTRACT. For a sample of 58 late type stars we analyse the relations between the soft X-ray flux density F_X, the Ca II H and K line-core flux density F_{H+K}, and parameters determining the global stellar structure. By analysing the soft X-ray spectra from 15 stars we determine the coronal temperatures T and specific emission measures per unit area ζ. We discuss the dependence of T on B-V, F_X and stellar radius R. The diagram of the specific emission measure ζ against the temperature T is interpreted in terms of a coronal model consisting of static loops. Also, a search for time variations in the X-ray flux has been performed.

1. OBSERVATIONS; TEMPORAL VARIABILITY

The X-ray observations (see Table 1 and Schrijver *et al.*, 1982) were obtained with the imaging proportional counter (IPC) onboard the HEAO-2 Einstein observatory (see Giacconi *et al.*, 1979). IPC counting rates are converted to X-ray fluxes f_X at Earth (0.15-4 keV) using a factor of 1.7 10^{-11} erg cm^{-2} ct^{-1} (see Schrijver *et al.*, 1982, Schrijver, 1982). X-ray and Ca II fluxes per unit area at the stellar surface are

TABLE 1. Characteristics of the stars used in the common-factor analysis. The number of stars in each class for which temperatures and emission measures are known is given between brackets.

Spectral type	Luminosity III	IV	V
F5 - F9			11 (3)
G0 - G4	1 (1)	1	10 (1)
G5 - G9	8 (2)	2	9 (3)
K0 - K4	7 (1)		6 (3)
K5 - K9			3 (1)

(Sources: see Schrijver, 1982)

J. O. Stenflo (ed.), Solar and Magnetic Fields: Origins and Coronal Effects, 205–210.

derived from fluxes at Earth using the bolometric flux at Earth and the stellar effective temperature.

An analysis in time and frequency (Fourier) domains did not reveal any significant variability for our sample of 12 stars (with typical observing times ranging from 1 to 3 ks and a total observing time of nearly 7 hours). The 3σ detection limit for variations on a 2 (15) min timescale ranges from 30 (10) % to 90 (30) % of the mean total X-ray intensity.

2. RELATION BETWEEN X-RAY AND CA II FLUXES

In Fig. 1 the stellar X-ray flux F_x (see Schrijver *et al.*, 1982; Schrijver, 1982; solar data from L. Golub, this conference) is plotted against the Ca II H+K line-core excess flux ΔF_{H+K} obtained by subtracting an observational lower-limi flux from the total line-core flux F_{H+K} (see Mewe *et al.*, 1981; Schrijv *et al.*, 1982; Schrijver, 1982; Skumanich and Eddy, 1981). It is seen tha these parameters are closely related the correlation coefficient is 0.82. A statistical common-factor analysis shows that they obey the relation (with fluxes in erg cm^{-2} s^{-1})

$$F_x = 2.4\ 10^{-3}\ F_{H+K}^{\ \ 1.4} \qquad (1$$

(Schrijver *et al.*, 1982). The relation does not depend on paramete determining the internal stellar structure (mass and radius). Upper limits for faint X-ray sources are consistent with the relation (1).

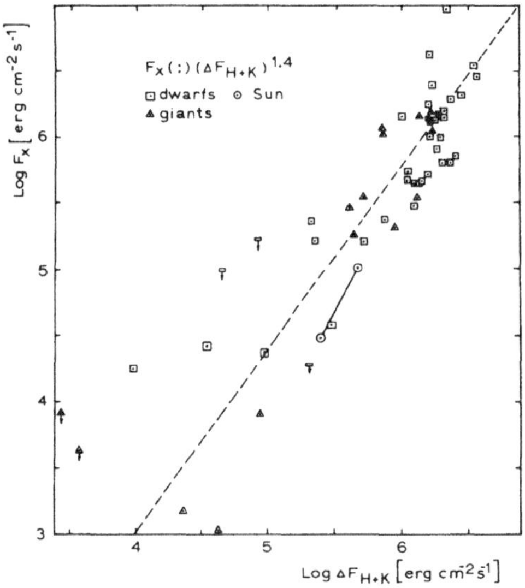

Fig. 1. Soft X-ray flux F_x vs the Ca II H and K excess flux ΔF_{H+K}. The line represents the relation resulting from the common-factor analysis.

3. SPECTRAL ANALYSIS: TEMPERATURES AND EMISSION MEASURES

In order to analyse the observed IPC X-ray spectra we convoluted theoretical line and continuum spectra (Mewe and Gronenschild, 1981) for cosmic abundances with the instrumental response function. A χ^2 analysis yields effective coronal X-ray temperatures T and emission measures $\varepsilon = \int n_e^2\, dV$. We have assumed negligible interstellar absorption. We found that the uncertainty due to the gain variations across the field of the IPC (≈7%) is generally much larger than the error due to counting statistics, so that only the gain uncertainty is considered.

In Fig. 2 T is plotted against (B-V). The vertical bar for the Sun indicates the range of temperatures for quiet and active regions (Vaiana

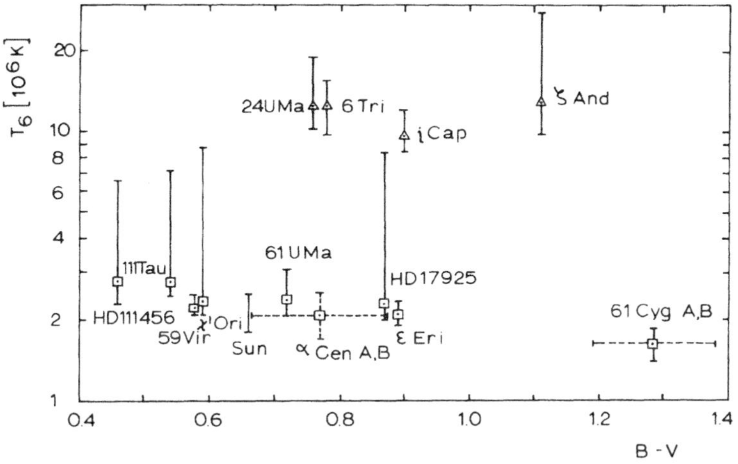

Fig. 2. Coronal X-ray temperature T vs color B-V (⊡ dwarfs; △ giants).

and Rosner, 1978). The datum for α Cen is taken from Golub *et al.* (1982). For our sample coronae around giants are significantly hotter (1 to 2 10^7 K) than those around dwarfs (1 to 4 10^6 K). Fig. 2 suggests a weak dependence of the coronal temperature on B-V for dwarfs.

We use the specific emission measure $\zeta = \varepsilon/(4\pi R^2)$ as an empirical parameter representing the stellar coronal activity. Mewe *et al.* (1981) have shown that ζ correlates strongly with the Ca II excess flux ΔF_{H+K} for the stars in our sample (see Table 1). In Fig. 3 we plot the obtained values of ζ_{27} (in units of 10^{27} cm^{-5}) against the X-ray temperature T_6 (in MK). The data for α Cen are taken from Golub *et al.* (1982). X-ray luminosities for the average Sun are taken from Manson (1977) and temperatures from Vaiana and Rosner (1978). The other solar data were derived from Skylab S054 observations (Pallavicini *et al.*, 1981).

In interpreting the spectral results we assume conditions in the outer atmospheres of the observed stars to be similar to those on the Sun: a corona highly structured by magnetic fields, in which the observed X-ray emission largely originates from plasma in closed magnetic loops. The model of a static, uniformly heated loop in local energy balance yields (Rosner *et al.*, 1978; Serio *et al.*, 1981):

$$T = 1400\ e^{-0.04L/H_p}\ (pL)^{1/3} \approx 1400\ (pL)^{1/3} \quad , \qquad (2)$$

where T is the (maximal) temperature, p is the (base) pressure, L is the halflength of the loop, and H_p is the pressure scale height (in cm) $H_p = 5\ 10^3\ T\ (g/g_\odot)^{-1}$. This scaling relation has been discussed by Pallavicini *et al.* (1981) on the basis of the Skylab observations. They found that, flare loops excluded, all features ranging from short, dense loops in active regions ($L \approx 3 \cdot 10^9$ cm, $p \approx 5$ dyn cm^{-2}; CARL in Fig. 3) to the long, tenuous loops connecting different active regions ($L \approx 2\ 10^{10}$ cm, $p \approx 0.2$ dyn cm^{-2}; LSL in Fig. 3) fit the scaling law well (see also Raymond and Rosner, 1981). Note that the point EARL in Fig. 3 corresponds to an 'equivalent' loop that matches the average of an active region on the Sun ($L \approx 9\ 10^9$ cm, $p \approx 1.5$ dyn cm^{-2}).

For a corona of N identical loops we find (Mewe *et al.*, 1980, 1982):

$$\zeta = G\, F\, p^2\, (2kT)^{-2} \ell \quad , \tag{3}$$

where G is a geometry factor (1-G is the fraction of the corona occulted by the star; we take G = 0.7), $F = 2NA/(4\pi R^2)$ is the coronal filling factor, A is the average cross section of the loops and ℓ is the emission scale length, which is limited either by the loop size ($\ell \approx L$ for $L \leqslant H_p/2$ with L the loops' halflength), or by the pressure scale height ($\ell \approx H_p/2$) (Mewe *et al.*, 1980, 1982). Combining Eqs. (2) and (3) with the limiting values for ℓ (with L_{10} in 10^{10} cm):

$$\zeta_{27} = 0.12\ (L_{10}/F)^{-1} T_6^4 \qquad (L \leqslant H_p/2) \quad , \tag{4a}$$

$$\zeta_{27} = 0.03\ (L_{10}/F)^{-2} F^{-1} (g/g_{\odot})^{-1} T_6^5 \qquad (L \geqslant H_p/2) \quad . \tag{4b}$$

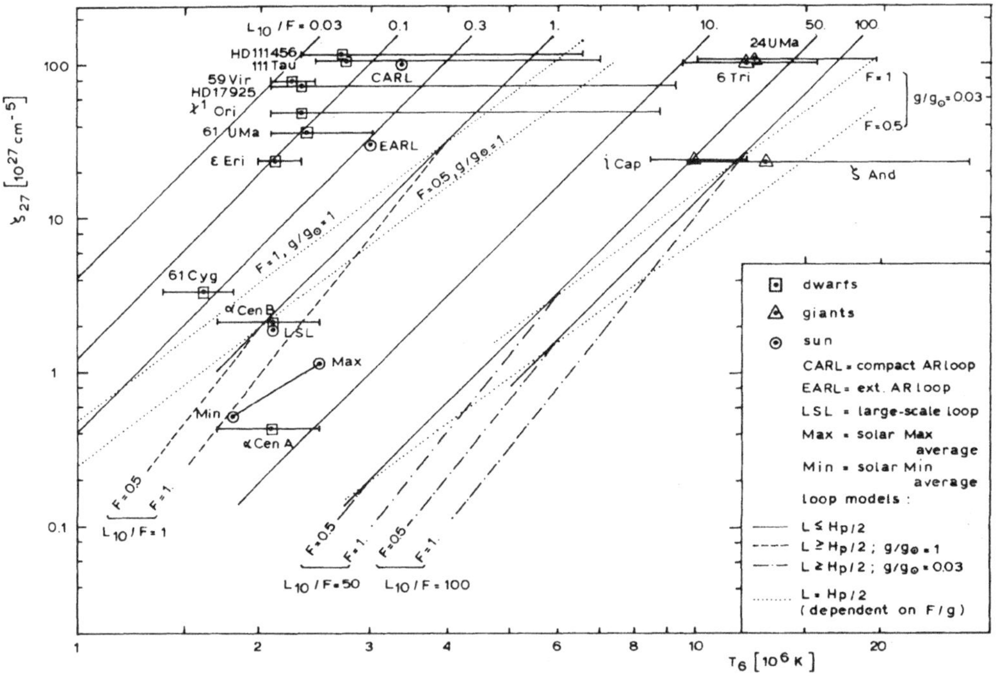

Fig. 3. Specific emission measure ζ vs coronal X-ray temperature T. The drawn curves represent theoretical relations for coronal loop models.

4. DISCUSSION

Curves calculated with Eqs. (4a,b) are drawn in Fig. 3. The dotted lines represent cases in which the loop halflength equals half the pressure scale height, with the filling factor F as a free parameter.

We see that the data for the average Sun and for α Cen A and B are compatible with a star covered for a large fraction with large-scale loops with $L \approx H_p/2 \approx 5\ 10^9$ cm ≈ 0.1 R. We know that the emission of these loops dominates the total solar emission because the filling factor for active regions (AR's) is small ($F \sim 10^{-3}$ to 10^{-2}, Vaiana and Rosner, 1978). The active dwarfs, however, lie in the region $L_{10}/F \approx 0.1$, where $L << H_p$. As $F \leqslant 1$ it follows that the majority of the loops on active dwarfs must be very compact, at least as compact as the loops in solar AR's. A search for rotation modulation in X-rays could give a clue on the values of the filling factors, and hence on loop lengths.

The active giants fall in the region $L_{10}/F \approx 50$, so we obtain as an upper limit $L \approx H_p/2 \approx R$. In the case of small filling factors the loops will be much smaller than H_p and R.

Thus, the observations indicate coronae with loops of different relative sizes. It is surprising that all these observations obey the same relation between F_x and ΔF_{H+K}.

ACKNOWLEDGEMENTS. We wish to thank H.M. Johnson for allowing us to analyse several of his observations. We also thank the staff members of the Center for Astrophysics in Cambridge who helped us in analysing the data, specifically D.G. Fabricant, E. Mandel and F.D. Seward. We are indebted to Ursule Smissaert for typing the manuscript.

REFERENCES

Giacconi, R. *et al.*: 1979, *Astrophys. J.* 230, p. 540.
Golub, L. *et al.*: 1982, *Astrophys. J.* 253, p. 242.
Manson, J.E.: 1977, The Solar Output and its Variation, ed. O.R. White, p. 261.
Mewe, R. *et al.*: 1980, Cool Stars, Stellar Systems and the Sun, ed. A.K. Dupree, S.A.O. Spec. Rep. 398, Cambridge, p. 107.
Mewe, R., Gronenschild, E.H.B.M.: 1981, *Astron. Astrophys. Suppl.* 45, p. 11.
Mewe, R., Schrijver, C.J., Zwaan, C.: 1981, *Space Science Rev.* 30, p. 191.
Mewe, R. *et al.*: 1982, *Astrophys. J.*, in press.
Pallavicini, R. *et al.*: 1981, *Astrophys. J.* 247, p. 692.
Raymond, J.C., Rosner, R.: 1981, Second Cambridge Workshop on Cool Stars, Stellar Systems, and the Sun, ed. M.S. Giampapa, L. Golub, S.A.O. Spec. Rep. 392, Vol. 1, p. 15.
Rosner, R., Tucker, W.H., Vaiana, G.S.: 1978, *Astrophys. J.* 220, p. 643.
Schrijver, C.J., Mewe, R., Zwaan, C.: 1982, Advances in Space Research, Proc. 24th COSPAR Conf., in press.
Schrijver, C.J.: 1982, in preparation.
Serio, S. *et al.*: 1981, *Astrophys. J.* 243, p. 288.
Skumanich, A., Eddy, J.A.: 1981, Solar Phenomena in Stars and Stellar Systems, ed. R.M. Bonnet, A.K. Dupree, D. Reidel, Dordrecht, p. 349.
Vaiana, G.S., Rosner, R.: 1978, *Ann. Rev. Astron. Astrophys.* 16, p. 393.

DISCUSSION

GIAMPAPA: Have you attempted coronal loop modelling with multiple components? That is, could you achieve some of the same results with a combination of "active" compact ($L < H_p$) loops and "quiet" loops ($L \gtrsim H_p$)?

MEWE: Not with the observations made with the imaging proportional counter (IPC) on the *Einstein* Observatory, because these are too inaccurate to allow such a procedure. However, on Capella, a giant with an activity comparable to that shown by the giants considered, we have performed high-resolution spectral measurements with the objective grating spectrometer (OGS) on *Einstein*. The resulting data are best fitted with two temperature components, which have the following characteristic parameters (assuming that Capella Ab is the X-ray source):
$T_1 = 5 \times 10^6 K$, $(L_{10}/F)_1 \approx 10$ (i.e., $L \ll H_p$), $(pF)_1 \approx 0.6$ dyn cm^{-2}, and
$T_2 = 10^7 K$, $(L_{10}/F)_2 \approx 100$ ($L \approx FH_p$), $(pF)_2 \approx 0.5$ dyn cm^{-2}.

VILHU: Are you able somehow (at least in principle) to determine the filling factor?

MEWE: A search for rotational modulation in X-rays (which has not yet been performed) could give us a clue to the values of the *coronal* filling factors (hence to the loop lengths). From the results for the magnetic area coverage factor of active dwarfs (G.W. Marcy, these proceedings) we may expect filling factors close to unity, which in that case would imply that modulations in X-rays are difficult to detect.

VAIANA: First a comment about the possibility of getting a filling factor from X-ray data. Of course any model that we have developed is able to give filling factors. There are two filling factors: one is the ratio of the area covered by coronal loops with respect to the total surface of the star; the other is a magnetic filling factor that is something like $\langle B_z \rangle / B_{equipart.} = f_m$. They are model dependent and will require independent determinations: The first from rotational modulation for instance, the second from direct magnetic field measurements.

Now I have a question. In your F_x vs. F_k figure, the scatter is acceptable only at high values of F_x and over a small region (perhaps one order of magnitude). At low fluxes the scatter is very large. Are you not worried that your correlation may be an artifact of this somewhat arbitrary subtraction technique?

MEWE: For stars with X-ray surface fluxes ranging from about 10^7 down to at least 10^5 erg cm^{-2} s^{-1} we had no problems in subtracting the Ca II H and K lower limit, but for fainter stars (with activity comparable to the sun or weaker) indeed this procedure becomes uncertain.

III. THEORY OF STELLAR MAGNETIC FIELD GENERATION

STELLAR DYNAMO THEORY

Manfred Schüssler
Universitäts-Sternwarte Göttingen
Federal Republic of Germany

ABSTRACT

Three topics of current interest are treated in this review:

a) The success and shortcomings of dynamo models for the solar cycle are explained, and oscillator models discussed briefly.

b) The intermittent (flux tube) nature of magnetic fields in convection zones leads to new conjectures about stellar dynamos. Arguments are given that the dynamo may operate in the overshoot region below a convective envelope. Mean-field theory for intermittent fields is illustrated.

c) I review nonlinear dynamo models and some attempts to interpret observational results concerning late-type active stars.

1. INTRODUCTION

Solar activity is no peculiarity of our parent star, no unique phenomenon. On the contrary, most late type stars show the signature of magnetic activity when observed in X-ray, UV, visible or radio wavelengths. Stellar activity cycles similar to the Sun's have been discovered by O.C. Wilson (1978) and since then many new exciting observational results have transformed "solar dynamo theory" into "stellar dynamo theory". This development opens - in principle at least - the opportunity to test competing dynamo models or other conjectures concerning the mechanisms of stellar activity using a whole set of examples with different physical parameters (e.g. rotation, effective temperature, age, multiplicity, luminosity, ...). The Sun remains the only star where the processes can be studied in detail, but predictions of any theory of solar activity have to meet the requirements set by observations of stars. Consequently, we are in a much better position than before: We can decide what is typical of stellar activity and what is peculiar, i.e. what can guide our search for basic principles and what is misleading. For comparison, theories for the formation of the planetary system still suffer from the fact that our system is the

J. O. Stenflo (ed.), Solar and Magnetic Fields: Origins and Coronal Effects, 213–236.

only one observed. Consequently, even the most unlikely process cannot be ruled out just because it may have happened once. Theory of solar/stellar activity is not in this situation: we have the Sun as a nearby laboratory for observation of the detailed plasma processes, interaction of magnetic and velocity fields - and we have active stars which enable us to test our hypotheses when the parameters are changed. The synthesis of the study of solar and stellar activity may therefore lead to a very fruitful stimulation of dynamo and related theories.

Recently, dynamo theory has reached the 'textbook stage', i.e. four monographs have been published, viz. Moffatt (1978), Parker (1979a), Krause and Rädler (1980), Vainshtein, Zel'dovich and Ruzmaikin (1980). The state of theory is described quite comprehensively in these books while the more recent developments have been reviewed by Cowling (1981), Gilman (1981), Stix (1981, 1982), Weiss (1981c, 1982) and Yoshimura (1982) from different points of view. It is clear from this wealth of literature that there is no need for another broad review of the field. In the following chapters, therefore, I will concentrate on three points of current interest and discuss them in some detail. Further information may be found in the books and review papers cited above.

2. THE SOLAR DYNAMO CONTROVERSY

The characteristic Ohmic decay time for large scale magnetic fields in the sun is $\tau_D = R_\odot^2 \cdot \eta_m^{-1}$ with the molecular magnetic diffusivity $\eta_m \sim 10^4 cm^2 s^{-1}$ for the deeper parts of the convective zone. τ_D is very large compared to the time scale of the solar cycle, i.e. 22 years. Consequently, induction processes have to be invoked to understand the temporal and spatial behaviour of solar magnetic fields. Two possibilities have found most interest in the literature so far:

a) Interaction of magnetic fields and velocity fields (e.g. differential rotation, turbulent convection) leading to dynamo action under certain circumstances: small seed fields grow and a sizable magnetic field is maintained against Ohmic dissipation ("dynamo models").

b) A primordial large scale magnetic field anchored in the radiative interior of the Sun (slowly decaying with time scale τ_D) is periodically deformed by a suitable oscillation and produces the activity phenomena through the interaction with differential rotation ("oscillator models", Layzer et al., 1979, Piddington, 1976, and reference therein).

While dynamo theory has been developed in considerable mathematical detail (cf. the 4 monographs mentioned in the introduction), the oscillator picture has not been worked out enough to make a detailed evaluation possible. On the other hand, the proponents of oscillator theories have criticised dynamo models (especially the turbulent dynamos). As far as the question of applicability of the theory to the solar case is concerned, the doubts seem to be partly justified: At least, the models have to be modified in order to include the filamentary structure of the field and the dynamical interaction with convection. This will be discussed in Sec. 3. The more formal criticism of the mean field concept has already been discussed by Stix (1981) and Cowling (1981).

Let us stress a few basic conceptual problems of oscillator models. In Piddington's version, a meridionally oscillating fossil dipole field is conjectured which is thought rooted in the radiative interior of the Sun. However, no sign of a steady (non-reversing) dipole component in the solar magnetic field has been found; on the contrary, polar fields reversals have been observed three times (1957/58, 1969/71, 1980/?) in full conformity with dynamo models (Howard and LaBonte, 1981; Stenflo, 1981). Apparently to avoid this problem, Layzer et al. (1979) talk about an "irregular large-scale field, largely confined to the non-convective core". But it is not clear what they mean: A large-scale field with small scale fluctuations? This would face the same problem as Piddington's, i.e. no permanent large-scale field is observed. Or a "highly tangled field" as the authors write in another section? But this is identical to a field with only very weak large-scale components, i.e. a chaotic field. Such a field would have decayed long ago due to its small typical length scale. However, even if we allow for this field as a source for toroidal fields produced through differential rotation, it is not clear how the large amount of order in the solar cycle (e.g. coherent polarity rules for both hemispheres) is possible.

Furthermore, for equipartition fields ($B \sim 10^4$ Gauß, a reasonable flux density for the deep parts of the solar convection zone, cf. Schüssler, 1977) the velocity of convective flows is comparable to the velocities associated with the oscillation (being essentially a standing Alfvén wave). Consequently, the influence of convection on the oscillation cannot be neglected, i.e. eddy viscosity has to be taken into account. Compare the Lorentz force as driving force for the oscillations with the viscous force (to order of magnitude) at the base of the convection zone:

$$\gamma \equiv \frac{F_{Lor}}{F_{visc}} \sim \frac{\dfrac{B_{pol} \cdot B_{tor}}{4\pi\rho d}}{\nu_t \cdot \dfrac{v_A}{d^2}} = \frac{B_{pol} \cdot B_{tor} \cdot d}{4\pi\rho \cdot \nu_t \cdot v_A} \qquad (2.1)$$

with poloidal field B_{pol}, toroidal field B_{tor}, density ρ, layer thickness d, Alfvén velocity v_A and eddy viscosity ν_t. For equipartition field strength, i.e. $B_{tor} \sim 10^4$ Gauß, $v_A = \frac{B_{tor}}{(4\pi\rho)^{1/2}}$ is equal to the convective velocity $v_c \sim 60$ m/s and, to order of magnitude, $\nu_t \sim v_c \cdot d$. Consequently, we get from (2.1):

$$\gamma \sim \frac{B_{pol}}{B_{tor}} << 1 \quad , \qquad (2.2)$$

since the poloidal field has to be very small compared to the toroidal field in a torsional oscillation, i.e. $\gamma \sim 10^{-2} \ldots 10^{-4}$. The viscous force due to convection dominates the magnetic force by a large factor and it is difficult to see how the torsional oscillation can be maintained.

On the other hand, there is no strong asymmetry between odd and even cycles; the energy dissipated during a cycle must be very small, because differential rotation as energy source (e.g. a tendency for $\Omega \sim r^{-2}$ due to convection) can only replenish the energy during, say, odd cycles which leads to asymmetry. But if the dissipation and the rate of energy replenishment is small, how can we reconcile this with the strong irregular fluctuations in the intensity of the cycles?

Thus, it seems not easy for the contemporary state of oscillator models to disprove the final statement of Plumpton and Ferraro (1955) after discussing torsional oscillations: "... a theory of sunspots based on torsional oscillations alone is likely to raise more difficulties than it resolves".

Let us now turn to the success and the problems of dynamo models for the solar cycle. Using the mean-field-approach, nearly all dynamo models lead formally to the same equation for the <u>mean</u> magnetic field $\langle \underline{B} \rangle$:

$$\frac{\partial \langle \underline{B} \rangle}{\partial t} = \underline{\nabla} \times (\langle \underline{u} \rangle \times \langle \underline{B} \rangle + \alpha(\underline{r})\langle \underline{B} \rangle) - \underline{\nabla} \times (\eta_t(\underline{r})\, \underline{\nabla} \times \langle \underline{B} \rangle) \qquad (2.3)$$

In the kinematical case, the fuctions $\alpha(\underline{r})$, $\eta_t(\underline{r})$ depend on the statistical properties of the fluctuating velocity field. η_t is the turbulent magnetic diffusivity, while α leads to a current parallel to the mean field (in the simplest case; generally, α and η_t are tensors).

Dynamo action is possible either through the combined effect of differential rotation (the $\langle \underline{u} \rangle \times \langle \underline{B} \rangle$-term) and the mean electromotive force of a random velocity field under the influence of rotation (the $\alpha\langle \underline{B} \rangle$-term) or through the α-term alone. The first kind of models ($\alpha\omega$-dynamos) with dominating influence of differential rotation predominantly yield oscillatory solutions, which can be compared with the observations of the solar cycle (Stix, 1976, 1978).

The following list shows observational results which have been reproduced by models based on Eq. (2.3) essentially:

- polarity rules of sunspot groups
- migration of the activity belt towards the equator ("butterfly diagram")
- polarity reversals of polar fields
- phase relationship between poloidal and toroidal fields consistent with direction of migration of the activity belt (Stix, 1976) and angular velocity increasing with depth as is indicated by observations of rotational splitting of 5-min-oscillations (Deubner et al., 1979; Claverie et al., 1981)
- rigid rotation of magnetic sector structure and coronal holes (Stix, 1974; 1977)

- zonal velocities associated with the activity belts observed by Howard and LaBonte (1980a): Since this flow has an 11-year-periodicity it is difficult to interpret as evidence for a torsional oscillation producing solar activity for which we would expect a 22-year period. It can be explained quite naturally by a wave of the (quadratic!) Lorentz force associated with a dynamo wave (Schüssler, 1981; Yoshimura, 1981; see LaBonte and Howard, 1982, for a diverging point of view).

Although the above list indicates that at least Eq. (2.3) describes the physics of the solar cycle to a great extent, we must not ignore the problems and weak points of the theory.

- Although the concept of positive turbulent diffusion of the mean field seems reasonable for the Sun (Kraichnan, 1976; Parker, 1979a, p. 584 ff.) so far there are no reliable means of calculating η_t (and α!) for the solar convection zone (Knobloch, 1978) because the First-Order-Smoothing-Approximation (FOSA) is hardly applicable. Essentially, this is due to our incomplete understanding of turbulent convection, in particular of MHD turbulence of high electrical conductivity (Cowling, 1981; Rädler, 1982).

- In view of the intermittent structure of photospheric magnetic fields, a kinematical approach which ignores the dynamical effects of concentrated magnetic fields can be questioned. The dynamical influence of convection which probably leads to the intermittent structures is not included (see Ch. 3).

- The cyclic variation of magnetic structures like the appearance of ephemeral active regions/X-ray bright points (Martin and Harvey, 1979; Golub et al., 1979) or sunspot intensity (Albregtsen and Maltby, 1981) cannot be described by the evolution of the mean field but may, nevertheless, contain important information on processes related to the cycle.

There is still the possibility that the success of dynamo models is possible just through the great number of free parameters or even free functions which can be tuned to give agreement with observations. It is very important to reduce this freedom for models by

- further detailed observation of solar surface magnetic and velocity structures

- reliable theory of the differential rotation $\Omega(r,\theta)$. Existing models depend very sensitively on boundary conditions (what about a fast rotating core?) and unknown parameters like the Prandtl-number (e.g. Schmidt, 1982) as well as on our insufficient understanding of the interaction between rotation and convection (Durney and Spruit, 1979; see also P. Gilman's review in these proceedings and Durney, 1981).

Possible alternatives to the classical type of $\alpha\omega$-dynamos are discussed in the next chapter.

3. DYNAMO ACTION AND CONCENTRATED FIELDS

As we have seen above, the $\alpha\omega$-models can describe the solar cycle rather well by parametrizing all effects of (turbulent) convection through the functions $\alpha(\underline{r})$ and $\eta_t(\underline{r})$ which appear in the equation for the mean field. The spatial two-scale approach (large scale: mean field and differential rotation, small scale: turbulent convection) loses its sense if large-scale convection ("giant cells") exists. Although observational confirmation is still lacking (Howard and LaBonte, 1980b), many numerical and analytical studies (e.g. Graham, 1975; Gilman, 1979; Gilman and Glatzmaier, 1981; Depassier and Spiegel, 1981) point clearly towards the existence of these structures. Such cells would be most efficient for field regeneration because they can be maximally influenced by rotation due to their long turn-over time ($\sim$ 1 month). Mean field dynamo equations may nevertheless be useful if time averages are invoked (Stix, 1976) because there is a big discrepancy between the time scale of the cycle (viz. 11 years) and that of giant cells ($\sim$ 1 month). Another possibility is averaging over longitude which leads to an axisymmetric mean field (e.g. Braginskij, 1964). Alternatively, large-scale convection can appear as a mean flow in spatially averaged dynamo equations while the traditional α-term arising from smaller scales can be neglected (but not η_t!).

Gilman and Miller (1981) have done such a calculation in the framework of a numerical study of Boussinesq convection in a spherical shell, in which also the differential rotation (the other ingredient of $\alpha\omega$-dynamos) is derived consistently through the action of the Reynold stresses exerted by global convection. The (not unexpected) result was that α-effect and ω-effect were of same order of magnitude, contrary to the assumptions for $\alpha\omega$-dynamos. The reason is that for convective eddies living about one turn-over time the differential rotation as well as the α-effect produced by them is of the order of the convective velocity itself. Estimating α from mixing length theory (e.g. Krause, 1967) already yielded values much too big for succesful simulation of the solar cycle with $\alpha\omega$-dynamos (Köhler, 1973). Consequently, as discussed by Gilman and Miller, they did not find the results of $\alpha\omega$-models but dynamos with randomly fluctuating fields, no oscillatory solutions. Recently, Gilman (1982) reports, that new calculations with reduced eddy viscosity give oscillating solutions near the threshold for dynamo action (weakly excited dynamo), but with toroidal fields migrating poleward, too small period and no clear dominance of solutions antisymmetric with respect to the equatorial plane (dipole parity, as suggested by the polarity rules). It is not clear whether a calculation for a compressible medium can remove these difficulties as expected by Gilman.

In some way, the influence of convection on the field has to be reduced, because the convective induction effect has to be smaller and the large-scale order in the solar cycle as expressed by the sunspot polarity rules is evidence against a dominating role of convection. How can the field escape? One possibility is that the fields are strong, at least of equipartition field strength, to be able to resist deformation by convective flows. For the photosphere of the Sun, this is indeed the case: The field is organized in strong flux tubes from big sunspots to the small facular points.

This "differentiation" in a magnetic and a convecting phase can be understood in terms of the high electrical conductivity of stellar plasmas which leads to an exclusion of magnetic fields from regions of closed streamlines (Galloway et al., 1978; Galloway and Weiss, 1981; Weiss, 1981a, b). These extensive studies of the nonlinear interaction of magnetic fields and cellular convection support the conjecture that the field throughout the convective zone of the Sun is organized in strong (at least equipartition field strength) flux tubes. Within and near such a tube convection is efficiently throttled leading to a decrease of the efficiency of field regeneration. Childress (1979) investigated an idealized case and showed that the α-effect may be reduced by a factor $R_m^{1/2}$, where R_m is the magnetic Reynolds number.

There are two paths on which the investigation of dynamos including the effects of concentrated fields proceeds: The "overshoot layer dynamo" which is throught to operate at the interface between the convective zone and the radiative interior and the "flux tube dynamo" which assumes the whole convective zone to be permeated by strong flux tubes. Those tubes are influenced by convection, rotation (and possibly meridional circulation) and lead to dynamo action through the action of Coriolis forces (the classical Parker loop) and reconnection plus dissipation. We shall discuss the two approaches in the following conceding that also "mixed models" are possible.

Dynamo Action in the Overshoot Layer

Assuming an organization of magnetic fields in equipartition flux tubes in order to resist convective deformation leads to a new problem: Strong flux tubes are buoyant and, therefore, are difficult to store within the convection zone for times comparable to the solar cycle (Parker, 1975). High turbulent viscosity reduces the rising velocity drastically (Unno and Ribes, 1976; Schüssler, 1977, 1979a; Kuznetsov and Syrovatskii, 1979) but on the other hand leads to strong coupling between flux tubes and convective down- or updrafts which may raise the flux to the solar surface within the convective timescale, viz. one month for global convection ("giant cells"). The only region where magnetic flux tubes could possibly be stored for times comparable to the solar cycle seems to be the lower boundary of the convective zone. In fact, a couple of mechanisms have been proposed which lead to an accumulation of flux there:

a) Topological pumping in a flow pattern which consists of a (multiple) connected network of downdrafts and isolated updrafts like the solar granulation (Drobyshevski and Yuferev, 1974; Drobyshevski et al., 1980)

b) Asymmetry between upflow and downflow velocities in an overshoot layer (van Ballegoijen, 1982; see discussion below)

c) The flux ejection dynamo discussed by Parker (1982a) produces magnetic flux at the bottom of a layer of convection cells subject to strong cyclonic rotation.

d) Van Ballegoijen (1982) proposes that toroidal flux tubes are transported equatorward by a meridional circulation. Conservation of angular momentum leads to a flow along the tubes opposite to the direction of rotation. The Coriolis force resulting from this flow is directed inwards and a flow velocity of a few m/s could stabilize the tube against buoyancy.

e) Trapping of flux tubes in large regions of closed streamlines (Parker, 1982e). A meridional circulation could trap toroidal flux in the deep convection zone.

f) A gradient in turbulent velocity (spatial dependence of turbulent diffusivity, "diamagnetism", see Zel'dovich, 1956; Spitzer, 1957; Rädler, 1968) is able to transport flux to the bottom of the convection zone (Ruzmaikin and Vainshtein, 1978). This effect is also evident from an $\alpha\omega$-dynamo calculation performed by Roberts and Stix (1972; see their Fig. 2).

g) Aerodynamic lift (Parker, 1979b) due to rotation with angular velocity increasing with depth is directed downward.

h) Convective instability of horizontal flux tubes leads to sinking of at least part of the tube (Spruit and van Ballegoijen, 1982).

i) Stable stratification in an overshoot layer below the formal boundary of the convection zone reduces the buoyancy force.

We see, many effects - which are all described by the MHD equations - lead to a downward transport of flux.

How may the field near the boundary of the convection zone look like? The downward forces compress the field and presumably lead to a more or less homogeneous magnetic layer which nevertheless may retain some flux tube structure imposed by convection. The field must have at least equipartition field strength B_e to resist severe deformation by convective flows. For the deep convection zone of the Sun, $B_e \sim 10^4$ Gauß. A value of $\phi_t \sim 10^{24}$ mx seems reasonable for the total toroidal flux produced during one half-cycle; consequently the hypothetical deep flux system located at the lower boundary of the convective zone consists of a magnetic layer with a minimum thickness between $d \sim 3000$ km and $d \sim 20\,000$ km depending on whether the whole latitude range or only a belt within ± 30° from the equator (where active regions appear) is taken (Schmitt and Rosner, 1982). Such a layer with equipartition field would interfere strongly with convective energy transport and we are led to the conclusion that it cannot be situated in a region where the main part of energy transport is by convection, because significant changes of luminosity would have to be expected during the cycle. A favourable site, however, is the overshoot layer below the "active" convection zone where energy is transported inward by convection and outward by radiation (the net flux being outward, of course; see Roxburgh, 1978; Cushman-Roisin, 1982). This has been proposed by Spiegel and Weiss (1980), van Ballegoijen (1982) as well as Schmitt and Rosner (1982). Earlier, somewhat related ideas have been put forward by Gokhale (1977). Van Ballegoijen (1982) estimates the depth of the overshoot zone and finds a few tenths of a scale height which is comparable with the depth d of the magnetic layer derived above (d $\sim$ 5 % - 30 % of scale height $\Lambda_e \sim 6 \cdot 10^9$ cm for the Sun). He also argues that downward flow is faster than upward flow in the overshoot layer because mechanical energy is converted into thermal energy (convective energy flux against temperature gradient). This effect leads to a net downward drag force: Take, for simplicity, a cylindrical flux tube with radius a. A flow with velocity v_c and density ρ_e leads to a drag force F_D per unit length of the tube (see e.g. Batchelor, 1967, p. 245 f.):

$$F_D = \rho_e v_c^2 \, a \, C_D \qquad (3.1)$$

with drag coefficient $C_D \sim 1$. Take a system of convective rolls with downflow velocity v_d and upflow velocity v_u, whose axis are perpendicular to that of the tube. Include vanishing net mass flux and compare the net downward (for $v_d > v_u$) drag force with the buoyancy force (assuming thermal equilibrium of the tube with the surrounding gas)

$$F_B = \frac{B^2 a^2}{8\Lambda} \ell$$

where ℓ is a length along the tube which contains an equal number of updrafts and downdrafts. A short calculation gives the condition for equilibrium:

$$\gamma \equiv \frac{1 - v_u/v_d}{1 + v_u/v_d} = \left(\frac{B}{B_e}\right)^2 \cdot \left(\frac{\pi a}{2\Lambda_e C_D}\right) \sim \left(\frac{B}{B_e}\right)^2 \cdot \left(\frac{a}{\Lambda_e}\right) \sim 0.1 \qquad (3.3)$$

for $B^2 \sim B_e^2 = 4u\rho_e v_d^2$ and a $\sim 10^9$ cm, a flux of $3 \cdot 10^{22}$ mx corresponding to an active region. From (3.3) follows $v_u/v_d \sim 0.8$, a value which seems not unreasonable for compressible convection (Graham, 1977). Because we have not taken into account the other downward forces listed above and the stable stratification we may safely conclude that fields of equipartition field strength can be stored in the overshoot layer. There they are amplified by differential rotation until buoyancy dominates or instabilities set in and flux tubes leave the region predominantly toroidal (Schüssler, 1980a) but possibly suffer from instabilities (Acheson and Gibbons, 1978) and dynamical fragmentation (Schüssler, 1979a; Tsinganos, 1980).

A caveat seems in order here: We have made our estimates so far using results (e.g. convective velocities) taken from mixing length models (Spruit, 1977). These velocities may be wrong and differ significantly from numerical simulations which take compressibility into account (Graham and Moore, 1978; Massaguer and Zahn, 1980).

Although we would expect an organization of the magnetic field in the overshoot layer in the form of individual flux tubes due to the interaction with convection, it has also been proposed that the field is more homogeneous ("diffuse") in the dynamo layer (J. Schmitt and Rosner, 1982), because only small-scale convection is thought to exist in the overshoot layer. These are favourable conditions for conventional dynamo theory, because First-Order-Smoothing would be applicable: Field fluctuations are small compared to the mean field. However, a kinematical theory is hardly applicable: Fields of equipartition field strength (or more) are likely to have a strong influence on the convective flow. Boundary layer methods as applied by Childress (1979) seem more appropriate.

A non-filamentary field can find a magnetohydrostatic equilibrium by changing the stratification and thus overcome the buoyancy problem. J. Schmitt and Rosner invoke double-diffusive instabilities to trigger the flux loss from the dynamo layer that leads to the formation of flux tubes and the surface appearance of active regions. Furthermore, slow magnetostrophic waves can be excited by unstable field configurations (Acheson and Gibbons, 1978). Those waves exhibit net helicity and can be important for the regeneration of the poloidal field (D. Schmitt, private communication).

Assuming a deep-seated solar dynamo and having evidence for a fast rotating solar core (Claverie et al., 1981), it is tempting to speculate that the differential rotation in the boundary layer between fast rotating core and slowly rotating convection zone is the main driving force for the dynamo. However, it is easily estimated that the rotational energy of a solar core rotating at ten times the surface value would be exhausted after only $\sim 10^7$ cycles, i.e. 10^8 years, a short episode compared to the main sequence lifetime of the Sun (from an 11- and 22-year cyclical variation of glacial varves in Australia, Williams (1981) concluded that the solar cycle has been in operation at least for the last $7 \cdot 10^8$ years). Furthermore, the accelerating torque exerted on the convection zone by the magnetic field would exceed the solar wind torque $D_S \sim 3 \cdot 10^{29}$ dyn cm/sterad (Pizzo et al., 1982) by a large factor which leads to a spin up of the envelope contrary to observation (Skumanich, 1972). We conclude that even a deep-seated solar/stellar dynamo must be driven by processes in the convective (and overshoot) region itself. Differential rotation and field regeneration (α-effect) have to be produced in situ by convection and rotation. Any influence of a possible "fossil" field located in the radiative interior on the generation of the solar cycle must be extremely small. Levy and Boyer (1982) come to a similar conclusion observing that a fossil field would induce a strong asymmetry between the two halves of the magnetic cycle which is not observed.

Flux Tube Dynamo

Assume an ensemble of strong (i.e. at least equipartition flux density) magnetic flux tubes permeating a non-magnetic convecting medium. Is such a system able to sustain a dynamo? In principle, this is a much more difficult question than the classical kinematic dynamo problem because of the possibility of a strong interaction between flux tubes and the convective pattern: While magnetic fields are excluded from regions of closed streamlines (Galloway, Proctor and Weiss, 1978, and references therein), existing flux concentrations can shape the convective pattern and possibly stabilize large cells like supergranules on the Sun (Parker, 1982b).

Two extreme pictures for the state of the flux tube/convection system are possible which lead to different consequences:

a) The "fibril state" with many ($> 10^6$) tubes which carry fluxes of $10^{17} \ldots 10^{18}$ mx and have diameters of a few hundred kilometers, similar to the observed photospheric flux concentrations.

b) The "flux rope state" with few ($\sim 10^2$) tubes with a flux of $\sim 10^{22}$ mx, equipartition field strength (or more) and diameters of some thousand kilometers.

While fibrils refer to the observed small flux concentrations, flux ropes contain the flux of a whole active region. Depending on which flux system we think is responsible for the dynamo, different physical processes are involved.

Due to their small diameter, fibrils are dynamically dominated by convective cells with scale Λ_e from a depth of 3000 km on down through the whole convection zone (Parker, 1982d). Consequently, a kinematic mean-field theory is adequate by averaging over a length scale λ_1 with $d \ll \lambda_1 \ll L$ where $L \sim 2 \cdot 10^{10}$ cm is the depth of the convection zone and $d \sim 10^7 \ldots 10^8$ cm is the fibril scale. The small-scale ($l \leq d$) interaction between flux tubes and convection is implied in such a model by assuming a differentiation between tubes and field-free convecting fluid (a related point of view using a three-scale analysis has been proposed by Weiss, 1981c). Defining $\underline{v}$ and $\underline{B}$ as average over λ_1 we obtain the mean momentum equation for a set of locally nearly aligned flux tubes (Parker, 1982c):

$$\frac{\partial v_i}{\partial t} + v_j \frac{\partial v_i}{\partial x_j} = \frac{1}{\rho'} \frac{\partial}{\partial x_j} \left[-\delta_{ij} p + m \frac{B_i B_j}{4\pi}\right] + \text{other forces} \qquad (3.4)$$

where p is the total pressure, ρ' a local mean fluid density (reduced within the tubes) and $m = b/B$ the ratio of the flux density b of the tubes to the mean field, the compression factor. Magnetic curvature forces are enhanced by the factor m as compared to a "diffuse" field of strength B. This is the essential difference to the simulation of mean field equations by Gilman and Miller (1981): The field is much less distorted than would be expected from a diffuse field; consequently, induction mechanisms are less efficient and the field cannot be tangled by the velocity until equipartition of the <u>total</u> kinetic and magnetic energy.

The tubes are bound to the nonmagnetic medium by means of the drag force, not by the high electrical conductivity as would be the case for a continuum field. Due to the curvature and buoyancy forces, the tube can slip through the medium with a velocity $\underline{u}$ derived from the equilibrium between curvature force $\underline{F}_C$, buoyancy force $\underline{F}_B$ and drag force $\underline{F}_D$:

$$\underline{F}_D(\underline{u}) + \underline{F}_B + \underline{F}_C = 0 \tag{3.5}$$

Because the fibrils are assumed to be locally aligned, this velocity has to be added to the mean fluid velocity $\underline{v}$:

$$\frac{\partial \underline{B}}{\partial t} = \underline{\nabla} \times [(\underline{u}+\underline{v}) \times \underline{B} + \underline{E}] + \eta_t \underline{\nabla}^2 \underline{B} \tag{3.6}$$

where η_t is the turbulent diffusivity. $\underline{E}$ stands for an electromotive force possibly arising from interaction of the fluctuating components of magnetic field and velocities. Such effects can be

a) The classical α-effect ($\underline{E} = \alpha \underline{B}$) which, however, may be drastically reduced or even absent for strong flux tubes because motions can be suppressed in the tube and the interaction restricted to a small boundary layer.

b) Helical waves driven by magnetic buoyancy in a rotating medium. Instability leads to slow magnetostrophic waves which propagate only in one direction with respect to the rotational velocity. Consequently, net helicity can be produced (Gibbons, 1977). D. Schmitt (1982, private communication) shows that this effect leads to an effective $\alpha \sim 10$ cm s^{-1} for parameters relevant for the lower solar convection zone.

Parker (1982c) shows that the slip velocity $\underline{u}$ can be neglected against $\underline{v}$ for fibrils if

$$\frac{a}{\Lambda_e} , \quad \frac{a}{L} << \frac{v^2}{v_A^2}$$

where v_A is the Alfvén velocity corresponding to the fibril field. Consequently, for sufficiently slender fibrils (3.6) takes the form of the "classical" kinematic mean field equation (2.3) by setting $\underline{u} = \underline{0}$.

As mentioned above, if giant cells exist, such a treatment only makes sense for temporal or longitudinal averaging, because only then these can be treated as part of the fluctuating velocities. In that case, α-effect and turbulent diffusion η_t result mainly from the giant cells but any calculation of these quantities has to take into account the fibril state of the field which leads to Eq. (3.4). First order smoothing may not be adequate then for determination of α and η_t. In particular, η_t may not necessarily be identical to the turbulent diffusivity of a scalar quantity. However, the global properties of a fibril dynamo may well be described by the mean field equation (3.6).

In view of the complications arising from flux tube dynamics when no averaging is performed, it is not surprising that no detailed fibril dynamo model has been presented in the literature so far. Only a few very idealized pictures have been proposed (e.g. Adams, 1977; Giovanelli, 1982).

The physics of "flux rope models" on the other hand, is quite different. Guided from the typical flux of active regions, the field is thought to be organized in a few ($\sim 10^2$) big flux tubes with fields of equipartition field strength ($\sim 10^4$ Gauß for the deep convection zone) and a flux $\phi \sim 10^{22}$ mx which leads to a diameter of about 10^9 cm. We have shown above that because of their large radius such tubes can avoid severe distortion from convective motions in the deep convective zone and can be maintained for a significant part of the solar cycle period in the overshoot layer below the solar convection zone. Following Babcock's (1961) original suggestion, Schmidt (1968) and Piddington (1976) have conjectured the existence of big flux tubes or "ropes", while Galloway and Weiss (1981), Spiegel and Weiss (1980), Golub et al. (1981) and Schüssler (1980b) have discussed some consequences of such a model for the theory of the activity cycle. The cyclic variation of sunspot brightness (Albregtsen and Maltby, 1981) and the related variation of appearance of X-ray bright points and ephemeral active regions (Maltby and Albregtsen, 1979; Golub et al., 1979) can be understood in terms of the age of the basic flux ropes. The flux rope models seem to lead necessarily to the idea of a dynamo operating in the overshoot layer which we have discussed above. The remarks made there are relevant for this model, too.

4. NONLINEAR DYNAMO THEORY AND APPLICATION TO STARS

Stimulated by the pioneering work of O.C. Wilson (1978) the investigation of stellar activity has become one of the most rapidly evolving fields in observational astrophysics (see reviews by Noyes, 1981, 1982; Zwaan, 1981, Dupree, 1981; Vaughan, this volume). The possibility to observe a large number of stars which differ in spectral type, evolutionary status, multiplicity, rotation rate, etc. is a big advantage for theory. Predictions from models can be compared with different stars and conclusions are not drawn solely from the solar case with its (possible) peculiarities. However, for predictions to be made, the models must have been worked out to a fair degree of sophistication, an important point for which is the inclusion of nonlinear effects (from the Lorentz force) which leads to a finite amplitude of dynamo action.

Nonlinear Dynamos

Several possibilities have been discussed in order to introduce nonlinearities into the dynamo formalism (all arising from the Lorentz force in the induction equation):

a) The cut-off-α-effect: The growing field $\underline{B}$ exhibits a stronger resistance against deformation by small-scale motions. The α-effect which results from this interaction therefore decreases as $|\underline{B}|$ increases. Models incorporating this kind of nonlinearity have been presented by Stix (1972), Rüdiger (1973), Jepps (1975), Ivanova and Ruzmaikin (1977), Yoshimura (1978), Kleeorin and Ruzmaikin (1981). The disadvantage of this procedure is the essentially arbitrary device of $\alpha = \alpha(|\underline{B}|)$.

b) Flows driven by the mean Lorentz-force: Inclusion of the force arising from the mean field $\underline{B}$, viz. $(\underline{\nabla} \times \underline{B}) \times \underline{B}/4\pi$ into the momentum equation for the mean velocity $\underline{v}$ leads to a flow which, according to Lenz's rule, limits the growth of the mean field. This approach has been pursued by Malkus and Proctor (1975), Proctor (1977), Hellmich (1978) for incompressible flow and Schüssler (1979b) for a compressible medium. A special case where only the magnetically induced change in differential rotation is considered has been given by Hinata (1982). These models suffer from the possibility that the mean Lorentz force from the fluctuating components of $\underline{B}$ (which is neglected) may possibly be much larger than the force arising from the mean field.

c) Magnetic buoyancy and convective transport: The amplitude of a dynamo working in a layer (as discussed in Sec. 3) can be limited by removal of magnetic flux out of the layer through the joint action of buoyancy and convection. Quantitative models are difficult to investigate because they involve uncertain details as diameter and flux density of magnetic flux tubes, turbulent viscosity and the dynamical interaction of flux tubes and convection. However, crude parametrizations have been tried beginning with Leighton's (1969) model, followed later by Köhler (1973), Yoshimura (1975), Schüssler (1980b), Durney and Robinson (1982) as well as Robinson and Durney (1982).

d) Hydromagnetic Dynamos: Models of this kind investigate simultaneously the generation of the velocity field (e.g. thermal convection driven by an unstable temperature gradient) and of the magnetic field (through dynamo action of the velocity field) allowing for nonlinearity through the Lorentz force in the momentum equation. Although suffering from not including possible small-scale effects, hydromagnetic dynamos have the advantage of treating generation <u>and</u> limitation of the field self-consistently. Childress and Soward (1972), Soward (1974) and Busse (1975) used perturbation methods (see also review by Soward, 1979) while full non-linear numerical solutions have been given by Baker (1978) for cartesian geometry as well as Gilman and Miller (1981) for a spherical shell, both employing truncated expansions.

It is well known that a non-linear system of equations can show a behaviour drastically different from the linear case. Beginning with early studies of aperiodic behaviour of a system consisting of two disc-dynamos (Allan, 1962), non-linear model equations for nonlinear dynamos have been investigated by several authors (Robbins, 1977; Jones, 1981; Ruzmaikin, 1981). Bräuer (1979, 1980) found subcritical solutions in an analytical study of a simplified nonlinear $\alpha\omega$-dynamo. Yoshimura (1978, 1979, 1980) introduced a time delay in the back-reaction term and received aperiodic behaviour which he used for an interpretation of the long-term variability of solar activity (Maunder minimum, "long cycle") and geomagnetic reversals. Fully developed MHD turbulence, inverse cascades and dynamo action have been investigated by Meneguzzi et al.(1981).

Observational Results

The main conclusions drawn from observations of active stars so far are the following:

a) The X-ray flux and the Ca^+-emission of late-type stars (taken as a measure of magnetic activity by analogy to the Sun) increase with the rotation rate Ω for stars of the same spectral type (Walter and Bowyer, 1981; Walter, 1981; Vaughan et al., 1981; Middelkoop, 1981).

b) Ca^+-Emission generally decreases with increasing age of the stars (Skumanich, 1972). In particular, there are indications for two distinct groups of active stars: Young, rapidly rotating stars with a high level of emission and strong temporal fluctuations but no cyclical activity; on the other hand, old, slowly rotating stars with a lower level of emission and solar-type cycles (Vaughan and Preston, 1980; Vaughan, 1980). The dependence of L_x/L_{bol} (X-ray luminosity divided by bolometric luminosity) on angular velocity is different for both groups (Walter, 1982). Solar-type cycle periods seem to be independent of Ω (Vaughan et al., 1981).

c) X-ray activity (L_x/L_{bol}) increases for stars with lower mass and deeper outer convection zones (Linsky, 1981). The high activity of close binaries (e.g. RS CVn systems) is attributed to synchronous rotation leading to high values for Ω, not to multiplicity itself (Walter and Bowyer, 1981). However, a variant of the Herzenberg (1958) two-sphere dynamo has been proposed by Dolginov and Urpin (1979) which takes into account the mutual induction between the members of close binaries.

Theoretical Interpretation

Comparison with models of magnetic field generation until now are all made in the framework of $\alpha\omega$-dynamos. Since both the α-effect (due to the Coriolis force $F_c \sim v_c \cdot \Omega$) and the differential rotation increase with Ω (Belvedere et al., 1980a) the dynamo number

$$N_D = \left(\frac{\alpha\Delta\Omega}{L}\right)^{1/2} \cdot \frac{L^2}{\eta_t} \tag{4.1}$$

($\Delta\Omega$: angular velocity difference within dynamo region of scale L) which is a measure for the excitation of the dynamo increases with Ω for otherwise identical stars. The ultimate level of magnetic energy increases with N_D in all nonlinear dynamo calculations presented so far and we expect an increase of stellar activity (as measured in Ca^+-emission and X-ray fluxes) with increasing rotation rate. This expectation is in full accord with the measurements summarized above. A more quantitative prediction of the expected level of activity is far more difficult because of all the uncertainties in our understanding of dynamo processes discussed in Chapters 2 and 3. All four quantities appearing in Eq. (4.1) are very uncertain and their detailed dependence on Ω even more. In spite of these difficulties Durney and Robinson (1982) and Durney et al. (1981) have tried to predict quantitatively the stellar activity level using simple estimates of α, $\Delta\Omega$ etc. guided by mixing length theory. Although the results are in qualitative agreement with observations, because of the many assumptions involved it is not clear how much weight should be attached to them. The same remark applies to the interpretation of X-ray data in terms of $\alpha\omega$-dynamos by Belvedere et al. (1981, 1982).

While the increasing activity for increasing Ω is straightforward to interpret, the growth of X-ray emission towards later spectral types calls for a more involved explanation. Several proposals have been put forward:

a) The convection zone of later stars extend into regions of higher temperature, scale height and density. Mixing length theory then predicts an increasing influence of rotation on convection expressed by decreasing Rossby number

$$Ro = \frac{v_c}{2\Lambda \cdot \Omega} \tag{4.2}$$

even if Ω decreases (Durney and Latour, 1978). This may lead to a larger α-effect for later stars. Preferred nonaxisymmetric modes ("starspots") for rapid rotation possibly can be understood by growing anisotropy of the α-tensor (Rüdiger, 1978, 1980). Furthermore, since the buoyant rise of flux tubes takes place at some fraction of the Alfvén velocity $v_A = B \cdot (4\pi\rho)^{-1/2}$ the higher density and scale height at the base of the stellar convection zone leads to a decreasing efficiency of buoyancy limitation of the field amplitude (Belvedere et al., 1980b).

b) In a model for differential rotation which invokes latitude-dependent heat transport, the differential rotation first decreases sharply from spectral type F5 to G5 and then increases from G5 to M0 (Belvedere et al., 1980a). Consequently, the induction by differential rotation (ω-effect) increases for stars later than G5.

c) The surface gas pressure Pg of the stars grows with decreasing effective temperature. Consequently, the field strength of flux tubes with $\beta = 8\pi Pg/B^2 \sim 1$ increases and - for the same total magnetic flux - the magnetic energy is a good deal larger (a factor $\sim$ 20 between G0 and M5, cf. Durney and Robinson, 1982, Table I). Magnetic heating of coronae by waves, currents or other mechanisms always strongly depends on the photospheric magnetic structure, i.e. flux tube geometry and peak field strength. For example, if the X-ray flux varies like B^2, the intensification of flux tubes due to stronger surface pressure alone could explain the increasing X-ray flux for later stars. The Ca^+ emission which at least for the Sun (Skumanich et al., 1975) is correlated with the total magnetic flux, seems to show a weaker dependence on the flux tube structure than the transition region (and coronal) emissions (Oranje et al., 1982). This may lead to the decrease of Ca^+-surface flux towards later stars while the X-ray flux increases. We should be very careful when interpreting the data in terms of dynamo efficiency while not knowing the surface structure of the fields and the implication for chromospheric and coronal heating.

There are different possibilities for the interpretation of the Vaughan-Preston gap in a plot of Ca^+ emission index S against (B-V), i.e. the bimodal distribution of active stars into old, slowly rotating stars with solar-type cycles (class C) and young, rapidly rotating stars with stronger but irregularly fluctuating activity (class I) (see also discussion in Durney et al., 1981):

a) Two different dynamo mechanisms: one operating for rapid rotation, one for slow rotation. No attempt has been made so far to investigate this proposal in detail.

b) A rapid change of rotation rate Ω: A star of class I is presumed to have a nearly closed corona and no efficient magnetic braking can take place. But as Ω decreases slowly, activity gets weaker and coronal holes can open. Now magnetic braking through high-speed stellar winds is far more effective, Ω decreases more rapidly, activity gets even weaker: a positive feedback. However, no such episode of rapid change of Ω with time is apparent from the observations (Skumanich, 1972).

c) As Ω decreases, the dynamo excitation decreases too and the magnetic field configuration may change from a state with many higher modes to a state where only the fundamental mode of the dynamo appears. Nonlinear interaction of the higher modes leads to the irregular fluctuations for class I. However, even the single-mode dynamo will operate in the non-linear domain (see the long-term fluctuations of the solar cycle) and it is not clear whether linear results (e.g. critical Reynolds numbers for higher modes) can be used under those circumstances. Subcritical phenomena (Bräuer, 1980) may appear.

d) Model equations for dynamos often show chaotic solutions in certain regions of parameter space (Jones, 1981; Ruzmaikin, 1981). The crude flux tube dynamo of Schüssler (1980b) changes suddenly from periodic to chaotic solutions if the excitation exceeds a certain threshold value (see also Model 2 of Jepps, 1975). The same behaviour appears for too small turbulent diffusivity. High rotation rate Ω favours strong helicity fluctuations which may lead to small (or even negative) turbulent diffusivity (Kraichnan, 1976). This may not necessarily choke the dynamo but lead to more unrelated variations in different parts of the convective zone. Hysteresis effects (see also e) below) may lead to the gap.

e) Knobloch et al. (1981) propose a change in the pattern of convection as Ω is increased associated with a hysteresis effect. This leads to a discontinuous dependence of the dynamo-built magnetic field on Ω. The two patterns of convection are cellular convection for low Ω and cylindrical convection rolls parallel to the rotation axis (as a consequence of the Proudman-Taylor constraint) for high Ω.

Cycle periods have been discussed by Belvedere et al. (1980b) who solve the linear dynamo equations with differential rotation derived from latitude-dependent heat transport and by Robinson and Durney (1982) discussing nonlinear model equations. Marginal dynamo action for the linear case is achieved if the period of the dynamo wave

$$\tau_w \sim \frac{L}{(\alpha\Delta\Omega)^{1/2}} \tag{4.3}$$

is equal to the magnetic diffusion time

$$\tau_d \sim \frac{L^2}{\eta_t} \tag{4.4}$$

$\tau_w \sim \tau_d$ means then that the dynamo number $N_D \sim 1$ (see also Hinata, 1982, for the marginally unstable case). Because L (depth of convection zone) increases towards later stars, τ_d increases and, consequently, Belvedere et al. find that the cycle period grows for later stars. However, this trend is also influenced by their assumption of a constant Prandtl number which leads to a decreasing turbulent diffusivity for stars with decreasing mass. On the other hand, Robinson and Durney with their nonlinear model equations obtain periods decreasing towards later stars (for moderate nonlinearity) and only a weak dependence on spectral type for strong nonlinearity. A moderate dependence of cycle period on excitation has also been found by Jepps (1975) over a large range of parameters.

It seems clear from the discussion above that the evaluation of observations of stellar activity for theoretical purposes is still at its infancy. Instead of fitting parameters for ad-hoc models to get a "reasonable" agreement, observations should be used as a guide for the development of physical concepts. From such a point of view we can ask "questions to nature". One example is the idea of dynamo action in the inner overshoot layer (Chapter 3): If the solar dynamo and those of stars with convective envelopes work on this basis, we should expect a totally different behaviour of very low mass stars which may be fully convective; see, however, Cox et al. (1981). In fact, Liebert et al. (1979) found no indication of chromospheric emission for some very late M dwarfs. Further research along these lines could lead to a better understanding of the physics of magnetic field generation in late type stars.

5. CONCLUSION

Stellar dynamo theory is a rapidly evolving field of research. In my opinion, this is due to the fact that we can draw information from two different reservoirs: The Sun enables us to study the basic nonlinear interaction of turbulent convection and magnetic fields with high spatial and temporal resolution, while active stars show us the dependence of the mechanism on different parameters. Any model has a hard life when confronted with these two sets of observational results; while solar data exclude the simple "on the envelope" ideas, stellar results should keep us from constructing models which are extremely "tuned" to fit the features of the solar cycle. However, the Sun remains the Rosetta stone in the yet unfinished puzzle of stellar activity - it is hard to imagine the state of our understanding of stellar activity without the information drawn from the Sun, but it is quite probable that the idea of a magnetic origin of the phenomena would be a mere speculation among a number of other conjectures and "scenarios".

ACKNOWLEDGEMENT

The author is grateful to Dr. M. Stix for comments on the manuscript and to Mrs. I. Schmidt for her careful typing.

REFERENCES

Acheson, D.J., Gibbons, M.P.: 1978, J. Fluid Mech. 85, 743
Adams, W.M.: 1977, BBSO preprint # 0163
Albregtsen, F. and Maltby, P.: 1981, Sol. Phys. 71, 269
Allan, D.W.: 1962, Proc. Camb. Phil. Soc. 58, 671
Babcock, H.W.: 1961, Astrophys. J. 133, 572
Baker, L.: 1978, Geophys. Astrophys. Fluid Dyn. 9, 257
Batchelor, G.K.: 1967, Fluid Dynamics, Cambridge University Press
Belvedere, G., Paternò, L., and Stix, M.: 1980a, Astron. Astrophys. 88, 240
Belvedere, G., Paternò, L., and Stix, M.: 1980b, Astron. Astrophys. 91, 328
Belvedere, G., Chiuderi, C., and Paternò, L.: 1981, Astron. Astrophys. 96, 369
Belvedere, G., Chiuderi, C., and Paternò, L.: 1982, Astron. Astrophys. 105, 133
Bräuer, H.: 1979, Astron. Nachr. 300, 43
Bräuer, H.: 1980, Astron. Nachr. 301, 203
Braginskij, S.I.: 1964, Sov. Phys. JETP 20, 726
Busse, F.H.: 1975, Geophys. J. R.A.S. 42, 437
Childress, S. and Soward, A.M.: 1972, Phys. Rev. Letters 29, 837
Childress, S.: 1979, Phys. Earth Planet. Inter. 20, 152
Claverie, A., Isaac, G.R., McLeod, C.P., van der Raay, H.B., and Roca Cortes, T.: 1981, Nature 293, 443
Cowling, T.G.: 1981, Ann. Rev. Astron. Astrophys. 19, 115
Cox, A.N., Shaviv, G., and Hodson, S.W.: 1981, Astrophys. J. 245, L37
Cushman-Roisin, B.: 1982, Geophys. Astrophys. Fluid Dyn. 19, 61
Deubner, F.L., Ulrich, R.K., and Rhodes, E.J.: 1979, Astron. Astrophys. 72, 177
Depassier, M.C. and Spiegel, E.A.: 1981, Astron. J. 86, 496
Dolginov, A.Z. and Urpin, V.A.: 1979, Astron. Astrophys. 79, 60
Drobyshevski, E.M., Yuferev, V.S.: 1974, J. Fluid Mech. 65, 33
Drobyshevski, E.M., Kalesnikova, E.N., and Yuferev, V.S.: 1980, J. Fluid Mech. 101, 65
Dupree, A.K.: 1981, Proc. 2nd Cambridge Workshop on Cool Stars, Stellar Systems and the Sun, eds. M.S. Giampapa and L. Golub, Smithsonian Astroph. Obs. Special Report 392, Vol. II, p. 3
Durney, B.R. and Latour, J.: 1978, Geophys. Astrophys. Fluid Dyn. 9,241
Durney, B.R. and Spruit, H.C.: 1979, Astrophys. J. 234, 1067
Durney, B.R.: 1981, in R. Dunn (ed.), Proceedings of the Symposium on Solar Instrumentation, Sacramento Peak Observatory
Durney, B.R., Mihalas, D., and Robinson, R.D.: 1981, Publ. Astr. Soc. Pac. 93, 537
Durney, B.R. and Robinson, R.D.: 1982, Astrophys. J. 253, 290
Galloway, D.J., Proctor, M.R.E., and Weiss, N.O.: 1978, J. Fluid Mech. 87, 243
Galloway, D.J. and Weiss, N.O.: 1981, Astrophys. J. 243, 945
Gibbons, M.P.: 1977, PhD dissertation, University of Oxford
Gilman, P.A.: 1979, Astrophys. J. 231, 284
Gilman, P.A.: 1981, in "The Sun as a Star", ed. S.D. Jordan CNRS/NASA SP-450, Paris/Washington, p. 231

Gilman, P.A.: 1982, Proc. 2nd Cambridge Workshop on Cool Stars, Stellar Systems and the Sun, eds. M.S. Giampapa and L. Golub, Smithsonian Astroph. Obs. Special Report 392, Vol. I, p. 165
Gilman, P.A. and Glatzmaier, G.A.: 1981, Astrophys. J. Suppl. 45, 351
Gilman, P.A. and Miller, J.: 1981, Astrophys. J. Suppl. 46, 211
Giovanelli, R.G.: 1982, preprint
Gokhale, M.H.: 1977, Kodaikanal Obs. Bull. Ser. A 2, 19
Golub, L., Davis, J.M., and Krieger, A.S.: 1979, Astrophys. J. 229, L145
Golub, L., Rosner, R., Vaiana, G.S., and Weiss, N.O.: 1981, Astrophys. J. 243, 309
Graham, E.: 1975, J. Fluid Mech. 70, 689
Graham, E. and Moore, D.R.: 1978, M.N.R.A.S. 183, 617
Graham, E.: 1977, in "Problems of Stellar Convection", eds. E.A. Spiegel and J.P. Zahn, IAU Colloquium 38, Lecture Notes in Physics, Springer, Berlin
Hellmich, R.: 1978, Geophys. Astrophys. Fluid Dyn. 10, 89
Herzenberg, A.: 1958, Phil. Trans. Roy. Soc. Lon. A250, 543
Hinata, S.: 1982, Astrophys. J. 256, L23
Howard, R. and LaBonte, B.J.: 1980a, Astrophys. J. 239, L33
Howard, R. and LaBonte, B.J.: 1980b, Astrophys. J. 239, 738
Howard, R. and LaBonte, B.J.: 1981, Sol. Phys. 74, 131
Ivanova, T.S. and Ruzmaikin, A.A.: 1977, Sov. Astron. 21, 479
Jepps, S.A.: 1975, J. Fluid Mech. 67, 625
Jones, C.A.: 1981, Geophys. Astrophys. Fluid Dyn., in press
Kleeonin, N.I. and Ruzmaikin, A.A.: 1981, Geophys. Astrophys. Fluid Dyn. 17, 281
Knobloch, E.: 1978, Astrophys. J. 225, 1050
Knobloch, E., Rosner, R., and Weiss, N.O.: 1981, M.N.R.A.S. 197, 450
Köhler, H.: 1973, Astron. Astrophys. 25, 467
Kraichnan, R.H.: 1976, J. Fluid Mech. 77, 753
Krause, F.: 1967, Habilitationsschrift, Univ. Jena
Krause, F. and Rädler, K.-H.: 1980, "Mean-Field Magnetohydrodynamics and Dynamo Theory", Pergamon, Oxford
Kuznetsov, V.D. and Syrovatskii, S.I.: 1979, Sov. Astron. 23, 715
Labonte, B.J. and Howard, R.: 1982, Sol. Phys. 75, 161
Layzer, D., Rosner, R., and Doyle, H.T.. 1979, Astrophys. J. 229, 1126
Leighton, R.B.: 1969, Astrophys. J. 156, 1
Levy, E.H. and Boyer, D.: 1982, Astrophys. J. 254, L19
Liebert, J., Dahn, C.C., Gresham, M., and Strittmatter, P.A.: 1979, Astrophys. J. 233, 226
Linsky, J.L.: 1981, in "Solar Phenomena in Stars and Stellar Systems", eds. R.M. Bonnet and A.K. Dupree, Reidel, Dordrecht, p. 99
Malkus, W.V.R. and Proctor, M.R.E.: 1975, J. Fluid Mech. 67, 417
Maltby, P. and Albregtsen, P.: 1979, Astrophys. J. 234, L147
Martin, S.F. and Harvey, K.L.: 1979, Sol. Phys. 64, 93
Massaguer, J.M. and Zahn, J.-P.: 1980, Astron. Astrophys. 87, 315
Meneguzzi, M., Frisch, U., and Pouquet, A.: 1981, Phys. Rev. Letters 47, 1066
Middelkoop, F.: 1981, Astron. Astrophys. 101, 295
Moffatt, H.K.: 1978, "Magnetic Field Generation in Electrically Conducting Fluids", Cambridge University Press

Noyes, R.W.: 1981, "Solar Phenomena in Stars and Stellar Systems", eds. R.M. Bonnet and A.K. Dupree, Proc. NATO ASI, Reidel, Dordrecht, p. 1
Noyes, R.W.: 1982, in "Reports on Astronomy", Transac. IAU 28a, ed. P.A. Wayman, Reidel, Dordrecht, p. 100
Oranje, B.J., Zwaan, C., and Middelkoop, F.: 1982, Astron. Astrophys., in press
Parker, E.N.: 1975, Astrophys. J. 198, 205
Parker, E.N.: 1979a, "Cosmical Magnetic Fields", Clarendon Press, Oxford
Parker, E.N.: 1979b, Astrophys. J. 231, 250
Parker, E.N.: 1982a, Geophys. Astrophys. Fluid Dyn. 20, 165
Parker, E.N.: 1982b, Astrophys. J. 256, 292
Parker, E.N.: 1982c, Astrophys. J. 256, 302
Parker, E.N.: 1982d, Astrophys. J. 256, 736
Parker, E.N.: 1982e, Astrophys. J. 256, 746
Piddington, J.H.: 1976, in V. Bumba und J. Kleczek (eds.), 'Basic Mechanisms of Solar Activity', IAU Symp. 71, 389
Pizzo, V., Schwenn, R., Marsch, E., Rosenbauer, H., and Mühlhäuser, K.-H.: 1982, Proc. IX-th Lindau Workshop: "The Source Region of the Solar Wind", Space Sci. Rev.
Plumpton, C., Ferraro, V.C.A.: 1955, Astrophys. J. 121, 168
Proctor, M.R.E.: 1977, J. Fluid Mech. 80, 769
Rädler, K.-H.: 1968, Z. Naturforsch. 23a, 1851
Rädler, K.-H.: 1982, Geophys. Astrophys. Fluid Dyn. 20, 191
Robbins, K.A.: 1977, Math. Proc. Camb. Phil. Soc. 82, 309
Roberts, P.H. and Stix, M.: 1972, Astron. Astrophys. 18, 453
Robinson, R.D. and Durney, B.R.: 1982, Astron. Astrophys. 108, 322
Roxburgh, I.W.: 1978, Astron. Astrophys. 65, 281
Rüdiger, G.: 1978, Astron. Nachr. 299, 217
Rüdiger, G.: 1980, Astron. Nachr. 301, 181
Rüdiger, G.: 1973, Astron. Nachr. 294, 183
Ruzmaikin, A.A. and Vainshtein, S.I.: 1978, Astrophys. Sp. Sci. 57, 195
Ruzmaikin, A.A.: 1981, Comm. Astrophys. 9, 85
Schmidt, H.U.: 1968, in "Structure and Development of Solar Active Regions", K.O. Kiepenheuer (ed.), IAU-Symposium 35, p. 95, Reidel, Dordrecht
Schmidt, W.: 1982, Geophys. Astrophys. Fluid Dyn., in press
Schmitt, J.H.M.M. and Rosner, R.: 1982, Astrophys. J., in press
Schüssler, M.: 1977, Astron. Astrophys. 56, 439
Schüssler, M.: 1979a, Astron. Astrophys. 71, 79
Schüssler, M.: 1979b, Astron. Astrophys. 72, 348
Schüssler, M.: 1980a, Astron. Astrophys. 89, 26
Schüssler, M.: 1980b, Nature 288, 150
Schüssler, M.: 1981, Astron. Astrophys. 94, L17
Skumanich, A.: 1972, Astrophys. J. 171, 565
Skumanich, A., Smythe, C., and Frazier, E.N.: 1975, Astrophys. J. 200, 747
Soward, A.M.: 1974, Phil. Trans. Roy. Soc. London A275, 611
Soward, A.M.: 1979, Phys. Earth Planet. Int. 20, 134
Spiegel, E.A. and Weiss, N.O.: 1980, Nature 287, 616
Spitzer, L.: 1957, Astrophys. J. 125, 525
Spruit, H.C.: 1977, Thesis, University of Utrecht
Spruit, H.C. and van Ballegoijen, A.A.: 1982, Astron. Astrophys. 106, 58

Stenflo, J.O.: 1981, in Jordan, C. (ed.) "Solar Activity", Proc. 3rd European Solar Meeting, Oxford, p. 11
Stix, M.: 1972, Astron. Astrophys. 20, 9
Stix, M.: 1974, Astron. Astrophys. 37, 121
Stix, M.: 1976, in V. Bumba und J. Kleczek (eds.), 'Basic Mechanisms of Solar Activity', IAU Symp. 71, 367
Stix, M.: 1977, Astron. Astrophys. 59, 73
Stix, M.: 1978, in S. Dumant und J. Rösch (eds.); "Plein feux sur la physique solaire", Proc. 2nd European Solar Meeting, Toulouse,p.37
Stix, M.: 1981, Sol. Phys. 74, 79
Stix, M.: 1982, in W. Fricke and G. Teleki (eds.), Proc. 6th European Regional Meeting in Astronomy, Reidel, Dordrecht, p. 63
Tsinganos, K.C.: 1980, Astrophys. J. 239, 746
Unno, W. and Ribes, E.: 1976, Astrophys. J. 208, 222
Vainshtein, S.I., Zel'dovich, Ya. B. and Ruzmaikin, A.A.: 1980, "Turbulent Dynamo in Astrophysics", Moscow (in Russian)
Van Ballegoijen, A.A.: 1982, Astron. Astrophys., in press
Vaughan, A.H.: 1980, Publ. Astr. Soc. Pac. 92, 392
Vaughan, A.H. and Preston, G.W.: 1980, Publ. Astr. Soc. Pac. 92, 385
Vaughan, A.H., Baliunas, S.L., Middelkoop, F., Hartmann, L.W., Mihalas, D., Noyes, R.W. and Preston, G.W.: 1981, Astrophys. J. 250, 276
Walter, F.M.: 1981, Astrophys. J. 245, 677
Walter, F.M. and Bowyer, C.S.: 1981, Astrophys. J. 245, 671
Walter, F.M.: 1982, Astrophys. J. 253, 745
Weiss, N.O.: 1981a, J. Fluid Mech. 108, 247
Weiss, N.O.: 1981b, J. Fluid Mech. 108, 273
Weiss, N.O.: 1981c, in Jordan, C. (ed.) "Solar Activity", Proc. 3rd European Solar Meeting, Oxford, p. 35
Weiss, N.O.: 1982, Proc. Workshop on Planetary and Stellar Magnetism, Budapest, Geoph. Astrophys. Fluid Dyn., in press
Williams, G.E.: 1981, Nature 291, 624
Wilson, O.C.: 1978, Astrophys. J. 226, 379
Yoshimura, H.: 1975, Astrophys. J. Suppl. 29, 467
Yoshimura, H.: 1978, Astrophys. J. 220, 692
Yoshimura, H.: 1979, Astrophys. J. 227, 1047
Yoshimura, H.: 1980, Astrophys. J. 235, 625
Yoshimura, H.: 1981, Astrophys. J. 247, 1102
Yoshimura, H.: 1982, in Trans. IAU 28a, Reports on Astronomy, ed. P.A. Wayman, Reidel, Dordrecht, p. 55
Zel'dovich, Ya.B.: 1956, Sov. Phys. JETP 31, 154
Zwaan, C.: 1981, in "Solar Phenomena in Stars and Stellar Systems", eds. R.M. Bonnet and A.K. Dupree, Proc. NATO ASI, Reidel, Dordrecht, p. 463

DISCUSSION

GARCIA DE LA ROSA: Apart from the zone of magnetic flux concentration at the overshooting layer, wouldn't you expect to observe another such zone at the bottom of the supergranular cells? As mentioned in your talk for fibrils, the buoyant force is not important in comparison with the hydrodynamic drag, so as soon as a "flux rope" decays into smaller elements, these can be moved down to the bottom of the supergranules where an eventual concentration of fibrils can give rise to a small tube, large enough to be buoyant and to produce a small active region. This may be the explanation for the observed emergence of several small active regions after the decay of a large active region.

SCHÜSSLER: This may well be true. As far as topological pumping is concerned, there are important new results obtained by W. Arter, D. Galloway, and M. Proctor.

ENDAL: Has anyone examined how the existence of an overshooting layer at the bottom of the convective envelope would affect the pulsation properties of the sun?

SCHÜSSLER: I don't know. Is there an expert in the audience who can comment on that?

MULLAN: Details of the transition from convection to radiative zone at the base of the convection zone are very important for determining the rate of emission of g modes. These are focussed towards the center of the sun and may contribute seriously to energy transport if the g-mode amplitudes become large. Hence the idea of flux loops at the convective-radiative interface may be significant in the discussion of g modes.

ROSNER: Your estimate of an upper bound on the solar angular momentum loss is derived from measurements in the solar wind in (or near) the ecliptic. Observations of the inner corona show however that much mass loss occurs out of the ecliptic (the exact amount being very uncertain); hence your angular momentum loss estimate is likely to be similarly uncertain.

SCHÜSSLER: Quite true, but the polar regions have a much lower specific angular momentum. However, having an upper limit of 10^{-5} G for the field penetrating the envelope, I do not mind discussing about two orders of magnitude.

GOKHALE: (1) I am aware that it is unfair to expect any theoretical model to be consistent with too large a number of observational constraints. However, is there any turbulent dynamo model that incorporates the 10 day time scale of emergence and replacement of solar magnetic flux (as observed by Dr. Howard) *in addition* to the observational constraints in your "hit list"? (2) If we were to introduce the inertial forces due to the mass flow within the fluxtubes, would it affect substantially the conclusions of the model?

SCHÜSSLER: (1) This phenomenon in my opinion is a surface phenomenon of appearance, pulling down and reappearance of flux, probably confined to the granular–supergranular layer. It may not be very important for the dynamo. (2) No. Substantial mass flows only occur during the final phases of the rise of a fluxtube. Near the surface ($z < 10^4$ km) the tube gets unstable to loop formation, and downflows may occur as observed.

MULLAN: Is there a minimum size of fibrils deep in the convection zone? Near the surface, Meyer et al. (1977) found $F > 10^{19}$ Mx. Is there a corresponding result deep down?

SCHÜSSLER: The minimum fibril size is determined by diffusion and stability. Apart from the paper you mentioned there are no detailed calculations available yet.

SODERBLOM: I wish to make two points and raise a question. First, if the chromospheric emission of the Vaughan-Preston survey (i.e., Ca II H and K emission) is corrected for the decline of the underlying continuum, the chromospheric emission *decreases* in going down the main sequence, it does not increase. I am referring to the normalized flux: H+K flux $/L_{bol}$. Second, the Vaughan-Preston gap may not necessarily be real. Consideration of the effects that place stars in the diagram in which the "gap" is so evident leads one to conclude that the gap need not be real — it may be only a statistical fluctuation. My question concerns your remark that the center of the sun is rotating rapidly. The observations I have seen to date are quite controversial (to put it mildly). Are we sure that the core is in fact spinning rapidly?

SCHÜSSLER: L_{HK}/L_{bol} decreases, but L_x/L_{bol} increases or at least is constant. However, that may have more to do with the structure of the surface fields than with dynamo efficiency. The two other points clearly have to be clarified — mainly by the observers!

MESTEL: I am not arguing against the genuine solar dynamo and in favour of amplification models which depend on an external source of flux. And of course I agree that a hypothetical primeval field in the solar core will, if coupled with the base of the convective envelope, tend to iron out the inferred differential rotation between core and envelope. However, we are still I think unclear on the interaction of turbulence and rotation, and I recall a suggestion by Gough that perhaps the effect is to expel vorticity, analogous to the expulsion of primeval magnetic flux. Is it completely ruled out that there may be a state which is both kinematically and dynamically steady, with convective expulsion of vorticity balanced by magnetic coupling, and turbulent resistivity limiting the generation of toroidal field by non-uniform rotation?

SCHÜSSLER: Such a model is possible, but the turbulent expulsion of vorticity must be a very powerful mechanism then, because magnetic tensions are very effective.

STIX: There is another reason why the observed solar mean field should be confined to the convection zone (including the overshoot layer). The field is periodic with 22 years and so cannot penetrate into the radiative interior because of the skin effect. This means that whatever part of the field participates in the cycle should be disconnected from any interior field.

A MOLECULAR WEIGHT GRADIENT BATTERY MECHANISM FOR STELLAR MAGNETIC FIELDS

L. Mestel
Astronomy Centre
University of Sussex
Brighton
BN1 9QH
U.K.

D. Moss
Mathematics Department
University of Manchester
Manchester
M13 9PL
U.K.

SUMMARY

Dolginov (1977) pointed out that a non spherical distribution of molecular weight μ can drive a Biermann type "battery" mechanism in a star and so generate a toroidal magnetic field. We consider uniformly rotating axisymmetric models of this type, containing a 'primeval' poloidal field, in which the currents maintaining the toroidal field flow along poloidal field lines, thus ensuring that the configuration is torque free. When the molecular weight distribution is chosen so as to produce a meridional circulation which exactly cancels the Eddington-Sweet circulation, then battery fields of order 1000 gauss can be maintained against ohmic decay in models appropriate to Ap stars near the main sequence with period of order 5 days.

1. INTRODUCTION

Under stellar conditions Ohm's law takes the form

$$\frac{\underline{j}}{\sigma} = \underline{E} + \frac{\underline{v}\times\underline{B}}{c} + \frac{\nabla p_e}{n_e e} - \frac{\underline{j}\times\underline{B}}{c n_e e} , \qquad (1)$$

in a standard notation. We also have

$$\frac{\partial \underline{B}}{\partial t} = - c\nabla \times \underline{E}, \qquad (2)$$

and the stellar material satisfies the equation of hydrostatic support

$$- \nabla p + \rho\nabla\phi + \rho\Omega^2 \underline{s} = \underline{0} , \qquad (3)$$

where $\Omega = \Omega(\underline{r})$ is the angular velocity, (s, z) are cylindrical polar coordinates, and we have neglected the Lorentz force. In a chemically homogeneous non-magnetic star, from (1) and (2)

J. O. Stenflo (ed.), Solar and Magnetic Fields: Origins and Coronal Effects, 237–242.

$$\nabla \times \underline{E} = - \nabla \times \left(\frac{\nabla p_e}{n_e e}\right) \propto \nabla \times \Omega^2 \underline{s} ,$$

and so $\nabla \times \underline{E} = \underline{0}$ for conservative rotation fields $(\Omega = \Omega(s))$. In this case the centrifugal force is balanced by a Coulomb electric field, maintained by an extremely small charge separation analogous to the "Pannekoek-Rosseland" field. But if $\Omega = \Omega(s, z)$ so that the centrifugal force is not irrotational, then $\nabla \times \underline{E} \neq \underline{0}$, and a toroidal magnetic field may be generated by Faraday's law (2). This was pointed out by Biermann (1950), who estimated that asymptotic fields of order 10^3 gauss could be built up in the sun, and much larger fields $(B_t \propto \Omega^2)$ were possible in rapid rotators. However Mestel and Roxburgh (1962) demonstrated that in the presence of even a weak poloidal field the requirement that the total field be torque free, together with isorotation, effectively 'short circuits' Biermann's battery.

In a chemically inhomogeneous star, even if it rotates uniformly, the battery term $\nabla p_e/n_e e$ may still have a non-zero curl (e.g. Dolginov, 1977). We pursue this idea by constructing axisymmetric torque-free mixed poloidal-toroidal fields, whose toroidal components are maintained against ohmic decay by the battery field generated by non-spherical μ-distributions that might plausibly arise spontaneously during stellar evolution.

2. THE QUASI-STEADY AXISYMMETRIC THEORY

We assume the presence of a "primeval" poloidal field $\underline{B}_p$. In a torque free steady axisymmetric system the currents $\underline{j}_p$ maintaining the battery field $\underline{B}_t$ must flow parallel to $\underline{B}_p$, and the torque free condition can be written (e.g. Lüst & Schlüter, 1954) as

$$\underline{j}_p = K\underline{B}_p, \ \underline{B}_p.\nabla K = 0. \tag{4}$$

If meridional circulation is neglected so that in equation (1) $\underline{v} = \Omega s\hat{\underline{t}} \times \underline{B}$, Mestel and Roxburgh (1962) showed that to a very good approximation Ω must be constant along poloidal field lines. We specialize by assuming Ω constant throughout the star. Then, integrating the poloidal component of equation (1) around poloidal field lines, assuming that a steady state has been reached, and using equation (4), we obtain

$$\oint \frac{\underline{j}_p.\underline{dr}}{\sigma} = K \oint \frac{\underline{B}_p.\underline{dr}}{\sigma} = \oint \frac{\nabla p_e}{n_e e} . \underline{dr} . \tag{5}$$

Thus the condition for the battery to operate is that the right hand integral in (5) be non zero. Suppose that the star consists wholly of hydrogen and helium, with molecular weight μ, so that

$n_e = \frac{(\mu+2)}{5\mu}\frac{\rho}{m_H}$. Thus in (5)

$$K \oint \frac{\underline{B}_p . d\underline{r}}{\sigma} = \frac{m_H}{e} \oint \frac{\nabla((\mu+2)p)}{\left(\frac{\mu+2}{\mu}\right)\rho} \cdot \underline{dr} \ . \qquad (6)$$

Note that if μ, p and ρ are equatorially symmetric, then the battery integral is only non zero if $\underline{B}_p$ has a component of even parity, for example if $\underline{B}_p$ is a quadrupole or decentred dipole field.

Introducing a 'stream function' P for $\underline{B}_p$, so that $\underline{B}_p = -\frac{\nabla P \times \hat{\underline{t}}}{r \sin\theta}$, where P is constant along lines of $\underline{B}_p$, then, using equation (4),

$$\underline{j}_p = \frac{c}{4\pi}\nabla \times \underline{B}_t = \frac{c}{4\pi}\nabla(r\sin\theta\ B_t) \times \frac{\hat{\underline{t}}}{r\sin\theta} = -K\frac{\nabla P \times \hat{\underline{t}}}{r\sin\theta} \ . \qquad (7)$$

Thus $r \sin\theta\ B_t = f(P)$, and so

$$\frac{df}{dP} = -\frac{4\pi}{c}K(P) . \qquad (8)$$

From (6) we can determine $K(P)$ by explicit integration along field lines, and then determine f and so $\underline{B}_t$ from (8), given that $f(0) = 0$. We note now that K will be greatly reduced on field lines which pass through low conductivity regions (equation (6)), and that with $\sigma \propto T^{3/2}$ the value of K on field lines which pass through the stellar surface will be sensitive to the value adopted for the temperature (T_c, say) of the surrounding regions. In the next section we present results for several values of this 'coronal temperature'.

3. ILLUSTRATIVE CALCULATIONS

Firstly we choose our μ distribution by the condition that the associated meridional circulation, $\underline{v}_\mu$ (Mestel 1953), exactly opposes the Eddington-Sweet circulation, $\underline{v}_\Omega$, driven by the stellar rotation throughout the radiative region of the star

$$\text{i.e.} \quad \underline{v}_\Omega + \underline{v}_\mu = \underline{0}.$$

The figures we present refer to a "Cowling model" of mass $2M_\odot$ and radius $1.6\ R_\odot$ with period approximately 10 days, so that the perturbation parameter $\lambda_\Omega = \frac{\Omega^2 R^3}{GM} \simeq 10^{-3}$. As the theory is linear the toroidal field scales as λ_Ω. The poloidal field $\underline{B}_p$ is that of the slowest dipole decay eigenmode, decentred by fractional amount d/R along the rotation axis. The salient results for the model are presented in Table I for 3 values of d/R and 3 choices of T_c.

$|B_t, (P_{max})|$ is the value of the toroidal field in gauss at the neutral line of the poloidal field, and $|B_{t, max,surface}|$ is the value where the first interior poloidal field line grazes the stellar surface.

Table I

d/R	T_c (^oK)	$\|B_t(P_{max})\|$	$\|B_{t,max,surface}\|$
0.05	10^4	175	0.3
	10^5	184	2.8
	10^6	192	8.5
0.10	10^4	316	0.6
	10^5	327	6.5
	10^6	348	14.0
0.20	10^4	419	0.9
	10^5	438	9.2
	10^6	473	25.0

Table I demonstrates the sensitivity of the surface field to the value of T_c, but the relative insensitivity of the interior field. Basically this arises from the small value of K (equation (6)) on field lines passing outside of the start when T_c is low. This results in f(P) remaining small until P corresponds to a field line totally within the star - see equation (7).

We studied this molecular weight distribution feeling that it might be an approximation to what happens as a rotating star evolves from the zero age main sequence (e.g. Mestel, 1981). More probably this asymptotic state will be approached slowly, and so we also investigated a model in which the molecular weight gradient was only present in the inner 65% by radius. Again the molecular weight distribution was determined by the condition that the total circulation, $\underline{v}_\Omega + \underline{v}_\mu$, be zero in this region. The resulting values of $|\underline{B}_t|$ were about 75% of those given in Table I. This agrees with the qualitative argument that most of the battery effect arises from high conductivity regions where $\nabla\mu \neq \underline{0}$. Provided that most field lines continue to pass through such a region, as they do here since the neutral line is at a fractional radius of about 0.45, the battery effect is not much reduced.

Clearly other sources of chemical inhomogeneity can be considered, such as the surface helium anomaly spots of Dolginov (1977). Trial calculations for plausible parameters again suggest that toroidal fields of around 1000 gauss are attainable, possibly with significantly larger surface fields than those of Table I. The extension of the theory outlined above to an oblique rotator model is difficult in detail, but estimates suggest that the asymptotic battery field is of a comparable order of magnitude to that found in the case of aligned rotational and poloidal magnetic axes.

4. CONCLUSIONS

It has been shown that the toroidal component of a torque-free, mixed poloidal -toroidal field can be maintained against decay in a uniformly rotating star with plausible non-spherical μ-distributions. The figures presented are for a modest value of λ_Ω , and larger fields are possible for shorter periods. Analogous processes may be occurring in the surface regions of chemically peculiar stars (Dolginov, 1977). Toroidal fields are necessary for the dynamical stability of poloidal fields (and v.v.). The potential observability of surface toroidal fields (Barker et al, 1981), and their possible importance for element diffusion (Alecian & Vauclair, 1981) are additional motivation for studying their generation and maintenance.

REFERENCES

Alecian, G., Vauclair, S.: 1981, Astron.Astrophys. 101, p.16.
Barker, P.L., Landstreet, J.D., Marlborough, J.M., Thompson, I., and Maza, J.: 1981, Astrophys. J. 250, p.300.
Biermann, L.: 1950, Zs. f. Naturforsch. 5a, p.65.
Dolginov, A.Z.: 1977, Astron.Astrophys. 54, p.17.
Lüst, R., and Schlüter, A.: 1954, Zs. f. Astrophys. 34, p.263.
Mestel, L.: 1953, Mon. Not. R. astr. Soc. 113, p.716.
Mestel, L.: 1981, In "Fundamental Problems in the Theory of Stellar Evolution", IAU Symposium 93.
Mestel, L., and Roxburgh, I.W.: 1962, Astrophys. J. 136, p.615.

DISCUSSION

KUPERUS: About 16 years ago I made a similar calculation on battery generation of toroidal magnetic fields in early-type stars. I found using the best available stellar models present at that time a peak value of a few hundred Gauss around the A type stars, which actually is the dominant spectral type of magnetic stars. The influence of ionization of H and He as well as the rotation was taken into account. It was published in *Bulletin of the Astronomical Institutes of the Netherlands* 1966 (P.S. The journal merged into *Astronomy and Astrophysics*).

MOSS: Did these models take account of the torque-free condition? Did they use a molecular weight gradient to drive the battery?

KUPERUS: No.

SPICER: How does the thermal electric field, which is proportional to the electron temperature gradient, compare with the electric field due to the electron pressure gradient?

IONSON: I agree with Spicer that you should include the thermoelectric field, $E_{thermo} = 0.7\nabla_{||}T_e/e$, which can be comparable to the electron pressure electric field, $E_{\nabla_{||}P_e} = \nabla_{||}P_e/me$.

MOSS: This term does not arise in the form of Ohm's law resulting from the mean free path approximation, which was used in the analysis presented. In any case it is a pure gradient term, and so can only give rise to a Coulomb electric field. As shown, this would not contribute to the battery mechanism.

ROXBURGH: Did you take a molecular weight distribution that cancelled the rotational circulation at all points in the star — or just in the central regions?

MOSS: Firstly we took a molecular weight gradient such that $v_\Omega + v_\mu = 0$ everywhere. This gave the fields in Table I. Another calculation with non-zero molecular weight gradients and $v_\Omega + v_\mu = 0$ in the inner 60% or so by radius gave fields about 0.75 of those of Table I.

TOPOLOGICAL MAGNETIC FLUX PUMPING REVISITED

Wayne Arter
Department of Applied Mathematics and Theoretical Physics,
Silver Street, Cambridge CB3 9EW, England

ABSTRACT

Calculations by Drobyshevski and Yuferev (1974) suggest that magnetic flux is 'topologically pumped' to the bottom of a convecting Benard layer as the electrical conductivity is increased. This is shown to be mistaken: however, compressible flows where motion is mainly around field-lines can confine flux near their base.

INTRODUCTION

Observations of the Sun reveal a magnetic field heavily distorted by the convecting plasma which forms its outer layers. The conditions there are typically too extreme to be duplicated in terrestrial laboratory experiments, so our understanding is much helped by the use of anlaytic and numerical techniques.

As a first approximation we neglect the dynamical back-reaction of the magnetic field on the plasma, and consider the effect of arbitrary, but hopefully realistic flows on field distribution. Rolls of incompressible fluid, where there is dependence on only two spatial coordinates, have been well studied, see e.g. the recent review by Proctor and Weiss, 1982 (PW). Fully three-dimensional calculations were performed by Drobyshevski et al. (1974 = DY, 1980).

This paper summarises a study of a number of 3-D patterns, including the original DY one, carried to more extreme values of R_m (a measure of the importance of advection relative to diffusion) than in DY. A fuller account is to be published elsewhere (Arter, 1983).

This work and a similar study by Galloway and Proctor (1983), show DY's description of high R_m behaviour to be misleading. The high-R_m regime is characterised by the production of spirals of field by flow about axes oblique, but not perpendicular, to the initial imposed field direction, and there is no significant vertical asymmetry in the absolute averaged magnetic flux distribution. However, compressible velocity patterns do produce marked asymmetry.

J. O. Stenflo (ed.), Solar and Magnetic Fields: Origins and Coronal Effects, 243–246.

BRIEF STATEMENT OF THE PROBLEM

We solve the dimensionless magnetic induction equation,

$$\frac{\partial \underline{B}}{\partial t} = R_m \, \text{curl}(\underline{u} \wedge \underline{B}) + \nabla^2 \underline{B}$$

for the magnetic field $\underline{B}$; t is time, $\underline{u}$ is a given velocity pattern with rectangular planform - see DY for further details. We note here only that the flux initially imposed in, say, the x-direction is conserved. In the anelastic approximation $\text{div}\{e(z)\underline{u}\}=0$ where the density $e(z)=\exp(cz)$ depends on the vertical co-ordinate z, and c is the compressibility factor. We take

$$\underline{u} = \text{curl curl } (S(x,y)\sin z \, \hat{\underline{z}}) + c(\text{grad}S(x,y))\sin z.$$

In DY c=0 and $S=\cos x + \cos y + \frac{1}{2}\cos x \cos y$. We also consider c=-1 with $S=\cos x + \cos 2y$ and $S=\cos 2x + \cos y$. These will be referred to as DY, $C_{\parallel}$ and $C_{\perp}$. DY has the topological property - isolated regions of ascending flow are surrounded by falling fluid - but $C_{\parallel}$ and $C_{\perp}$ are roll-like with motion chiefly in planes parallel or perpendicular to the field respectively.

Furthermore, our R_m is defined using $|u_z|$, thus the largest R_m studied here, $R_m=200$, corresponds to $R_m=66.7$ in DY. The numerical techniques also differ. This paper, after Roberts and Weiss (1966), employs a primitive variable finite difference scheme, which turns out to preserve $\text{div}\underline{B}=0$. The mesh size is 24×24×24 and integration to a steady solution typically takes about 3 minutes using an ICL DAP.

SUMMARY

At lower R_m the results here agree with DY's, but for $R_m>50$, measures of horizontal flux become negative in $z>\frac{1}{2}$ (Fig.1) indicating a preponderance of reversed field. Reconnection takes place at flow cell boundaries (Fig.2) rather than inside the eddies: contrast Fig.3 with e.g. Fig.1 of PW. However, the formation of persistent isolated loops of field in Fig.3 is due to a symmetry caused by the absence of net field in the y-direction: in general reconnection will involve at least two or more field-lines producing spirals. Either way, the unsigned flux is enhanced by a factor $R_m^{1/2}$ throughout most of the layer.

The topological property of $\underline{u}$ results in the formation of a flux-tube at the cell bottom, as opposed to a sheet at the top. Thus from PW, as $R_m\to\infty$ all the magnetic energy goes to the base. It is not clear how significant this is for dynamo action.

$C_{\parallel}$ (and $C_{\perp}$) are topologically different. $C_{\parallel}$ has flow mostly around the field-lines and we find B_x stays positive. It is enhanced where $\text{div}\underline{u}=0$, in falling fluid, and 80% of the flux resides at the layer bottom. Interestingly in $C_{\perp}$ flux is trapped at the top by the velocity shear in x

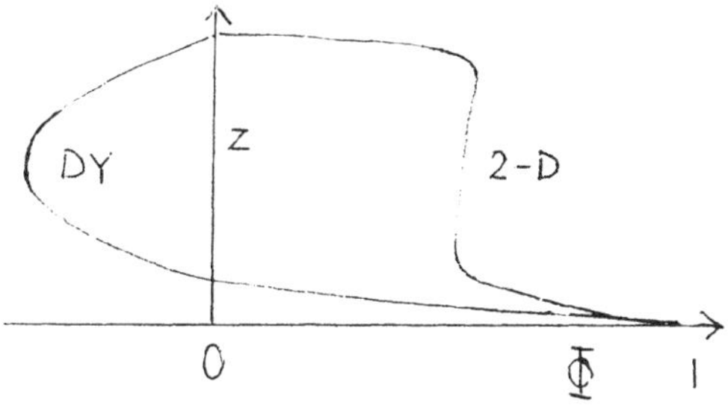

Fig.1 Sketch of $\Phi(z)=\int_V B_x\,dV$, where V is volume $1>\zeta>z$, for typical 2-D and 3-D flows, $R_m=150$.

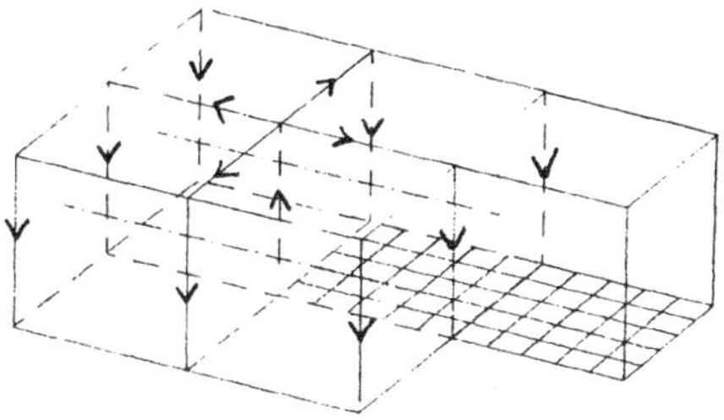

Fig.2 Schematic view of flow DY and field direction (dash-dot line).

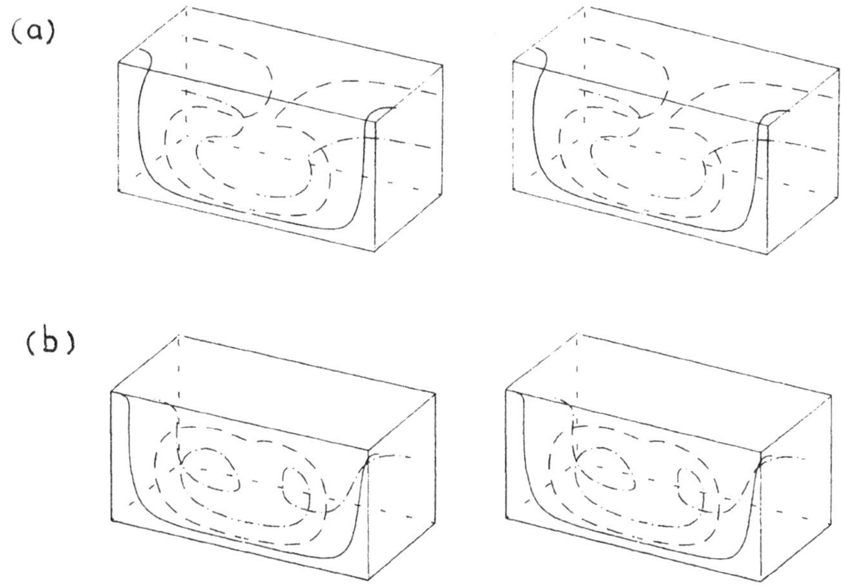

Fig.3 Stereo pairs showing field-lines reconnecting at $R_m=150$: time t increases from (a) to (b). The base is drawn hatched in Fig.2.

Now eddies in a strong dynamic magnetic field might look like $C_{\parallel}$, while others in the solar context might resemble $C_{\perp}$ due to the effect of rotation. Thus we have a mechanism which holds down strong field and expels weak field from a convective layer (cf. magnetic buoyancy). This needs further study, especially as the boundary conditions on $\underline{B}$ may strongly affect these results (Parker, 1975; Arter et al., 1983).

ACKNOWLEDGEMENTS

These are made to Drs. M.R.E. Proctor, N.O. Weiss and D.J. Galloway, and to the SERC.

REFERENCES

Arter, W.: 1983, in preparation.
Arter, W., Galloway, D.J. and Proctor, M.R.E.: 1982, submitted to Mon. Not. R. astr. Soc.
Drobyshevski, E.M., Kolesnikova, E.N. and Yuferev, V.S.: 1980, J. Fluid Mech. 101, pp.65-78.
Drobyshevski, E.M. and Yuferev, V.S.: 1974, J. Fluid Mech. 65, pp.33-44.
Galloway, D.J. and Proctor, M.R.E.: 1983, in preparation.
Parker, E.N.: 1975, Astrophys. J., 202, pp.523-527.
Proctor, M.R.E. and Weiss, N.O.: 1982, Rep. Prog. Phys., in press.
Roberts, K.V. and Weiss, N.O.: 1966, Math. Comp. 20, pp.272-299.

DISCUSSION

PROCTOR: Dr. Galloway and myself have done similar experiments only in a hexagonal geometry: we find similar results. It was not clear to me, though, how the compressibility of the field would affect the nature of the pumping in a general case.

GALLOWAY: The point of these calculations is the creation of large amounts of negative flux at the top of the layer (the input flux being positive), for high R_m. This looks nothing like the sun, where emerging flux is mainly vertical. Recently I have been looking at the effect of changing the top boundary condition so that the convecting layer lies underneath a current-free atmosphere. This implies that the *average* horizontal field has to vanish at this boundary, more like the sun. Preliminary results suggest that the flux is initially pumped to the bottom on a turnover timescale T, and subsequently decays on a longer timescale $R_m^{1/2}T$ as flux leaks out of the top (there is no steady state in this problem). So perhaps flux can be trapped at the base of the convection zone for an appreciable part of the solar cycle if the effective R_m is high enough.

DYNAMOS OF THE SUN AND STARS, AND ASSOCIATED CONVECTION ZONE DYNAMICS

Peter A. Gilman
High Altitude Observatory
National Center for Atmospheric Research

1. Introduction

I see my assignment in this talk as being to focus on the interaction between the magnetic fields produced by dynamo action and the dynamics of the fluid flow which drives this dynamo, and to make some connections to the solar and stellar dynamo problems. To do this really requires we start with a fluid dynamical model that satisfies relevant laws of fluid dynamics, and in which the flow can actually respond to the induced magnetic field. Thus the so-called kinematic dynamo models are not enough for our purposes, and we must address the full MHD dynamo problem in a self-consistent way. For example, we do not allow ourselves the license to vary independently the convection and differential rotation, as is commonly done in kinematic dynamo calculations, because the laws of physics do not allow that. We have been attempting to do self consistent MHD dynamo modeling at Boulder for the past several years, starting from a nonlinear fluid dynamical model for convection in a rotating spherical shell. This model we believe is physically complete in itself, with a minimum of ad hoc assumptions. It is much simpler than the real sun, but contains a lot of the physics we consider most relevant to the solar and stellar dynamo problem. I like to view the model more as an analog to the solar or a stellar convection zone, rather than as an approximation--much as a laboratory rat or monkey is used as an analog to a human being in many medical experiments. Each exists and is physically complete, and much, though not all, of their biochemistries are the same or are closely related.

Our model is for fully three-dimensional, nonaxisymmetric, nonlinear convection in a deep shell initially in uniform rotation and heated uniformly from inside. (Details of the computational approach are given in Gilman, 1975, 1977.) It has a central gravity, and constant viscosity and thermal conductivity (which we are forced to compare to eddy viscosities and thermal diffusivities on the sun). Stress free

---

The National Center for Atmospheric Research is sponsored by the National Science Foundation

J. O. Stenflo (ed.), Solar and Magnetic Fields: Origins and Coronal Effects, 247–270.

velocity boundary conditions are applied at top and bottom boundaries along with constant heat flux bottom and constant temperature top. The fluid is currently incompressible but a compressible version is now undergoing testing and should be operational this fall. (And we do have some results for linear compressible convection in a deep rotating spherical shell, see, eg. Gilman and Glatzmaier (1981), Glatzmaier and Gilman (1981a,b,c, 1982).

My plan in this talk is first to comment on the observational constraints provided by the sun which seem most relevant to the MHD dynamo problem, and then summarize the hydrodynamic results bearing on those constraints. Then I will discuss what kind of dynamo behavior we actually get with the model and contrast that with the observations. Finally, I will generalize and extrapolate some to a more general picture of rotating convectively driven spherical shell dynamos, and comment on possible implications of these generalizations for stellar dynamos.

2. Observational constraints

You all know the principal observational characteristics of solar cycles--the migration of spot zones toward the equator, Hale's sunspot polarity law, field reversals at the poles near cycle maximum, amplitude variations from cycle to cycle, and between north and south hemispheres, etc. In addition, the surface differential rotation of the sun is well known. As Robert Howard has told you, there also appear to be small amplitude ($\pm \sim$ 5m/sec) torsional oscillations on this mean rotation profile, linked in phase to the solar cycle, but extending to much higher latitude. The search for giant cell convective velocities has so far revealed only fleeting glimpses, and an upper limit of roughly 10m/sec per longitudinal wave number (LaBonte, Howard and Gilman, 1981, and earlier references cited therein). There is some indication of other rotation rate changes with time both within a cycle and from cycle to cycle, of a few percent. For example, Eddy et al (1978) reported from sunspot measurements a declining rotation rate in the first half of the 20th century, followed by a leveling off and possible increase. There is a suggestion in these data of an inverse correlation between amplitude of the cycle envelope, and low-latitude rotation, which may be a significant clue regarding the long-term workings of the solar dynamo.

The most important missing element in our knowledge of solar rotation for the dynamo problem is its variation with depth. The faster rotation rate of spots compared to the plasma has long been interpreted as evidence of an increase of angular velocity with depth (Foukal, 1972). Angular velocity increases with depth near the surface were suggested by Deubner et al. (1979) from frequency shifts in 5 minute oscillations, but the errors are large and more recent measurements at Kitt Peak (unpublished) do not confirm this result. More startling are new observations by Hill and colleagues (to be published; see also Gough, (1982)) indicating interior rotation rates several times the surface rate. It is not clear yet how much of the proposed increase occurs within the convection zone itself, but such a result, if confirmed, will be very important for solar differential rotation and dynamo theory.

This is because, as many of you know, kinematic dynamo theory applied to the sun favors angular velocity increasing with depth to give migration of the zone of sunspot formation toward the equator with time, while global convective models, at least those in which the strong influence of rotation upon convection in the deep layers of the convection zone is taken into account, favor angular velocity constant on cylinders and decreasing with depth.

As you also know, there is now evidence of luminosity changes associated with sunspot passage across the solar disk (e.g. Willson, et. al., 1981) which imply the magnetic field is modifying the energy output of the sun and causing intermittent storage of energy in the convection zone. Dynamo models will ultimately have to address this question, though none appears capable of doing so now, at least not on sunspot spatial scales.

3. Results from spherical shell convection model relevant to the dynamo problem

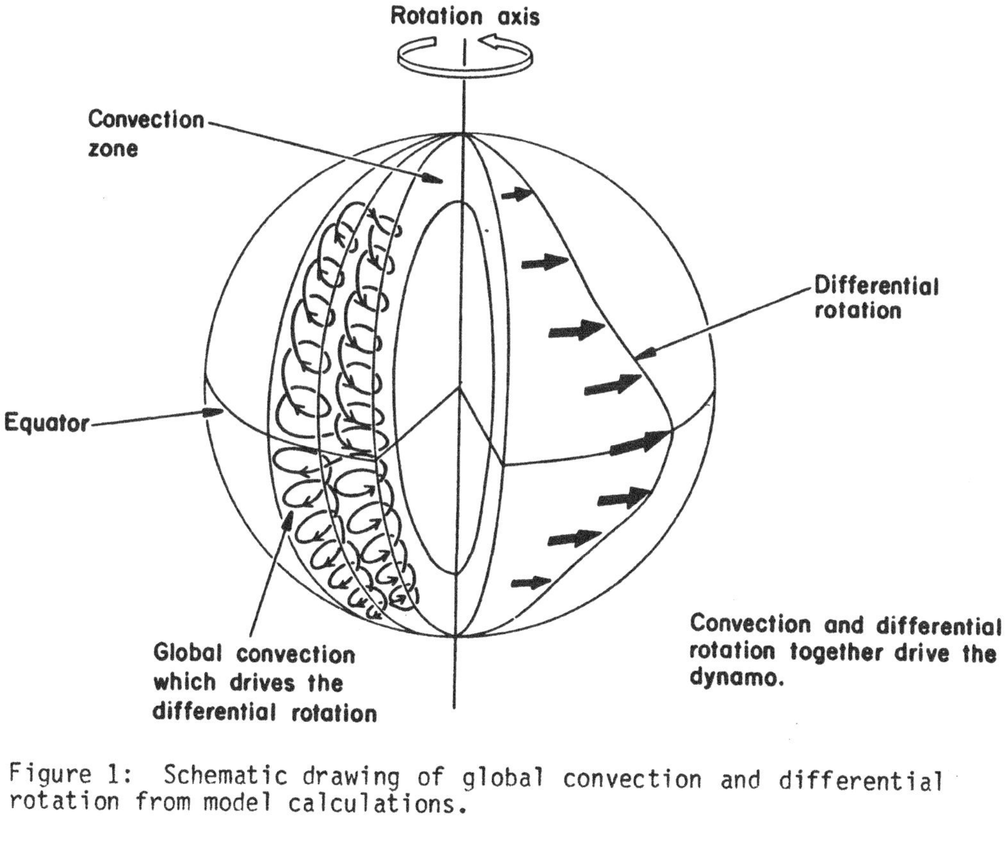

Figure 1: Schematic drawing of global convection and differential rotation from model calculations.

We have now performed a large number of numerical experiments for nonlinear convection of an incompressible fluid in a rotating spherical shell, varying the convection zone depth, heating rate, viscosity and rotation rate. One general conclusion from all these calculations is that in order for the convection to sustain a large amplitude, broad equatorial acceleration such as the sun has, two conditions must be met: First, the convection zone must be relatively deep, roughly one-third of the radius or more. Second, the influence of rotation upon the convection must be strong so that the turn over time for the convection is not short compared to the rotation time. This condition should be met in the deep part of the solar convection zone. In shallow convection zones the equatorial acceleration has small extent in latitude, and if the influence of rotation upon the convection is weak, the differential rotation is weak, and is in the form of an equatorial deceleration even slightly above critical for convection to occur.

A schematic picture of the global convection and differential rotation is shown in Figure 1. Convective rolls with a north-south axis are preferred due to rotation (if the rotational influence is very strong, the roll axis will tend to line up with the rotation axis, rather than bend with latitude as seen here). The horizontal velocity vectors in these rolls are tilted with respect to the east-west direction (due to the coriolis force) in such a way that fast moving particles are also moving toward the equator. This results in a Reynolds stress and net angular momentum flux toward the equator from high latitudes, which is how the equatorial acceleration is produced. The strong influence of rotation upon the convection leads to a strong preference for north-south rolls with this property, allowing a large amplitude differential rotation to build up. Also, with strong influence of rotation, fluid particles moving radially do not conserve angular momentum but are instead subject to longitudinal pressure torques that keep the particle paths from deviating far from the radial direction -- this is a manifestation of the near "heliostrophic" balance between pressure gradient and coriolis forces, which allows the convection to continue to release potential energy and transport heat efficiently. As a consequence radial angular momentum transport is either outward, or weakly inward. In either case, a surface equatorial acceleration is maintained.

In addition, also due to the constraint of rotation, the angular velocity tends to be constant on cylinders (the Taylor - Proudman theorem), and therefore decreases with depth. This is also a quite general result, modified only when some additional force is large enough to compete with pressure gradients and coriolis forces. It does not depend, for example, on the detailed properties of turbulent transport, unless such transport is large enough to upset the heliostrophic balance. In our model calculations, this can be made to occur by increasing the buoyancy force by increasing the heating rate. Then inward radial angular momentum transport due to departures from heliostrophic balance begins to dominate. But the first consequence of this is to wipe out the surface equatorial acceleration, and then to change the radial gradient so angular velocity increases inward. Because of the loss of equatorial acceleration, the solution ceases to be of interest for the sun, but may be for other stars (see e.g. Gilman, 1980).

The relative amplitudes of differential rotation and the convection which drives it are functions of, among other things, the fluid viscosity. Generally speaking, the lower the viscosity, the larger the differential rotation that can be produced. In an earlier paper (Gilman, 1980) I reported that it appeared that an upper limit to differential rotation energy was perhaps 40% of the total kinetic energy. That estimate was erroneous as we have now found, by reducing the viscosity, examples for which the differential rotation is as much as 80% of the total kinetic energy, the convection only 20%. I mention this because this ratio of energies is very important for determining the kind of dynamo behavior we get, for example, whether magnetic cycles occur or not. Solutions with relatively low amplitude global convection are also easier to fit within the observational upper limits for giant cells.

If we return to Figure 1, we can see the elements needed for dynamo action: differential rotation, and a helical structure to the convection, such that in the northern hemisphere, fluid moving toward the pole has a clockwise rotation, fluid moving toward the equator a counterclockwise rotation. This results in left-handed or negative helicity (scalar product of velocity and vorticity) in the northern hemisphere, right handed or positive in the southern. The shearing due to differential rotation will generate toroidal field from poloidal field, while the transporting and twisting of fields by the helicity will generate new poloidal field. The relative strengths of these two processes clearly depends on the relative strengths of convection and differential rotation.

We can also see from Figure 1 a reason why one would like to do the MHD rather than the kinematic dynamo problem: the helicity arising from the spiral structure of the flow, and the angular momentum transport arising from the tilt of the velocity vectors, are obviously closely linked. A change in helicity due to a change in the convection is likely to be accompanied by a change in the rate of angular momentum transport, and therefore in the differential rotation.

How much of what we have just said concerning convection of an incompressible fluid in a rotating spherical shell is likely to carry over to the compressible case? The answer is that we do not know for certain, because no comparably nonlinear calculations have yet been done. However, from the work of Glatzmaier and myself for linear compressible convection, extended to the initial tendency for nonlinear convection in Glatzmaier and Gilman (1982), we can make preliminary statements. In particular from compressible convective modes strongly influenced by rotation that do extend the full depth of the convection zone, we still get angular momentum transport toward the equator from high enough latitudes to produce a broad equatorial acceleration. The helicity of these modes also looks to be similar to that of the incompressible convection. However, the radial gradient of angular velocity they produce may change sign at mid-depth, with angular velocity constant on cylinders only in the deep part of the convection zone where the influence of rotation upon the convection is strongest. A typical angular velocity profile might look like that of Figure 2. This result does depend upon what is assumed for both the relative magnitude, and depth dependence, of the viscosity and thermometric diffusivity. Its reality and magnitude (and generality) in the fully nonlinear case

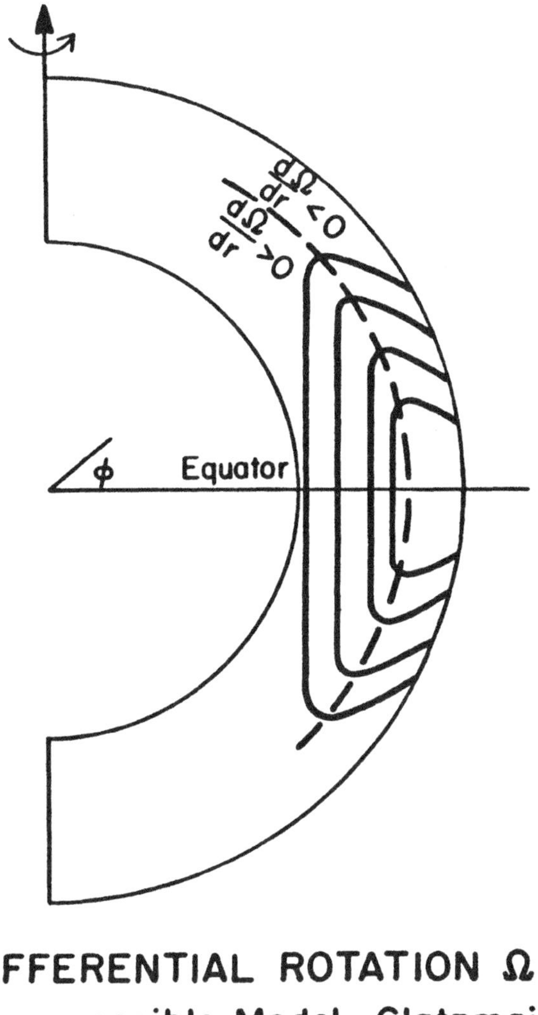

DIFFERENTIAL ROTATION $\Omega(r,\phi)$ (Compressible Model - Glatzmaier & Gilman Initial Tendency Calculation)

Figure 2: Schematic profile of differential rotation estimated from first calculations of compressible convection in a rotating spherical shell of Glatzmaier and Gilman (1982) when convective modes extend to full depth of convection zone.

needs to be tested. If such a radial angular velocity gradient reversal is present somewhere in the middle of the solar convection zone, it would have important consequences for solar dynamo models, because it could result in opposite directions of migration in latitude of the toroidal field above and below the level of gradient reversal. We return to this point after discussing our dynamo results from the incompressible model.

4. Dynamo results from an incompressible convection model

a) Early solutions

Our numerical experiments to study the dynamo action of convection in a rotating spherical shell are usually begun by first establishing a finite amplitude statistically stationary state for the convection and the differential rotation it drives, and then adding a small seed magnetic field and following its subsequent growth and evolution as well as the full nonlinear feedbacks on the inducing motions. We concentrated on hydrodynamical solutions for which the differential rotation at the outer boundary was about equal in amplitude and profile to the observed solar differential rotation. The first experiments, described in Gilman and Miller (1981) were for hydrodynamic states in which the convection contained initially at least two-thirds of the total kinetic energy of the flow, the differential rotation no more than one-third. To our dismay, we were not able to find any field reversing dynamos, but only more random dynamo action which maintained a broad magnetic spectrum. The reason was evident from a study of the energetics of the solutions: the differential rotation was not large enough, compared to the convection, to contribute the dominant mechanism for maintenance of the toroidal field. In this case, the helicity of the convection predominated in maintaining all components and scales of field. In the language of turbulent dynamo theory, our system behaved more like an " α^2" dynamo than an " $\alpha - \omega$ " dynamo.

b) Solutions with lower viscosity

The only way around this difficulty was to look for hydrodynamic states in which differential rotation was a much larger fraction of the total kinetic energy of the system. These we found by reducing the viscosity and thermometric diffusivity of the system each by a factor of 10 (still within the range of uncertainty of mixing length estimates, when comparing to the sun). The reason this works is that viscous dissipation is what limits the amplitude of differential rotation in the model, so a smaller viscosity allows a smaller Reynolds stress, and therefore smaller convection amplitude, to maintain the same amplitude differential rotation. Smaller viscosity also implies the influence of rotation upon the convection is stronger. For solar amplitude differential rotation, these new solutions have a kinetic energy spectrum for the convection which falls near the upper limit from observations as calculated in LaBonte, Howard and Gilman (1981).

With these new solutions, we have found several examples of cyclic dynamos, together with a variety of electromagnetic feedbacks. A typi-

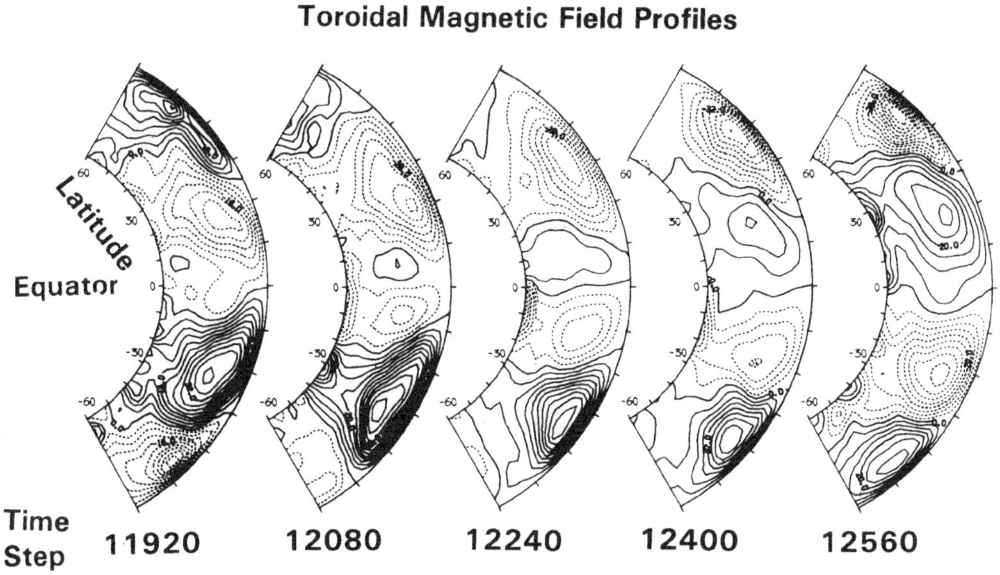

Figure 3: Contours of toroidal magnetic field amplitude for a sequence of time steps from typical solution of the convectively driven dynamo model. Solid contours represent positive field (into page) dashed contours, negative field.

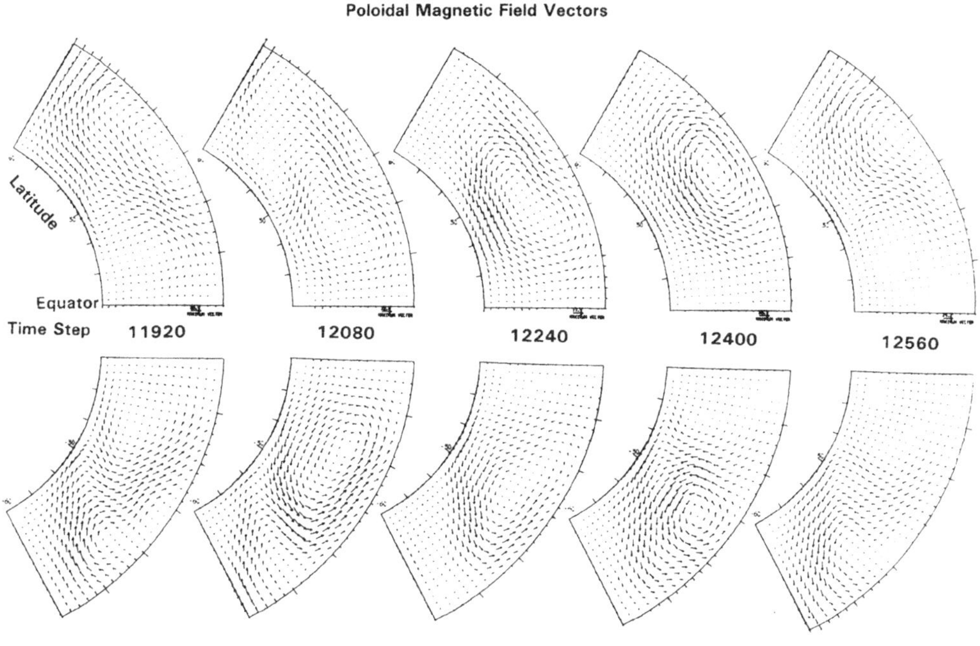

Figure 4: Poloidal field vectors for same time steps as in Figure 3.

cal example of toroidal and poloidal fields from such a solution is shown in Figures 3 and 4. Solid and dashed contours represent toroidal field amplitudes of opposite signs. (The calculations are cut off at 60 latitude for computational stability, an artifact that needs to be removed in subsequent models). From Figure 3, one can see that a new magnetic cycle in toroidal field begins deep in the convection zone near the equator, then grows in amplitude and expands upward and toward the poles, reaching the outer boundary in mid-latitudes. In Figure 4, the poloidal field migrates along with it, producing the strongest surface poloidal fields near the high latitude end of the toroidal field cycle. In a qualitative sense, the migration is as predicted from $\alpha - \omega$ dynamo theory.

I call your attention to another detail, namely the tendency for the poloidal field vectors in high latitudes in the interior of the convection zone to line up parallel to the rotation axis, and therefore with the contours of differential rotation, since they are cylindrical. This tendency toward "iso-rotation" appears to be important for limiting cyclic dynamo action to lower latitudes, particularly the region outside the tangent cylinder to the inner boundary. That is, a poloidal field parallel to the rotation contours is not sheared out into a new toroidal field. One can see from Figure 3, particularly in the southern hemisphere, a tendency for the toroidal field to avoid polar regions at deep levels. We will comment on this point further in talking about dynamos in stars generally. The reason the effect occurs in the model is that, at high latitudes, the convective flow itself including particularly the axisymmetric meridional circulation, tends to be parallel to the rotation axis, causing the poloidal field lines they induce to have a similar orientation. This is the opposite of low latitudes, where the dominant convective flow is a rotational one about the roll axis, which is parallel to the rotation axis..

c) Energetics of a typical solution

Rather than look at a lot of details of these complex dynamo solutions, I think it is instructive to instead concentrate on their energetics. Figure 5 shows time traces of the various kinetic and magnetic energies of the same solution as shown in Figures 3 and 4, in dimensionless form. The magnetic cycles show up most clearly in the toroidal field energy -- max to min in each cycle is a change of a factor of two or so. By contrast, the poloidal field energy is much noisier. This is because it is maintained by convection, which itself is much noisier than the differential rotation responsible for maintaining the toroidal field. This is obvious from the kinetic energy traces for these quantities seen near the top of the figure.

From the dimensional time scale drawn in the middle of Figure 5, one can see the cycles this model produces are at least an order of magnitude shorter than for the sun, even though differential rotation has solar amplitude. We return to a discussion of this point later.

The toroidal field energy trace also shows an "envelope" to the magnetic cycle, whose typical time scale is several cycles. This envelope amplitude appears to be connected to long term changes in dif-

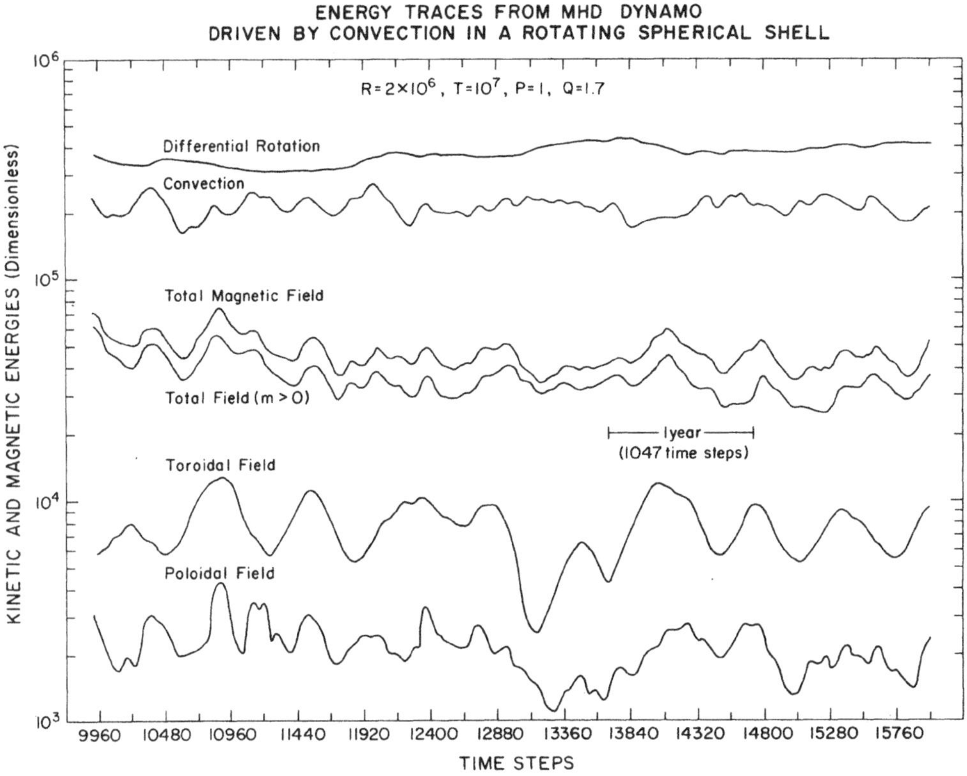

Figure 5: Kinetic and magnetic energies (dimensionless) as a function of time for several thousand time steps from the model solution shown in Figures 3 and 4, illustrating cycle amplitudes and envelope changes. Note the dimensional time scale of 1 year on the figure.

ferential rotation amplitude, in the sense that a declining sequence of cycles is associated with a generally rising differential rotation energy (which is experiencing a much smaller percentage change). This is what we would expect if the j x B force is modulating differential rotation amplitude. Relatively quick change in differential rotation, such as the sharp drop following step 13800, is due to a combination of a sharp rise in j x B force due to previous induction when convection was larger, combined with a drop in Reynolds stress due to a subsequent drop in convection amplitude. This event is followed by a new magnetic cycle of considerably weaker amplitude in toroidal field, a result of the weaker differential rotation.

Thus we can see that in this model, cycle envelope changes are a natural outcome of the interplay between the dynamo and its drivers, as well as the more independent fluctuations in convection itself. The differential rotation does not appear to respond much to the change of j x B force in an individual cycle. This is probably because the cycles

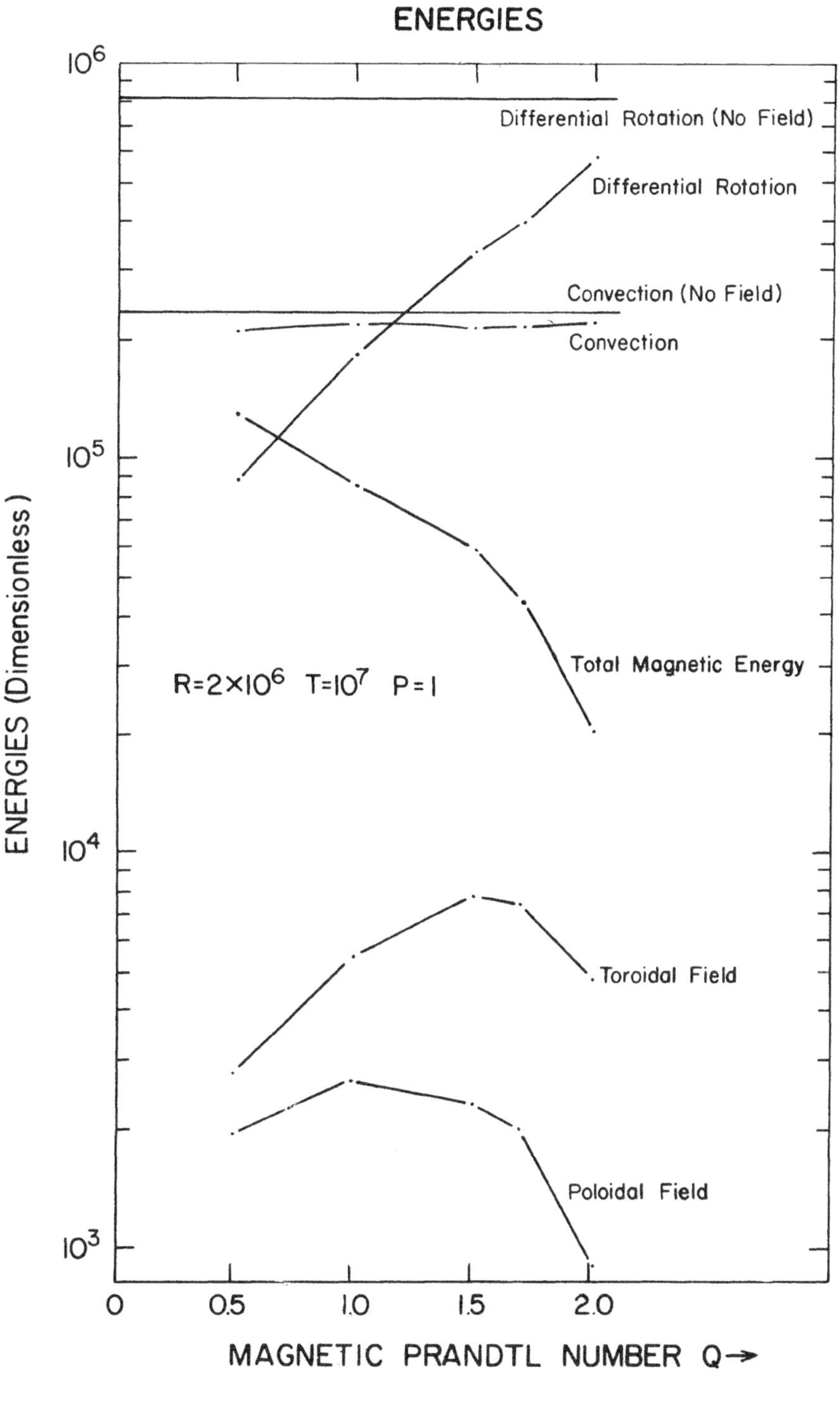

Figure 6: Average kinetic and magnetic energies from various dynamo solutions as a function of magnetic Prandtl number $Q = \eta / \kappa$, in which η is the magnetic diffusivity, and κ the thermometric diffusivity.

are relatively short, only a few convective turnover times, while the differential rotation changes more on its own internal "spin up" time, which is, for the viscosity chosen, considerably longer. Were the induced magnetic cycles much longer, we would expect to see a single cycle modulation of differential rotation energy as well.

It is worth noting that the differential rotation changes we find correspond to changes in absolute rotation velocity at the equator of only 1 - 2%, so they are in the range discussed from observations.

d) Effects of decreasing magnetic diffusivity: increasing feedbacks

We have found it particularly instructive to study changes in behavior of the dynamo as a function of the assumed magnetic diffusivity η. We measure that quantity in our model by a magnetic Prandtl number $Q = \eta / \kappa$, in which κ is the thermometric diffusivity. (In virtually all our calculations $\kappa = \nu$, the kinematic viscosity, so the ordinary Prandtl number is unity). Then the smaller is Q, the stronger dynamo action we should get, since the magnetic Reynolds number, which is proportional to Q^{-1}, is rising. So by varying Q for the same initial hydrodynamic solution, we can examine how the dynamo changes as we rise above the threshold for dynamo action. In our particular case, this threshold for small amplitude magnetic fields falls between Q = 1.7 and 2.0.

Figure 6 displays average kinetic and magnetic energies for several solutions of the system as a function of Q. Several effects are evident. First, as Q is decreased, the total magnetic energy rises, as we would expect, but the toroidal and poloidal field energies first rise, and then fall off, so that by Q = 0.5 together they constitute only 3% or so of the total magnetic energy of the system, as compared to 30% at Q = 2.0. Through this range of Q, the convection energy hardly changes at all, indicating very little feedback on it; by contrast, the differential rotation has dropped steeply, to the point that it becomes much smaller than the convection which drives it. (Incidentally, the dynamo solution at Q = 2.0 represents a finite amplitude dynamo field not reachable from a small amplitude initial field, which would decay away -- we found it instead by using a Q = 1.7 solution as an initial state.) The change in surface profile of differential rotation is shown in Figure 7, indicating it remains an equatorial acceleration as Q is decreased, but is simply of smaller amplitude (its cylindrical profile with depth is also retained).

What is happening is that as Q is decreased, $j \times B$ forces take over from viscous forces in braking the differential rotation, as can be seen in Figure 8, which plots the dimensionless rates at which work is done to buildup (+) or brake (-) the differential rotation energy. These rates are also normalized with respect to the viscous dissipation rate (always -1 unit). At Q = 2.0, viscosity dominates, while by Q = 0.5, the balance is principally between Reynolds stress and Maxwell stress, with viscous dissipation only about 1/3 of the total. (Note that coriolis forces acting on the meridional circulation do not play a significant role at any Q).

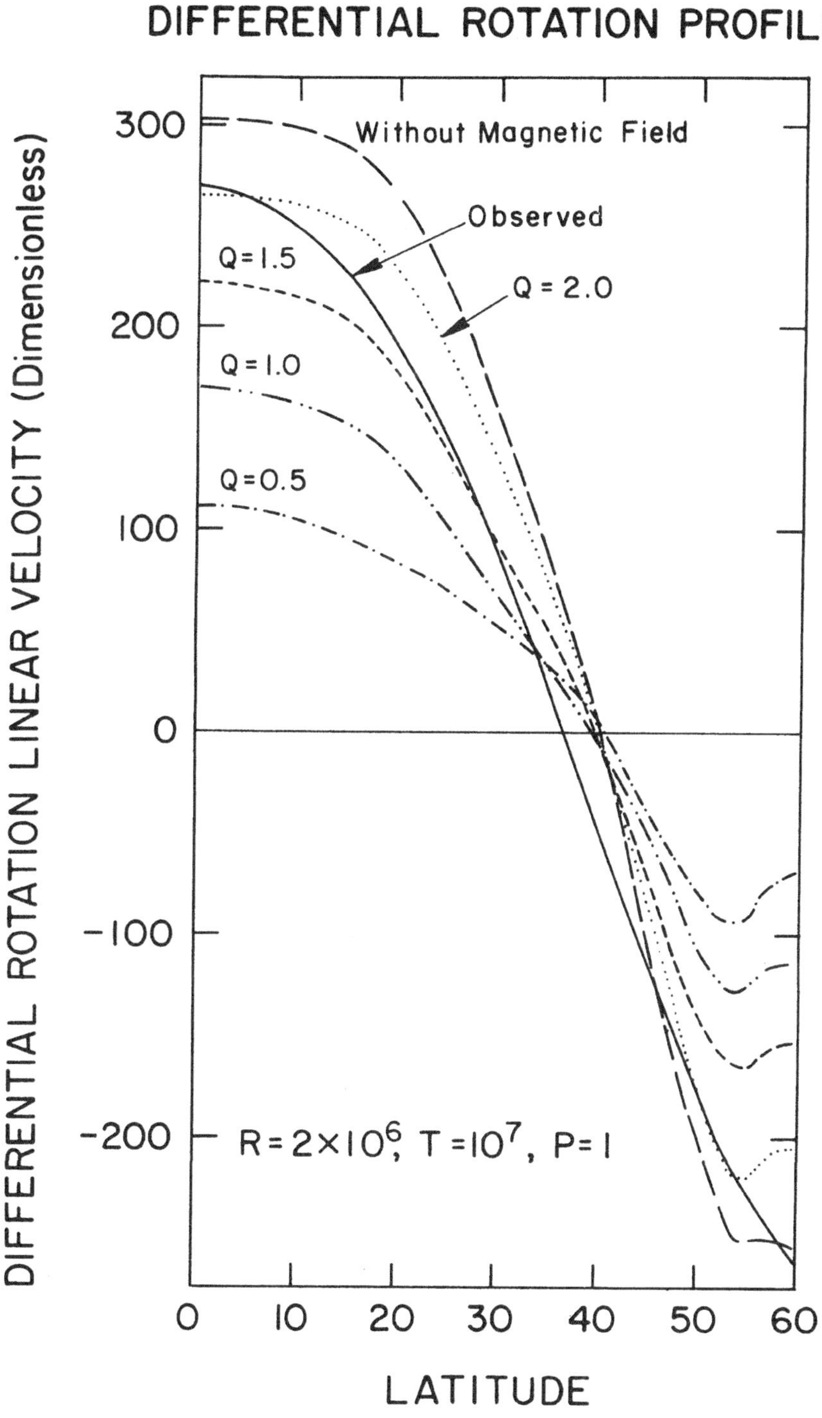

Figure 7: Dimensionless surface differential rotation profiles from the model for several values of Q, illustrating the damping by j x B forces. A typical average differential rotation for the sun is plotted for reference in the same units.

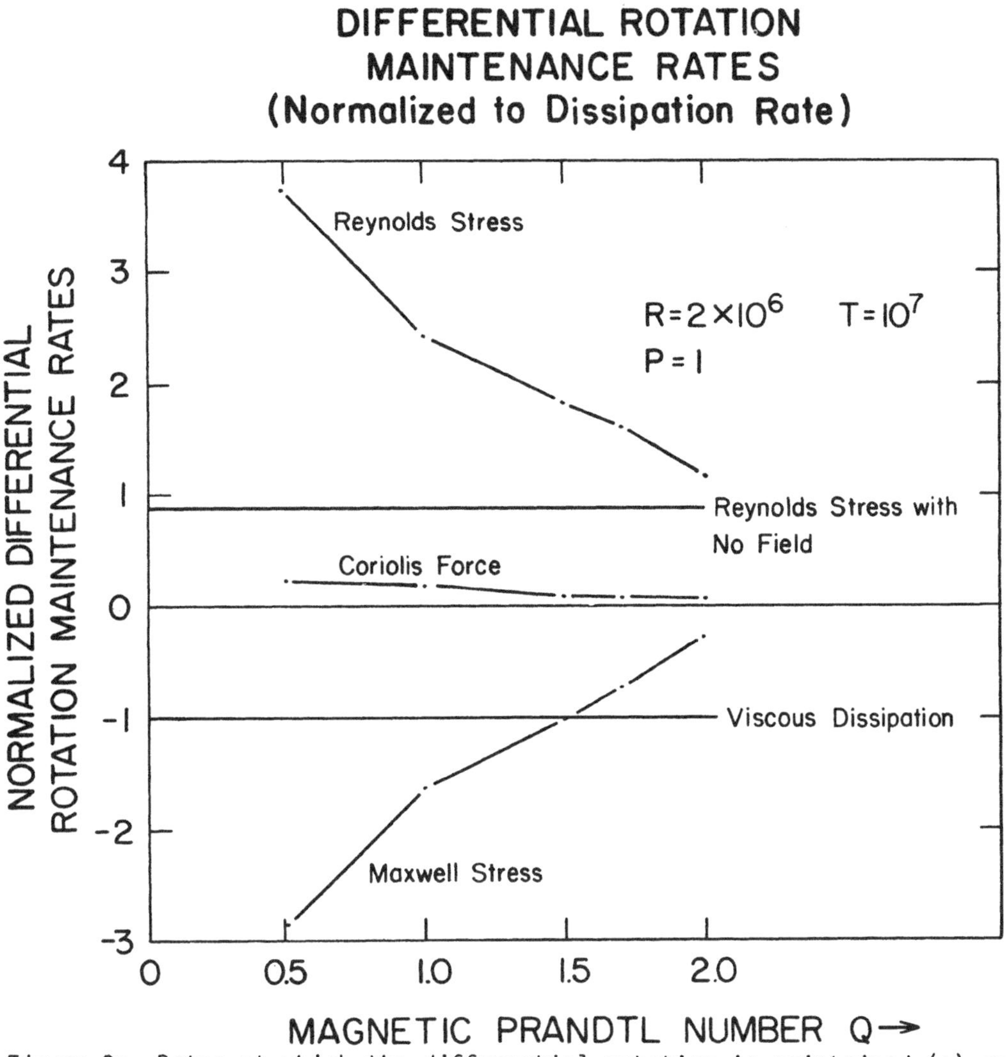

Figure 8: Rates at which the differential rotation is maintained (+) or braked (-), by Reynolds stresses, Maxwell stresses, and coroilis forces acting on the meridional circulation as functions of the magnetic Prandtl number. All rates are integrals over the whole volume of the fluid, and are normalized with respect to the viscous dissipation rate.

While the j x B force on the differential rotation is large enough to take over from viscosity in braking it, it is not so large that it interferes with the basic "heliostrophic" balance between pressure gradient and coriolis forces, so the Taylor-Proudman constraint is still satisfied and the differential rotation remains constant on cylinders. The j x B force affects the differential rotation at higher Q more than it does the convection because the viscous dissipation of differential rotation is relatively small. The profile is much larger scale than is that of the convection which has much fine structure. There must be a value of Q below which convection, too, is damped, thus limiting the total magnetic energy, but we have not investigated this regime.

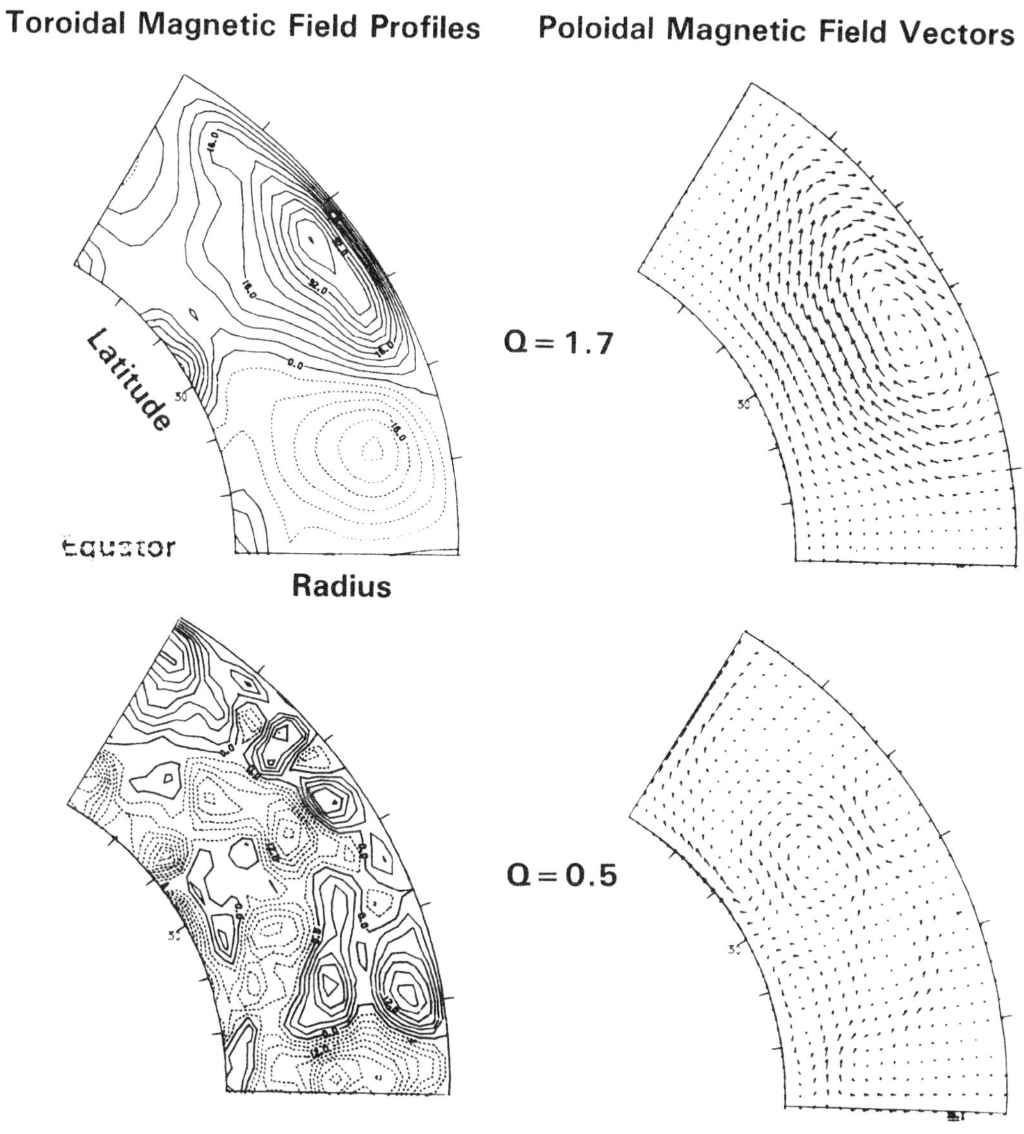

Figure 9: Sample toroidal and poloidal field profiles for a high Q and low Q case, showing break up of organized global field into much more chaotic form as the differential rotation drops.

As a result of this damping of differential rotation with decreasing Q, less toroidal field is induced from poloidal field, and the dynamo changes its character. Typical toroidal and poloidal field profiles are shown in Figure 9. By Q = 0.5, both poloidal and toroidal fields have lost their global coherence, and have just as much fine structure as the small scale fields at high longitudinal wave number in the model. Furthermore, the magnetic cycle has disappeared. It does not seem to be hidden in the noise of the instantaneous toroidal field profile for Q = 0.5, it is really not present at all. If one looks at the rates of maintenance of toroidal field shown in Figure 10, one can see what has happened. At Q = 2.0, the dominant mechanism of toroidal field maintenance was shearing of the poloidal field by differential rotation. The effect of helicity, or the " α effect", which is the sum of shearing and transport of field by the convection, is actually negative. But by Q = 0.5, the differential rotation has decreased to the point that transport and shearing by convection is now the dominant mechanism in maintaining the toroidal field. We are back closer to the " α^2 dynamo" case, rather like what we found with higher viscosity (although our "mean field" is not smooth and coherent on a global scale). Thus, the dynamo is capable of shutting off its own cycles; and for it to retain a cycle requires the sum of poloidal and toroidal field be a substantial fraction of the total magnetic energy of the system (in these model solutions, nearly 10%). This also illustrates how important a strong differential rotation is to produce cycles in the first place, something long known from kinematic dynamo theory.

It is clear from our results that the dynamo is very sensitive to its feedbacks, because all of these changes occur within a factor of 4 of the threshold magnetic diffusivity for any dynamo action at all.

e) Other feedback effects.

We have already seen that j x B feedbacks can profoundly affect the differential rotation, including its average amplitude and modulations from cycle to cycle. The Q = 2.0 solutions illustrate that a finite amplitude magnetic field can allow a dynamo to be sustained when a weak field would die out. In addition, Gilman and Miller (1981) showed that even a weak magnetic field can completely change the history of convection patterns. Two time dependent solutions for the flow, initially identical but one with a small magnetic field, will diverge away from each other in their amplitude and phase histories of the various convective modes in the spectrum. This precludes long term predictions, and will contribute to a certain randomness in the cycle envelope, and in field amplitudes with time within a given cycle.

One feedback effect we have looked for but not found is magnetically driven torsional oscillations. It turns out the model does have torsional oscillations in it, but these are present even without a magnetic field. They do migrate in the same direction as the toroidal field in the model but a rate which is faster by 40% or so. They have a similar amplitude to that observed on the sun. They appear to be a form of axisymmetric inertial oscillations in which the restoring force is the coriolis force. It is entirely possible magnetically driven tor-

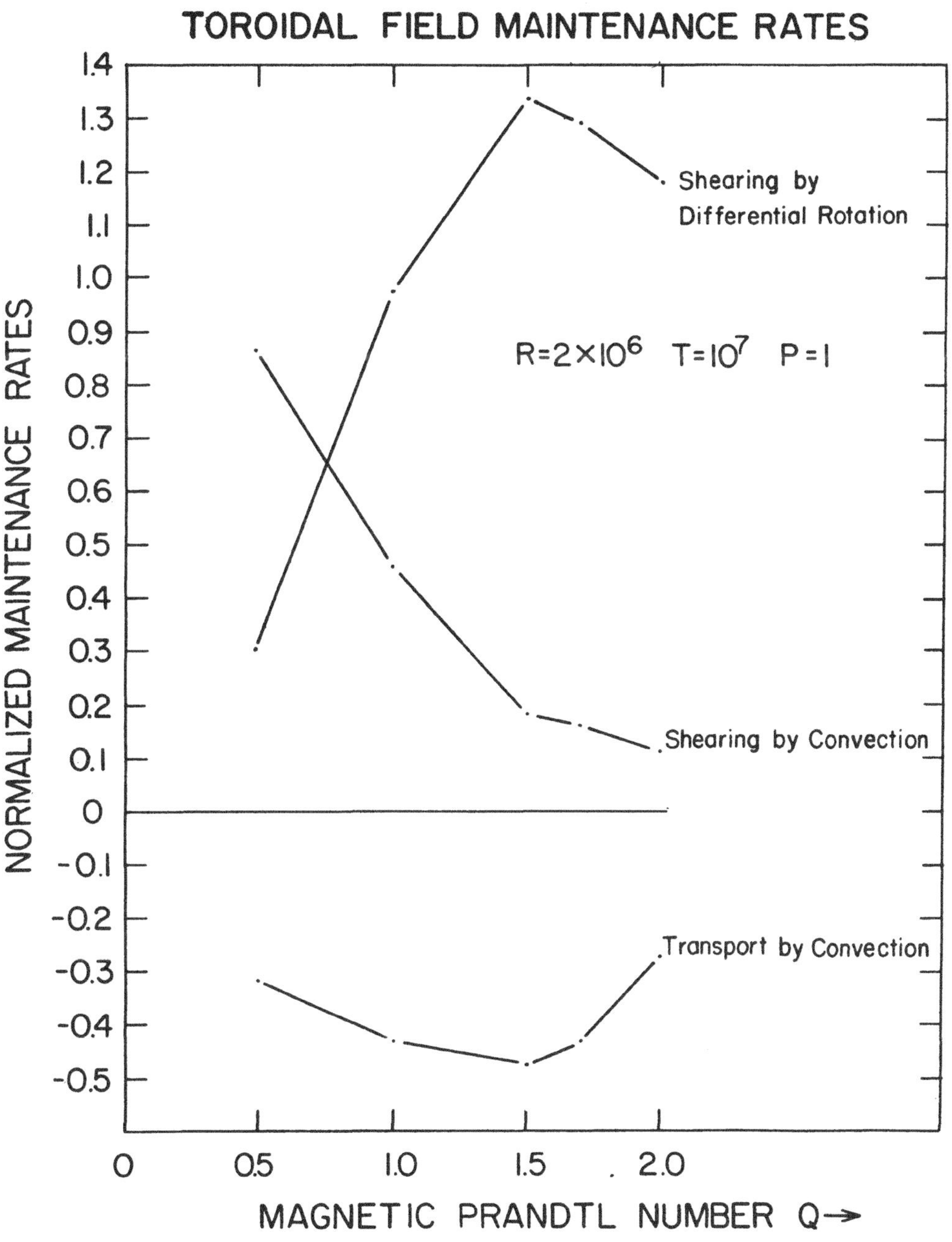

Figure 10: Rates at which toroidal field is maintained by various processes, normalized with respect to the magnetic field dissipation rate, as functions of magnetic Prandtl number.

sional oscillations will be produced in the model in solutions whose dynamo period is much longer than we presently find.

f) Symmetry of the magnetic field about the equator.

In our solutions, we have found that, if the initial magnetic field has one dominant symmetry about the equator (for example, toroidal field either symmetric or antisymmetric about the equator), this symmetry is likely to be preserved over many magnetic cycles (several thousand model time steps) and initial growth of the field for either symmetry is virtually the same. However, due to departures from one particular symmetry in the motion field, the second symmetry does come into the magnetic field with time, and there are occasions when the dominant symmetry changes, and becomes locked into the opposite form for several cycles. This phenomenon seems to be easier to produce with the model than is a "maunder minimum", apparently the opposite of the sun.

It is easy to see how a symmetry change can come about if one just allows the convection to be temporarily a little more vigorous in one hemisphere than the other. A stronger poloidal field is then induced in that hemisphere, earlier in the cycle, which the differential rotation then amplifies into a stronger toroidal field also at an earlier phase in the cycle. A small succession of these events leads to the phase of the cycle in one hemisphere getting well ahead of that in the other, leading to the opposite symmetry of field about the equator. It would seem that in the sun, the two hemispheres are more closely linked than in the model.

5. Generalizations and applications

a) Application to the sun

Obviously, our dynamo solutions fail to reproduce the solar dynamo in some important respects. Kinematic dynamo theory so far has done better, but its practitioners have had more free parameters and functions at their disposal. Our problem is how to get the correct direction of migration of toroidal field, and cycle period, while still staying within the constraints of fluid dynamics. In the former problem, compressibility may help greatly. If the schematic differential rotation profile shown in Figure 2, with an angular velocity maximum at mid-depth, proves to be the rule in the compressible case, then we might generate magnetic cycles which start near the equator deep in the convection zone, for which the toroidal field then migrates roughly parallel to the axis of rotation toward the surface and higher latitudes. Upon reaching the reversed gradient of angular velocity there, it then migrates back toward the equator. In the sunspot cycle, all we would then be seeing is this upper branch of the cycle.

This picture offers some real advantages. For example, it would explain why spots first appear in mid latitudes, and the equatorward moving branch in each hemisphere would keep the two hemispheres more closely linked together. A disadvantage would be that it is not clear how the high latitude part of the torsional oscillations would fit in.

Such a hybrid dynamo would prove unnecessary if we were to find that with several density scale heights in the convection zone, the convective modes maintaining the equatorial acceleration also produced an angular velocity increasing inward all the way to the bottom of the convection zone. This would be more consistent with what is emerging from the oscillations measurements, but there is no evidence from what compressible convection calculations we have done so far that it will happen.

Either way this problem is solved, it would provide resolution of the conflict between differential rotation and dynamo models over the sign of the radial gradient of angular velocity.

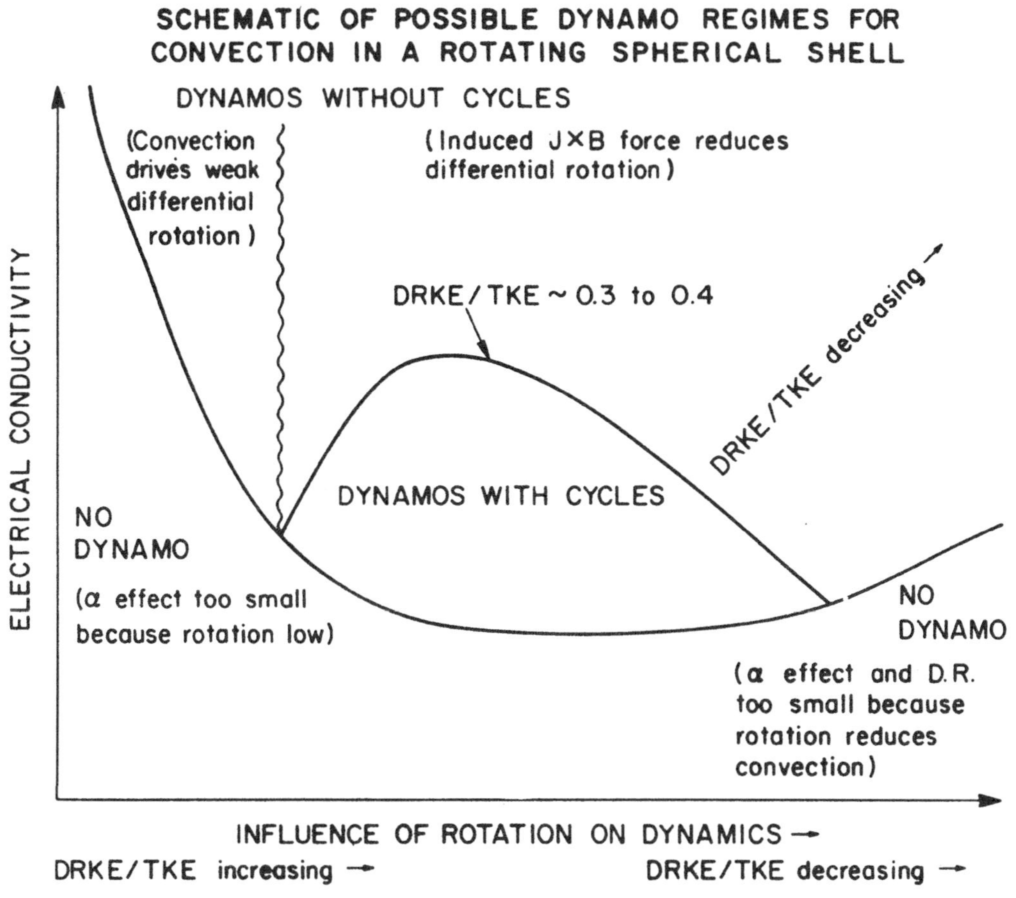

Figure 11: Schematic regime diagram for kinds of dynamos expected from model as functions of the conductivity and influence of rotation upon the dynamics.

As for the short dynamo period we find, we see at least two possibilities. One is that the global convection driving the differential rotation of the sun is even weaker than in our current model -- for which rms. horizontal velocities are already no more than 10m/sec per wave number. The other is that, due to the highly intermittent magnetic field structure in the solar convection zone, not all of the helicity present in the convection is felt by the field. For example, a field largely confined to the boundaries of convection cells would not feel the helicity of the cells' interior. If this is the case, it will be very difficult to deal with in global dynamo models, because of practical limits on spatial resolution, and will require sophisticated parameterization of the unresolved small scale interaction of velocities and magnetic fields to capture its global effects. The role of magnetic buoyancy also needs to be considered further.

b) Generalizations from our dynamo results.

Based on our experience with the dynamo model so far, as well as other physical arguments, it is possible to construct a plausible, but very schematic, "regime diagram", illustrating what kind of dynamo action to expect. This may help us think about dynamos in other stars. Such a diagram is shown in Figure 11. In Figure 11, the vertical axis is the electrical conductivity for the model (proportional to η^{-1}. The horizontal axis is an unspecified measure of the influence of rotation upon the dynamics. It could, for example, be the ratio of rotation frequency to buoyancy growth rate or convective turnover rate, although, particularly at low and moderate influence of rotation, viscosity also enters in. Below the horizontal axis, we indicate that, without magnetic fields, as the rotational influence grows, the differential rotation energy DRKE grows as a fraction of the total kinetic energy TKE of the system. But eventually, the rotation has such a strong influence on the convection it nearly suppresses it. This drives the differential rotation down at an even faster rate, because it is driven by Reynolds stresses, which are second order in convective velocities.

Figure 11 illustrates the obvious point that, for any amount of influence of rotation upon the dynamics, there will be an electrical conductivity below which there is no dynamo action. At low influence of rotation, this is because while convection has its full amplitude, the helicity or α effect is too small to overcome the dissipation. At high influence, it is because rotation is suppressing convection and differential rotation. At higher conductivities but weak rotational influence, we get dynamos, but they are not field reversing, because the differential rotation is too weak. This is the region to the left of the vertical wavy line. To the right of that line, we get cyclic dynamos in an enclosed pocket inside of which the differential rotation exceeds some substantial fraction of the total kinetic energy of the system. From our model, that fraction appears to be in the range 0.3 to 0.4. Above and to the right of that pocket, we still get dynamos, but the cycles are lost, this time because j x B forces reduce the differential rotation to the point where shearing of the poloidal field by it is not the principal means by which toroidal field is maintained.

The various boundary lines we have drawn may not be sharp or unique, but may depend on the direction from which they are approached in the parameter space -- that is, hysteresis may be present, with memory of the previous state of the dynamos that were used as initial conditions for the next calculation. We have already seen an example of this in our solutions at Q = 2.0.

Our calculations so far have essentially been along two vertical lines in this diagram, one to the left and one to the right of the wavy vertical line. To really quantify such a diagram and verify its various parts, we obviously need many more calculations, but these are quite expensive with our large model. Such calculations would be worthwhile and much more practical with a simplified version, if a good one can be found.

c) Applications to stars

Even given uncertainties in the regime diagram for the dynamo model, there are additional uncertainties in extrapolating from it to the sun and other stars. For example, to make the connection, we must identify the electrical conductivity with a turbulent conductivity for a convection zone. But that may be affected by the influence of rotation upon convection, in which case the horizontal and vertical axes would not be independent. This would be a problem mostly on the far right-hand side of the diagram.

But let us suppose that the regime diagram in Figure 11 does apply to stars, at least in some form. Stars with cycles like the sun such as observed in Calcium emission by Wilson (1978) obviously would fall inside the enclosed "dynamos with cycles" region, at intermediate influence of rotation on the dynamics. This would be consistent with the observations of Vaughan, et al. (1981) that stars with cyclic calcium emission all have rotation periods of 20 days or longer. Stars with shorter rotation period have no cycles, and higher emission. On Figure 11, these stars would lie to the right of the upper boundary of the cyclic region. For some distance to the right of this boundary, the magnetic field amplitude ought to grow, beyond which it drops as the convection is more suppressed by rotation. Clearly other stars with strong but non-cyclic emission could simply lie well above the whole "cycles" region on the diagram. For these, even at intermediate rotation, the conductivity was sufficiently high that the induced j x B force shut down the differential rotation, eliminating the cycle. The calcium emission should be larger than for the cyclic stars, which is what is observed.

With respect to cycle periods, we presumably would have the same problem with applying our model to other stars as we have had for the sun, and the solution may be the same and therefore equally difficult -- better representation of the small-scale interaction between velocities and magnetic fields. With respect to the direction of migration of the toroidal field, we have no information from other stars, but it would obviously be extremely valuable input to dynamo models. I would speculate that, the stronger is the influence of rotation on the dynamics of a particular star that has a magnetic cycle, the more likely the dominant migration will be from low to high latitudes, as in our dynamo calculations. I would think this would be particularly true for cyclic red giants in which the convection zone is geometrically very deep but still contains very few scale heights.

The effects of convection zone depth are hidden in the regime diagram in Figure 11. In general, the shallower the depth (characteristic of early type main sequence stars) the shorter is the turnover time in the convection zone. As, for example, Gilman (1980) showed, even though the rotation rate of these stars is higher than the sun, ultimately the influence of rotation on the dynamics is less, reducing the probability of cyclic dynamo action in these stars, and, for the earliest, eliminating dynamo action altogether (see also Durney and Latour, 1978).

But there is an even more subtle effect here. If the suggestion from our calculations is correct that the seat of cyclic dynamo action is principally in low latitudes outside the tangent cylinder to the inner boundary of the convection zone, then the shallower the convection zone in a star with a cyclic dynamo, the more the variable Calcium emission should be confined to low latitudes. If such a star is being observed pole-on, it might be very difficult to measure cyclic variability, even if present. On the other hand, if its axis is nearly perpendicular to the line of sight, even a convection zone only 13% of the radius would produce activity over a belt ± 30° latitude, or half of the observed area. It would be very instructive if the amplitude of cyclic emission could be correlated with orientation of the star's axis, for a range of convection zone depths.

I thank Ron Gilliland for reviewing the manuscript, Betsy Alves and Bobbie Morse for typing it, and Jack Miller and Mike Kuhn for programming the calculations.

References

Deubner, F.-L., Ulrich, R.K. and Rhodes, E.J.: 1979, Astron. Astrophys. 72, 177.

Durney, B.R. and Latour, J.: 1978, Geophys. Astrophys. Fluid Dyn. 9, 241.

Eddy, J.A., Noyes, R.W., Wohlbach, J.G. and Boornazian, A.A.: 1978, Bull. Am. Astron. Soc. 10, 400.

Foukal, P.: 1972, Astrophys. J. 72, 177.

Gilman, P.A.: 1975, J. Atmosph. Sci. 32, 1331.

Gilman, P.A.: 1977, Geophys. Astrophys. Fluid Dyn. 8, 93.

Gilman, P.A.: 1980, in Proc I.A.U. collogu #51 "Stellar Turbulence", Gray, D.F. and Linsky, J.L., eds. (Springer Verlag, New York) p 19.

Gilman, P.A. and Glatzmaier, G.A.: 1981, Astrophys. J. (Suppl.) 45, 335.

Gilman, P.A. and Miller, J.: 1981, Astrophys. J. (Suppl.) 46, 211.

Glatzmaier, G.A. and Gilman, P.A.: 1981a, Astrophys. J. (Suppl.) 45, 351.

Glatzmaier, G.A. and Gilman, P.A.: 1981b, Astrophys. J. (Suppl.) 45, 381.

Glatzmaier, G.A. and Gilman, P.A.: 1981c, Astrophys. J. (Suppl.) 47, 103.

Glatzmaier, G.A. and Gilman, P.A.: 1982, Astrophys. J. 256, 316.

Gough, D.O.: 1982, Nature (in press)

LaBonte, B.J., Howard, R. and Gilman, P.A.: 1981, Astrophys. J. 250, 796.

Vaughan, A.H., Baliunas, S.L., Middlekoop, F., Hartmann, L.W., Mihalas, D., Noyes, R.W., and Preston, G.W.: 1981, Astrophys. J. 250, 276.

Willson, R.C., Gulkis, S., Janssen, J., Hudson, H.S., and Chapman, G.A.: 1981, Science, 211, 700.

Wilson, O.C.: 1978, Astrophys. J. 226, 379.

DISCUSSION

IONSON: Dr. Golub has been stressing for years that the total magnetic flux is essentially constant throughout the solar cycle. What appears to be cycling is the magnetic flux power spectrum. That is, during solar minimum the flux is found primarily in bright points, during maximum in larger scale active regions. Has this very important empirical constraint been confirmed by contemporary dynamo theories?

GILMAN: Not really, and I am not sure we should expect them to, since such a spectrum shift may well depend strongly on the details of surface dynamics of active regions, which dynamo models do not usually account for.

SCHÜSSLER: First I have a comment on Dr. Ionson's question: In a model of the solar dynamo on the basis of fluxtubes, the anticorrelation of ephemeral active regions and big active regions is achieved if ephemeral active regions are assumed to be the shredded parts of fluxtubes under the influence of convection (cf. *Nature* **288**, p. 150).

Next I have a question to Dr. Gilman: You get a ratio Toroidal Field Energy/Poloidal Field Energy ≈ 3 for your model. Don't you think this is too small for the solar case?

GILMAN: It may be. I would note that whatever is happening on the sun that allows its magnetic fields not to sense the full helicity of global convection could also be the cause of the low poloidal field, since in the dynamo theory helicity is responsible for it. So if we solve the problem of the short dynamo period, we may also have solved the problem of poloidal field amplitude. But that remains to be seen.

WEISS: It is possible to include magnetic buoyancy within the Boussinesq approximation, and it would be interesting to see how this affected your results. In order to explain small-scale magnetic structures at the solar surface (and thus satisfy the observers) one would need to include intermittent magnetic fields and magnetic buoyancy in the theoretical description.

GILMAN: I agree, but this task is not an easy one. I decided to wait to look at magnetic buoyancy effects until the compressible version of our model was running, in which buoyancy would occur naturally.

SPRUIT: You mentioned that concentration of fields at the boundaries of cells will reduce the effect of helicity on the field. This suggests that the amount of dynamo action obtained should depend on the numerical resolution. Is this seen?

GILMAN: This would be true only if the resolution were inadequate to resolve the finest structure the model was capable of producing at the magnetic Reynolds number R_m of the flow. In our model, R_m for the smallest resolvable flow patterns is of order unity, so a large increase in resolution should not change the dynamo action much.

FRISCH: First, I would like to stress that to the best of my knowledge, the Gilman-Miller dynamo is the first example of a numerical nonlinear fluid dynamo where no violence is done to the equation of fluid dynamics. They have clearly been able to capture the first non-trivial magnetic bifurcation. Now, from nonmagnetic convection experiments (and calculations) we have learned that there are generally many bifurcations leading to quite different statistical states. It seems hard to rule out similar possibilities for the magnetic bifurcations. For example I would not rule out the possibility of getting field reversals just by increasing Reynolds numbers (and resolution).

GILMAN: Since differential rotation plays such a prominent role in determining whether there are cycles or not, I would be quite surprised if cycles reintroduced themselves for higher magnetic Reynolds numbers of the convection, while rotation and other parameters are kept fixed. It would have to come from a different cause than stretching of the poloidal field into the toroidal direction by differential rotation.

MULLAN: What happens to the dynamo as the convection zone becomes deeper? In particular, does the dynamo mode change radically (or at all) when the convection occurs throughout the star, rather than in a shell only?

GILMAN: I doubt our results would change much for deeper convection zones, since these calculations are already for a depth of 40% of the outer radius. The convective scale, and therefore the magnetic patterns would get somewhat larger, but by no more than a factor of 2. However, if the convective layer were made much shallower, the dynamo would change, because we would no longer have a broad equatorial acceleration.

GIOVANELLI: The calculations shown give very impressive results on flow patterns. I am not so happy with the magnetic fields, which seem to me to have been parameterized to the extent that results of various types can be obtained by varying the parameters. But this makes it very difficult to produce results which can be compared in detail with what we will observe at the surface. Yet this is what the observer expects from any theory. We really want to find out the physical relationships between what we observe and what causes it. There still seems a long way to go.

GILMAN: There is a long way to go, and I think you are asking too much of the theory in terms of details. To make lasting progress on a problem of this difficulty, we need to stick fairly close to the laws of physics and build, step by step.

HELICAL AND NON-HELICAL DYNAMOS

U. Frisch A. Pouquet
CNRS, Observatoire de Nice
B.P. n° 252
06007 NICE CEDEX, France

M. Meneguzzi
CNRS, DPHEP-CEN Saclay
B.P. n° 2
91190 GIF sur Yvette, France

Regeneration of magnetic fields by dynamo effect in the presence of turbulence is of interest both for astro-geophysical applications and because it may produce unwanted effects in large scale fast breader reactors. In such machines magnetic Reynolds numbers R^M of the order of 100 can be reached, possibly in excess of the critical value R_c^M for dynamo action.

Closure-based studies have revealed that the critical magnetic Reynolds number R_c^M is of the order of thirty. They also show that R_c^M can be lowered when the flow is helical, especially when there is a scale separation between the energy-carrying eddies and the largest scale of the flow (Léorat et al., 1981).

Direct numerical simulations of the three-dimensional MHD equations must be performed with a high spatial resolution in order to compute in a reliable way at (kinetic and magnetic) Reynolds numbers of 30 and/or include a good scale separation. Computing with a CRAY 1 on a grid of 64^3 points, one may reach kinetic and magnetic Reynolds numbers up to 100; such a computation takes of the order of 20 hours. Some studies may be performed at a lower resolution (and cost) on a 32^3 grid, for example by using like Gilman and Miller a magnetic Prandtl number in excess of one (Gilman, 1983).

Salient features of the results are now given. For kinetic and magnetic Reynolds numbers of 100 with non-helical forcing, a seed magnetic field with a magnetic to kinetic energy ratio β of 2% grows. After a few eddy turn-over times a statistically stationary state obtains, in which the ratio β fluctuates around 10%. The spatial structure of the magnetic field is strongly intermittent, which may be due to proximity to the transition. Other simulations with helicity gave rise to magnetic fields, well ordered in the large scales and whose energy is at least twice that of the kinetic eddies (Meneguzzi et al., 1981).

We benefitted from a contract with the National Center for Atmospheric Research (Boulder, Colorado), sponsored by the N.S.F.

REFERENCES

Gilman, P.A.: 1983, these Proceedings.
Léorat, J., Pouquet, A., and Frisch, U.: 1981, J. Fluid Mech. **104**, p. 417.
Meneguzzi, M., Frisch, U., and Pouquet, A.: 1981, Phys. Rev. Lett. **47**, p. 1060.

J. O. Stenflo (ed.), Solar and Magnetic Fields: Origins and Coronal Effects, 271.

A QUANTITATIVE STUDY OF MAGNETIC FLUX TRANSPORT ON THE SUN

N.R. Sheeley, Jr.
E.O. Hulburt Center for Space Research
Naval Research Laboratory
Washington, DC 20375
J.P. Boris and T.R. Young, Jr.
Laboratory for Computational Physics
Naval Research Laboratory
Washington, DC 20375
C.R. DeVore
Berkeley Research Associates
Springfield, VA 22150
K.L. Harvey
Solar Physics Research Corporation
Tucson, AZ 85716

Abstract: A computational model, based on diffusion, differential rotation, and meridional circulation, has been developed to simulate the transport of magnetic flux on the Sun. Using Kitt Peak magnetograms as input, we have determined a best-fit diffusion constant by comparing the computed and observed fields at later times. Our value of 730 $\pm$ 250 km^2/s is consistent with Leighton's (1964) estimate of 770-1540 km^2/s and is significantly larger than Mosher's (1977) estimate of 200-400 km^2/s. This suggests that diffusion may be fast enough to account for the observed polar magnetic field reversal without requiring a significant assist from meridional currents.

This paper presents the initial results of a project to simulate the transport of solar magnetic flux using diffusion, differential rotation, and meridional circulation. The study concerns the evolution of large-scale fields on a time scale of weeks to years, and ignores the rapid changes that accompany the emergence of new magnetic regions and the day-to-day changes of the supergranular network itelf.

Our initial objective was to determine the value of an effective diffusion constant that would provide the best fit between the computed and observed fields. To our knowledge, no fully quantitative determination has been attempted using modern computational techniques and high-quality observations. A resolution of the discrepancy between Leighton's (1964) model-dependent estimate of (770-1540 km^2/sec) and Mosher's (1977) semi-quantitative estimate of (200-400

J. O. Stenflo (ed.), Solar and Magnetic Fields: Origins and Coronal Effects, 273–278.

km^2/sec) would help to determine the relative importance of diffusion and meridional circulation in transporting flux to the Sun's poles. With a value as low as Mosher's, diffusion would require a substantial assist from meridional flow to produce a timely reversal of the polar magnetic field, whereas with a value as high as Leighton's, diffusion could accomplish the reversal alone.

We assume that the photospheric field's radial component satisfies a continuity equation containing the transport effects of Leighton diffusion, differential rotation, and meridional circulation. In spherical coordinates this equation is:

$$\frac{\partial B}{\partial t} = \frac{1}{\tau_1}\left[\frac{1}{\sin\theta}\frac{\partial}{\partial\theta}\left(\sin\theta\frac{\partial B}{\partial\theta}\right) + \frac{1}{\sin^2\theta}\frac{\partial^2 B}{\partial\varphi^2}\right] + \frac{1}{\tau_2}\left[\cos^2\theta\frac{\partial B}{\partial\varphi}\right] + \frac{1}{\tau_3}\left[\frac{1}{\sin\theta}\frac{\partial}{\partial\theta}\left(B\sin\theta\sin 2\theta\right)\right] ,$$

where the time constants are $\tau_1 = R^2/K$ (K is the diffusion constant), $\tau_2 = 1/\omega_0$ (ω_0 is the differential rotation rate), and $\tau_3 = R/V_0$ (V_0 is the amplitude of the meridional flow speed). Typical values are τ_1 = 20 years (for K = 800 km^2/sec), τ_2 = 21 days (for a Newton and Nunn rate of 2.77 deg/day), and τ_3 = 1 year (for a meridonal circulation amplitude of 20 m/sec (Duvall, 1977)).

Our study has progressed through several increasingly difficult and realistic phases. First, to test our modeling procedures, we used photographic prints of Kitt Peak magnetograms to estimate the location, pole separation and strength of new bipolar magnetic regions during 1976-1981 as the present sunspot cycle evolved. Using these doublet sources, and an assumed initial global dipole field of 1 Gauss, we calculated the evolution of the large-scale fields and displayed the fields in Carrington synoptic charts which we updated at 27-day intervals to include the contributions of flux from newly emerging regions.

Second, to estimate the diffusion constant, we selected four relatively isolated magnetic regions which were observed on several consecutive solar rotations. For each of these regions, our objective was to deposit the Kitt Peak digital magnetic measurements into the simulator during the initial solar rotation, and to compute the evolving large-scale magnetic fields at elapsed times of 1, 2, and 3 rotation periods for a range of values of the diffusion time τ_1. In practice, we also considered other combinations such as depositing the flux during its second disk appearance and comparing the computed and observed fields on the third rotation. This procedure not only provided additional tests of our measurement consistency but also helped to avoid possible magnetic field calibration errors associated with the presence of sunspots during the initial rotation. In this second phase of our study, we deposited digital data for only a limited area including the source region. The surrounding area contained the typical, but unrealistic, background field computed from the doublet sources of phase one. To minimize the unknown errors associated with this unrealistic background field we limited our comparisons to the immediate area of the expanding bipolar region.

For each region, the best fit was obtained numerically by minimizing the sum of the squared errors between the computed and observed fields and by allowing for a possible bias in the zero level of the Kitt Peak measurements. A relatively broad minimum was always obtained probably due to the unknown influence of the weak background fields. However, in half of these cases the minimum solution also provided close agreement between the observed and computed values of the centroids and the total surviving flux of the magnetic region.

The results for the case of V_o=0 are shown in the following table:

TABLE 1 - DIFFUSION CONSTANT MEASUREMENTS

Region	A	B	B	C	C	C	D	D
Deposition Rotation	1	0	2	0	0	1	0	2
Comparison Rotation	2	1	3	1	2	2	1	3
Diffusion Constant	470*	680*	700*	1160	1140	1220	1690	1070*

Summary: K = 1050 $\pm$ 400 km^2/s (all 8 measurements)
K = 730 $\pm$ 250 km^2/s (4 best measurements)

* most reliable determinations, based on considerations of the location and total flux of the evolving region.

The eight determinations in Table 1 yield an average diffusion constant of 1050 $\pm$ 400 km^2/sec, but the four most reliable measurements give the smaller value of 730 $\pm$ 250 km^2/sec. Both are significantly larger than Mosher's (1977) 200-400 km^2/sec, and are consistent with Leighton's (1964) 770-1540 km^2/sec. In future phases, we plan to study the effects of including the observed background field as well as a non-zero value of meridional circulation speed.

Figure 1 illustrates some of our procedures graphically. The upper left frame shows a Kitt Peak magnetogram on December 30, 1976. A large bipolar magnetic region (BMR) is visible in the southern hemisphere on this otherwise quiet day near sunspot minimum. The upper middle frame shows these fields after suppressing the short wavelength components of their Fourier representation. Obviously on this day the Kitt Peak zero level was biased in the negative (blue) direction and the positive-polarity north polar field is not visible. The upper right frame, displaying the normal component rather than the line-of-sight component of field, shows the same BMR deposited within the artificial background field. Relatively few BMRs have contributed to this background field at this early stage of the new sunspot cycle.

The lower frames in Figure 1 show the situation 28 days later on January 27, 1977. The lower left and center frames show the raw and smoothed Kitt Peak measurements, respectively. The lower right frame shows the BMR and its surroundings as they would have evolved with a diffusion constant of 680 km^2/sec. At this time, a newly emerged BMR, visible in the magnetogram's northern hemisphere, has not yet been deposited in the simulator. A recently deposited doublet is visible west of the large BMR in the southern hemisphere. Our measurements are based on a comparison of the flux distributions only in the immediate vicinity of the evolving BMR and not in the surrounding areas where disagreement is both obvious and expected.

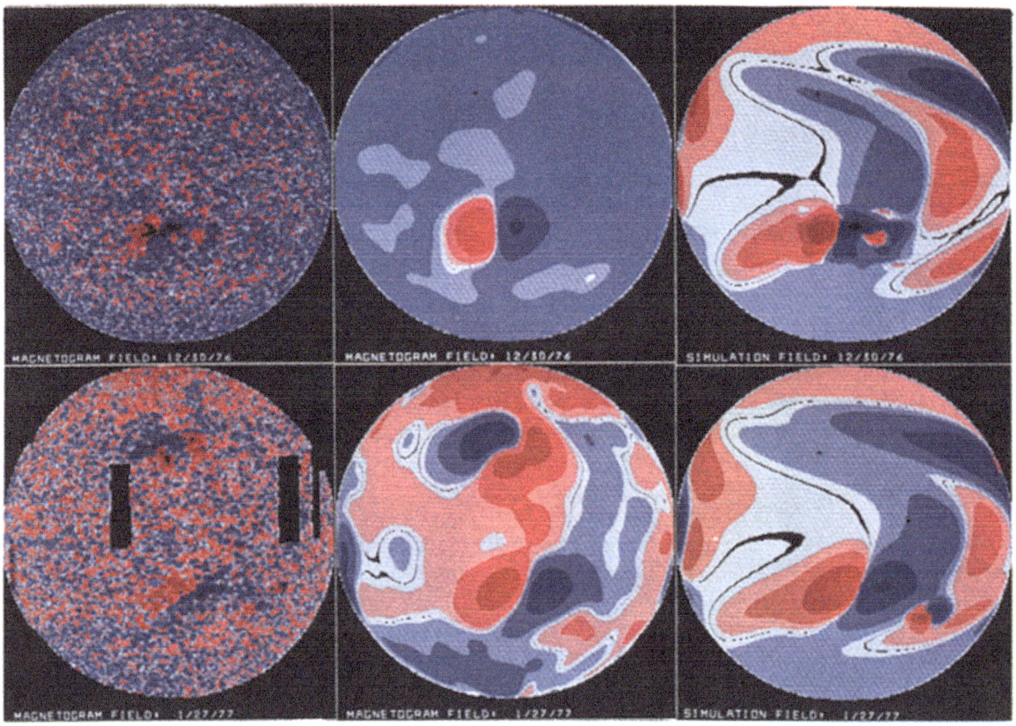

Fig. 1: Observed BMR (upper left and center) is deposited into simulated field (upper right), evolved (lower right), and compared with later observed field (lower left and center).

In conclusion, we should like to emphasize that our study is just reaching the point where most of the technical problems have been overcome. While our results to date are tentative, they are encouraging and in particular give a diffusion constant that is consistent with Leighton's original estimate.

We are grateful to C. Keppel for technical support, and to J.W. Harvey for Kitt Peak magnetograms. This work was supported by NASA (DPRs W 14, 429 and W 14,777), and by AFGL (FY7121-82-03062).

REFERENCES

Duvall, Jr., T.L.: 1979, Solar Phys. 63, 3.
Leighton, R.B.: 1964, Astrophys. J. 140, 1547.
Mosher, J.M.: 1977, The Magnetic History of Solar Active Regions, Ph.D. Thesis, Calif. Inst. of Tech.

DISCUSSION

IONSON: **Could you give us some idea of the physics hidden in your black box?**

SHEELEY: We begin with a continuity equation for the variation of the normal component of the magnetic field. We assume that the magnetic flux is transported by the combined effects of random-walk diffusion, differential rotation, and meridional circulation. Then by continually entering new sources of flux as they occur, we calculate the expected evolution of the large-scale fields.

GIOVANELLI: This is a highly important investigation. I hope that it will be possible to study flux migrations on a shorter time scale than one rotation, particularly when dealing with the fate of flux originating near the equator, which I suspect is of vital importance for the cycle mechanism.

SHEELEY: Eventually we do plan to consider a short time scale, perhaps using daily magnetograms or magnetograms every few days. Meanwhile to establish a realistic background field for the future short-term study we will try to use 27-day synoptic data and run the program for several years into the present sunspot cycle.

GOKHALE: This is a very interesting and instructive way of studying the magnetic flux transport on the Sun. Won't it be useful to include source and sink terms in the equations to account for the emergence of new flux and for disappearance of flux either by dissipation or by fluxtubes going above the surface?

SHEELEY: In fact we do include the eruption of new sources in our study of the evolution of the large-scale field during the sunspot cycle. We provide a sink by adding the positive and negative fields algebraically so that they annihilate when they overlap.

KOUTCHMY: I would like to know if the method of "smearing" you used can reproduce the neutral lines we observe over the sun, in particular through tracers like filaments?

SHEELEY: Neutral lines are clearly visible in our synoptic plots of the large-scale field. In fact, they evolve dramatically as the sunspot cycle advances, the wavy equatorial neutral line being the most prominent example. Incidentally, we have also seen the evolution of saddle-point-like field configurations associated with coronal holes.

LOW: It seems that you have constructed your equation in analogy with the conservation law for mass flow, with diffusion added. We do have, already, the hydromagnetic equation for magnetic field transport. Have you tried to show from first principles that your constructed equation is derivable from the hydromagnetic equation? This is important, because a missing or additional term can overwhelm otherwise physically real effects.

NORDLUND: Concerning the possibility of deriving your equation from first principles: To the extent that the photospheric magnetic fields are vertical, I think your equation follows from the induction equation. The vertical component is transported by horizontal flows as a scalar quantity.

SHEELEY: Yes, our equation follows directly from the induction equation $\partial\mathbf{B}/\partial t = \nabla \times (\mathbf{v} \times \mathbf{B})$, provided that $\mathbf{B}$ is purely radial, and the velocity field is given by $\mathbf{v} = (-K\nabla\mathbf{B}/B) + \mathbf{v}_{differential\ rot.} + \mathbf{v}_{meridional\ flow}$.

GAIZAUSKAS: **In the examples of bipolar regions selected for your simulation, how sure are you that there were not just one but multiple injections of flux?**

SHEELEY: Starting near sunspot minimum in 1976, we looked systematically for those regions whose evolution was both isolated and free from subsequent eruptions of flux. Only in one case did we find an anomalous field distribution, which subsequently was proved to be a secondary eruption.

VAN BALLEGOOIJEN: Can polar field reversals be explained with this model, or is it necessary to involve additional forces on the surface fluxtubes that are polarity dependent (poleward for following flux, equatorward for leading flux)?

SHEELEY: Polar field reversals can be explained with this model. Diffusion provides a meridional component of transport, and the meridional component of the doublet sources insures that the net poleward flow has the "following" polarity in each hemisphere.

ON THE STELLAR ROTATION-ACTIVITY CONNECTION

Robert Rosner
Harvard-Smithsonian Center for Astrophysics
Cambridge, MA 02138 USA

ABSTRACT

I discuss the connection between the rotation rate of late-type dwarf stars and their level of surface "activity" (as deduced from visible, UV, and X-ray observations) from both theoretical and observational perspectives.

1. INTRODUCTION

Understanding the connection between stellar activity and intrinsic stellar parameters (such as stellar mass, age, and rotation rate) is, strictly speaking, a rather formidable problem because it presupposes that one has a full understanding of the various theoretical components that are thought to be involved: stellar interior and evolution theory, stellar dynamo theory and, finally, the theory of chromospheric and coronal formation. On a qualitative level, much is of course known: that is, after all, part of the point of this symposium. Nowadays, it is common wisdom that the production of stellar magnetic fields in stellar interiors largely determines the level of stellar surface "activity" (see, for example, discussions in Bonnet & Dupree 1980). It is on the detailed, quantitative, level that the difficulty manifests itself: given a star with known mass, age (or radius), and surface rotation rate, can we predict the expected level of surface activity? In trying to address this question, three major stumbling blocks arise:

(i) It is not obvious that the given parameters are sufficient to uniquely determine stellar surface emission levels. In addition to the (trivial) fact that activity levels vary substantially during the course of the solar cycle (during which the Sun's age and mass hardly can be thought to vary), one must keep in mind the much more profound problem that stellar surface rotation rates may be a very poor guide to the actual rotation profile in the stellar interior (which is presumably a major determinant of activity levels). I will not concern myself with this basic difficulty here.

(ii) It is not obvious in what sense current theories of stellar magnetic field production and coronal activity are predictive; one must carefully examine not only the limitations of the theoretical tools used, but also scrutinize the limitations of the physics which enters current calculations. I will address this problem first.

J. O. Stenflo (ed.), Solar and Magnetic Fields: Origins and Coronal Effects, 279–300.

(iii) It is not obvious what theory is to account for. Before theorists invest much effort in detailed modeling of specific "rotation-activity" correlations which are claimed by observers, it is well to ask for the basis of these correlations; as I will try to show, although the sought-for connection between rotation and activity (for example) can be made very precise, it is another matter whether observations are presently able to establish this connection.

Unsurprisingly, there are thus both observational and theoretical conundrums to puzzle over; in the following, I will try to discuss these, keeping in mind that much of the basic details of both the observations and theory are summarized here by others (see especially Gilman 1983; Golub 1983; Schussler 1983; Vaiana 1983).

2. ON THEORY

Can one predict the variation of stellar surface "activity" with the stellar rotation rate Ω and other intrisic stellar parameters, such as stellar mass and age? To begin the discussion, it is worthwhile to distinguish the several possible theoretical approaches[1], and the roles that these play in answering the above question (see Schussler 1982 for further details):

(i) *Linear (kinematic) dynamo theory* is exploratory, but not quantitatively predictive. Linear kinematic dynamo theory is based on the supposition that Lorentz forces are negligible, so that in order to determine the evolution of (mean) stellar magnetic fields, it is sufficient to consider the field as kinematically transported by a given fluid flow (which is determined by some other theory, such as a model for interior stellar convection). Such theories (which by far dominate what one usually refers to as "dynamo theory") cannot provide information concerning the amplitude of the magnetic fields built up during the course of dynamo action; this is so because the equations are linear in the very variable one would like to have information about, the magnetic field. However, such theories do provide a physical picture for many aspects of the dynamo process, motivate the construction of more sophisticated models, and allow the calculation of growth rates, for example, for unstable dynamo waves (cf. Moffatt 1978, Parker 1979). It therefore makes little sense to attempt to link the *amplitude of stellar activity* with the measures of dynamo activity derived from linear kinematic theory (such as the growth rate or dynamo number). Naively, one might of course hope that such a connection does exist; indeed, the literature abounds with such hopes. But it remains a disheartening fact that linear theory has little, if any, predictive power regarding the behavior of solutions to the full non-linear equations (which include, for example, the back-reaction of the magnetic fields on the fluid flow). An instructive example to convince the skeptics in this regard are the recent calculations of simple non-linear extensions of standard dynamo equations by N. O. Weiss and collaborators (Weiss 1983; see also Robbins 1977, Jones 1981, and Ruzmaikin 1981); it is very evident from their work that many of the essential features of the non-linear solutions (such as quasi-periodic and chaotic behavior) are entirely absent in the linear solutions. This is of course not to say that linear theory ought to be forgotten: the point is only that the essence of the present problem -- understanding the effects of magnetic field production in the non-linear regime -- is virtually by definition outside the domain of linear dynamo theory (whose more modest aims are more in the nature of demonstrating the possibility of dynamo action, rather than in providing detailed models of observed dynamos). I suspect that it is the success of linear theory in successfully "modeling" the gross features of the solar cycle, such as the sunspot butterfly diagram and the polarity laws, by parameter fitting that has obscured the fact that such modeling actually says little regarding the governing physics (viz., Gilman 1982, 1983).

(ii) *Model non-linear dynamos* fall in a very distinct class of dynamo calculations in which the equations used are approximations of the full non-linear equations of motion (through, for example, some truncation process and/or simplifications of geometry or of the dimensionality of the system); these approximations are generally made in the spirit of, first, preserving what are believed to be the essential non-linear interactions thought to occur in the physical system and, second, making the problem mathematically tractable. The emphasis here is on capturing the basic properties of the physical processes likely to be occuring in a computationally feasible way; because these calculations are not full simulations (see below), they are not likely to provide a quantitatively-accurate account of the competition between various non-linear effects in the actual physical system (i.e., the solar convection zone). Recent calculations of magnetoconvection (see Weiss 1983 in this volume) fall into this category of dynamo modeling.

(iii) *Dynamo simulations*, that is, "full" numerical simulations of the magnetohydrodynamic equations, seek to model the dynamo behavior of the (thermally-driven) convective flows which are ultimately responsible for dynamo action; the calculations of Gilman (1981, 1982a, b), reported in this volume, are exemplary of this kind of approach. In this case, a serious attempt is made to solve the full set of equations in a realistic geometry, so that one would think that limitations of the simplified non-linear models to have been overcome. As Gilman (1983) has noted, there remain, however, several major limitations: present calculations are not fully compressible, do not take magnetic buoyancy into account, and are limited in spatial resolution. The latter difficulty manifests itself in the necessity to impose artificial (turbulent) diffusivities (because the diffusive range is not simulated), and in the fact that the formation of localized flux ("flux ropes"), as in the classic calculations of Weiss (1966), cannot be taken into account. An interesting further limitation of present calculations (which is likely to be relatively easily disposed of) is the fact that the effects of fluid helicity (i.e., $\langle v \cdot \mathrm{curl}\ v \rangle$) are derived only from fluid flow on the resolved scales (unlike the eddy diffusivities, which do take into account the "macroscopic" effects of turbulent motions on unresolved scales); thus, the turbulent diffusivities and the "α-effect" due to unresolved turbulence are not taken equivalent account of. Finally, it is argued by some (see Frisch 1982 in this volume) that the very introduction of eddy diffusivities into a simulation negates the purpose of the simulation; although, strictly speaking, this point of view has merit, I believe that for the purposes of modeling stellar dyanamo behavior, the introduction of eddy diffusivities is benign (the limitations imposed by the neglect of compressibility being much more serious).

[1] It is useful to note at this point that there do exist dynamo calculations which fall into none of the above categories; specifically, these are models in which account is taken of possible non-linearities, but without providing a systematic basis for deriving these effects (so that the non-linearities are imposed *ad hoc*). At times, such models are very enlightening (the best example perhaps being the original dynamo model of Leighton 1964, in which an artificial flux eruption and loss term was inserted into the induction equation); but, in general, I do not see such models as useful contributions because, in a profound sense, these models are not testable.

It is an interesting fact that magnetic buoyancy plays a very subsidiary role in most of the above approaches to dynamo theory: it is either altogether neglected, or enters only by defining a critical time scale (i.e., the typical rise, or crossing, time for magnetic flux generated in the deep convection zone to reach the solar surface; cf. Robinson & Durney 1982 and Schussler 1980). This neglect is very likely seriously in error. Acheson (1978) showed that, if stellar rotation is taken into account, magnetic buoyancy changes from a problem of lack of steady equilibrium to a classic instability problem; hence, a new time scale enters the problem, namely the inverse growth rate of the relevant instability (see Schmitt & Rosner 1982 for a full discussion of the various possible instabilities and associated growth rates). This insight led a number of theoreticians to re-consider the possible location of the magnetic flux generating region of the dynamo (Rosner & Vaiana 1979; Spiegel & Weiss 1980, 1981; Golub *et al.* 1980; Knobloch *et al.* 1981); and, in particular, to consider the possibility that this generating region actually lies below the convection zone proper, i.e., in the "overshoot" boundary layer that must separate the fully-convective zone from the stably-stratified radiative core, leading to the so-called "shell dynamo" (Rosner 1980).

It must be noted that this new approach to the flux generation problem is conceptually distinct from Parker's (1975) reanalysis of the rate of rise of magnetic flux due to magnetic buoyancy. In this calculation, Parker showed that reasonable estimates of the rise rate of buoyant flux tubes lead to the difficulty that, unless flux tubes are produced in the lower region of the solar convection zone, the rise time to the surface is too short to allow the "α-effect" (resulting from the action of the Coriolis force on rising flux "bubbles") to function effectively; hence, Parker argued, it must be the case that dynamo action does not occur everywhere within the solar convection zone, but rather occurs preferentially in its lower depths. In this case (as in the case of Schussler 1979, 1980, Spruit 1981 and, more recently, van Ballegooijen 1982), the focus is on the motion of flux tubes; and, in the case of Parker (1975, 1979) and others, the assumption is that the relevant dynamics occurs within the convection zone proper.

This second approach of course does not attempt to answer the question of how these flux ropes came to be, regards this problem as distinct from the buoyancy problem and, typically, appeals to (for example) the calculations of N. O. Weiss (1966) and collaborators in order to account for magnetic flux intermittency (cf. Schussler 1983). In the alternative approach which J. Schmitt and I have been following, the initial focus is on the stability of *diffuse* magnetic flux lying just below the base of the convection zone; i.e., we assume that in the region of toroidal magnetic flux amplification (which is assumed to coincide with the boundary layer separating the solar convection zone proper from the radiative core), the magnetic field has not as yet filamented into flux ropes, and is therefore "diffuse". One then investigates the following formal problem: consider the MHD stability of an electrically conducting and differentially rotating gas in the presence of a toroidal magnetic field **B**, an external, constant gravitational field, and radiation pressure, with the effects of viscosity (ν), magnetic field diffusion (η), and heat diffusion (κ) to be included. Thus posed, the problem can be addressed by application of classic linear (and local) analysis, and was first studied in depth analytically (without inclusion of radiation pressure) by Acheson (1978, 1979); the full linear analysis, covering the plausible range of available parameters (the most important of which are the (thermal) Prandtl number ν/κ and the (magnetic) Prandtl, or Schmidt, number η/κ), and presenting growth rates for both low and high frequency modes, has been recently given by Schmitt & Rosner (1982). In the following, I will briefly summarize the principal conclusions from these studies that directly impinge on the subject at hand.

To orient ourselves, we chose cylindrical coordinates (r, ϕ, z) such that the local Ω-vector points in the z-direction, the local magnetic field in the ϕ-direction and local gravity g^* is a vector in the meridional plane; Ω and B are functions of the meridional coordinates; viscous forces, radiation pressure and, of course, gravity are regarded as external forces. The basic "equilibrium" state is assumed to be axisymmetric and stationary; no motions other than the prescribed rotation rate $\Omega(r,z)$ are assumed to be present (thus one ignores flows whose time scales are of order the Eddington-Sweet circulation). Each of the diffusive processes (of heat, magnetic field, and momentum) defines a characteristic time scale: we have ν/λ^2, η/λ^2, and κ/λ^2, where λ is some scale length. These time scales may be vastly different; this possibility is of considerable importance because it is this disparity in diffusive time scales that allows for the presence of so-called doubly-diffusive instabilities (Stommel, Arons & Blanchard 1956; see review of Huppert & Turner 1981).

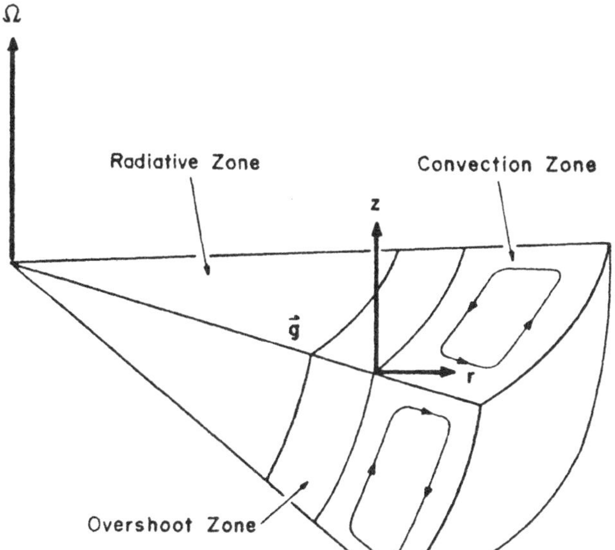

Figure 1: Sketch of the magnetic field geometry envisaged for the convectively-stable boundary layer separating the radiative core from the outer convection zone (from Schmitt & Rosner 1982).

Because the phenomenon of double-diffusive instability may not be universally familiar, I digress for a moment. Consider the classic case of cold, fresh water overlying warm salty water (Stern 1960) such that the density (as well as the salinity and temperature) decreases with height. Naively, one would think that because the density profile is statically stratified, the system ought to be also stably stratified; this is incorrect. Focus on the upward displacement of a "bubble" of water from its equilibrium position: the dense, hot, and salty bubble will try to equilibrate with the ambient less dense, cooler, and fresher water by means of molecular diffusion. Suppose that the heat diffusivity is larger than the salt diffusivity (as is indeed the case for water); the bubble's temperature will then equilibrate with its surroundings more rapidly than its salinity does, so that the bubble, upon its return to its equilibrium position, will be relatively more dense than its surroundings at this position. It therefore sinks past its initial equilibrium, and the above process repeats; hence the bubble will execute overstable oscillations which are driven by the unstable temperature gradient (i.e., in the absence of a salinity gradient, the fluid would be unstably stratified as denser cold water would overlies less dense warm water). Note, however, that the key to the instability is the inequality of the two governing diffusivities (of heat and salinity), with the third diffusivity in the problem, the viscosity, simply acting to damp the bubble's motion. Of course, more complex cases can be considered, the best-known of these being warm, salty water overlying cold, fresh (and denser) water; in this case (which is commonly encountered in oceans), the water develops highly horizontally-structured and vertically-enlongated "salt fingers", again as a result of the higher diffusion rate of heat than salinity (so that a descending, initially hot and salty "bubble" becomes yet denser than its surroundings, and continuous to descend). Where does the connection to the magnetic buoyancy problem lie? Here the appropriate analogies are between the density, temperature, and salt distributions in water and the temperature, entropy, and magnetic field in the magnetofluid, respectively; analogous to the saline water case, the magnetic diffusivity is substantially smaller than the heat diffusivity, so that overstable oscillations and, presumably, analogues of salt fingers become possible. It is the latter possibility that led J. Schmitt and me to consider the possibility that the magnetic buoyancy problem and the magnetic flux tube problem in the deep convection zone may have a common solution.

As an example of the possible behavior of the magnetofluid, consider the physically-simplest case, namely that corresponding to the axi-symmetric solutions of the full dispersion relation (studied first by Acheson and Gibbons; see Acheson 1978) for a stably-stratified region. Three different inertial frequencies must be considered: the rotation rate Ω, the Brunt-Vaisala frequency N (i.e., the frequency of adiabatic buoyancy oscillations), and a "magnetic" Brunt-Vaisala frequency M (defined analogously to N, but based on the magnetic field, rather than the entropy, gradient), which we assume to be much smaller than $\max\{N, \Omega\}$; we hence require for stability that the frequency of buoyancy oscillations far exceeds the rotation and magnetic Brunt-Vaisala frequencies. We then adopt the (astrophysically interesting) assumption that the entropy gradient is stable, but that the magnetic field gradient is unstable -- as could occur in the overshoot zone lying below the solar convection zone. The diffusivity (η) corresponding to the destabilising ingredient (the magnetic field) then obeys the ordering $\nu < \eta < \kappa$, and is thus bracketed by the diffusivities corresponding to the stabilising ingredients (rotation and entropy). In that case, buoyancy modes will be damped by efficient heat conduction on shorter length scales; and one obtains doubly-diffusive instability as long as the modulus of the ratio of the normal and magnetic Brunt-Vaisala frequencies does not significantly exceed η/κ. What actually happens? For these (axisymmetric) modes, the fluid motions must lie in the r-z plane; because of incompressibility, modes whose wave vector lies in the r-direction correspond to motions along z, and vice versa. Consider first motions along z (which are not subject to Coriolis forces, but only to buoyancy effects); because there is no effective restoring force, these modes will grow exponentially. Next, consider what happens as we rotate the wave vector in the r-z plane: as we cross the direction in which wave motions lie in the plane perpendicular to the buoyancy forces (i.e., such that $\mathbf{g}\times\mathbf{k} = 0$), we obtain stability (because there are no driving forces for these motions). When $\mathbf{k}\times\mathbf{z} = 0$, the corresponding fluid motions along the r-direction will be subject to both Coriolis and buoyancy forces; hence, Coriolis forces may stabilise buoyancy effects (by "pushing" in the azimuthal direction). However, if instability occurs, it will be oscillatory, with a frequency determined by the rotation rate Ω if the magnetic Brunt-Vaisala frequency is less than Ω. In that case, doubly-diffusive effects will enter again: the influence of Coriolis forces is now damped by diffusion of momentum (controlled by ν), while the influence of buoyancy is damped by diffusion of heat (controlled by κ) and/or magnetic fields (controlled by η).

A basic feature of these doubly-diffusive instabilities is that they occur only on "intermediate" length scales, that is, on scales lying between the short wavelengths which are heavily damped, and are thus stabilised, and the long wavelengths which are assumed to be stable; one can thus calculate cut-off wavenumbers, such that damping dominates at larger wavenumbers, by invoking the condition of marginal stability; we find that, to leading order (and for the ordering $\nu \ll \eta \ll \kappa$), the cutoff wave number s depends very strongly upon κ

$$s^4 \sim -[\,\eta\kappa\,]^{-1} \cdot N^2[\,M^2 + \Omega^2 + (\eta/\kappa)N^2\,] \,/\, [\,M^2 + \gamma(\eta/\kappa)\Omega^2 + (\eta/\kappa)N^2] \qquad (2.1)$$

(where γ is the ratio of specific heats), in contrast to Acheson's (1979) result obtained for an unstratified atmosphere.

It is evident from the above that the actual behavior of the magnetic layer depends crucially upon the actual values of the various diffusivities, as well as on the actual value of the superadiabatic gradient and mean (diffuse) field strength in the unstable layer. Consider first the diffusivities. In the radiative interior, typical values for the Prandtl (ν/κ) and Schmidt (η/κ) numbers are

$$\nu/\kappa \sim 2\times10^{-6}, \qquad \eta/\kappa \sim 7\times10^{-4}, \qquad (2.2)$$

and thus we are most certainly in the parameter regime $\nu \ll \eta \ll \kappa$ assumed by Acheson (1978): application of this theory to the radiative interior of the Sun thus seems to be on safe grounds (but is not particularly interesting). The situation is however more problematical in the solar convection zone: this region is turbulently convective (the Reynolds number is very large), so that if one regards the physical variables in the MHD equations of motion as averages (i.e., as variables defined on spatial and temporal scales large when compared to the integral scales of the turbulent flow), then the *eddy* (rather than the molecular) diffusivities are appropriate; these we take to be isotropic. This latter assumption is probably reasonable because, as the scales modeled by us are much smaller than the local pressure scale height, and as the scales of turbulent motions which provide the eddy diffusivities occur on yet smaller scales, the relevant Rossby number is likely to be large; hence rotation is unimportant for the scales of the turbulence responsible for the eddy diffusivities.

As alluded to above, there are a myrad of other obstacles to (numerically) evaluating the growth rates for the unstable modes: we do not know the value of the superadiabatic gradient in the boundary layer separating the convection zone from the radiative interior (this is related to the classic problem of calculating the extent of the overshoot region which must underlie the level at which the Schwarzschild criterion is met); and we do not know with any certainty exactly where the azimuthal (toroidal) magnetic flux is really built up (the earlier arguments not withstanding). But let us suppose that magnetic flux storage *does* occur below the fully-developed convection zone (cf. Arter 1983). In order to scale our results, we adopt the parameter values

$$B_{mean} \sim 10{,}000 \text{ gauss}; \quad (2.3a)$$

$$|B|/|\nabla B| = 0.2\,\rho/|\nabla\rho|; \quad (2.3b)$$

$$\Omega = 10 \times \Omega_{\odot}; \quad (2.3c)$$

$$\nabla - \nabla_{adiabatic} \sim -10^{-4}. \quad (2.3d)$$

The values for the density, the density and pressure scale heights, the sound speed, the curvature radius, the gravitational acceleration, and the adiabatic index γ are relatively well established, and are kept constant throughout our calculations. The radiation pressure is neglected, and the surface angular velocity is adopted as the rotation rate (although the rotation rate very likely varies radially); we thus neglect local effects of differential rotation.

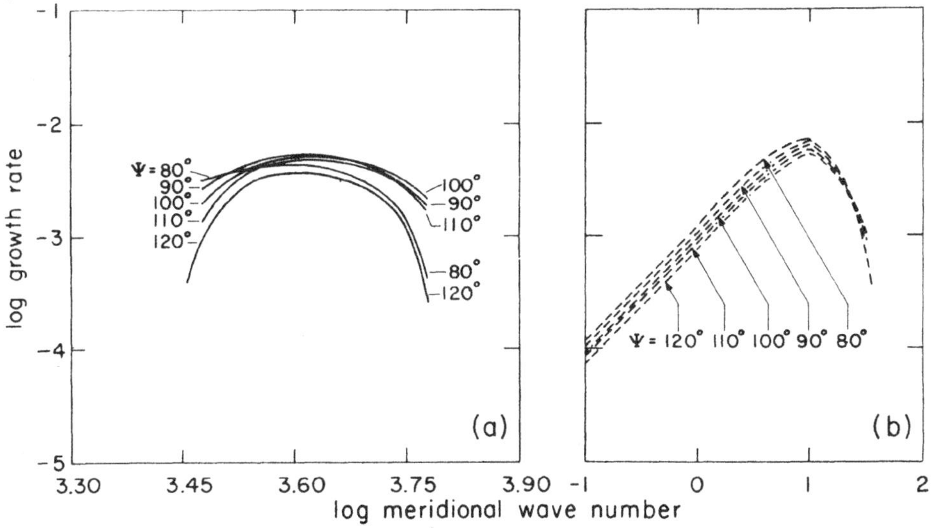

Figure 2: Growth rate versus wave number in the nonaxisymmetric case, with azimuthal wavenumber m = 3, rotation rate $\Omega = 10 \times \Omega_{sun}$, for (typical) values of the superadiabatic gradient $\nabla - \nabla_{ad} = -10^{-4}$, field strength $B = 10^4$G, and magnetic field gradient scale $H_B = H_\rho/5$; ψ is the angle between the direction of wave propagation and the radial direction at colatitude $\theta = 60^o$. Except for the singular case $\theta = 0^o$, the results are insensitive to the choice of colatitude. We assume that there is no magnetic field gradient along the z-axis, and plot the growth rate for various values of ψ. There are two principal branches, corresponding to (a) inertial ("fast") waves, whose growth rates peak at small spatial scales (large wave numbers); and (b) magnetostrophic ("slow") modes, whose growth rates peak at large spatial scales (small wave numbers) (from Schmitt & Rosner 1982).

Typical results for the growth rate of the unstable modes are shown in Figure 2 for the simplest non-axisymmetric case, in which there are no magnetic field gradients along Ω (so that there is no buoyancy driving force in the direction in which fluid motions are unaffected by the Coriolis force); we arbitrarily adopt an azimuthal wave number m = 3. Under these assumptions, motions along Ω will be stable, and motions along the magnetic field gradient will be subject to the full stabilisation of rotation. The most unstable meridional wavevectors lie in the radial direction, as shown in Figure 2; in particular, the "slow" (magnetostrophic mode) instability occurs on the longer scales (smaller wavenumbers), and the "fast" (inertial mode) instability occurs on shorter scales (it turns out that axisymmetric modes, with m = 0, have no "slow" mode instability). Because the maximum growth rates are almost independent of the assumed Brunt-Vaisala frequency if $|M^2/N^2| < \eta/\kappa$, we conclude that, qualitatively, the assumed magnetic field configuration does not seem to be critical for the appearance of instability.

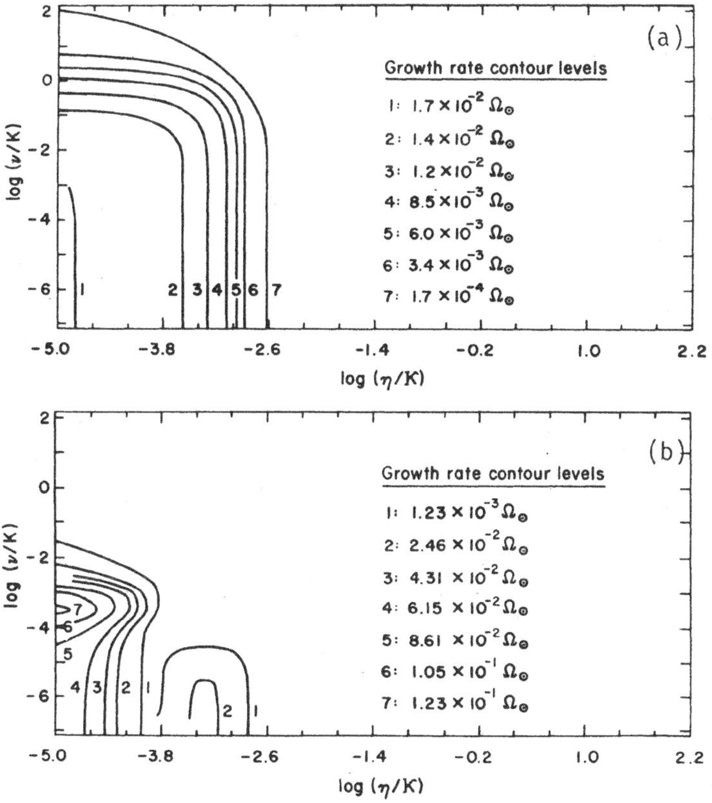

Figure 3: Variation of the maximal growth rate versus the Prandtl and Schmidt numbers, for the parameter values used in Figure 2; we fix ψ = 160°, and show results for two different wave number regimes, $\kappa s^2/\Omega$ = (a) 10 ("slow" mode regime); (b) 10^3 ("fast" mode regime). Note that if turbulent eddy diffusivities apply (such that both the Prandtl and Schmidt numbers are of order unity), the magnetic field configuration is stable (from Schmitt & Rosner 1982).

Now, suppose that the diffusivities were given instead by their eddy values; does the behavior of the instabilities change? On the scales of interest, the turbulent viscosity will be given by $\nu_{turb} \sim u(\lambda)\lambda$, where u is a typical velocity on the scale λ. If we assume u to be given by typical velocities in the lower part of the convection zone, we have $u \leq 10^3$ cm sec^{-1}; setting $\lambda \leq 10^5$ cm, we then obtain $\nu_{turb} < 10^8$ cm^{2}sec^{-1}. We assume that the energy flux is largely carried by radiation, so that $(\nu/\kappa)_{eff} \sim O(1)$; and, because the magnetic Prandtl number and Schmidt numbers are not well established, we shall explore the range of values $10^{-4} << \eta/\kappa << 10^{+2}$.

In order to carry out this exploration of the parameter space of the (thermal) Prandtl and Schmidt numbers, we fix the wave numbers s and m, and calculate growth rates as a function of η/κ and ν/κ. Figure 3a shows the result of such a calculation for the case $\kappa s^2/\Omega = 10$, m = 3, assuming again a spherically symmetric mean magnetic field configuration. The "slow" mode is evidently unstable for $\nu/\kappa \leq 10$, $\eta/\kappa \leq 10^{-2}$. Similarily, we obtain the growth rates for the most unstable mode for the case $\kappa s^2/\Omega = 10^3$ (so that one expects to obtain the "fast" mode instability), shown in Figure 3b. These figures clearly show that the region of instability in the η/κ - ν/κ plane is a function of both parameters, although the growth rate, once it exceeds zero, depends only weakly on η/κ and ν/κ.

What are some of the implications of the above formal results? Here we tread on far less certain grounds. Doubly-diffusive processes characteristically give rise to intrinsically preferred length scales; thus, in our case, maximum growth rates typically occur on (dimensionless) scales $\kappa s^2/\Omega \sim 1 - 1000$ for the range of subadiabatic gradient values considered by us, corresponding to meridional wavelengths $\lambda_{mer} \sim 10 - 400$ km for $\kappa \sim 10^8$ cm^{2}sec^{-1}. One interesting possibility is that this instability (in the non-linear regime) gives rise to *flux tubes* of diameter $\sim \lambda_{mer}$ (cf. Spiegel & Weiss 1980; Hughes 1982); simple-minded scaling arguments for the change in flux tube diameter during the course of its rise to the surface then yield flux rope scale sizes of ~ 10,000 km or more, not an outrageous number. Thus, it may be that flux tube formation may be a natural (nonlinear) consequence of doubly-diffusive instability.

Our growth rate calculations also show that in a not too strongly stabilised region, *both* "slow" hydromagnetic waves *and* "fast" inertial waves may become unstable; these instabilities occur with comparable growth rates, but rather different characteristic spatial scales. Radiation is sufficiently efficient to cause these waves to lose their thermal buoyancy, but magnetic field diffusion does not occur rapidly enough to remove their magnetic buoyancy; the energy driving the instability therefore must come from the magnetic field free energy. It is not evident which modes are likely to be excited; however, experience with other systems exhibiting convectively unstable and stable zones (viz., thermosolutal convection) suggests that gravity modes will indeed be efficiently excited (see Press 1981).

Perhaps the most intruiging consequence of these doubly-diffusive instabilities is the implication for the so-called "shell" dynamo (Rosner & Vaiana 1979). As argued above, the overshoot convection boundary layer is thought to be the site for toroidal magnetic flux generation by the ω-dynamo. Because of the presence of turbulent flows, the Schmidt (or magnetic Prandtl) number will be large [i.e., O(1)]. Hence, the ratio of the magnetic and non-magnetic Brunt-Vaisala frequencies will be smaller than the Schmidt number, and the doubly-diffusive instabilities discussed above will be inhibited. Thus, toroidal magnetic flux accumulation is possible in this boundary layer. However, once the toroidal magnetic field exceeds some critical threshold, turbulent motions will be suppressed; the eddy diffusivities will hence decrease (and so the Schmidt number will become small), and the system will become unstable. Because the instability growth rates are large, toroidal magnetic flux ejection in this model would then be episodic on a recurrence time scale fixed by the rate of toroidal flux amplification by the ω-dynamo. We conjecture that the formation of active region complexes is a consequence of this episodic eruption process, and that the spatial scale of the erupting magnetic flux is determined by the doubly-diffusive mechanism we have discussed above.

Where does this theoretical work leave us? It is not hard to notice the many assumptions adopted in the above discussion; although these are necessary in order to progress in the present instability calculations, one would really prefer to know the actual parameter regime applicable to the Sun. The overshoot problem is just now being seriously addressed (cf. Hulbert, Toomre, & Massaguer 1981; van Ballegooijen 1982; Marcus, Press & Teukolsky 1982; Schmitt & Rosner 1983), so that there is some hope of understanding the stratification of the deep convecton zone and overshoot boundary; and the crucial problem of the non-linear development of the doubly-diffusive modes is now also beginning to be attacked. But whether a definitive statement regarding the quantitative connection between stellar rotation rate and "activity" can be made is somewhat doubtful.

3. ON OBSERVATIONS.

I now turn to a far more observationally-oriented aspect of the stellar "activity-rotation connection"; and ask to what extent we have observational knowledge of the existence of such a "connection" and, more specifically, to what extent one should -- as a theoretician -- pay heed to the precise nature of the correlations found by observers to connect stellar rotation rates and the various measures of stellar surface activity.

The most basic question to be addressed is, quite clearly, why one ought to expect difficulties in the first place. Certainly, it must be that, as the volume of data has grown (virtually exponentially), application of standard data reduction and error analysis techniques, together with careful attention to the possible existence of sample biases, will avoid gross error; this ought certainly hold for correlation analyses of *complete* samples (that is, samples containing every star meeting some *a priori* selection criteria or, somewhat more weakly, containing a truly random selection of such stars). In the following, I will show that such *standard* analyses, when applied to *currently-available* stellar data, will very likely lead to error.

To be more precise, consider a volume-limited sample of stars, for which we have measurements of stellar parameters such as rotation rates, mass, effective temperature, x-ray luminosity, Ca II index, etc. Let us assume for the moment that for every star in the sample (of N stars), *all* data are available; that is, every star in the sample is characterized by M distinct parameter measurements $F \equiv \{f_1, f_2, \ldots, f_M\}$, so that the full sample is completely characterized by the N x M matrix of measured parameters. Classically, one can describe the sample by specifying the joint probability distribution function $\psi(f_1, f_2, \ldots, f_M)$, such that

$$\psi(\cdots)\, df_1 \cdots df_M \qquad (3.1)$$

is the probability that a given star has its characteristic parameters lying in the range (f_1, $f_1 + df_1$; ... ; f_M, $f_M + df_M$). Now, it is always possible to write ψ in the form (with $f_1 \equiv L$, L some passband luminosity, and $f_2 \equiv \Omega$, Ω the stellar rotation frequency)

$$\psi(L, \Omega, \cdots) = \Theta(\Omega) \cdot \Phi(L|\Omega, \cdots), \tag{3.2}$$

so that the first moment of the marginal (or conditional) distribution function Φ yields the *regression curve*

$$\langle L(\Omega, \cdots)\rangle = \int dL\, L\, \Phi(L|\Omega, \cdots). \tag{3.3}$$

The aim of standard correlation analysis (in its various guises) is in fact to construct this regression curve, which functionally relates the *mean* luminosity to the remaining stellar parameters. Least-squares, or minimum-variance, fitting and common factor analyses all fall into this generic class; in the most sophisticated versions, one attempts to construct a minimal set of stellar parameters (with $M' < M$) that fully describe the sample, and so "factor out" those parameters which are not truly independent (the functional dependence of these "factored-out" parameters on the remaining stellar parameters is one of the principal goals of such analyses). All of this is subject to two major restrictions:

(i) In general, we do not know $\Phi(L \mid \Omega, \ldots)$. The common remedy is to *assume* some *a priori* functional form for $\langle L\rangle$ which contains several adjustable parameters θ_i, and then to minimize the variance of the fit of $\langle L\rangle$ to the data by variation of these parameters. This is the aim of classical regression analysis, and has been used to describe, for example, the correlation between stellar rotation rate and stellar x-ray emission (cf. Pallavicini 1980, 1982; Walter & Bowyer 1981; Walter 1981, 1982); in this case, the adjustable parameters θ_i are the coefficient and exponent of the power law usually assumed to define the *mean* variation of L_X, $\langle L_X\rangle$, with Ω.

(ii) In general, one may not have measurements of *all* M of the parameters characterizing a given star; more specifically, some of the stars in the sample may have only upper bounds on their (X-ray) luminosity or rotation rate. This lack of information is only troubling insofar as standard regression analysis cannot deal with upper bounds. For example, classical least squares fitting can only be performed if one has a measurement to compare theory with. It is unfortunately common to deal with this difficulty by simply ignoring it; typically, one finds that stars whose parametric description is incomplete are simply removed from the sample under study. Naively, one might think that removal of incomplete information is benign, but in fact upper bounds do provide information concerning the distribution of a given parameter within the test sample.

To see how both detections and upper bounds contribute information about the sample population, consider the luminosity function $\phi(L)$,

$$\phi(L) = \int d\Omega \cdots \Phi(L|\Omega, \cdots), \tag{3.4}$$

obtained by integrating Φ from Eq. (3.2) over all the remaining stellar parameters; $\phi(L)$ has the property that the probability for a given star drawn from the sample population to have a luminosity L in the range [L, L + dL] is just

$$\phi(L)\, dL$$

The basic idea, first discussed by Avni *et al.* (1980; also Avni 1981) in the context of quasar luminosity functions and quasar evolution, is that upper bounds impose integral constraints on the luminosity function which provide information rather analogous to that provided by actual detections. More specifically, consider the following simplified heuristic argument. Suppose we bin our detections and upper bounds as shown in Figure 4; and let us suppose that we really know the true distribution of L, that is, we know the bin probabilities ϕ_i, i = 1, ... , M (where M is the total number of luminosity bins), defined by

$$\phi_i \equiv \int_{L_i}^{L_i + \delta L} dL\, \phi(L) . \tag{3.5}$$

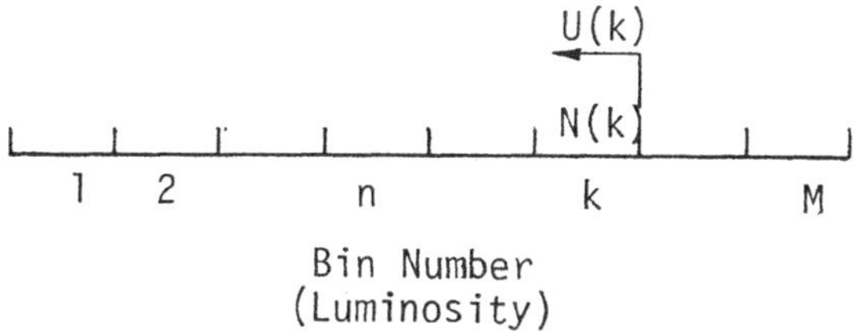

Figure 4: Sketch of luminosity binning arrangement; the horizontal luminosity axis is consists of discrete bins, each characterized by an as yet unknown bin probability that a given observed star will have a luminosity in the luminosity range spanned by the bin. See text for further details.

Now denote the number of detections in bin k by N(k), and the number of upper bounds in bin k by U(k). Then the probability that a given upper bound in bin k (say) "really" belongs in bin n (that is, the star really has a luminosity that would place it in bin n, if we had sufficient sensitivity to observe it), is

$$\phi_n \;/\; \sum_{k'=1}^{k} \phi_{k'} . \tag{3.6}$$

Thus, the expected number of stars in our sample that have upper bounds on L which placed them in bin k, but which really "belong" in bin n, is just

$$U(k) \cdot \phi_n \;/\; \sum_{k'=1}^{k} \phi_{k'} . \tag{3.7}$$

Thus, the "effective" number of stars which belong in bin n is just the sum of the stars with measured luminosities which fell into bin n, plus the number obtained from the upper bounds, Eq. (3.7):

$$N_{effective}(n) = N(n) + \sum_{k=n}^{M} U(k) \cdot \phi_n \; / \; \sum_{k'=1}^{k} \phi_{k'}. \tag{3.8}$$

Now, in the limit of large samples, we expect that the ratio $N_{effective}(n)/N_t$ (N_t the total number of stars in the sample) will approach the bin probability ϕ_n; hence, with a bit of algebra, we can solve for ϕ_n in Eq. (3.8):

$$\phi_n = N(n) \; / \; [N_t - \sum_{k=n}^{M} U(k) \; / \; [1 - \sum_{k'=k+1}^{M} \phi_{k'}\,]\,]. \tag{3.9}$$

This result gives, in closed iterative form, the differential luminosity function for the sample, and takes the upper bounds into account; given the set of values {N(k), U(k)}, one starts with n = M (the upper-most luminosity bin) and work downwards to smaller n. *Note that this result is significantly different from what would be obtained if upper bounds were not taken into account (i.e., if U(k) = 0 for all k in Eq. 3.9).* For some purposes, it is more convenient to work with the integral luminosity function F(L), defined simply as

$$F(L) \equiv \int_L^{\infty} dL \; \phi(L); \tag{3.10}$$

an example of a result of such a calculation is shown in Figure 5, the integral x-ray luminosity function for dwarf M stars (Rosner *et al.* 1981).

Now let us return to the original problem of determining the functional form of the marginal distribution $\Phi(L \mid \Omega, \ldots)$, which contains the crucial information linking activity and the classical stellar parameters, such as rotation rate. Following the approach just described for deriving the stellar luminosity function, we will now outline the method for constructing Φ; the parametric form presented here is due originally to Y. Avni (Avni & Tananbaum 1982; Tananbaum *et al.* 1982).

To begin, we shall adopt three major assumptions:

1. We suppose that the problem can be described parametrically, that is, we shall assume that the mean variation of the stellar (X-ray) luminosity $\langle L \rangle$ (defined in Eq. 3.3) can be expressed as an analytic function of the remaining stellar parameters,

$$\langle L \rangle = f(\Omega, \cdots ; \theta_1, \theta_2, \cdots \theta_m), \tag{3.11}$$

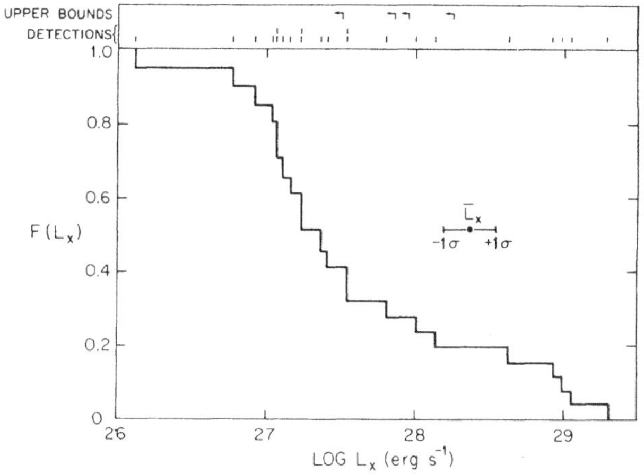

Figure 5: Integral x-ray luminosity function for a volume-limited sample of M dwarfs, as derived from *Einstein Observatory* data; F(>L) indicates the fraction of the entire population of dwarf M stars whose luminosity exceeds L (from Rosner *et al.* 1981).

where $\theta_1, \dots, \theta_m$ are the m regression parameters to be determined. For example, it is commonly assumed that one can connect $\langle L \rangle$ with Ω using the relation

$$\log \langle L \rangle = \theta_1 + \theta_2 \log \Omega, \tag{3.12}$$

where θ_1 and θ_2 are the fitting (or regression) parameters. It is not essential to assume a specific functional form for the connection between $\langle L \rangle$ and, say, Ω; indeed, it is straightforward to apply this method non-parametrically (so that no assumptions regarding the functional form of $\langle L \rangle$ must be made), as a student at Harvard, J. Schmitt (1982), has recently done.

2. We assume that all stellar parameters have been determined, with the sole exception of L, for which we have measurements as well as upper bounds; we regard each value of L (or upper bound on L) as an independent random variable whose distribution we seek to determine. Again, it is straightforward to extend this method to the case in which upper bounds for some of the other stellar parameters also exist (most crucially, for the rotation rate Ω); J. Schmitt has carried this extension out as well.

3. We assume that the residuals $\{(L^o - f^{best})_i\}$, where L^o is the observed luminosity and $f^{best} = \langle L \rangle$ is the best-fit mean relation, have a Gaussian distribution with zero mean and variance σ. This assumption is essential, but may be invalid: whether this assumption is applicable can only be determined *a posteriori*.

The method is now quite straightforward: we first construct the likelihood function Λ (Kendall & Stuart 1976)

$$\Lambda(X \mid \Theta) = \Phi(L_1 \mid \Theta)\, \Phi(L_2 \mid \Theta) \cdots \Phi(L_n \mid \Theta) \qquad (3.13)$$

where $X = (L_1, \dots, L_n)$ is the set of detections and upper bounds for L, and $\Theta = (\theta_1, \dots, \theta_m)$ is the set of m independently varied parameters in the fitting distribution function (in our case, the distribution function is assumed to be a Gaussian, and only one parameter enters, namely the variance σ). Note that the likelihood function is nothing but the joint probability density function for the independent random variables L_i. It is customary (and algebraically helpful in the following) to work with the negative natural logarithm of Λ, $S \equiv -2 \ln \Lambda$; for the case at hand, we find that S is given by

$$S = 2N_d \ln \sigma + \sigma^{-2} \sum_{i=1}^{N_d} (L^d_i - f)^2 - 2 \sum_{i=1}^{N_u} \ln G[\,(L^u_i - f)/\sigma\,], \qquad (3.14)$$

where

$$G(t) \equiv \int_t^{\infty} dt'\, (2\pi)^{-1/2} \exp(-t'^2/2); \qquad (3.15)$$

As an example of this method, I have applied the above formalism in order to obtain the conditional x-ray luminosity function

$$\Phi \equiv \Phi(L_x \mid \Omega) \qquad (3.16)$$

for single late-type main sequence stars (using the data of Pallavicini *et al*. 1982); this distribution function basically measures the variation in dispersion of the stellar x-ray luminosity about the mean $\langle L_x \rangle$ as a function of the stellar rotation rate Ω. The outcome of our calculation, cast in terms of the cumulative distribution F (scaled to unit variance)

$$F(L_x) = \int_0^{L_x} dL_x'\, \Phi(L_x' \mid \Omega), \qquad (3.17)$$

is shown in Figure 6. The most striking result is that the residuals of the best-fit relation connecting $\langle L_x \rangle$ with Ω do seem to follow a Gaussian distribution of zero mean and unit variance, as we *assumed* in the beginning; that is, the distribution of L_x about its mean $\langle L_x \rangle$ for fixed rotation rate is consistent with a random distribution; this suggests that:

(i) the variation of $\langle L_x \rangle$ with Ω is the dominant systematic contribution to the observed range in x-ray luminosities;

(ii) the remaining scatter in luminosity (associated with the residuals $L^d_i - \langle L_x \rangle$) may be due to intrinsically stochastic processes; the obvious possibilites are a superposition of contributions from short-term stellar variability (for example, from flares) and from stellar activity cycle variations (which are randomly sampled in phase).

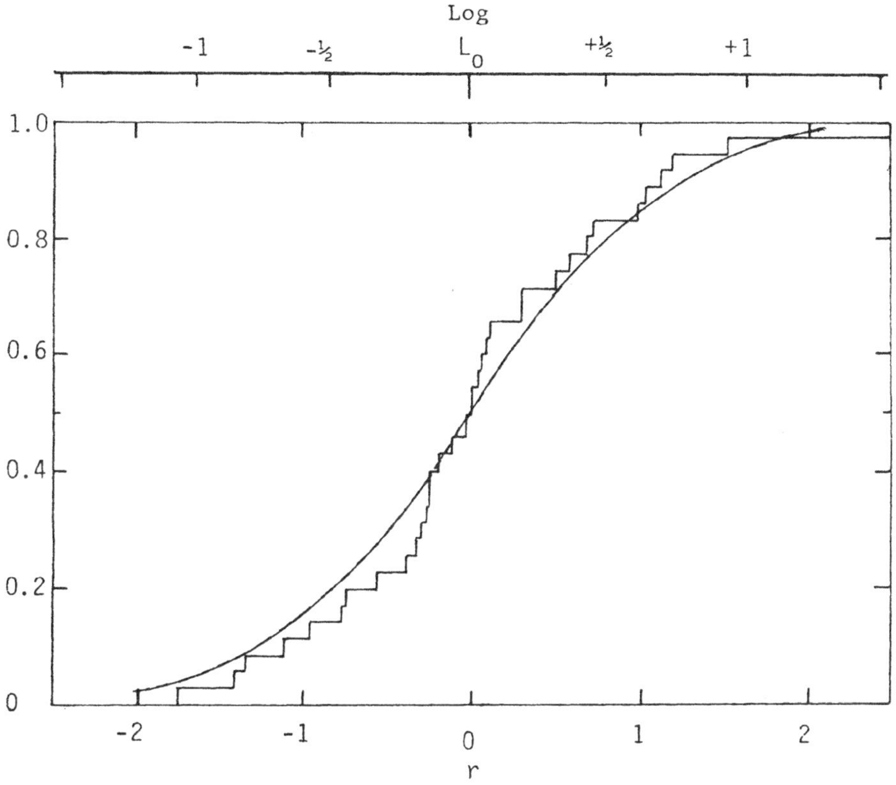

Figure 6: Cumulative x-ray luminosity distribution function $F(L \mid \Omega)$, obtained from a sample of single, late-type dwarf stars for which x-ray luminosities (or upper bounds) and rotation rates were available (from Pallavicini *et al.* 1982). Φ gives the distribution of stellar X-ray luminosity about the regression curve $\log \langle L_x \rangle = \theta_1 + \theta_2 \log \Omega$. Also shown is the predicted distribution function (solid curve) for a Gaussian random variable of zero mean and variance 1; note the close correspondence between the observational result and the theoretical curve.

It is interesting to note that, if the residuals are dominated by stellar cycle variability, then the observed variance of the residuals for fixed Ω (given by Figure 6) *immediately gives the amplitude of stellar activity as a function of rotation rate*; hence, long-term monitoring may not be required to study this aspect of activity cycles (long-term monitoring is of course crucial to studies of, for example, cycle periods).

4. SUMMARY.

I have outlined some of the limitations of present attempts to "model" the connection between stellar activity and intrinsic stellar parameters, such as the rotation rate; and have discussed some remedies. It appears that at present, it may be premature to ask theory to account for reported correlations between, for example, x-ray luminosity and stellar rotation rate, on a quantitative level; but the time seems ripe to apply more rigorous analysis techniques in order to establish the observational *quantitative* connection between intrinsic stellar parameters and "activity" characteristics. Rather than "reinventing the wheel", we have taken advantage of recent work on quasar luminosity functions (in which many of the observational problems encountered in stellar research -- upper bounds on detections, incomplete samples, etc., come up as well), and have applied this methodology (with some modification) to analysis of stellar x-ray and rotation data; these methods could be easily extended to similar work with other "activity" indicators, such as Ca II data.

ACKNOWLEDGEMENTS.

I would like to thank Y. Avni, J. Schmitt, and N. O. Weiss for many extensive discussions and suggestions. This work was partially supported by NAGW-79 and NAG8-445 at Harvard University.

REFERENCES.

Acheson, D. J.: 1978, *Phil. Trans. Roy. Soc. Lond.* **A, 289**, 459.
Acheson, D. J., and Gibbons, M. P.: 1978, Appendix to Acheson (1978).
Acheson, D. J.: 1979, *Solar Phys.*, **62**, 23.
Arter, W.: 1983, in **Solar and Stellar Magnetic Fields**, ed. J. O. Stenflo, this volume.
Avni, Y.: 1981, private communication.
Avni, Y., Soltan, A., Tananbaum, H., and Zamorani, G.: 1980, *Ap. J.*, **238**, 800.
Avni, Y., and Tananbaum, H.: 1982, preprint.
Bonnet, R. M. and Dupree, A. K.: 1981, **Solar Phenomena in Stars and Stellar Systems**, editors (Dordrecht: Reidel).
Durney, B. R., and Spruit, H. C.: 1979, *Ap. J.*, **234**, 1067.
Galloway, D. J., and Weiss, N. O.: 1981, *Ap. J.*, **243**, 945.
Gilman, P. A.: 1970, *Ap. J.*, **162**, 1019.
Gilman, P. A.: 1981, in **The Sun As a Star**, ed. S. Jordan (NASA SP-450), p. 231.
Gilman, P. A.: 1982, in **Cool Stars, Stellar Systems, and the Sun**, ed M. S. Giampapa and L. Golub (SAO Report 392).
Gilman, P. A.: 1983, this volume.
Golub, L.: 1983, this volume.
Golub, L., Rosner, R., Vaiana, G. S., and Weiss, N. O.: 1981, *Ap. J.*, **243**, 309.
Hulbert, N., Toomre, J., and Massaguer, J. M.: 1981, *BAAS*, **13**, 912.
Huppert, H. E., and Turner, J. S.: 1981, *J. Fluid Mech.*, **106**, 299.
Jones, C. A.: 1981, *Geophys. Astrophys. Fluid Dyn.*
Kendall, M., and Stuart, A.: 1976, **The Advanced Theory of Statistics, Vol. 2**, (McMillan).
Knobloch, E., and Rosner, R.: 1981, *Ap. J.*, **247**, 300.
Knobloch, E., Rosner, R., and Weiss, N. O.: 1981, *M. N. R. A. S. (Comm.)*, **197**, 45P.
Marcus, P. S., Press, W. H., and Teukolsky, S. A.: 1982, *Ap. J.*, in press.

Moffatt, H. K.: 1978, **Magnetic Field Generation in Electrically Conducting Fluids** (Cambridge: Cambridge Univ. Press).
Pallavicini, R., Golub, L., Rosner, R., and Vaiana, G. S. 1982, in **Cool Stars, Stellar Systems, and the Sun (Vol. I)**, ed. M. S. Giampapa and L. Golub, pp. 77.
Parker, E. N.: 1975, *Ap. J.*, **198**, 205.
Parker, E. N.: 1979, **Cosmical Magnetic Fields** (Oxford: Clarendon Press).
Press, W. H.: 1981, *Ap. J.*, **245**, 286.
Roberts, P. H., and Stewartson, K.: 1977, *Astr. Nachr.*, **298**, 311.
Robinson, R. D. and Durney, B. R.: 1982, *Astron. Ap.*, **108**, 322.
Rosner, R.: 1980, in **Cool Stars, Stellar Systems, and the Sun**, ed. A. K. Dupree (SAO Report No. 389), p. 79.
Rosner, R., and Vaiana, G. S.: 1979, in **X-ray Astronomy**, ed. R. Giacconi and G. Setti (Dortrecht: Reidel), p. 129.
Rosner, R., *et al.*: 1981, *Ap. J. Letters*, **249**, L5.
Ruzmaikin, A. A.: 1981, *Comm. Astrophys.*, **9**, 85.
Schmitt, J. H. M. M.: 1982, in preparation.
Schmitt, J. H. M. M., and Rosner, R.: 1982, *Ap. J.*, in press.
Schmitt, J. H. M. M., and Rosner, R.: 1982, in preparation.
Schussler, M.: 1979, *Astron. Ap.*, **71**, 79.
Schussler, M.: 1980, *Nature*, **288**, 150.
Schussler, M.: 1983, this volume.
Spiegel, E. A., and Zahn, J.-P.: 1970, *Comments Astrophys. Space Phys.*, **2**, 178.
Spiegel, E. A., and Weiss, N. O.: 1980, *Nature*, **287**, 616.
Spiegel, E. A., and Weiss, N. O.: 1981, preprint (Columbia Univ./Astronomy #A10).
Spruit, H. C.: 1976, Ph. D. Thesis, Utrecht.
Spruit, H. C.: 1981, preprint.
Spruit, H. C. and Ballegooijen, A. A.: 1982, *Astron. Ap.*, **106**, 58.
Stern, M. E.: 1960, *Tellus*, **12**, 172.
Stommel, H., Arons, A. B., and Blanchard, D.: 1956, *Deep-Sea Res.*, **3**, 152.
Tananbaum, H., Wardle, J. F. C., Zamorani, G., and Avni, Y.: 1982, preprint.
Vaiana, G. S., *et al.*: 1981, *Ap. J.*, **245**, 163.
Vaiana, G. S.: 1983, this volume.
van Ballegooijen, A. A.: 1982, *Astron. Ap.*, **106**, 43.
Walter, F. M.: 1981, *Ap. J.*, **245**, 677.
Walter, F. M.: 1982, *Ap. J.*,, in press.
Walter, F. M., and Bowyer, S.: 1981, *Ap. J.*, **245**, 671.
Weiss, N. O.: 1966, *Proc. Roy. Soc. Lond.*, **293**, 310.
Weiss, N. O.: 1981a, *J. Fluid. Mech.*, **108**, 247.
Weiss, N. O.: 1981b, *J. Fluid. Mech.*, **108**, 273.
Weiss, N. O.: 1983, this volume.

DISCUSSION

GILMAN: Parker's argument that magnetic buoyancy requires the solar field to be retained mostly in the deep layers of the convection zone for the dynamo to work must assume a magnitude for the α effect consistent with the observed magnetic cycle length. But in the global MHD dynamo models, the problem is that α is much too big to give the right period. Perhaps the presence of magnetic buoyancy would not eliminate the dynamo from the bulk of the convection zone, but rather would simply reduce the effective α to the point where a more nearly correct period comes out. What do you think about this?

ROSNER: I think that in the context of your global model, in which the dynamo processes occur throughout the convection layer, your suggestion is very plausible, and in fact is yet another good argument for including magnetic buoyancy in the model.

SPICER: If you look at the question of magnetic buoyancy as an MHD stability problem, you find that magnetic buoyancy is just a magnetized form of the Raleigh-Taylor instability called the Kruskal-Schwarzschild instability. Now it is possible to stabilize the K-S instability by shearing the magnetic field by adding a parallel current, albeit a weak parallel current. Hence, shearing the field may help keep the field from rising through the convection zone too quickly allowing amplification of the field.

ROSNER: The relation between magnetic buoyancy and the Kruskal-Schwarzschild instability is well-known. Indeed, there is a poster presentation by D. Hughes at this symposium on the interchange instability (which is what you are addressing), which discusses the geat richness of the possible instabilities. Now, if one knows the particular instability to be suppressed, you are quite right that one can in principle stabilize the system by imposing suitable current flows (as is done in the laboratory). However, in the case of the Sun, it is very hard for me to see how the Sun manages to arrange for stabilizing current flows for the entire zoo of MHD instabilities which follow from the full dispersion relation. Perhaps some modes are stabilized by the kind of effect you propose, but at most this will just change the dominant mode of instability.

SPRUIT: In general by adding complications to the magnetic configuration, like shear, one increases the free energy. Though certain instabilities may be reduced, the overall instability tends to get worse.

SPICER: In response to the comment that adding an additional parallel current may suppress the K-S instability but causes other instabilities, I must state that it is a question of the magnitude of the magnetic field associated with the parallel current relative to the ambient field. Only a weak shearing field is necessary to stabilize the K-S instability, while large fields, associated with the parallel currents, give rise to the instabilities Spruit refers to — generically referred to as kinks. Perhaps fields are kept at the base of the convection zone by shearing but brought up only after the current density gets large enough to cause a kink.

GARCIA DE LA ROSA: According to the current ideas on magnetic buoyancy, the velocity of ascension for a fluxtube is roughly proportional to the magnetic flux content of the tube. If you consider a unique layer of departure for magnetic fluxtubes travelling towards the surface, how can you fit the obvious observation of the simultaneous coexistence on the solar surface of large and small active regions corresponding to fluxtubes with a difference of flux content of several orders of magnitude.

ROSNER: Your question is quite apt, and in fact similar arguments led L. Golub, G. Vaiana, N.O. Weiss, and myself to a phenomenological picture of flux emergence, in which the small-scale fluxtubes indeed do come from higher layers, i.e., they are produced as large flux ropes rise through the convection zone and are "shredded" by turbulence (1981, *Astrophys. J.* **243**, p. 309).

BASRI: I just wanted to point out that the RS CVn stars are the best current sample of stars which satisfy the theoretical properties Dr. Rosner has indicated are desirable. Observation of a volume limited sample has the best chance of having all detections in various diagnostics (no upper limits) coupled with well determined values of Ω . One must then just be careful to know whether results from this sample are generally applicable to a larger class of stars.

ROSNER: I agree. In fact, as I pointed out, because the sample of stars considered by Walter and Bowyer contained only detections (i.e., every star looked at was detected), their "rectified" L_X/L_{bol} distribution is in fact identical to the one derived by maximum likelihood techniques. I would however caution that in those cases in which periods of RS CVn stars are established purely by considering the orbital period of the system (i.e., arguing that because of tidal coupling, the rotation period is identical to the orbital period), there may be a difficulty: as discussed by J.-L. Tassoul (*Theory of Rotating Stars*, p. 36ff), synchronism does *not* always obtain. For this reason, one should try to obtain the photometric period (as has in fact been done for many of the stars in this sample).

GIOVANELLI: (1) Observations tell us a great deal about the origin of the small-scale flux elements. These *have* their origins in sunspots, and their disintegration has been studied in great detail. This is no mere possibility — it is a fact. (2) We have been able also to follow the transport of flux elements to polar regions. Since the tubes of force do not break, we have long subsurface tubes joining the polar flux elements back to their origins in the sunspot flux ropes. There must be a deeper flow back to the equator, for continuity, and these meridional circulations carry tubes of force downwards, where they are concentrated greatly *unless* gas can cross *into* the tube. I don't know whether this is a vital problem. But the main result is that the fluxtubes are *always* strong deep in the convection zone. I believe that the theoreticians have been concentrating on the wrong thing: they have been concerned too much with building up strong fields deep in the convection zone, whereas the problem is rather in dealing with strong fields, and explaining how these strong fields can survive without floating.

ROSNER: (1) My comments were directed not at the problem of forming small magnetic features *at* the surface, but rather at the problem of explaining the emergence of magnetic features (such as fields associated with X-ray bright points) which are already small upon emergence. In this case, Golub et al. (1980) argued that their origin might well be in the "shredding" of larger flux elements. I thus do not believe we are in disagreement. (2) I believe that we are not discussing the same issue. I agree that the ultimate fate of fluxtubes is an important problem, but it is not the problem I discussed. I addressed the question: How might fluxtubes be formed in the first place? Thereby, J. Schmitt and I considered the stability of an equilibrium field configuration at the base of the convection zone to doubly-diffusive modes.

IONSON: What is the rotation-dependent force that competes with the buoyancy force? If one builds this effect into, for example, an α-ω dynamo, is the cycle time modified?

ROSNER: The Coriolis force is the competing element. Although it is of course hard to specify without doing the full calculation, qualitatively the answer is yes, the cycle time *is* modified. E. Knobloch, N.O. Weiss, and myself in fact recently looked at this very question (1981, *Monthly Not. Royal Astron. Soc.* **197**, p. 45P), and suggested that it is the buoyant loss process from the flux-producing layer that controls the time scale of field eruption.

FRISCH: You have rightly stressed the basic difference between ad hoc non-linear theories, which have little predictive power, and full non-linear calculations. Among the limitations of presently achievable calculations you mention problems relating to *turbulent diffusivities*. This, however, is a concept that belongs to ad hoc modelling (like dimensional analysis). The only known case where a systematic justification can be given for the use of turbulent diffusivities is when there is a clear-cut separation of scales. Otherwise, if we were to take seriously the concept of turbulent viscosity we would predict that high and moderate Reynolds number flows have essentially the same large scale features and differ only in the small scales. This however contradicts experimental evidence (e.g. from Taylor-Couette flows) that by increasing Reynolds numbers we may have a transition from chaotic large scales to *more* organized (coherent) large scales. Such *relaminarization* phenomena are likely to play a role both in dynamo theories and in stellar variability theory.

ROSNER: I agree with your basic point, but I note that, as a practical matter, one requires some artificial dissipation at small scales in current simulations, which are very strongly limited in spatial resolution. In the absence of "turbulent" transport coefficients, energy cascading to large wavenumbers could not be dissipated at the smallest calculated scale length (where classical diffusion is still negligible).

VAN BALLEGOOIJEN: It was pointed out by Acheson (1978) that a small radial gradient of angular velocity brings about a new instability. This instability is more important than the usually considered buoyant instability for $v_A \ll 2\Omega H$ (this corresponds to $B \ll 6 \times 10^4$ G in the deep layers of the solar convective zone). An outward decrease of angular velocity makes the field unstable, while an outward increase stabilizes the non-axisymmetric modes of the tube.

ROSNER: Quite right. In our work, we explicitly avoided looking at these modes, except for the $\mathbf{B} = 0$ case, in which case we again showed (cf. Acheson, 1978) that one can retrieve the well-known Goldreich and Schubert result. As far as "importance" is concerned, I note that we use $B \approx 10^4$ G (which is, in light of our total ignorance, as reasonable a number as any), so that the inequality is barely satisfied. Furthermore, my comments regarding the inertial mode apply here as well: we really ought to do the non-linear calculation before deciding which modes are really important.

GALLOWAY: In your linearized model you took a prescribed magnetic field. In practice the field must be generated by a dynamo, and that needs motions. The convective overshoot motions themselves could do it, but they would presumably also bring the flux into the convection zone, where it would be removed very quickly. Do you have any other ideas about how to generate the field?

ROSNER: Your question aptly contrasts our idealized model with reality. Of course, we would like the penetrative flow into the boundary layer to carry flux down (to be amplified within the boundary by the ω-effect), but these very motions can also carry flux outwards. The point is that the actual situation is likely to be very far from a simple stationary equilibrium, so that there may never be a time during which linear analysis applies. That is yet another reason to look at the non-linear regime, as we are now doing.

SCHÜSSLER: We should distinguish between buoyancy-related instabilities, which depend on an influence of the magnetic field on the overall stratification holding up more material than would be there without a field, and genuine buoyancy of an isolated fluxtube, which simply is a non-equilibrium phenomenon. Furthermore, we must not forget convection: small fluxtubes are dominated by the drag force due to convective flows, while big flux ropes are influenced by buoyancy. Thus the two types of fluxtubes may behave quite differently.

ROSNER: You are quite right on both counts. In our paper (J. Schmitt and R. Rosner: 1982, *Astrophys. J.*, in press) we in fact go to great lengths to distinguish between the instability problem we deal with, and the buoyancy of "isolated" fluxtubes. As far as convection is concerned, I am a bit worried about available theories for the rise of buoyant fluxtubes. The problem may be very much complicated by processes such as recently discussed by Parker (1982), in which he shows that fluxtubes lying parallel to the axis of convective rolls can be "trapped" within regions of downflow, executing a stable motion with no net upward component. This seems to say that the actual rise rate of fluxtubes in a convection zone may be very sensitive to the details of the convective flow, and hence be a very difficult problem to solve generally.

AMPLIFICATION OF MAGNETIC FIELDS BY COMPRESSIBLE CONVECTION

M.R.E. Proctor
Dept. of Applied Mathematics and Theoretical Physics,
University of Cambridge, England

Abstract: Previous work on the concentration of magnetic field by cellular convection in a Boussinesq fluid is extended to a perfect gas, so that the gas pressure is reduced in the presence of a strong magnetic field. Attention is focussed on 2-dimensional flows and on the case of small layer depths, so that the pressure is effectively uniform in field free regions. It is shown that if the field is sufficiently strong, flux sheets with significantly reduced pressures and densities may form. Criteria are established which measure the relative importance of this effect to the more familiar 'magnetic drag' which acts to prevent the concentration of field by generating counter vorticity.

In recent years, high resolution observations of the solar photosphere have shown the existence of vertical magnetic fields confined to thin flux tubes, with peak fields of up to 2000G (see e.g. Beckers 1981). Near the surface, these fields approach the maximum permitted by the requirement that the gas pressure must be small inside the tubes. If the fluid is regarded as a perfect conductor, so that the field lines are 'frozen in', it is difficult to produce such high fields by dynamical means (see e.g. Parker 1979). A complete theory must involve the action of diffusion (either molecular or turbulent) which allows flow across field lines. In this view, the size of a flux tube is determined by a balance between advective effects (due to the convective motions) and diffusion (Galloway et al. 1977). Previous theories of the dynamical effects of flux concentration (Galloway et al. 1978) have ignored the compressibility of the fluid, supposing it to be Boussinesq. Thus the evacuation of the tube cannot be modelled. In this paper the latter effect is investigated in the simple case of 2-dimensional convection, making use of the 'magneto-Boussinesq' approximation (Spiegel & Weiss 1982). It is shown that the effects of compressibility are less than might have been expected and that although the tubes can become evacuated, the amplification mechanism is essentially the same as for the Boussinesq case, at least in the parameter range considered.

J. O. Stenflo (ed.), Solar and Magnetic Fields: Origins and Coronal Effects, 301–305.

ANALYSIS

We consider steady 2-dimensional convection in a perfect gas in the presence of an imposed vertical magnetic field $B_o z$; the convecting layer has depth d. The physics is simplified by supposing large thermal conductivity, so that the temperature T can be taken as fixed, and the energy equation dispensed with. Motion is then caused by horizontal density and pressure gradients, and this confines the magnetic field to thin boundary layers (see e.g. Proctor & Weiss 1982). If we write the magnetic field $\underline{B} \equiv \nabla_\wedge (A\underline{y})$, where $\underline{y}$ is at right angles to the convection rolls, the dimensionless equations describing the system are,

$$\nabla.(\rho\underline{u}) = 0 \qquad (1) \qquad R_m(\underline{u}.\nabla A) = \nabla^2 A \qquad (2)$$

$$Re(\underline{u}.\nabla\underline{u}) + \frac{1}{\varepsilon\beta}\nabla A\nabla^2 A + \left[\nabla^2\underline{u}+\frac{1}{3}\nabla(\nabla.\underline{u})\right] - \frac{1}{\varepsilon}\nabla(\rho T) + \rho\underline{z} = 0 \qquad (3)$$

where $\underline{u}$ is velocity, ρ density, and the dimensionless parameters are

$$\varepsilon = gd/RT_o, \qquad \beta = \frac{\mu_o R T_o \rho_o}{B_o^2}, \qquad R_m = \rho_o g d^3/\mu\eta, \qquad Re = \rho_o{}^2 g d^3/\mu^2 \qquad (4)$$

if g is gravitational acceleration, R the gas constant, μ_o permeability, μ viscosity and ρ_o, T_o typical values of ρ, T. R_e is the Reynolds and R_m the magnetic Reynolds number, β is the ratio of initial gas pressure to magnetic pressure and ε the ratio between the layer depth and a pressure scale height. Previous work (Galloway et al. 1977, 1978) has concentrated on the case $\varepsilon \to 0$ (Boussinesq fluid). In this case the pressure is effectively uniform and the role of the Lorentz forces is to generate vorticity opposite in sense to that of the flow in the absence of field. If $R_m >> 1$, the field is confined to a boundary layer of (scaled) width $O(R_m^{-1/2})$, and in this case the curl of (3) may be used to show that this 'magnetic drag becomes significant (reduces the amplification) if $\beta \lesssim R_m^{1/2}\varepsilon^{-1}$. If R_m is so large that $R_m^{1/2}\varepsilon >> 1$, however, then β can be $O(R_m)$ or less before the former inequality can be satisfied, and in this case the gas pressure in the flux sheet will be reduced below its ambient value since the total pressure must be constant, and the peak field B^* is $O(R_m^{1/4})B_o$. If we suppose Re not too large, the effects of dynamic pressure may be ignored. Then because of continuity the velocity field will change to accommodate the horizontal density contrast, and this will change the shape of the flux sheet, which is Gaussian for very large β. There is also a generation of vorticity due to horizontal density gradients, but this is always small compared with the original vorticity. We can therefore ignore the effect of magnetic drag (which requires curved field lines) and consider a simple model in which $u = (R_m^{-1/2}\hat{u}(\xi), z)$, where $\xi = R_m^{1/2} x$ is the boundary layer coordinate for a flux sheet near x=0 (Fig.1). This causes a vertical sheet in which $A = A(\xi)$, $|\underline{B}| = R_m^{1/2}\hat{B}(\xi) = R_m^{1/2} dA/d\xi$. The boundary layer equations are then (if $T \simeq 1$ near x=0 and $\rho=1$ outside the boundary layer)

$$\rho + \frac{R_m}{2\beta}\hat{B}^2 = 1; \qquad \frac{\partial}{\partial\xi}(\rho\hat{u}) = -\rho; \qquad \hat{u}\hat{B} = d\hat{B}/d\xi \qquad (5)$$

with $\partial\rho/\partial\xi = \partial\hat{B}/\partial\xi = \hat{u} = 0$ at $\xi=0$, and $\rho\to 1$, $\hat{B}\to 0$, $\hat{u}\to -\xi$ as $\xi\to\infty$, with $\int_0^\infty \hat{B}d\xi = \frac{1}{2}$,

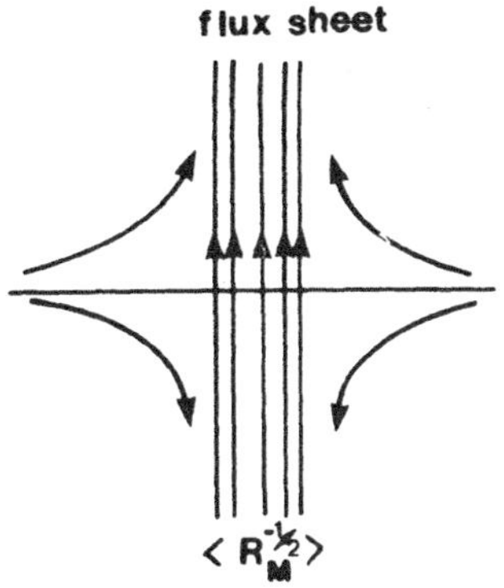

Figure 1. Geometry of the model problem: converging flow creates a thin flux sheet

say, since the flux is fixed. This system can be reduced to a first order o.d.e. which has been solved numerically. Fig.2 shows $\hat{B}$ and ρ as a function of ξ for 3 values of $\gamma \equiv \beta/R_m$. The sheet becomes flattened as γ decreases, with much more flux in the wings; it seems that the central density $\rho_{centre} \to 0$ as γ tends to a finite value (Fig.3) (though of course the assumptions of the theory are no longer valid then.)

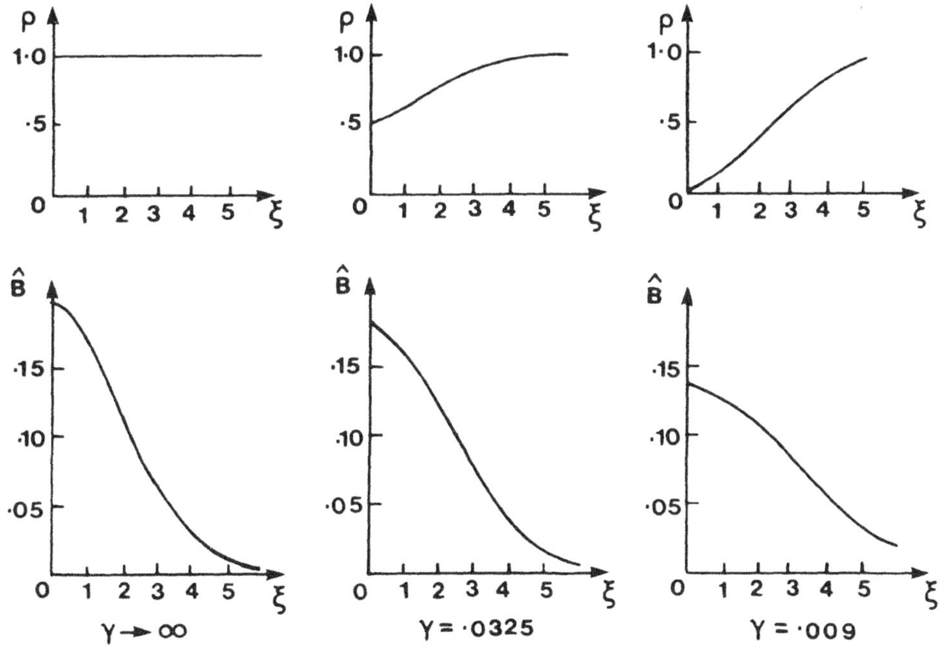

Figure 2. ρ and $\hat{B}$ as functions of ξ for three different values of γ.

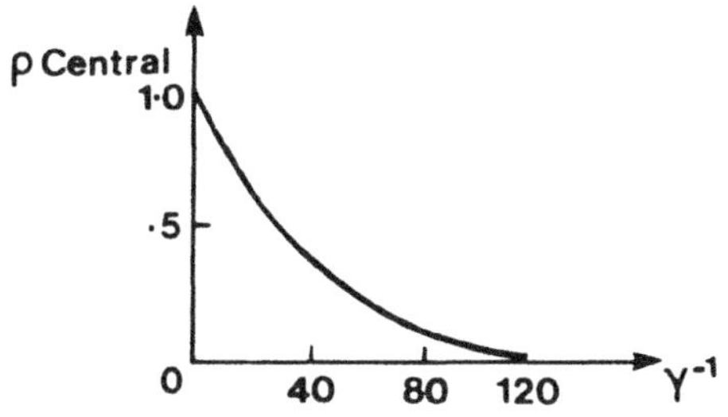

Figure 3. Central density as a function of γ^{-1}.

DISCUSSION

A start has been made in assessing the effect of compressibility on the concentration of magnetic fields by convection. The small ε results here complement those for large ε obtained numerically by Cattaneo (1982), which show qualitatively similar effects. They show that provided diffusion plays a role in fixing the size of the flux tubes (itself a controversial matter, see e.g. Spruit 1981), evacuated tubes can be supported almost as easily as non-evacuated ones. Further work is in progress to include the effects of an energy equation, and to consider the interaction between evacuation and magnetic drag effects.

ACKNOWLEDGEMENT: I am grateful to F. Cattaneo for helpful discussions.

REFERENCES

Beckers, J.M.: 1981, in "The Sun as a Star", ed. S. Jordan, NASA Report SP-450, pp.11-65.
Cattaneo, F.: 1982, "Compressible magnetoconvection", in preparation.
Galloway, D.J., Proctor, M.R.E. and Weiss, N.O: 1978, Nature 266, pp.686-9.
Galloway, D.J., Proctor, M.R.E. and Weiss, N.O.: 1978, J. Fluid Mech. 87, pp.243-261.
Parker, E.N.: 1979 "Cosmical Magnetic Fields", Oxford University Press.
Proctor, M.R.E. and Weiss, N.O.: 1982, "Magnetoconvection", Rep. Prog. Phys. (in press).
Spiegel, E.A. and Weiss, N.O.: 1982, Geophys. Astrophys. Fluid Dyn. (in press)
Spruit, H.C.: 1981, in "The Sun as a Star", ed. S. Jordan, NASA Report SP-450, pp.385-413.

DISCUSSION

GIOVANELLI: (1) Can you give me any idea of the depth of a granule? One suggestion has been 1/4 granule diameter. Others have spoken of 1500 km to perhaps 10 000 km. If they are shallow, then of course the idea of a structure small compared with the scale height is OK, but otherwise this is not so. (2) The observations have shown increasing downflow velocities *inside* magnetic elements. Most of the downflow observations have been made by my colleagues and myself. These observations were done before we realized that the fields splayed out so rapidly with height. I believe that the downflow observations should now be redone taking cogniscance of our present improved knowledge on tube structure. Jack Harvey and I are attempting this, but I hope that it will be possible to interest others in similar studies. It is obviously of great theoretical interest.

PROCTOR: (1) A few hundred km is the most commonly accepted level for granules, and one imagines that the appropriate depth is one scale height (though if the granulation takes place in strongly stably stratified regions the vertical scale may be rather less). A full theory should certainly take account of larger layer depths. However, Cattaneo's numerical work seems to show similar features, even though layer depths there are large. (2) I heartily concur with the need for new downflow observations.

CRAM: Recent calculations of the electrical conductivity of the solar photosphere give a minimum value of about 1 Siemens m^{-1}. This value is too large to lead to significant Joule dissipation. If "classical" diffusivity is not operative, what is the picture of "turbulent" diffusivity that we should think about in relation to magnetic structures that are only a few hundred km across?

PROCTOR: "Turbulent diffusion" acts at both scalar and vector fields both to reduce scales below the resolution threshold and, more importantly, to such scales that molecular or radiative effects can operate on the time scale available. For scalars the effective diffusivity does not depend significantly on the molecular value, but there are certainly problems in understanding the diffusion processes when they are strongly affected by the anisotropy introduced by the Lorentz force. I would expect the existence of very small-scale fluctuating flows down to the scale of a few (perhaps a few tens) of km, as the end product of the cascade due to the turbulent convection.

SPICER: I have a comment with respect to Cram's remark on the electrical conductivity of the plasma at the temperature minimum. As there are flow fields which appear to flow perpendicular to the ambient magnetic field, perpendicular currents will result. Hence the Pedersen conductivity can lead to significant dissipation via $\eta_{Peder} J_{\perp}^2$.

PROCTOR: In highly conducting fluids the field lines move with the fluid, so there is no perpendicular flow! One does not want *too much* dissipation anyway, since otherwise the very intermittent magnetic field structure would not be seen.

WEISS: In modelling fluxtubes, we poor theoreticians have to rely on what the observers tell us, and it is confusing when even Australian observers seem to disagree with each other. As I understand it, Dr. Giovanelli still maintains that there is a flow of gas across the field lines and downwards into the fluxtube. Dr. Cram insists that ohmic diffusion will not permit this flow. So we have to postulate the existence of a different diffusion process.

PROCTOR: It is certainly important not to be content with "kinematic" turbulent diffusivity theories but to include the effects of Lorentz forces. Perhaps we can then understand how we can obtain flows "down under" the tubes.

PERIODIC AND APERIODIC BEHAVIOUR IN STELLAR DYNAMOS

F. Cattaneo*, C.A. Jones** and N.O. Weiss*
*Department of Applied Mathematics and Theoretical Physics, University of Cambridge, England
** School of Mathematics, University of Newcastle-upon-Tyne, England

Abstract: We have constructed a simple parametrized mean field dynamo model that includes the dynamical interaction between the magnetic field and differential rotation. This system of seven coupled nonlinear ordinary differential equations has finite amplitude oscillatory solutions (corresponding to Parker's dynamo waves) when the dynamo number D>1. We have studied two regimes. In the first, the velocity shear is reduced by the Lorentz force and there are stable periodic solutions for all D>1. In the second there is a transition from strictly periodic oscillations to aperiodic (chaotic) behaviour as D is increased. This simple example shows that nonlinear hydromagnetic dynamos can produce aperiodic cycles, with Maunder minima, as observed in the sun and other late-type stars.

The solar cycle provides a convincing example of a strange attractor (Ruzmaikin 1981). There are several routes from regularity to chaos (Eckmann 1981) and in order to discover which is relevant to the sun it is natural to investigate the simplest model that captures the essential physics of a nonlinear stellar dynamo. We have therefore constructed a system of equations that describes a parametrized $\alpha\omega$-dynamo in one space dimension, including the dynamical effect of the Lorentz force, which acts to reduce differential rotation. The solutions are nonlinear dynamo waves (Parker 1979).

This seventh order system is given, in dimensionless form, by the equations

$$\dot{A} = 2DB - A \tag{1}$$
$$\dot{B} = i(1+\omega_0)A - \tfrac{1}{2}iA^*\omega - B, \tag{2}$$
$$\dot{\omega}_0 = \tfrac{1}{2}i(A^*B-AB^*) - \nu_0\omega_0, \tag{3}$$
$$\dot{\omega} = -iAB - \nu\omega. \tag{4}$$

The poloidal flux function A and the toroidal field B are complex, so that $B=B_1+iB_2$ etc., and the change in the velocity shear is separated into a real, spatially uniform component ω_0 and a complex component with twice the spatial frequency of A and B; ν and ν_0 are real constants.

J. O. Stenflo (ed.), Solar and Magnetic Fields: Origins and Coronal Effects, 307–310.

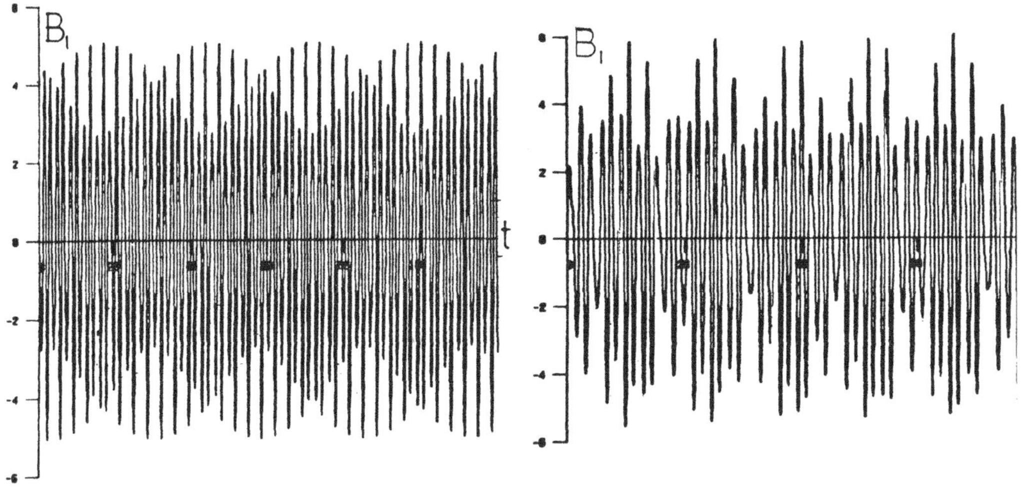

Figure 1: Nonlinear dynamo waves: the toroidal field B as a function of time for the sixth order system with ν=0.5. (a) Doubly periodic behaviour for D=3.0. (b) Triply periodic motion for D=3.5.

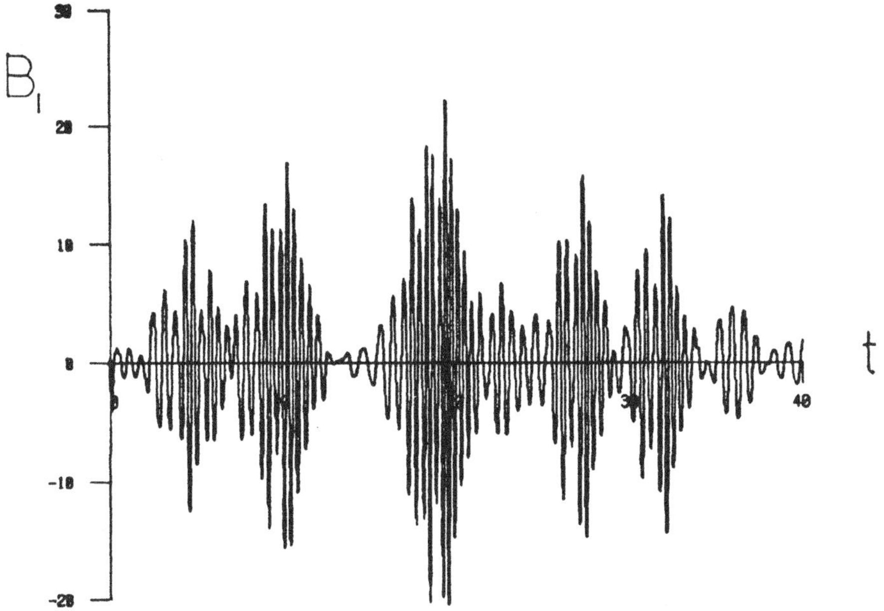

Figure 2: Chaotic behaviour with episodes of reduced activity. As Fig.1 but with D=16.

The dynamo number D is a stability parameter for this problem and there is a Hopf bifurcation from the trivial solution A=B=ω=0 at D=1(Parker 1979).

We have studied two cases. In the limit $\nu\to\infty$, $\omega\to 0$ equations (1)-(3) yield a fifth order system with an exact periodic solution: A and B vary sinusoidally and ω_0 is a (negative) constant. This solution is stable for all D. Thus the Lorentz force reduces the velocity shear, giving a lower effective dynamo number, as in Gilman's (1982) numerical experiments.

When $\nu_0\to\infty$, $\omega_0\to 0$, equations (1),(2) and (4) give rise to a sixth order system which exhibits more interesting behaviour (cf. Jones 1982). Again there is an exact periodic solution for D>1, with A,B,ω varying sinusoidally in time. We have located successive bifurcations from this solution as D is increased for ν=0.5. The exact solution loses stability at D$\approx$2.07 and Fig.1(a) shows $B_1(t)$ for a doubly periodic solution at D=3.0. The next bifurcation leads to triply periodic solutions, shown in Fig.1(b) for D=3.5. There follows a transition to chaotic behaviour around D$\approx$3.84. Fig.2 shows a typical solution for D=16. The cycles are aperiodic and are separated by intervals of stasis, with drastically reduced activity. (This pattern is typical of all runs made for D$\geq$4.) Mathematically, there is a bifurcation to a 2-torus at D$\approx$2.07, followed by a bifurcation to quasiperiodic motion on a 3-torus at D$\approx$3.47. Two of the three frequencies subsequently lock and, after a cascade of period doubling bifurcations, the flow eventually becomes chaotic.

The variable ω describes variations in differential rotation with twice the frequency, in space and in time, of the mean magnetic fields. Corresponding behaviour has been found by Howard and LaBonte (1980,1982) on the sun. Dynamical coupling with ω is sufficient to produce recurrent episodes of reduced activity (Maunder minima) even in our simple model. We expect that similar behaviour can be found in a variety of models (cf. Childress and Spiegel 1981) and also in computations like those of Gilman (1982). Provided that the initial bifurcation leads to an oscillatory magnetic field (as in $\alpha\omega$-dynamos) and that the dynamics are sufficiently complex to allow the development of a chaotic regime, this pattern of behaviour is likely to occur. Thus Maunder minima are a characteristic feature of a wide class of nonlinear dynamos.

REFERENCES

Childress, S. and Spiegel, E.A.: 1981, in 'Solar Constant Variations', ed. S. Sofia, NASA, Washington.

Eckmann, J.P.: 1981, Rev. Mod. Phys. 53, pp.643-654.

Gilman, P.A.: 1982, in "Cool Stars, Stellar Systems and the Sun', ed. M.S. Giampapa and L. Golub, pp.165-179, SAO Special Report 392.

Howard, R. and LaBonte, B.J.: 1980, Astrophys. J. 239, pp.L33-L36.

Jones, C.A.: 1982, in 'Planetary and Stellar Magnetism', ed. A.M. Soward, Gordon and Breach, London.

LaBonte, B.J. and Howard, R.: 1982, Solar Phys. 75, pp.161-178.

Parker, E.N.: 1979, 'Cosmical Magnetic Fields', Clarendon Press, Oxford

Ruzmaikin, A.A.: 1981, Comments on Astrophys. 9, pp.85-96.

DISCUSSION

GILMAN: (1) I am confused as to the role of torsional oscillation in your model. (2) Could you simulate bifurcations and strange attractor behaviour with respect to magnetic field symmetry? My own model calculations show examples of symmetry persistence over several cycles, followed by occasional switchovers to the opposite symmetry.

WEISS: (1) In our model the torsional oscillations are driven by the Lorentz force, and the variations in differential rotation then affect the generation of the toroidal magnetic field. This feedback is sufficient to produce chaotic behaviour (though the amplitude of the torsional oscillations is then embarrassingly large). (2) We have not yet attempted to model bifurcations leading to *spatial* asymmetry, but it is certainly possible to do so.

ROSNER: I would like to reemphasize that, in my view, so-called ad hoc non-linear dynamo models *are* perfectly appropriate for exploring dynamo behaviour, with appropriate approximations (viz. truncations), which retain the essentials of some physical process in question. However, I argue that it is not meaningful to ask such models to *quantitatively* account for observed correlations between, for example, activity diagnostics and rotation.

WEISS: Of course it is also important that the qualitative predictions of such models should be checked by comparison with further calculations, like those that Gilman has described.

SPRUIT: Are your aperiodic solutions always of the Maunder-minimum type, and what features of the model are responsible for producing that particular form of aperiodicity?

WEISS: No, we have obviously looked for chaotic behaviour, and we only found it for the sixth-order system in a restricted parameter range ($\nu < 1$). However, once the solutions become chaotic they exhibit Maunder minima. What is needed for this pattern is irregular behaviour such that trajectories in phase space approach the neighbourhood of the origin and are attracted towards the stable manifold of the origin before being flung out on its unstable manifold. This is likely to be found in a wide variety of models.

GIOVANELLI: (1) Are the Maunder-minimum type solutions fairly periodic, e.g. do they have similar periods between successive minima, and similar harmonics? (2) Can you give any idea as to the way the Maunder-minimum type solutions may vary with solar or stellar evolution?

WEISS: (1) Between the "Maunder minima" the model exhibits oscillations that are fairly regular but nevertheless aperiodic. The average period is well defined, at least in the limit of large dynamo number, but the phase is not preserved through the "Maunder minima". (2) As a star evolves it is spun down, and the dynamo number D decreases. In the model, this leads to more regular behaviour, eventually becoming periodic as D approaches unity. One might expect a rapidly rotating star to vary in a more chaotic manner, but I doubt whether the model can explain the lack of cyclic behaviour in younger, more rapidly rotating stars.

IV. THE ROLE OF MAGNETIC FIELDS IN THE STRUCTURE AND ENERGY BALANCE OF STELLAR ATMOSPHERES

THE ROLE OF MAGNETIC FIELDS IN STELLAR CHROMOSPHERES AND TRANSITION REGIONS

Jeffrey L. Linsky[1]
Joint Institute for Laboratory Astrophysics National Bureau of Standards and University of Colorado, Boulder, Colorado 80309 USA

ABSTRACT

In this review based largely on observations with the IUE and Einstein satellites, I will summarize the different roles that magnetic fields play in controlling the structure and energy balance in the chromospheres and transition regions of late-type stars. Solar observations clearly show that magnetic flux tubes are the dominant structural element in the solar atmosphere, but the rotational modulation of plages (structures that are bright in ultraviolet emission lines) that overlie dark starspots provide strong evidence that magnetic flux tubes are the dominant structural elements in late-type stellar atmospheres as well. The wide range of radiative loss rates (and thus heating rates) observed in chromospheric and transition region emission lines also provides evidence for the importance of magnetic fields, but it is not yet clear whether the most active stars can be understood in terms of a large fractional coverage by solar-like magnetic flux tubes or whether brighter flux tubes are needed. I propose that the existence of a boundary between solar-like stars and those with little or no hot plasma, as well as the different types of G-K giants and supergiants, can be understood in terms of the fractional surface coverage by closed magnetic structures. Transition region downflows, the chromospheric heating mechanism, and the relative heating rates at different layers can be simply explained by the control of the energy balance by magnetic fields. Finally, I will intercompare models computed for active and quiet regions on the Sun with similar models computed for active and quiet stars, that is stars with intrinsically bright or weak emission lines.

[1]Staff Member, Quantum Physics Division, National Bureau of Standards.

J. O. Stenflo (ed.), Solar and Magnetic Fields: Origins and Coronal Effects, 313–338.

I. INTRODUCTION

The organizers of this IAU Symposium have asked me to review a very difficult topic. While I will do my best to rise to this challenge, I feel that I should describe the inherent difficulties of the topic before we proceed further.

From the point of view of an observer, one can say very little about the role magnetic fields play in stellar atmospheres, because we have very few measurements of magnetic fields in cool stars and even these measurements are spatially averaged whereas the stellar fields are likely inhomogeneous and filamentary. Other speakers have reviewed what we know about the role magnetic fields play in the solar atmosphere, and it seems reasonable to assume that magnetic fields play similar roles in solar-like stars, but we do not know a priori which stars are solar-like and which are very different. Furthermore, we have not yet resolved individual magnetic flux tubes in the solar atmosphere, and as a result we have no definitive information yet on the role played by magnetic fields even in this case.

From the point of view of a theoretician, the situation may be worse. The introduction of magnetic forces and magnetic flux tube geometries into the hydrodynamic equations makes it almost impossible to solve the equations for nonradiative heating rates and the steady state response of the flux-tube embedded atmosphere to this heating. One must either ignore the magnetic force terms, simplify the geometry, or make some other drastic approximations in order to make any progress. I suspect that the past emphasis on acoustic wave heating, ignoring the magnetic fields that are located at the sites of the greatest heating in the solar atmosphere, was due to difficulties in solving the MHD equations for realistic geometries.

In view of these difficulties I cannot be rigorous but must instead be speculative, and will follow solar analogies in trying to explain the interesting results coming from IUE and Einstein. There is a danger in this approach as a solar analogy may be a poor guide for explaining phenomena on stars with parameters very different from the Sun. Nevertheless, let us pursue solar analogies in a self-consistent manner, because they should be reliable when we understand the basic physical processes underlying the solar phenomena.

It is important to ask what the properties of stellar chromospheres and transition regions (TRs) would be for stars without magnetic fields. According to the Vogt-Russell theorem (cf. Lang 1980), the mass, age, and initial chemical composition of a star determine its effective temperature, radius, and gravity, and thus its location in the H-R diagram. Thus for nonmagnetic stars, there are unique values for the luminosity and convective zone properties at each point in the H-R diagram. Furthermore, the atmospheric structure should be relatively uniform across the surface of these stars, except for small variations due to gravity darkening at the poles, meridional circulation, and the spatial variations due to the cellular nature of convection. The nonradiative heating by acoustic waves or other nonmagnetic waves generated by convective motions should still produce chromospheres, TRs, and coronae in late-type stars, but the heating rates should be relatively small. We therefore expect only quiescent outer atmospheric layers that are weak, steady emitters in the ultraviolet and X-rays, are spatially homogeneous, and show few differences between stars of similar effective temperature and gravity.

By contrast with this rather boring picture of idealized nonmagnetic stellar atmospheres, we know that magnetic flux tubes in the solar atmosphere are the basic structure responsible for atmospheric inhomogeneity, are regions of enhanced nonradiative heating, are the locations of time-variable phenomena on time scales of seconds to months, are regions of important systematic flows, and are characterized by very different energy balance and atmospheric properties than regions of weak fields. The main theme of this review is to describe similar phenomena in late-type stars and to summarize the various roles that magnetic fields likely play in modifying the chromospheres and TRs of these stars. I will conclude by comparing atmospheric models for stars believed to have large and small areas covered by strong magnetic fields, and by suggesting some unsolved problems for future work. Some recent reviews relevant to this topic include Dupree (1981a,b;1982), and Linsky (1980;1981a,b,c, d;1982).

II. INHOMOGENEITY AND VARIABILITY IN STELLAR CHROMOSPHERES AND TRANSITION REGIONS

a) Spatial Inhomogeneity

Earlier in this symposium Marcy and Spruit reviewed the observations and theory of photospheric magnetic fields in late-type stars, and Zwaan and Vaiana reviewed the evidence for stellar magnetic structure using chromospheric and coronal indicators. I do not wish to convince you again that magnetic fields are responsible for atmospheric structure, but rather to point out how this may be true.

The best evidence we have for the existence of large scale bright structures in the chromospheres and TRs of stars comes from measurements of periodic variations in the intensity of specific spectral features at the presumed rotational period of the star. Vaughan et al. (1981), for example, monitored the chromospheric Ca II H and K lines to derive rotational periods of 19 stars, mainly dwarfs of spectral type F6-M0. These data argue conclusively for the inhomogeneous distribution of brightness across the chromospheres of these stars and solar analogy argues that the structures primarily responsible for this patchiness are bright plage regions that are inhomogeneously distributed in longitude and that are long-lived compared to a rotational period. If a star were completely covered with plages or the plages were uniformly distributed in longitude, then there would be no modulated intensity signal to indicate a rotational period. Hallam and Wolff (1981) followed a similar technique using IUE to determine rotational periods of three dwarf stars -- 111 Tau (F8 V), ε Eri (K2 V), and 61 Cyg A (K5 V). They observed rotational modulation of intensity in lines of Lyα, Si II 1812 Å, and Mg II 2796 Å. Several monitoring programs are under way, in some cases with coordinated magnetic field observations, to confirm the hypothesis that the plage regions have strong magnetic fields.

Close binary systems with cool components typically show bright ultraviolet emission lines and photometric variability indicative of rotational modulation of dark star spots (e.g. Kunkel 1975; Hall 1981). One therefore expects that the ultraviolet emission line flux should vary with rotational phase such that emission line maximum (maximum coverage by plages) corresponds to photometric minimum (maximum coverage by dark star spots). Baliunas and Dupree (1982) did this experiment on the long-period RS CVn system, λ And (G8 III-IV + ?), confirming that the emission lines are strong at photometric minimum and weak at photometric maximum. Marstad et al. (1982) monitored three RS CVn systems (HR 1099, II Peg, AR Lac) and the prototype BY Dra system, and found clear evidence for periodic variability in HR 1099 and II Peg. For HR 1099, the data cover three rotational periods (P = 2.8 days) and are consistent, indicating that the plages are long-lived. For II Peg, the data reveal important clues concerning stellar plages:

(1) The strength of all the II Peg emission lines increases rapidly at orbital phase 0.45, is roughly constant for one-half the period, and then decreases just as rapidly at phase 0.95 (see Fig. 1). This is strong evidence for the rotational modulation of a relatively compact plage group centered at phase 0.70. The compactness of this

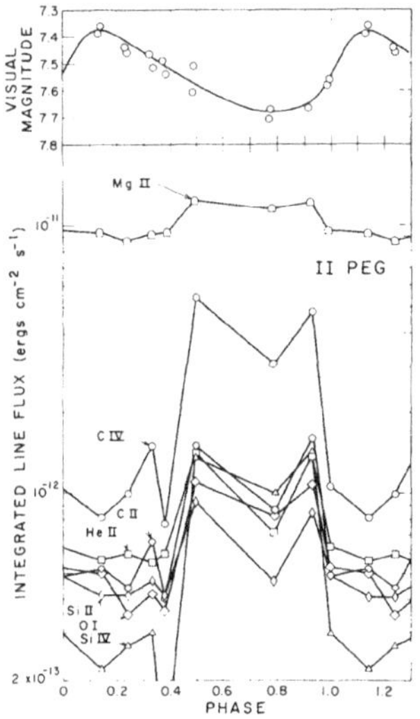

Fig. 1. Observed flux (from Marstad et al. 1982) in the Mg II λ2800, C IV λ1550, He II λ1640, C II λ1335, Si II λ1812, O I λ1304, and Si IV λ1400 features for II Peg as a function of phase. Also given at the top are FES visual magnitudes obtained simultaneously with the IUE spectra.

plage group and the absence of large intensity changes during the 3.3 day half period are important new information.

(2) The increase in emission line strength corresponds to the decrease in photometric brightness (see Fig. 1) measured simultaneously by the FES on IUE, indicates that the plage group overlies dark star spots and thus the plages are regions of strong magnetic fields.

(3) The spectrum of II Peg at plage maximum differs considerably from that at plage minimum as can be seen in Figure 2, in the sense that the high temperature TR lines (C IV 1549 Å, Si IV 1400 Å, and C II 1335 Å) are enhanced by much larger factors than the chromospheric lines (Si II 1812 Å, C I 1657 Å, and Mg II 2800 Å). This phenomenon is seen in solar plages and throughout the IUE data, as we shall see; therefore, it is an important consequence of magnetic fields in stellar atmospheres that requires an explanation.

b) Time Variability

According to our corollary to the Vogt-Russell theorem, nonmagnetic stars emit nearly constant flux, and flux variability, therefore, should be a direct consequence of magnetic fields. Variability

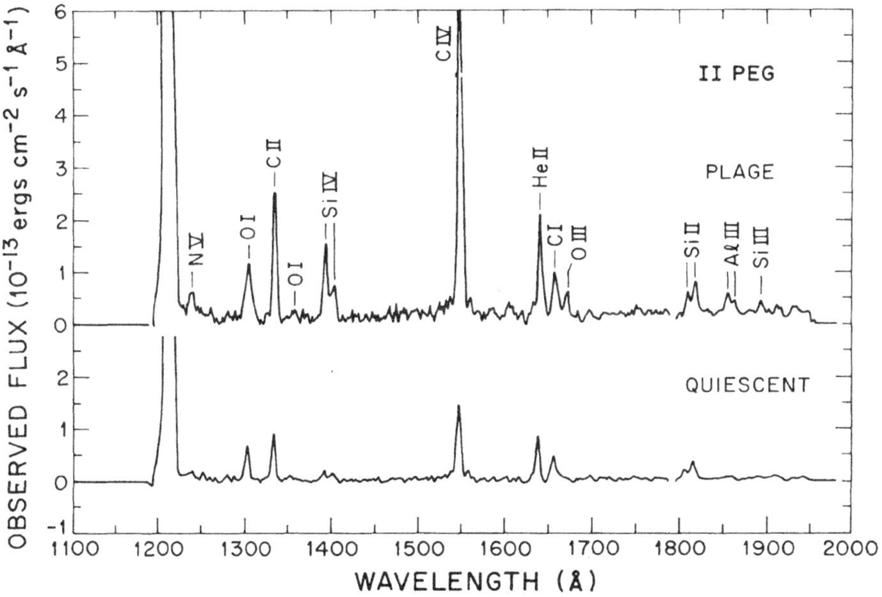

Fig. 2. Composite IUE spectra from Marstad et al. (1982) of the RS CVn-type system II Peg when the plage is on the disk (sum of three spectra) and when the plage is not present (sum of five spectra).

on many time scales has already been detected from the chromospheres and TRs in late-type stars. For example, Wilson (1978) discovered periodic variations in the Ca II flux from G-K dwarf stars with periods ~10 years, which he ascribed to magnetic cycles. White and Livingston (1981) found similar changes in the solar Ca II line flux between minimum and maximum in the solar cycle, and they argued that these changes are due to differences in the fractional area coverage by plages and not to any changes in the chromospheric network. We believe that the stellar data can be interpreted in a similar way.

Flares have been detected in IUE spectra of the RS CVn systems UX Ari (Simon, Linsky and Schiffer 1980a) and HR 1099 (Marstad et al. 1982), as well as the dMe flare stars GL 867A (Butler et al. 1981) and Proxima Centauri (Haisch et al. 1982). Also, Baliunas et al. (1981) detected short time scale variations in the Ca II lines in α Tau (K5 III), λ And (G8 III-IV + ?), and ε Eri (K2 V). Such flares are probably also magnetic in character, with the dMe star flares similar to solar flares and the RS CVn flares perhaps involving reconnection between flux tubes of the two stars. In addition to the expected enhancement of the emission lines, Butler et al. (1981) found continuous ultraviolet emission during a flare on GL 867A. One property seen in both the dMe and RS CVn flares is the relative enhancement of the TR lines compared to the chromospheric emission lines. This is illustrated by comparing the IUE spectra of HR 1099 during a flare and at quiescent times immediately before and after the flare (see Fig. 3). This important property will be discussed in §VI below.

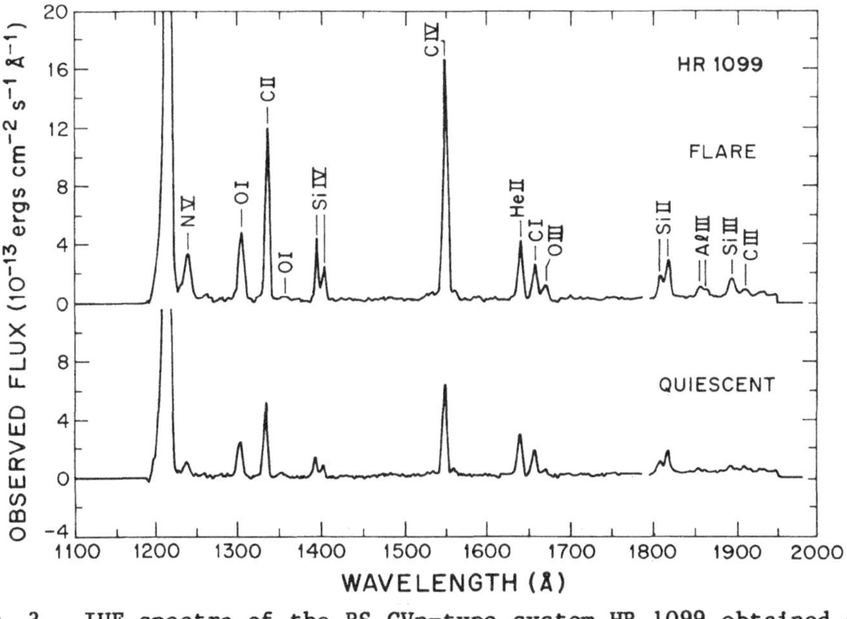

Fig. 3. IUE spectra of the RS CVn-type system HR 1099 obtained during a flare and when the system was quiescent.

III. ENHANCED HEATING

A vital role played by the magnetic field is to enhance the local nonradiative heating rate in the chromosphere and higher layers over that which occurs in regions of weak or no fields. This conclusion is based on the tight spatial correlation of bright chromospheric emission with photospheric magnetic field strength (e.g. Skumanich, Smythe and Frazier 1975). Stein (1981) has summarized the arguments why slow mode MHD waves are the most likely heating mechanism for those regions in stellar chromospheres that have strong magnetic fields.

It is important to go beyond this general conclusion to a quantitative evaluation of the nonradiative heating rates at different atmospheric layers in different types of stars. One way of comparing these rates in different stars is by plotting emission line surface fluxes, the flux per unit area of the star, as a function of mean temperature of formation of each line. Linsky et al. (1982) compared line surface fluxes for nine active chromosphere dwarf stars and six quiet chromosphere dwarf stars of spectral types G2-M5.6e using IUE observations. The surface fluxes of these stars divided by the mean quiet Sun values are shown in Figure 4. Several conclusions can be drawn from these data:

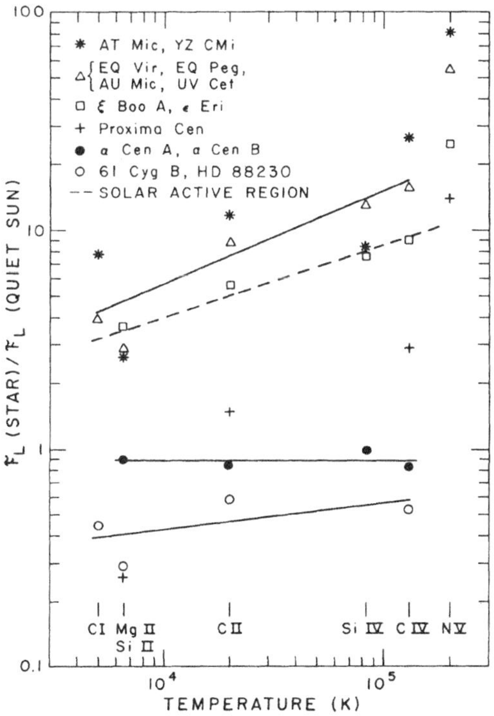

Fig. 4. Ratios of emission-line surface fluxes from Linsky et al. (1982) for six groups of stars and a bright solar active region relative to the quiet Sun. The stars are grouped according to their C IV ratios into three groups of dMe stars (AT Mic and YZ CMi; EQ Vir, EQ Peg, AU Mic, and UV Cet; Proxima Centauri), the dM stars (61 Cyg B and HD 88230), the active G-K dwarfs (ξ Boo A and ε Eri), and the quiet G-K dwarfs (α Cen A and α Cen B). Straight lines are drawn to indicate rough trends in the data for several groups of stars. The data for ξ Boo A and ε Eri are similar to those for a solar active region.

(1) The surface fluxes of the M dwarf stars with the weakest emission lines are about a factor of three below the mean quiet Sun values. These stars likely have very weak magnetic fields with few plage regions. Since 61 Cyg B (K7 V) shows rotationally modulated Ca II emission (Vaughan et al. 1981), the plage regions on these stars are intrinsically fainter, and hence the heating rates lower, than for the Sun.

(2) The two stars most like the Sun [α Cen A (G2 V) and α Cen B (K1 V)] have surface flux ratios close to unity independent of temperature, indicating magnetic fields similar to those on the Sun.

(3) The active chromosphere stars, including the dMe stars, have surface flux ratios that increase from 3-8 in the chromosphere to 10-100 at 2×10^5 K in the TR. Similar trends in the surface flux ratios are detected in other active stars including: (a) G giants and supergiants (e.g. Hartmann, Dupree and Raymond 1982; Stencel et al. 1982a,b) for which the ratios increase from 1 in the chromosphere to about 10 in the TR, and (b) RS CVn systems (e.g. Simon and Linsky 1980) for which the ratios increase from 10-20 in the chromosphere to as large as 600 in the TR. T Tauri stars (e.g. Imhoff and Giampapa 1982) also show very large ratios.

To what extent can we explain these large surface flux ratios by assuming that the active stars are mostly covered by solar-like plage regions? Included in Figure 4 is a dashed line indicating the surface flux ratios for a bright solar plage (Vernazza and Reeves 1978). Since the active M dwarf stars, RS CVn binaries, and T Tauri stars lie above this line, even complete coverage by solar-like plages cannot explain the large surface flux ratios. Furthermore, there is evidence described in the previous section (see Fig. 1) that these active stars also have a patchy distribution of emitting regions.

There is another important point, however, that must be considered. Solar plages are not homogeneous regions, but instead consist of many individual magnetic flux tubes that are not resolved in the existing data as their widths are much smaller than 1 arcsec (Frazier and Stenflo 1972; Stenflo 1973). By comparison, the solar plage data in Figure 4 were obtained by the EUV spectrometer-spectroheliometer on *Skylab*, which had a spatial resolution of 5×5 arcsec^2 (Vernazza and Reeves 1978). The High Resolution Telescope Spectrograph (HRTS) experiment (cf. Brueckner, Bartoe and Van Hoosier 1977; Basri et al. 1979) obtained ultraviolet solar spectra with a resolution of 1×1 arcsec^2. These data show a range of intensities between the brightest portions of plages and the darkest quiet regions that is a factor of 100 in the C IV 1548 Å line (Brueckner, private communication). Schindler et al. (1982) used HRTS data to show that the He II 1640 Å line in the brightest plage regions is a factor of 50 brighter than the mean quiet Sun value. Even in these data, the magnetic flux tubes are not resolved, and therefore we cannot yet say whether the observed flux ratios for chromospheric and TR lines in the active dwarf stars,

RS CVn systems, and T Tauri stars can be explained by solar-like flux tubes covering portions of their surfaces.

IV. THE EXISTENCE, LOCATION, AND WIDTH OF THE TRANSITION REGION BOUNDARY IN THE H-R DIAGRAM

Prior to IUE and Einstein, one could only speculate concerning which portions of the H-R diagram contain stars with outer atmospheric layers and winds similar to the Sun and which portions contain stars with very different outer atmospheres and winds. The rich data from these two spacecraft have partially answered this question and also, I believe, told us that the question was improperly posed. Since magnetic fields control the energy balance in the outer layers of a star, there must be a range of outer atmospheric properties for stars in each location in the H-R diagram. Thus, the answer to the question is not straightforward, since magnetic fields play a crucial role.

Near the beginning of IUE operations, Linsky and Haisch (1979) noted a trend in the ultraviolet spectra of cool giants and supergiants in which the warmer stars with $V-R < 0.80$ (the yellow giants) show emission lines formed at all temperatures up to 10^5 K, whereas the cooler stars with $V-R > 0.80$ (the red giants) show only chromospheric emission lines. On this basis they proposed a nearly vertical dividing line in the H-R diagram near $V-R = 0.80$ (see Fig. 5) separating the yellow giants that typically have TRs from the red giants that do not. Subsequently, Ayres et al. (1981a) showed that the Einstein soft X-ray observations are consistent with the typical presence of hot coronae in stars to the left of a similar boundary and the absence of coronae in single stars to the right (see Fig. 5). Also Stencel (1978), and Stencel and Mullan (1980a,b) presented evidence for the onset of massive cool winds (observed by asymmetries in the Mg II lines) in stars lying to the right of a similar boundary.

The idea of a boundary as proposed by Linsky and Haisch has been criticized on the basis of the few stars in the original data sample, the absence of detected TR emission lines in some yellow giants (Hartmann, Dupree and Raymond 1982), and the existence of some red giants, the so-called hybrid stars, that show evidence for cool winds and C IV emission lines (Hartmann, Dupree and Raymond 1980,1981; Reimers 1982). The hybrid stars were proposed to have winds that are cool far from the star, on the basis of blue-shifted absorption features in the Mg II lines, and 10^5 K outflowing plasma closer to the star, on the basis of the observed broad C IV emission lines. These valid criticisms led to a reexamination of the existence of a boundary by Simon, Linsky and Stencel (1982) on the basis of a much larger sample of 39 single stars. They found (see Fig. 5 and subsequent data in Simon and Linsky 1982, and Hartmann et al. 1982) that the yellow giants show a wide range of C IV normalized fluxes (i.e. $f_{C\ IV}/\ell_{bol}$), including stars with small upper limits, that they ascribed to a mixed evolutionary status of these stars. Some stars have just recently

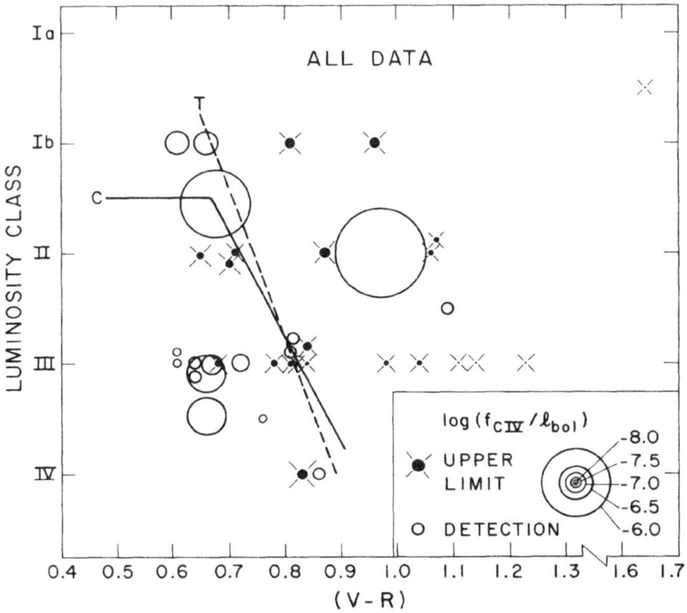

Fig. 5. H-R diagram showing measured ratios of the C IV λ1550 flux to the apparent stellar bolometric luminosity from Simon, Linsky and Stencel (1982). Open circles are detections and filled circles are upper limits. The line marked T is that originally proposed by Linsky and Haisch (1979) to separate stars with (to the left) and without (to the right) 10^5 K plasma. The line marked C was proposed by Ayres et al. (1981a) to separate stars that generally show soft X-ray emission (to the left) from stars that generally do not (to the right). Two of the three detections to the right of the line marked T are previously unknown binary systems.

evolved from the upper main sequence and could be relatively rapid rotators with strong dynamo-amplified magnetic fields and bright emission lines, whereas some may be post-helium flash stars that are slow rotators or stars that were slow rotators on the main sequence. Among the 18 red giants in their sample, only three stars have detectable C IV emission, but two of these three show ultraviolet continuum emission indicative of previously unknown companions (56 Peg and α UMa) and the third is the hybrid star α TrA (Hartmann et al. 1981). In a separate IUE study of 15 red giants, Strickland and Sanner (1981) also found no evidence for C IV emission. Simon et al. (1982) therefore concluded that the boundary originally proposed by Linsky and Haisch is real in that single stars to the right, with the exception of one hybrid star, contain significantly less 10^5 K plasma than typical single stars to the left of the boundary.

The question of a boundary remains open, however, for the following reasons. First, for the red giants, fluorescence in the fourth positive system of CO can produce emission features near the C IV 1550 Å and C II 1335 Å lines (Ayres, Moos and Linsky 1981b). Thus, high dispersion spectra are needed to estimate reliably upper limits for these emission line strengths. Second, the C IV lines detected in such hybrid stars as ι Aur (K3 II) and θ Her (K2 II) (cf. Reimers 1982) are at the level of some noise features. T. Simon recently re-observed θ Her with a 175 min SWP low-dispersion exposure, and saw no evidence for C IV emission. Thus, C IV emission in some of the hybrids is variable or not definitively detected. Third, the existence of a boundary demands a physical explanation.

A clear picture is now emerging from the observations of the G and K giants and supergiants on both sides of the boundary. I believe that three types of stars are present in this group as a consequence of different magnetic field strength and geometry:

(1) There are active stars, of which β Dra (G2 Ib) is a prototype, which show bright TR emission lines emitted by a geometrically thin region, bright X-ray emission, no evidence for any outflow of material, and redshifted emission lines (see next section). These are stars for which closed magnetic flux tubes dominate their outer atmospheres, and are responsible for enhanced heating and the suppression of outflowing gas. These stars probably have large magnetic fields either because they have just recently evolved from the upper main sequence (β Dra may be an example) or because they are members of close binary systems that are forced to rotate synchronously (the cool components of RS CVn systems and the W UMa stars are examples).

(2) There are quiet stars, of which α Boo (K2 III) is a prototype, which show no evidence of TRs or hot coronae to very small upper limits (Ayres, Simon and Linsky 1982a), have cool winds with significant mass loss, and geometrically extended chromospheres as determined from IUE studies of the C II 2325 Å intersystem lines (Stencel et al. 1981). These stars probably have no, or very few, closed magnetic flux tubes but rather have outer atmospheres with magnetically open topologies like coronal holes, and little or no hot plasma. These stars are probably slow rotators. Precisely how the decay of magnetic fields can lead to a star changing from active to quiet remains to be worked out in detail.

(3) I believe that the hybrid stars, of which α TrA (K4 II) and α Aqr (G2 Ib) are prototypes, are hybrid but in a different sense than originally proposed. These stars show weak and likely variable TR emission lines, no detected X-ray emission, and evidence for a cool wind. I believe that there is no real evidence for 10^5 K winds in these stars, since broad C IV emission lines occur in β Dra which shows no other evidence for a wind, but Doppler shift measurements are needed to answer this point. I would describe these stars as hybrid

in the sense that their outer atmospheres contain mostly open field lines along which the cool wind flows, but they do contain a few closed magnetic flux tubes from which the 10^5 K lines are emitted. Rotational modulation of these few flux tubes could explain the variability. These tubes may also contain 10^6 K plasma emitting soft X-rays, but the X-rays are absorbed by the surrounding cool plasma.

We need to ask why there are apparently no active single stars to the right of the boundary near (V-R) ≈ 0.80. I suspect that the explanation may be very simple. Gray (1981) measured the rotational velocities of five G5 III stars; four are slow rotators (v sin i ≈ 4 km s^{-1}) and one, ε UMi, is a quickly rotating (v sin i = 24 km s^{-1}) RS CVn system. These data led him to propose that as stars evolve to this location in the H-R diagram from the upper main sequence, the coupling of dynamo-generated magnetic fields and mass loss rapidly decreases the stellar rotation. The absence of active stars among the single red giants is likely due to the weak magnetic fields resulting from their slow rotation.

V. SYSTEMATIC DOWNFLOWS OF TRANSITION REGION PLASMA

Perhaps the most exciting and unexpected discovery by _IUE_ concerning cool stars is the very recent evidence concerning flows of the 10^5 K plasma. Stencel et al. (1982a,b) measured line centroid velocities for 18 lines in the SWP high dispersion spectrum of the supergiant β Dra (G2 Ib) by fitting least-squares Gaussians to the observed profiles. They estimated the velocity at the base of the chromosphere using eight subordinate or intersystem lines of C I, O I, S I, and Cℓ I. A measurement of the mean velocity of ten high excitation lines of He II, C III, C IV, N V, O III, Si III, and Si IV then gave a relative motion of the high and low excitation lines of 20±4 km s^{-1}. The TR plasma is thus flowing down into the star and we are observing a _stellar "antiwind."_

This was an unexpected discovery, because we are accustomed to studying outflows (winds) in luminous late-type stars with mass loss rates that generally increase towards the upper right-hand corner of the H-R diagram. To confirm this result, Ayres et al. (1982b) then reexamined high dispersion _IUE_ observations of active chromosphere stars and found that β Ceti (K1 III), α Aur Ab (F9 III), λ And (G8 III-IV+?), and ε Eri (K2 V) also show net redshifts of TR emission lines relative to chromospheric lines (see Fig. 6). They also reobserved α Aur Ab during conjunction (when both components of the α Aur A system have the same radial velocity) through the _IUE_ small aperture to obtain an absolute velocity scale, with the result (see Fig. 6) that the chromospheric lines have an absolute redshift of about 5 km s^{-1} and the TR lines have an absolute redshift of 17 km s^{-1}.

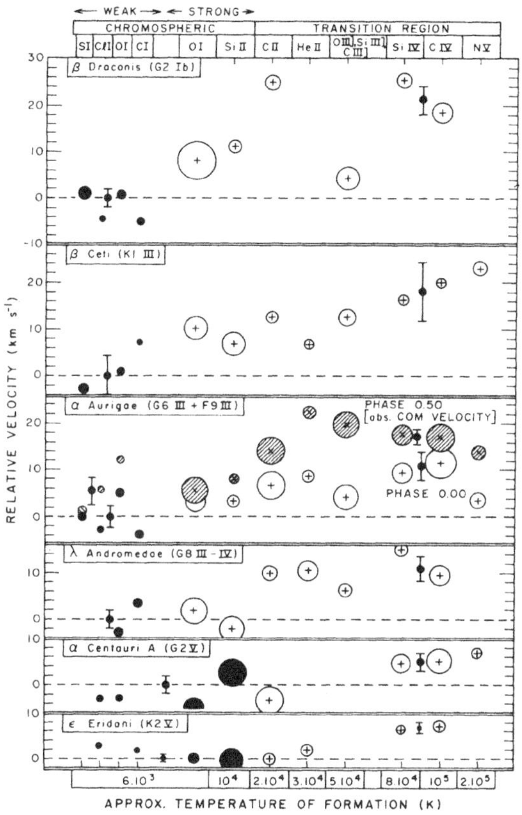

Fig. 6. A comparison of the line-of-sight velocities of high and low excitation lines obtained by Ayres et al. (1982b). The size of the bubbles indicates the total relative flux of the lines of a given ion. The filled circles are narrow chromospheric lines used in obtaining the flux-weighted mean zero velocity, and the standard error of the mean is indicated by the error bars to the left. The open circles indicate the mean velocities of higher excitation lines, and the error bars to the right, indicate the flux-weighted mean velocity of the four C IV and Si IV lines and the error of this mean (including the error of the zero velocity determination). The partially filled bubbles for Capella indicate velocities obtained from a small aperture observation at phase 0.50. These velocities were placed on an absolute center of mass velocity scale by comparison with a platinum lamp exposure obtained after the Capella exposure. Note that the velocity difference between high and low excitation lines is the same for both Capella data sets, but the small aperture data indicate that the chromospheric lines also appear to exhibit a small red shift.

At first sight, the idea of downflows ("antiwinds") in luminous stars, seems preposterous, however, solar downflows of 10-20 km s^{-1} are typically seen in such lines as C IV and Si IV in the OSO-8 (Roussel-Dupré and Shine 1982), Skylab (Doschek, Feldman and Bohlin 1976; Feldman, Cohen and Doschek 1982), and HRTS (Brueckner 1981; Dere 1982) data. These downflows are best seen in observations of the chromospheric network and plages, where magnetic flux tubes are located, but the downflows are even detected in integrated light because they are bright in ultraviolet emission lines and thus make large contributions to the integrated light line profiles. An interesting result seen in the high spatial resolution HRTS data is that the downflow velocities increase with temperature from 10^4 to 10^5 K (Dere 1982), but Doschek and Feldman (1977) measured small downflow velocities ($\approx$5 km s^{-1}) even for the chromospheric Mg II lines in the solar supergranulation network. Ayres et al. (1982b) detected the same increase in absolute downflow velocities between the chromosphere and TR of α Cen Ab.

Several explanations have been proposed. The flows may be produced by coronal material that is cooling, condensing, and falling back down to the chromospheric footpoints of magnetic loops following the interruption of the internal heating source (Rosner, Tucker and Vaiana 1978). The downflows may be part of a circulation pattern within large flux tubes for which the upleg portion of the circulation is too cool (spicules, for example) to be visible in C IV (Pneuman and Kopp 1977). It is even possible that material is flowing upward at C IV temperatures more rapidly than it is flowing downward, such that the decrease in density required by mass conservative flow will greatly reduce the visibility of the upward moving gas (since $F_L \sim n_e^2$), resulting in a net redshift (Doschek, Feldman and Bohlin 1976). In any case, the appearance of redshifts is clear evidence for the existence of strong, closed magnetic field structures in a stellar atmosphere.

There are several important consequences of redshifts, which we interpret as true downflows of material in magnetic flux tubes:

(1) The existence of downflowing gas implies an enthalpy flux of heat from the corona that must be included in any study of the energy balance of a stellar TR. Pneuman and Kopp (1977), for example, demonstrated that the enthalpy flux in a typical solar downflow exceeds the thermal conductive flux.

(2) An important result from the stellar redshift measurements that cannot be obtained from the solar data alone is the trend of redshift velocity (from the C IV and Si IV doublets) with stellar gravity. Ayres et al. (1982b) found an inverse relation of the form,

$$v_* \approx 8\ (g_*/g_\odot)^{-0.15}\ \text{km s}^{-1} \quad .$$

This relation allows one to estimate a crude lower limit on the typical flux tube height using energy conservation,

$$h_{min} \geq \frac{v_*^2}{2g_*} \approx 10^2 \ (g_*/g_\odot)^{-1.3} \ \text{km} \quad ,$$

which predicts a rapid increase with decreasing stellar gravity.

(3) Since many stars should have both downflows and upflows and since the downflowing regions probably emit brighter emission lines, one should be especially careful in estimating stellar mass loss rates from integrated disk Doppler shifts. The Sun does have a wind!

VI. ENERGY BALANCE IN STELLAR CHROMOSPHERES AND TRANSITION REGIONS

a) The Heating Mechanism for Late-Type Stellar Chromospheres

Linsky and Ayres (1978) showed, on the basis of Mg II fluxes, that the chromospheric radiative loss rate per unit surface area of a star shows no dependence on stellar gravity. This implies that the heating rate is also independent of gravity, contrary to computations for the dissipation of shocks produced by nonmagnetic acoustic waves, which imply a g^{-1} dependence and a T_{eff} dependence different than observed. This result was modified slightly by Stencel et al. (1980), who showed that IUE observations of cool supergiants are consistent with a small increase ($\sim g^{-1/4}$) in the heating rate as the gravity decreases. These data and the large range in heating rates for stars of similar T_{eff} and g, provide important clues on the chromospheric heating mechanism.

Stein (1981) and Ulmschneider and Stein (1982) derived approximate scaling laws for how the emitted flux from a chromosphere (F_{chromo}) depends on T_{eff}, g, and B for four wave heating modes (acoustic, Alfvén, acoustic slow, and magnetic fast). Only the acoustic slow mode for weak or equipartition magnetic fields predicts

$$\frac{F_{chromo}}{\sigma T_{eff}^4} \approx g^{-0.192} \ T_{eff}^{2.13} \quad ,$$

consistent with the above data. In addition, the chromospheric heating rate will depend on the fractional surface coverage by the magnetic flux tubes, which can explain the range of radiative loss rates for stars of similar T_{eff} and g. Thus stellar observations have played a major role in guiding theoretical calculations by pointing out the important role played by magnetic fields.

b) Relation Bewteen Heating Rates in Stellar Chromospheres, Transition Regions, and Coronae

Using SWP low dispersion observations of 28 cool stars, Ayres, Marstad and Linsky (1981c) showed that the emission line fluxes of chromospheric and TR lines are not linearly correlated (see Fig. 7). Instead, as one goes to stars with brighter chromospheric emission (i.e. $f_{Mg\ II}/\ell_{bol}$), the TR lines brighten even faster such that $(f_{C\ IV}/\ell_{bol}) \sim (f_{Mg\ II}/\ell_{bol})^{1.5}$ and the coronal X-ray flux brightens faster yet. Walter, Basri and Laurent (1982) and Oranje, Zwaan and Middlekoop (1982) found similar results. This phenomenon was previously noted in the comparison of the II Peg plage to quiescent spectra (Fig. 2) and the flare to quiescent spectra for HR 1099 (Fig. 3) and the dMe star Proxima Centauri (Haisch et al. 1982). It is also seen by comparing solar plage to quiescent spectra and thus must be a general property of stellar atmospheres.

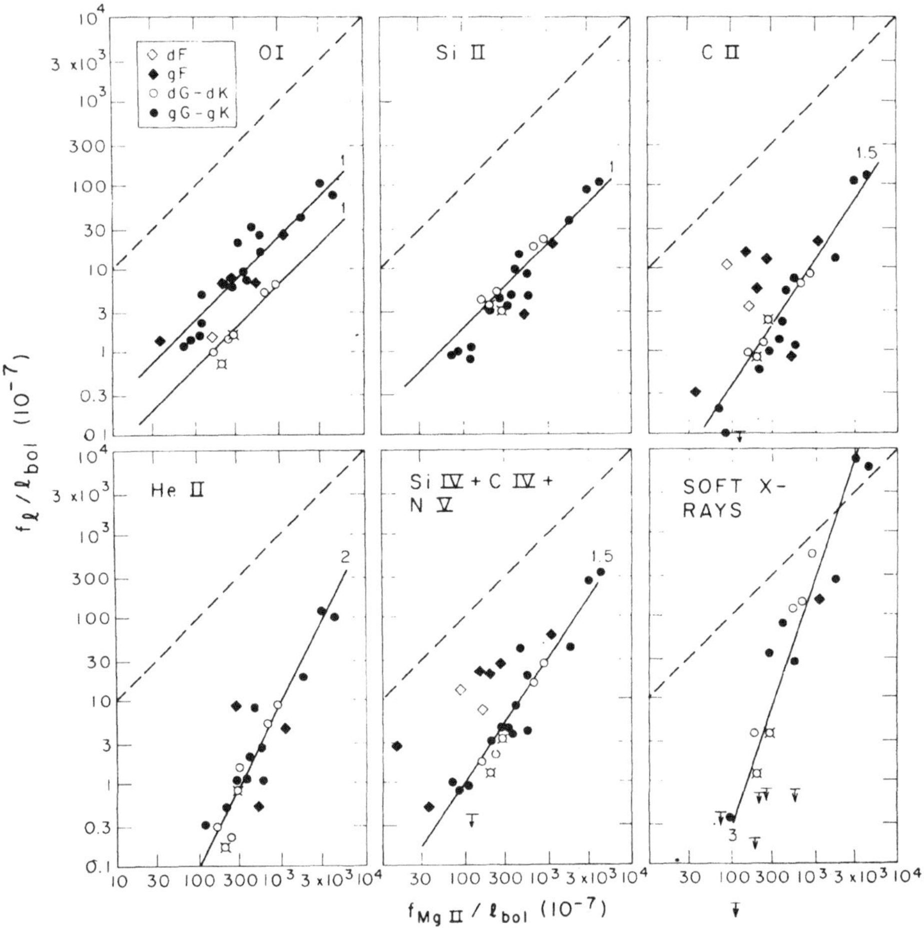

Fig. 7. Correlation plots of chromospheric, TR, and coronal fluxes compared to the Mg II line relative flux (Ayres, Marstad and Linsky 1981c).

Hammer, Linsky and Endler (1982) proposed an explanation for this phenomenon. They pointed out that the radiative loss rate in TR emission lines for realistic magnetic flux tube models (e.g. Rosner, Tucker and Vaiana 1978) depends on pressure to a higher power than the corresponding radiative loss rate in chromospheric emission lines (e.g. models in Vernazza et al. 1981). Thus, with increasing mechanical energy flux, the location of the base of the TR (intersection point of the curves in Fig. 8) moves to larger pressures and the TR lines brighten by a larger factor than the chromospheric lines.

There are additional factors that must be included in realistic future energy balance studies of stellar chromospheres and TRs:

(1) For their sample of ten active dwarf stars (including solar plages), Linsky et al. (1982) showed that the fraction of the stellar flux emitted by the chromosphere increases a factor of 5 as T_{eff} decreases from 5770 to 3200 K, but the fraction emitted by the TR increases a factor of 100. Thus the relative heating rates in different atmospheric layers may depend on T_{eff} as well as the magnetic field.

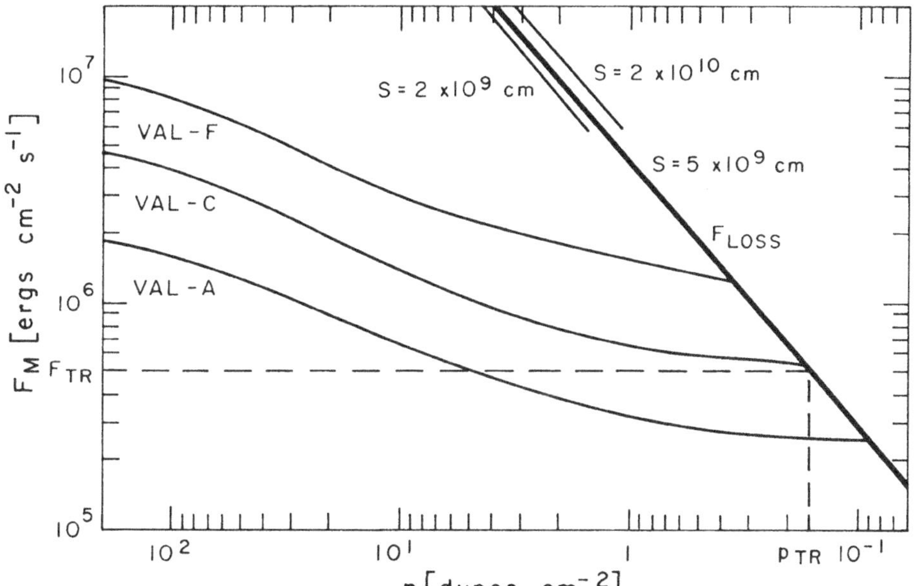

Fig. 8. Total mechanical energy flux F_M as a function of the pressure p for the chromospheric models A, C, and F of Vernazza, Avrett and Loeser (1981). The transition region lies at the intersection point with a curve (heavy line) that gives the total energy losses F_{Loss} of the TR and corona as a function of the base pressure and the semilength S of the coronal loops (cf. Rosner, Tucker and Vaiana 1978). From Hammer, Linsky and Endler (1982).

(2) The enthalpy flux can be an important heating source for stellar TRs. Thus models should include dynamic phenomena -- downflows and also upflows (e.g. spicules).

(3) Cram (1982) called attention to the importance of heating of dMe star chromospheres by coronal X-rays. Since solar coronal X-ray emission is almost entirely confined to magnetic loops, even the nonmagnetic regions of dMe star chromospheres may be heated indirectly by magnetically controlled plasma.

VII. MODELS FOR ATMOSPHERIC LAYERS WITH AND WITHOUT MAGNETIC FIELDS

Until now we have summarized the different qualitative roles played by magnetic fields, but in addition, several investigators have built detailed models of solar and stellar atmospheres for regions in which fields are important. Chapman (1981), Ulmschneider and Stein (1982), and Linsky (1980) have reviewed this work. No models are wholly satisfactory as yet due to gross simplifications in the assumed geometry, treatment of the radiative transfer equation, and the strong dependence of the magnetic flux tube parameters on the assumed filling factor, but I will mention those trends in the models that are likely to be valid despite these simplifications.

a) One-Component Atmospheric Models

The simplest approach is to construct a one-component model atmosphere to match line profiles and fluxes for the brightest regions of a solar plage or the integrated flux of a star with very bright emission lines. Since the data are spatial averages including magnetic flux tubes (the active component) and nonmagnetic (quiet component) plasma, the derived differences between the active model and the reference quiet Sun or star model are lower limits to the true differences between the properties of magnetic flux tubes and the nonmagnetic regions. Thus we need to extrapolate these trends to the case of complete filling of the aperture by flux tubes.

One-component models of solar plages include the work of Shine and Linsky (1974) using the Ca II lines, Morrison and Linsky (1978) and Kelch and Linsky (1978) using the Mg II lines, Basri et al. (1979) using Lα, and Vernazza et al. (1981) using a number of continuum and emission line features. These models have a number of elements in common that are illustrated in Figure 9:

(1) The minimum temperature (T_{min}) reached between the photosphere and chromosphere is enhanced by several hundred degrees (270 K in the Vernazza et al. Model F and about 200 K in the Morrison-Linsky models) compared to the quiet Sun. There is also considerable enhancement of temperature and thus nonradiative heating in the upper layers of the photosphere.

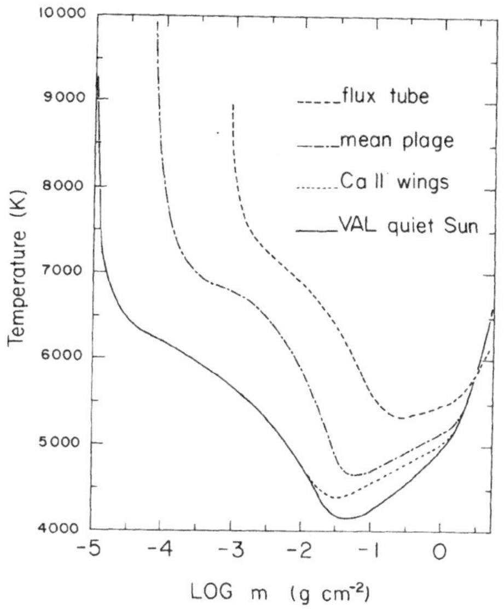

Fig. 9. Plage, flux tube, and quiet Sun models. The solid line is the VAL quiet Sun model. The short dashed lines (Ca II wings) represent a modification of the VAL designed to reproduce the Ca II H and K damping wings. The dash-dot curve is a plage model based on H I Lα, Ca II K, and Mg II k data obtained by the French (LPSP) experiment on OSO-8. The long dashed (higher) curve represents a flux tube model with a chromospheric portion matching the OSO-8 plage profiles with a 20% filling factor. The photospheric portion ($m > 0.3$ g cm^{-2}) is similar to the class of flux tube models advocated by Chapman (1977). From Chapman (1981), courtesy of Colorado Associated University Press.

(2) The location of T_{min} is displaced inward to larger values of column mass density (m); for example, T_{min} is displaced inward about 40 km and $m(T_{min})$ increases from 0.05 to 0.08 g cm^{-2} between the quiet Sun (Model C) and Model F in the Vernazza et al. grid.

(3) Similarly, the location of the steep rise in temperature beginning at 8000 K and extending through the TR is also displaced inwards to larger values of mass column density. For example, the grid point corresponding to 10,000 K is located 40 km deeper and $m(T = 10^4$ K) increases from 6×10^{-6} to 1.2×10^{-5} g cm^{-2} between the mean quiet Sun and Model F in the Vernazza et al. grid. In the Shine and Linsky (1974) grid the range in $m(T = 10^4$ K) is nearly a factor of 40. Since the temperature rises steeply in all TR models, the TR pressure ($P_{TR} \approx m(T = 10^4$ K)g) is nearly constant for each model. Thus a sequence of models with increasing P_{TR} typically produces a sequence of emission lines with increasing flux.

(4) Different authors have assumed different functional forms of the temperature structure, T(m), between the temperature minimum and the top of the chromosphere ($T \approx 10^4$ K), but a general result is that T(m) and the local electron density at each column mass density are larger for models with increasing m($T = 10^4$ K). Thus typical chromospheric lines like Ca II H and K, Mg II h and k, Lα, and Si II λλ1808, 1816 brighten as m($T = 10^4$ K) increases.

(5) Avrett (1981) computed radiative loss rates for the five Vernazza et al. models, the Basri et al. (1979) plage model, a flare model from Machado et al. (1980), and two stars (α Boo and λ And). For the Vernazza et al. models and the plage model, Ca II and Mg II are the dominant emitters, although hydrogen becomes important at the base of the TR. Thus, the observed Ca II and Mg II surface fluxes in the Sun and solar-like stars should be accurate diagnostics of the chromospheric heating rate as had been assumed previously, but, for the dMe stars, the Balmer lines are more important chromospheric emitters (Linsky et al. 1982).

To what extent do models of active chromosphere stars show properties similar to the models of solar plages? Linsky (1980) and Ulmschneider and Stein (1982) have reviewed this question. There are three groups of models that can be used to answer it: models of F-K dwarf stars (Kelch, Linsky and Worden 1979; Simon, Kelch and Linsky 1980b), models of dM and dMe stars (Giampapa, Worden and Linsky 1982), and models of RS CVn-type active subgiants (Simon and Linsky 1980; Baliunas et al. 1979). Evidence for enhanced heating in the upper photosphere and temperature minimum region is clearly shown in the dMe models and perhaps also in the active F-K dwarfs. In addition, the dMe stars and the active G-K dwarfs (see also Ayres et al. 1982c) show broader base widths of the Ca II and Mg II resonance lines, indicating that the column mass density at the temperature minimum is systematically larger in stars with enhanced heating. Finally, all of these active stars models have m($T = 10^4$ K) similar to solar plages, so that TR pressures are enhanced over quiet stars with similar T_{eff} and gravity. Thus the differences between solar plage and quiet models are repeated in the active and quiet late-type stars.

b) Two-Component Atmospheric Models

Chapman (1981) conclusively argued that no one-component model can accurately determine the properties of unresolved flux tubes, but instead the true models for flux tubes must be extrapolations of the one-component plage models. Three approaches have been followed in estimating flux tube properties:

(1) One approach is to estimate the factor by which flux tubes fill the aperture and then to solve for the flux tube parameters assuming that the remaining portion of the aperture is filled by a quiet atmosphere model. Chapman (1981) derived such a flux tube model from OSO-8 plage spectra of the Lα, Ca II, and Mg II lines assuming a 20%

filling factor (see Fig. 9). His solar flux tube model has T_{min} enhanced by 1200 K, $m(T_{min})$ displaced inwards from 0.04 to 0.3 g cm^{-2}, chromospheric temperatures enhanced by nearly 2500 K, and $m(T = 10^4$ K) displaced inwards from 1×10^{-5} to 1×10^{-3} g cm^{-2}.

(2) A second type of model, typified by the work of Chapman (1977,1979), assumes a diverging flux tube in horizontal pressure equilibrium (gas and magnetic forces) with the surrounding nonmagnetic atmosphere. Again one tries to match spatially averaged spectra by assuming a quiet model and solving for the parameters of the flux tubes assuming a value for the base magnetic flux.

(3) A third approach is to assume a geometry with isolated flux tubes embedded in a nonmagnetic medium and then solving the transfer equation in two dimensions including the horizontal flow of radiation. Two examples are the work of Stenholm and Stenflo (1977) and Owocki and Auer (1980). This approach increases the computational complexity, but it is probably a necessary complication in the chromosphere where most lines are effectively thin and thus photoexcitation from adjacent structures may be important.

VIII. SOME SUGGESTIONS FOR FUTURE WORK

I began this talk by mentioning some of the inherent difficulties in this topic, and I suspect that you now appreciate what I meant. While we have made real progress recently in identifying the roles played by magnetic fields in stellar chromospheres and TRs, we now recognize how little we know. Let me conclude by stating how we will probably make real progress in this area:

(1) We must resolve individual flux tubes on the Sun in order to derive their physical properties in a definitive way. I anticipate that this will be the most important accomplishment of the Solar Optical Telescope (SOT) when it flies at the end of this decade.

(2) I suspect that future high spectral resolution stellar observations with IUE and Space Telescope will reveal new phenomena and trends with stellar parameters that will point towards the important physical processes that future theoretical studies must include.

(3) The basic physical processes responsible for downflows in flux tubes and the existence of the TR and coronal boundaries in the H-R diagram remain unknown.

(4) Even though heating processes will probably remain poorly understood for some time, it is nevertheless important to model magnetic flux tubes properly, taking into account the energy balance, dynamics, and radiative transfer for parameterized heating rates. It is important to study the stability of such models and to look for unique signatures of the location and mechanism of the heating.

This work was supported in part by NASA grants NAG5-82, NGL-06-003-057, and NAG5-199 through the University of Colorado. I would like to thank my colleagues T. R. Ayres, P. L. Bornmann, S. Drake, R. Hammer, N. C. Marstad, M. Schindler, T. Simon, R. E. Stencel, and F. M. Walter for stimulating discussions and for permission to describe unpublished work at this time.

REFERENCES

Avrett, E.H.: 1981, in R. M. Bonnet and A. K. Dupree (eds.), "Solar Phenomena in Stars and Stellar Systems," Reidel, Dordrecht, p. 173.
Ayres, T.R., Linsky, J.L., Simon, T., Jordan, C., and Brown, A.: 1982c, in preparation.
Ayres, T.R., Linsky, J.L., Vaiana, G.S., Golub, L., and Rosner, R.: 1981a, Astrophys. J. 250, p. 293.
Ayres, T.R., Marstad, N.C., and Linsky, J.L.: 1981c, Astrophys. J. 247, p. 545.
Ayres, T.R., Moos, H.W., and Linsky, J.L.: 1981b, Astrophys. J. (Letters) 248, p. L137.
Ayres, T.R., Simon, T., and Linsky, J.L.: 1982a, Astrophys. J. (submitted).
Ayres, T.R., Stencel, R.E., Linsky, J.L., Simon, T., Jordan, C., Brown, A., and Engvold, O.: 1982b, in preparation.
Baliunas, S.L., Avrett, E.H., Hartmann, L., and Dupree, A.K.: 1979, Astrophys. J. (Letters) 233, p. L129.
Baliunas, S.L., and Dupree, A.K.: 1982, Astrophys. J. 252, p. 668.
Baliunas, S.L., Hartmann, L., Vaughan, A.H., Liller, W., and Dupree, A.K.: 1981, Astrophys. J. 246, p. 473.
Basri, G.S., Linsky, J.L., Bartoe, J.-D.F., Brueckner, G.E., and Van Hoosier, M.E.: 1979, Astrophys. J. 230, p. 924.
Brueckner, G.E.: 1981, in F. Q. Orrall (ed.), "Solar Active Regions," Colorado Assoc. Univ. Press, Boulder, p. 113.
Brueckner, G.E., Bartoe, J.-D.F., and Van Hoosier, M.E.: 1977, in E. Hansen and S. Schaffner (eds)., "Proceedings of the OSO-8 Workshop," LASP, Boulder, p. 380.
Butler, C.J., Byrne, P.B., Andrews, A.D., and Doyle, J.G.: 1981, Monthly Notices Roy. Astron. Soc. 197, p. 815.
Chapman, G.A.: 1977, Astrophys. J. Suppl. Ser. 33, p. 35.
Chapman, G.A.: 1979, Astrophys. J. 232, p. 923.
Chapman, G.A.: 1981, in F. Q. Orrall (ed.), "Solar Active Regions," Colorado Assoc. Univ. Press, Boulder, p. 43.
Cram, L.E.: 1982, Astrophys. J. 253, p. 768.
Dere, K.: 1982, Solar Phys. 77, p. 77.
Doschek, G.A. and Feldman, U.: 1977, Astrophys. J. Suppl. 35, p. 471.
Doschek, G.A., Feldman, U., and Bohlin, J.D.: 1976, Astrophys. J. (Letters) 205, p. L177.
Dupree, A.K.: 1981a, in R. M. Bonnet and A. K. Dupree (eds.), "Solar Phenomena in Stars and Stellar Systems," Reidel, Dordrecht, p. 407.
Dupree, A.K.: 1981b, in C. Chiosi and R. Stalio (eds.), "Effects of Mass Loss on Stellar Evolution," Reidel, Dordrecht, p. 87.

Dupree, A.K.: 1982, in "Advances in Ultraviolet Astronomy: Four Years of IUE Research" (in press).
Feldman, U., Cohen, L., and Doschek, G.A.: 1982, Astrophys. J. 255, p. 325.
Frazier, E.N. and Stenflo, J.O.: 1972, Solar Phys. 27, p. 330.
Giampapa, M.S., Worden, S.P., and Linsky, J.L.: 1982, Astrophys. J. (in press).
Gray, D.F.: 1981, Astrophys. J. 251, p. 155.
Haisch, B.M., Linsky, J.L., Bornmann, P.L., Stencel, R.E., Golub, L., and Antiochos, S.K.: 1982, Astrophys. J. (submitted).
Hall, D.S.: 1981, in R. M. Bonnet and A. K. Dupree (eds.), "Solar Phenomena in Stars and Stellar Systems," Reidel, Dordrecht, p. 431.
Hallam, K.L. and Wolff, C.L.: 1981, Astrophys. J. (Letters) 248, p. L73.
Hammer, R., Linsky, J.L., and Endler, F.: 1982, in "Advances in Ultraviolet Astronomy: Four Years of IUE Research" (in press).
Hartmann, L., Dupree, A.K., and Raymond, J.C.: 1980, Astrophys. J. (Letters) 236, p. L143.
Hartmann, L., Dupree, A.K., and Raymond, J.C.: 1981, Astrophys. J. 246, p. 193.
Hartmann, L., Dupree, A.K., and Raymond, J.C.: 1982, Astrophys. J. 252, p. 214.
Imhoff, C.L. and Giampapa, M.S.: 1982, in "Advances in Ultraviolet Astronomy: Four Years of IUE Research" (in press).
Kelch, W.L. and Linsky, J.L.: 1978, Solar Phys. 58, p. 37.
Kelch, W.L., Linsky, J.L., and Worden, S.P.: 1979, Astrophys. J. 229, p. 700.
Kunkel, W.W.: 1975, in V. E. Sherwood and L. Plaut (eds.), "Variable Stars and Stellar Evolution," Reidel, Dordrecht, p. 15.
Lang, K.R.: 1980, "Astrophysical Formulae," 2nd ed., Springer-Verlag, Berlin, p. 510.
Linsky, J.L.: 1980, Ann. Rev. Astron. Astrophys. 18, p. 439.
Linsky, J.L.: 1981a, in R. M. Bonnet and A. K. Dupree (eds.), "Solar Phenomena in Stars and Stellar Systems," Reidel, Dordrecht, p. 99.
Linsky, J.L.: 1981b, in C. Chiosi and R. Stalio (eds.), "Effects of Mass Loss on Stellar Evolution," Reidel, Dordrecht, p. 187.
Linsky, J.L.: 1981c, in I. Iben, Jr. and A. Renzini (eds.), "Physical Processes in Red Giants," Reidel, Dordrecht, p. 247.
Linsky, J.L.: 1981d, in "X-ray Astronomy in the 1980's," NASA Technical Memorandum 83848, p. 13.
Linsky, J.L.: 1982, in "Advances in Ultraviolet Astronomy: Four Years of IUE Research" (in press).
Linsky, J.L. and Ayres, T.R.: 1978, Astrophys. J. 220, p. 619.
Linsky, J.L., Bornmann, P.L., Carpenter, K.G., Wing, R.F., Giampapa, M.S., and Worden, S.P.: 1982, Astrophys. J. (in press).
Linsky, J.L. and Haisch, B.M.: 1979, Astrophys. J. (Letters) 229, p. L27.
Machado, M.E., Avrett, E.H., Vernazza, J.E., and Noyes, R.W.: 1980, Astrophys. J. 242, p. 336.
Marstad, N. et al.: 1982, in "Advances in Ultraviolet Astronomy: Four Years of IUE Research" (in press).
Morrison, N.D. and Linsky, J.L.: 1978, Astrophys. J. 222, p. 723.

Oranje, B.J., Zwaan, C., and Middelkoop, F.: 1982, Astron. Astrophys. (in press).
Owocki, S.P. and Auer, L.H.: 1980, Astrophys. J. 241, p. 448.
Pneuman, G.W. and Kopp, R.A.: 1977, Astron. Astrophys. 55, p. 305.
Reimers, D.: 1982, Astron. Astrophys. 107, p. 292.
Rosner, R., Tucker, W.H., and Vaiana, G.S.: 1978, Astrophys. J. 220, p. 643.
Roussel-Dupre, D. and Shine, R.A.: 1982, Solar Phys. 77, p. 329.
Schindler, M., Kjeldseth-Moe, O., Bartoe, J.-D.F., Brueckner, G.E., and Van Hoosier, M.E.: 1982, Astrophys. J. (submitted).
Shine, R.A. and Linsky, J.L.: 1974, Solar Phys. 39, p. 49.
Simon, T., Kelch, W.L., and Linsky, J.L.: 1980b, Astrophys. J. 237, p. 72.
Simon, T. and Linsky, J.L.: 1980, Astrophys. J. 241, p. 759.
Simon, T. and Linsky, J.L.: 1982, in "Advances in Ultraviolet Astronomy: Four Years of IUE Research" (in press).
Simon, T., Linsky, J.L., and Schiffer F.H. III: 1980a, Astrophys. J. 239, p. 911.
Simon, T., Linsky, J.L., and Stencel, R.E.: 1982, Astrophys. J. 257, p. 225.
Skumanich, A., Smythe, C., and Frazier, E.N.: 1975, Astrophys. J. 200, p. 747.
Stencel, R.E.: 1978, Astrophys. J. (Letters) 223, p. L37.
Stencel, R.E., Linsky, J.L., and Ayres, T.R.: 1982a, in "Advances in Ultraviolet Astronomy: Four Years of IUE Research" (in press).
Stencel, R.E., Linsky, J.L., Ayres, T.R., Jordan, C., and Brown, A.: 1982b, Astrophys. J. (submitted).
Stencel, R.E., Linsky, J.L., Brown, A., Jordan, C., Carpenter, K.G., Wing, R.F., and Czyzak, S.: 1981, Monthly Notices Roy. Astron. Soc. 196, p. 47P.
Stencel, R.E. and Mullan, D.J.: 1980a, Astrophys. J. 238, p. 221.
Stencel, R.E. and Mullan, D.J.: 1980b, Astrophys. J. 240, p. 718.
Stencel, R.E., Mullan, D.J., Linsky, J.L., Basri, G.S., and Worden, S.P.: 1980, Astrophys. J. Suppl. 44, p. 383.
Stenflo, J.O.: 1973, Solar Phys. 32, p. 41.
Stenholm, L.G. and Stenflo, J.O.: 1977, Astron. Astrophys. 58, p. 273.
Stein, R.F.: 1981, Astrophys. J. 246, p. 966.
Stickland, D.J. and Sanner, F.: 1981, Monthly Notices Roy. Astron. Soc. 197, p. 791.
Ulmschneider, P. and Stein, R.F.: 1982, Astron. Astrophys. 106, p. 9.
Vaughan, A.H., Baliunas, S.L., Middelkoop, F., Hartmann, L.W., Mihalas, D., Noyes, R.W., and Preston, G.W.: 1981, Astrophys. J. 250, p. 276.
Vernazza, J.E., Avrett, E.H., and Loeser, R.: 1981, Astrophys. J. Suppl. 45, p. 635.
Vernazza, J.E. and Reeves, E.M.: 1978, Astrophys. J. Suppl. Series 37, p. 485.
Walter, F.M., Basri, G.S., and Laurent, R.: 1982, in "Advances in Ultraviolet Astronomy: Four Years of IUE Research" (in press).
White, O.R. and Livingston, W.C.: 1981, Astrophys. J. 249, p. 798.
Wilson, O.C.: 1978, Astrophys. J. 226, p. 379.

DISCUSSION

KOUTCHMY: In your analysis of the global solar activity compared with the stellar activity, why did you not incorporate more ***transient*** events like events connected with *flares* or, even, the so-called super-energetic events as they are almost permanently occuring over the surface of the Sun, at least at the time of the maximum of the solar activity cycle? They produce a lot uf UV and X-ray radiation probably connected with some kind of dissipation or change of the magnetic field. So, should they not play a significant role for G-K stars as well?

LINSKY: At present we have no way of distinguishing between continuous heating processes and heating by a large number of microflares in stars.

LAGO: In your graph showing the dividing line in the H-R diagram, where would you place T Tauri stars that have massive winds and yet strong line emission from transition regions. For instance in RU Lupi, C IV is very strong, and yet the IUE data we obtained suggest a P Cygni-type profile for the C IV line.

LINSKY: The proposed dividing line was based entirely on observations of evolved stars, for which the processes of angular momentum loss and dynamo magnetic field regeneration have greatly modified the initial properties of a stellar atmosphere. Thus I would not expect the T Tauri stars to fit at all into the concept of a dividing line.

UCHIDA: In relation to the evolutionary aspects you mentioned in your talk, I would like to mention that Dr. Bappu and myself have recently proposed a mechanism for the revival of the magnetic activity in giants after they have lost their magnetic activity in the long life on the main sequence. This is, in our proposal, due to the contraction, in the evolution to the giant stage, of the core and its spin-up. The convection develops to deeper layers as the envelope expands and is cooled, and as the convection reaches deep down, a dynamo layer reappears in the bottom region of the envelope and leads to the revival of the dynamo. This suggests that not only the surface rotation, but the rotation in the invisible interior could be essential.

HARTMANN: (1) In your discussion of downdrafts in transition-region lines, you neglected to mention that even the optically thin lines exhibit very large widths, much larger than the redshifts, and much larger than in the sun. This seems to me to be a very important fact. (2) If you have a general, roughly spherically symmetric downflow in an optically thick line, you will tend to get an inverse P-Cygni profile, with blueshifted emission. (3) Open magnetic-field regions in the sun are sources of hot wind. If we are to use the solar analogy, we would expect this to be true of stars as well. I don't understand why the opening up of magnetic field lines will necessarily result in a cool wind.

LINSKY: (1) Most of the stars with redshifted lines detected to date are giants and supergiants with transition region line widths greater than for the Sun. I suspect that these stars have upflows and downflows of greater magnitude than seen in the Sun. One indication of this is that the inferred net downflows are larger. (2) It is correct that *optically thick* emission lines formed in an expanding extended atmosphere can exhibit a net redshift due to the absorption or scattering of radiation in the blue side of the line profile by expanding gas in front of the star. However, we can test this possible explanation for the observed transition region redshifts by looking to see if the redshift decreases or disappears as one goes to lines of decreasing optical depth in the sequence C IV $\lambda1551$, $\lambda1548$, Si IV $\lambda1393$, $\lambda1402$, N V $\lambda1238$, $\lambda1242$, and finally the O III, Si III, and C III intersystem lines, which must be optically thin. We do not find such a trend in our data, except perhaps for β Dra. On the other hand, the intersystem lines for the small aperture Capella data show a larger redshift than the C IV lines. Thus the opacity explanation does not appear to be valid. (3) If it is valid to think in terms of Parker-type critical points for late-type giant winds, then the critical temperature should decrease with decreasing stellar gravity. For the Sun we know that the coronal gas in closed magnetic regions is hotter than for open regions, presumably because of increased heating and the absence of wind expansion and conduction to space as cooling mechanisms. On the basis of solar analogy I would argue that the open field regions of late-type giants would be cooler than the closed field regions and much cooler than solar coronal holes.

ZWAAN: (1) One comment in passing: You seemed to suggest that plages overlie sunspots. However, in the solar case plages border sunspots, and that may be important in modelling solar-type stars. (2) As to your category of quiet giants: also the *chromospheric* emission of these stars is quite low. During my talk last Monday I have discussed indications that the chromospheric flux attains a lower-limit value for these stars. There is one category of single K-type giants showing an enhanced Ca II H and K emission: many of the intrinsically brighter ($M_V < +1$) K-type giants do. These may correspond to your hybrid stars. (3) You have shown inferred heating rates as a function of surface gravity. However, how do you disentangle heating rates and filling factors?

LINSKY: (1) The II Peg data I showed imply a general spatial correlation of starspots with plages, but the data are insufficient to argue whether or not these two types of structures are cospatial in detail. (2) Your point is well taken. Based on the limited data available, the Mg II surface fluxes for the hybrids are similar to the quiet Sun, whereas for the prototype quiet giant α Boo, the Mg II surface flux is 0.15 times that of the quiet Sun. Thus the chromospheric surface fluxes for the single K giants may be correlated with the hybrid stellar characteristics. (3) The inferred heating rates are based on the measured radiative loss rates independent of filling factors.

NOYES: Two comments: (1) (relevant to Linsky's response to Zwaan). The star HD 1835 (G2V) shows perhaps 30 $\times$ as great photometric modulation (i.e. spots with 30 $\times$ as much area), but only a few times greater chromospheric (Ca II) emission. This reduction in plage/spot ratio may be less in younger stars. (2) Note that the *minimum* chromospheric flux of "active chromosphere" stars (e.g. HD 1835, but also many others) is much larger than the maximum integrated chromospheric flux from the sun. This suggests that dynamo generation of quiet chromosphere (magnetic-field related) emission also evolves downward with time. The study of spectra from "quiet" stars (i.e., of minimum activity) can thus lead to additional and independent information on the origins of magnetic activity.

GAIZAUSKAS: With regard to the spatial relation between plages and sunspots, I wish to point out that plages overlie a spot *only* when the spot is just emerging. When seen near the limb, the bright Hα emission overlying newly emerging spots has a lot of local vertical structure, extending to heights of say 10^4 km. Once the spot has fully emerged, in about 2 days, it is free of overlying plage. At that stage the plage covers the typical collection of trailing pores and is much dimmer. At the limb, a mature region shows much less vertical structure in its plage. Now if your stellar observations demand that the plage and sunspot emissions be cospatial, an analogy with the sun would suggest that a great deal of flux is continually emerging near your starspots.

LINSKY: The II Peg data permit us to say only that the plage and spots are cospatial to within $\pm 20°$ in longitude or so, which is too coarse for detailed solar comparison. It is likely, however, that there is continuously emerging flux near the spots in RS CVn-type stars.

THE MAGNETIC FIELD CONFIGURATION IN SOLAR AND STELLAR CHROMOSPHERES

U. Anzer and D.J. Galloway
Max-Planck-Institut für Astrophysik
Karl-Schwarzschild-Str. 1, 8046 Garching, FRG

ABSTRACT

Calculations are presented for the inhomogeneous magnetic field structure above a stellar photosphere which has magnetic flux tubes located at the downdraughts of its supergranulation pattern. Regions can be delineated where the ambient magnetic energy density is large or small compared with the thermal energy density derived from a model atmosphere. This enables the relative importance of magnetic versus non-magnetic heating mechanisms to be assessed. For the quiet Sun, over half the chromospheric emission must be supplied non-magnetically, whilst the network and active regions require a magnetic supply. For other late-type stars, a simple working rule suggests that when the magnetic field is strong enough to be directly observable, the chromosphere will be magnetically dominated.

1. INTRODUCTION

The quiet Sun's photospheric magnetic field is concentrated into thin tubes which are most commonly located in the regions of convective downdraught at the boundaries between supergranulation cells. In the chromosphere above, the field fans out and flux from the different tubes merges to give a field which is more uniform by the time the corona is reached. In this paper we give a three-dimensional calculation of this fanning-out structure based on the hypothesis that the field in the atmosphere is current-free and use it to decide where and when in the chromosphere magnetic versus non-magnetic heating mechanisms are important. Two-dimensional calculations have already been given by Kopp and Kuperus (1968) and Gabriel (1976); less detailed 3-D models are shown in the paper on coronal plumes of Newkirk and Altschuler (1968), and models for various isolated tubes are given in Simon et al. (1982). The work here considers a horizontally periodic atmosphere with tubes stationed at strategic points in a hexagonal supergranulation pattern.

The calculated models are static with simple field configurations, and are thus best suited for treating quiet regions. For the Sun, Avrett

J. O. Stenflo (ed.), Solar and Magnetic Fields: Origins and Coronal Effects, 339–344.

(1981) has worked out in this case the run of chromospheric radiative losses as a function of height, basing his calculations on the model atmospheres of Vernazza et al. (1981). The losses peak at a height of around 850 km, and fall off exponentially above 1000 km. In a steady state, these losses must be supplied by some form of mechanical heating; Ulmschneider (1981) advocates non-linear acoustic waves, and the divergence of the flux of these waves appears to be able to reproduce Avrett's loss curves. Ulmschneider and Stein (1982) have also argued that in the network an essentially magnetic mechanism must operate, and indeed the large excess emission there and in active regions indicates that such a mechanism is far more efficient when given a chance. It is thus important to work out where in the chromosphere the field dominates, and where by contrast it is so weak that wave motions are essentially acoustic or gravitational in nature. We shall do this numerically for the Sun, and then provide an approximate working rule for use with other stars. A full description of this work will be available in Anzer and Galloway (1982).

2. OUTLINE OF THE MODEL

The bottom boundary of the model is taken to be a plane consisting of a periodic pattern of hexagons representing supergranulation cells. The magnetic field at this boundary is concentrated into individual tubes whose centres are located at strategic points on the lines separating the hexagons, i.e. in the convective downdraughts. Each tube is small compared with the supergranule half-width L, and the vertical component of field B_z at its bottom boundary is uniform out to some radius R, and thereafter zero. All hexagons are identical. Various different configurations have been run; the one illustrated in this paper has tubes at each vertex (figure 1).

The field above the boundary z=0 is calculated on the assumption that it is current-free. The resulting potential problem can then be solved by Fourier series; for a $\underline{B}$-field which is even in the horizontal directions x and y, the solution can be written

$$B_z = \sum_{m=0}^{\infty} \sum_{n=0}^{\infty} B_{mn} \cos(m\alpha x)\cos(n\beta y)\ e^{-\gamma_{mn} z}$$

together with corresponding sine/cosine expressions for B_x and B_y, where $\alpha = \pi/\sqrt{3}\,L$, $\beta = \pi/L$, $\gamma_{mn} = (m^2\alpha^2+n^2\beta^2)^{1/2}$, and the B_{mn}'s are calculated from the distribution of B_z at z=0. When the tube is centred at (X,Y), and has a radius R and a uniform B_z of 1 at z=0, the Fourier coefficients can be calculated analytically; the resulting formulae for the B_{mn}'s are given in terms of Bessel functions in AG 1982. Several such tubes are then superposed, and the field at any desired (x,y,z) is found by summing the Fourier series on a computer. In two dimensions a full analytic solution is possible (see AG 1982).

To describe the importance of the magnetic field in various regions of the chromosphere, we will consider the 'equipartition surface' (EPS), defined as the level z(x,y) where magnetic and gas pressures are equal; thus z solves

$$B^2(z)/8\pi = p(z)$$

and is easily found numerically for any (x,y) once the B-distribution is known and a model atmosphere (normally VAL (1981) model C) is prescribed. Above this surface the ambient magnetic energy density dominates over the thermal energy density, while below it the opposite is true. In consequence energy transport is likely to be magnetically controlled in the regions above, but to proceed essentially non-magnetically in those below.

3. RESULTS

For the quiet Sun, we have considered a variety of cases with average field strengths B_0 up to 10 G, assorted tube layouts, and a supergranule length scale 2L of 20 000 or 30 000 km. For reasons which will become clear in a minute, only the value of B_0 is really important in determining the height of the EPS away from the immediate neighbourhood of the flux concentrations. Thus here we show only one example, in figure 1, where the EPS is displayed for one hexagon of a planform with L = 15 000 km and a B_0 of 10 G concentrated into tubes of strength 1500 G at each vertex, so that the tubes each have radius 909 km.

Various other calculations were run, and some can be found in AG 1982. In all appropriate cases, away from the immediate surroundings of the flux tubes the EPS lies several scale heights above those regions where, according to Avrett (1981), the dominant chromospheric losses occur. This means there are large areas of quiet Sun requiring a non-magnetic heating mechanism (and, correspondingly, smaller areas in the network that require a magnetic one.) To quantify this, we note that for figure 1, some 76% of the EPS lies above 1000 km (61% above 1300 km), whereas Avrett's (1981) radiative losses peak at a height of 800 km, and decrease approximately exponentially above 1000 km. Though the non-magnetic regions radiate less than the network, their greater area means they actually contribute slightly more than half of the quiet Sun's total chromospheric emission. Taking a B_0 which is less than the rather high value 10 G increases the height of the EPS even further. Moreover, test cases where 'hidden flux' (Stenflo, 1982) is included in the form of small bipolar regions located away from the sides and vertices of the hexagon, show little difference. One must include substantially more net flux before the EPS comes down as low as the radiative losses. For instance, using a B_0 of 110 G, we have been able to obtain results similar to those of Giovanelli and Jones (1982), who found observationally a value of around 500 km for what they call the 'canopy', a concept corresponding loosely to our EPS. Their study was for active regions, which, not surprisingly, are primarily magnetically heated.

For other stars, some less elaborate recipe is helpful. For unipolar regions it turns out that the EPS lies roughly half above and half

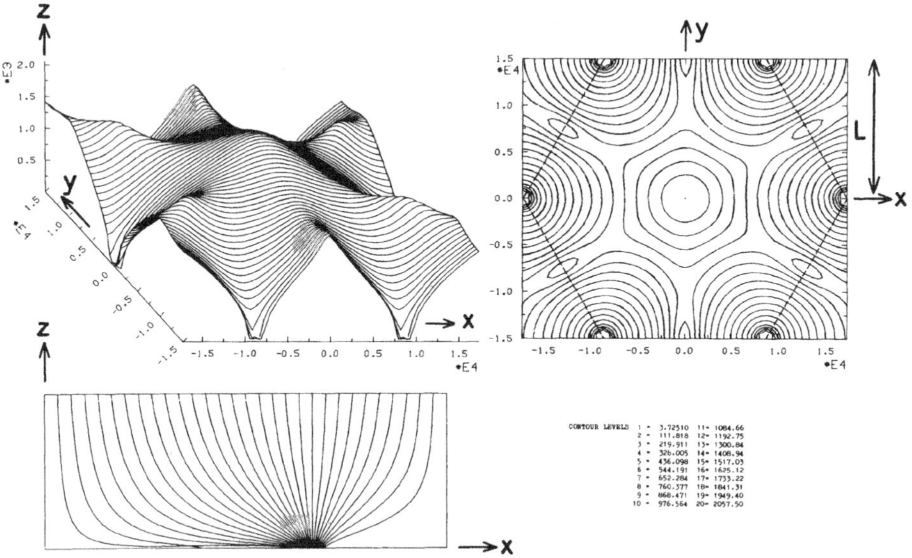

Fig. 1: Perspective plot and contours of the EPS, for the run in the text. Also shown are the field lines in the vertical plane y=0, continued to $x=L\sqrt{3}$. Distances in km.

below that level where $B_0^2/8\pi=p$. For the analogous 2-D problem the result is exact, as is proved analytically in AG 1982. Observational information is rather scanty to apply this result to other stars: either B_0 or p(z) can sometimes be inferred, but both together are known only for the Sun. In AG 1982 we used fields in ξ Boo A and 70 Oph A measured by Robinson et al. (1980), and the model of Arcturus due to Ayres and Linsky (1975), to suggest that late-type stars with directly observable fields should have magnetic chromospheres.

REFERENCES

Anzer, U. and Galloway, D.J.: 1982, to appear in Mon.Not.Roy.Astron.Soc.

Avrett, E.H.: 1981, in "Solar Phenomena in Stars and Stellar Systems", ed. R.M. Bonnet and A.K. Dupree (Reidel), p. 173.

Ayres, T.R. and Linsky, J.L.: 1975, Astrophys.J. 200, p. 660.

Gabriel, A.H.: 1976, Phil.Trans.Roy.Soc.London A281, p.339.

Giovanelli, R.G. and Jones, H.P.: 1982, preprint.

Kopp, R.A. and Kuperus, M.: 1968, Solar Phys. 4, p. 212.

Newkirk, G., Jr. and Harvey, J.W.: 1968, Solar Phys. 3, p. 321.

Robinson, R.D., Worden, S.P., and Harvey, J.W.: 1980, Astrophys.J. 236, p. L155.

Simon, G.W., Weiss, N.O., and Nye, A.: 1982, preprint.

Stenflo, J.O.: 1982, Solar Phys. 80, p. 209.

Ulmschneider, P.: 1981, in "Solar Phenomena in Stars and Stellar Systems", ed. R.M. Bonnet and A.K. Dupree (Reidel), p. 239.

Ulmschneider, P. and Stein, R.F.: 1982, Astron.Astrophys. 106, p. 9.
Vernazza, J.E., Avrett, E.H., and Loeser, R.: 1981, Astrophys.J.Suppl. 45, p. 635.

DISCUSSION

SPRUIT: The surface you are talking about is not the same as the interface between the field and the field free atmosphere, since you left out the corresponding currents in your calculations. Could this be the cause of the discrepancy?

GALLOWAY: We estimated that Giovanelli's procedure corresponds loosely to finding the surface $B^2/8\pi = 0.5p$ in our model. This is indeed lower than the EPS, but not by enough to explain the apparent discrepancy between these results and the latest measurements of Giovanelli and Jones on the poster downstairs.

CAMPOS: I think that the concept of the equipartition surface (EPS) is a magnificent one; it is defined by equality of gas and magnetic pressure. It corresponds to the sound speed being equal to the Alfvén speed (apart from a factor of $2/\gamma$, which is close to unity). Thus this surface can be interpreted in terms of waves: a magnetosonic wave behaves below it as a hydrodynamic wave dominated by compressibility, and above it as a hydromagnetic wave dominated by the magnetic field. The EPS is a critical layer separating these two different regimes, and could be a level of wave absorption.

Now I have a question: The surface (EPS) reaches up to 1300 km for a mean magnetic field $B_0 = 10$ G. What is the influence of different values of B_0 on the height of the EPS? Can it reach into the transition region? Or can you say that, for realistic values of B_0, it does not reach the transition region?

GALLOWAY: If the field is unipolar, parts of the EPS begin to penetrate the transition region when the field average B_0 gets down to about 2 G. When the polarity is mixed there are smaller intrusions for higher B_0, in those regions where the field cancels. I do not know if such regions have any particular significance on the Sun.

BASRI: I have become somewhat confused about what can be learned about heating rates from your results. If the chromosphere is concentrated within fluxtubes, then the tubes would still be confined at 850 km (where the heating rate has a maximum). If indeed most of the heating occurs outside these tubes, why has this conference concentrated so much on fluxtubes?

GALLOWAY: According to Avrett (1981), the non-network parts of the chromosphere have appreciable radiative losses and thus require a heating mechanism. Although of course the losses are greater in the network regions, their smaller area means that they contribute only about half the chromospheric total emission. We have shown here that the *non*-network regions are essentially non-magnetic, so you have to supply their losses some other way, by sound or gravity waves. I appreciate that this goes against the current fashion for magnetic fluxtubes, but that is just what comes out of the calculation. I can only say in apology that one must still heat the network with fluxtubes, and that they will be the dominant mechanism for stars more active than the Sun, which seems to be a borderline case in this respect.

NORDLUND: The crucial assumption in your calculations is that the chromosphere is current free; thus the gas pressure is assumed to be only a function of height (uniform horizontally), and you plot the level where the magnetic pressure is equal to the gas pressure. You therefore get results similar to those of Spruit, who assumes a negligible gas pressure inside the fluxtubes and derives heights for equilibrium between the tube magnetic pressure and the external gas pressure.

If we consider the real situation, there are significant sources of gas pressure (above hydrostatic) inside the magnetic fluxtubes; spicules shoot up, are heated above Hα temperatures, and "rain down" through transition-zone temperatures. With enhanced internal gas pressure, equilibrium can only be achieved with lower canopy heights, such that the external pressure can balance the *sum* of the magnetic pressure and the internal gas pressure. Thus the *determining factor* for the height is the strength of the source of extra gas pressure inside the tubes! Do you agree?

GALLOWAY: It depends on how cynical you want to be about the model atmospheres. One gets the impression reading VAL that the things you are suggesting do not occur with sufficient vigour to affect the derivations, and without any more information one just has to assume that the job is being done properly. However, it is certainly true that there are spicules in the network, and that these are *very* far from hydrostatic equilibrium. If there are indeed corresponding supersonic motions filling the regions above supergranule centres, I agree that the picture given here will have to be altered.

GIOVANELLI: Mathematicians can get exact answers to problems where the boundary conditions are prescribed, but these do not necessarily tell us anything about the real Sun. It is the latter that we should be concerned about. Indeed, the Sun is extremely complicated, with all sorts of chromospheric structures — including the spicules and the fibrils, both of which originate in the magnetic network. They are magnetic structures.

Dr. Jones and I have been concerned with trying to find out how the magnetic field expands with height in the Sun, not in a model. It is necessary to study in detail how the magnetograph responds to canopy-type fields, where the field is not uniform through the region of line formation. It is necessary to use non-LTE excitation and make proper allowance for magnetic pressure. We have explored as full a range of atmospheric models as we could find, e.g. models A-F of Vernazza, Avrett, and Loeser. The conclusions we derive (1982, *Solar Phys.* **79**, p. 247) seem to give canopy heights with uncertainties of the order of ± 50, perhaps ± 75 km, but the field strengths are not well determined.

EMPIRICAL SCALING LAWS FOR CORONAL HEATING

Leon Golub
Smithsonian Astrophysical Observatory
Cambridge, MA 02138, USA

ABSTRACT

We review the origins and uses of scaling laws in studies of stellar outer atmospheres, with particular emphasis on the properties of coronal loops. The evidence for a fundamental structuring of the Solar corona is reviewed and a discussion of thermodynamic scaling laws is presented. In order to intercompare different theories for coronal formation and heating, it is necessary to recast the theories in terms of observable quantities. As an example, we present a discussion of magnetic field-related heating and scaling laws which can be obtained relating coronal pressure, temperature and magnetic field strength; available data are shown to be consistent with scaling laws obtained in this way. However, some parameters of the theory must be treated as adjustable at the present time and it is necessary to examine data from other stars in order to determine whether these are true parameters in coronal heating. We examine some of the difficulties involved in using unresolved stellar data when dealing with loop atmospheres, by first treating the Sun as an unresolved source. Using the detailed observations now available we examine the limits of applicability of single-loop models. The possibilities and limits of stellar data are then discussed.

1. THE USES OF SCALING LAWS

The outstanding problem in coronal physics for over four decades has been the explanation of the corona's high temperature. The realization that some flux of mechanical energy is needed to maintain the million degree plasma has led in recent years to a plethora of theories, whose number is limited primarily by the ability of observations to disagree with specific predictions of the various theories.

It is clearly recognized now (see Vaiana and Rosner 1978) that a major difficulty which arises in attempting to specify the coronal heating mechanism is that the fraction of the Solar luminosity which needs to go into maintaining the corona is extremely small, of order 10^{-6} Thus there are many mechanisms which can be proposed and for each one a serious evaluation would involve a detailed and sometimes difficult search for observable signatures which could then be compared with available data, or which may necessitate new instrumental developments.

J. O. Stenflo (ed.), Solar and Magnetic Fields: Origins and Coronal Effects, 345–361.

The other major difficulty in studying coronal heating is that there may be more than one mechanism operating, depending on which part of the corona is under consideration. That is, the closed regions of strong magnetic fields in an active region core may be heated by an entirely different process than that responsible for heating the large scale, weak field regions or the open field regions with their large energy losses to the escaping Solar wind. It is also unclear at this time whether the more sudden and violent release of energy in a Solar flare occurs by the same process that provides the quasi-steady heating of the more quiescent corona; the possibility also exists that in the two different types of flare, the compact flare and the two-ribbon flare (Pallivicini, Serio and Vaiana 1977), there are again different processes responsible for heating the coronal plasma.

Although the difficulties just listed may appear to represent an unsatisfactory situation, the exact opposite is the case. The complexities of present work on coronal heating are a sign that our approach is more realistic than in previous studies and that we are likely to be approaching a better understanding of coronal formation and heating. In the following, I will attempt to provide a small piece of the story by reviewing briefly some of the steps which have led us away from the simple acoustic theories, with particular emphasis on the role played by magnetic fields. This discussion will be followed by an examination of some ways in which we can use scaling laws in combination with the available data for critical comparison with heating theories; we will then conclude with some discussion of the uses of stellar observations in extending the Solar results which have thus far been obtained. The present discussion covers only a small part of the total picture and should be viewed as complementary to the presentations by Vaiana, Rosner, Ionson and Chiuderi in these Proceedings.

a) Coronal Structure

The structuring of the corona is a fact of fundamental importance in understanding the physics of coronal formation and heating. The existence of some structure and also the importance of understanding the role of magnetic fields in active regions was recognized before high resolution x-ray images were available (Billings and Saito 1964). However, the availibility of on-disk observations of the corona in x-rays revealed the ubiquitous loop structuring of the emitting regions and the close connection with regions of strong surface magnetic fields (Vaiana Krieger and Van Speybroeck 1970). A series of rocket flights led to semi-empirical modelling of coronal loop structures and coronal magnetic fields (Vaiana, Krieger and Timothy 1973) and the long duration *Skylab* missions permitted study of the temporal evolution of the emission regions in the corona and the quantitative link between magnetic field strength and coronal emission, as we will discuss below.

Examples of the observed loop structure in the corona are shown in figure 1, which illustrates a typical view of the corona as seen from *Skylab*. In this image we find most of the typical non-flare coronal structures and we also see the boundary of a large N-S oriented coronal hole which represents an open magnetic field topology. The key point to notice, one which has been well documented by detailed studies (Van Speybroeck, Krieger and Vaiana 1970; Vaiana, Krieger and Timothy 1973; McIntosh *et al.* 1976), is that all of the coronal emission comes from closed structures (cf. however Maxson and Vaiana 1977). In all of the following discussion we will assume that high temperature plasma, i.e. material hot enough that its primary radiation is x-ray emission, on solar-type stars is predominantly confined in closed loop structures as is the case in the Solar corona.

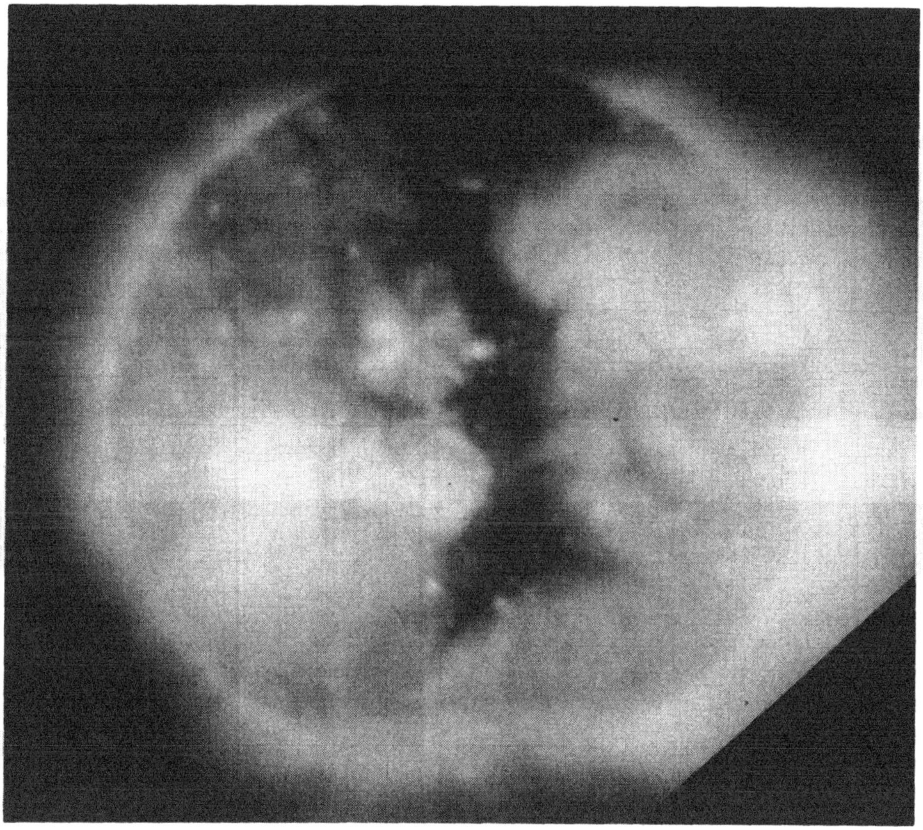

Fig. 1. The Solar x-ray corona as seen from Skylab on 1 June 1973. This photo shows all of the typical loop structures seen in the corona outside of flares and it illustrates the dominance of closed loops in controlling the emission. A large open structure (coronal hole) is clearly visible near disk center as a region of low emission.

The connection between photospheric magnetic fields and coronal emission regions is illustrated in figure 2, which shows a series of xray exposures of increasing duration and a standard photospheric magnetogram. The xray images show successively larger and fainter loop structures, ranging from the bright inner core loops of the active reions to the large-scale loops interrconnecting separated regions. The diffusion of the magnetic fields across the surface after they have emerged is also evident by comparing the small, newly emerged region to the older, larger one.

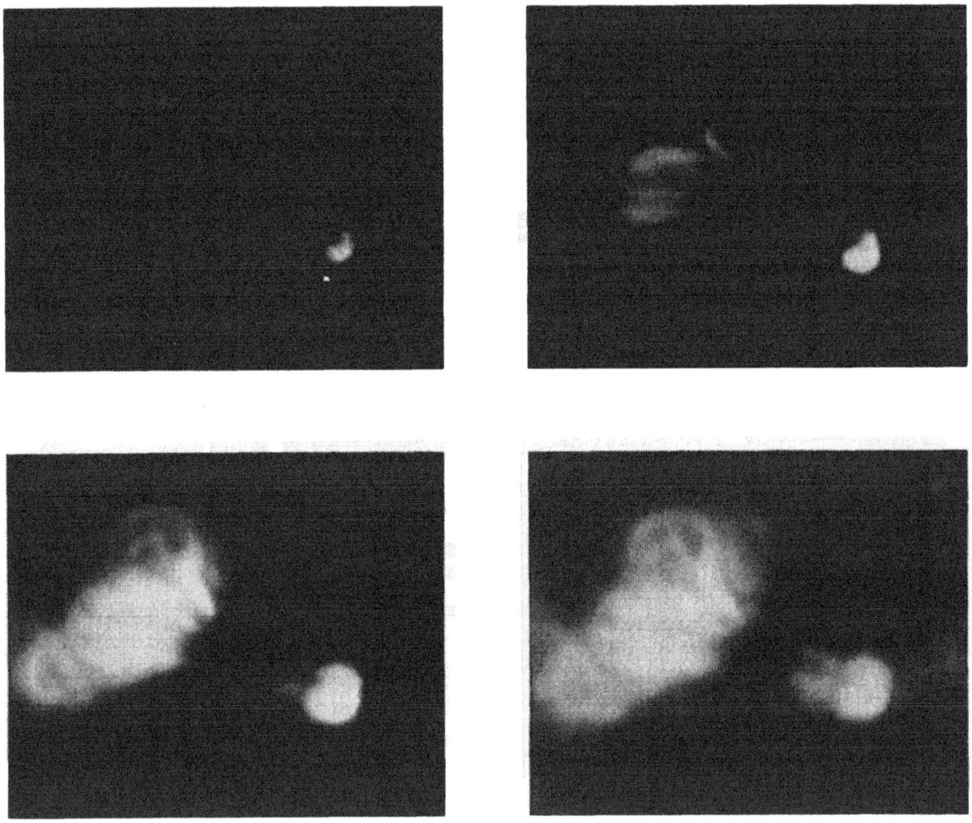

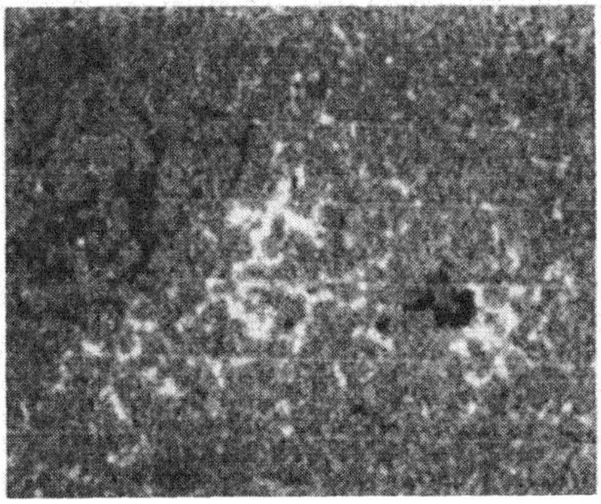

FIG. 2—Four X-ray photographs of a pair of active regions; the larger is one rotation old, and the smaller is 4 days old at the time of these photographs (1973 June 20, 0900 UT). X-ray exposure durations increase by successive factors of 4 from top left to bottom right, showing successively larger and fainter loop structures surrounding the bright cores. Note also that the energy density in the younger region is more than an order of magnitude greater than that of the older region.

The rapid variability and possible complexity of active region loop structures are illustrated in figure 3, which shows ~ 3 days in the life of an active region. The large number of loops and the interaction between different active regions is evident and indicative of the xray corona during times of moderately high activity.

The importance of structuring in coronal physics was put on a firm theoretical foundation by Rosner, Tucker and Vaiana (1978) and by Rosner and Vaiana (1978), who showed that each individual loop structure could be considered as a relatively independent "mini-corona". The true Solar corona, as observed at any given time, is to be viewed as an ensemble of these building blocks having a mix of statistical properties which are determined by the stochastic eruption of magnetic flux from the Solar interior. This work led directly to a *testable* scaling relation involving observable quantities and it forms the basis for most of the subsequent efforts at formulating scaling relations by which coronal heating theories may be tested.

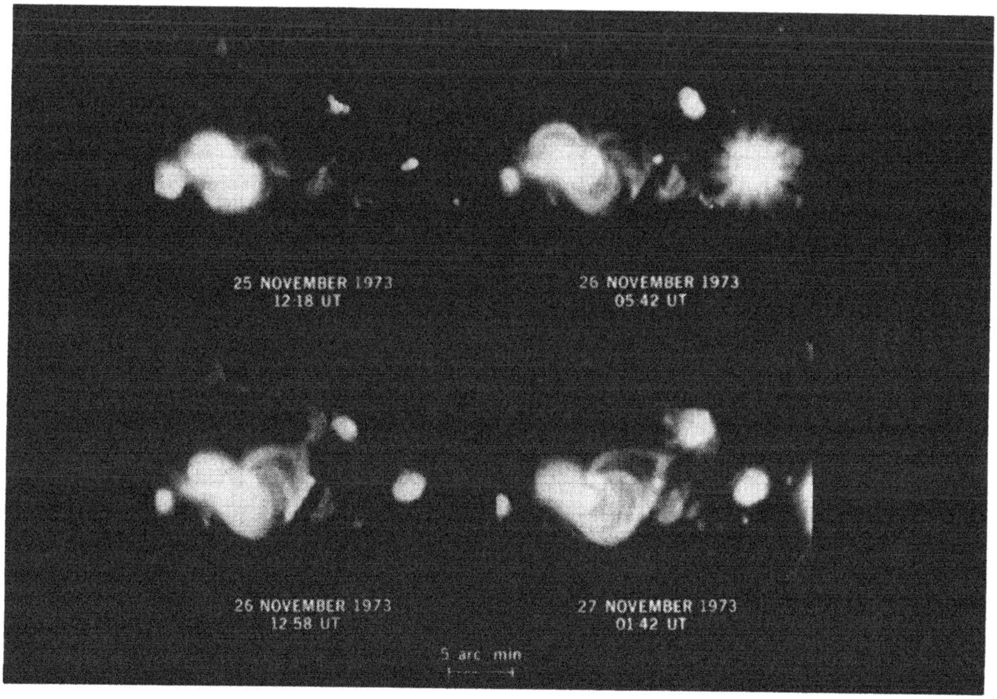

Fig. 3. The rapid and complex development of active regions loops seen in x-rays, as shown by two days' history of a large well-developed region and two newly emerging regions.

b) Testing the Role of Magnetic Fields

In the present discussion we will concentrate on observational tests of magnetic field-related heating theories. This work has been motivated by the types of observations described in the preceding section, which argue for an active role for magnetic fields in coronal physics. That is, the Solar surface fields play a passive role in determining the coronal topology as we have shown, but they may also play an active role by providing the direct means for dissipation of mechanical energy flux to heat the coronal plasma *in situ*.

Given the conditions under which astronomical observations are made, it appears unlikely that any particular theory can ever be *proven*. The best that we can hope to do is to search for observational implications of an hypothesis and determine whether the data are or are not consistent with the theory's predictions. We have begun this type of work for the general case of magnetic heating, in which the energy for coronal heating is supplied by stressing of the magnetic field and subsequent dissipation of this nonpotential field energy.

We will start by examining the types of scaling laws which have been proposed for coronal loop structures without consideration of specific heating mechanisms, i.e. scalings of the thermodynamic kind. However, we will then concentrate on those scalings which are more directly related to a specific mechanism for coronal heating. In particular, we will describe a simplified but general loop model which specifically incorporates the magnetic field in an active role, leading to scaling laws relating observable quantities to the magnetic field strength, finding that significant tests can be performed over a wide range of the observable parameters. However, there are parameters which appear in the theory which cannot be varied when observing only the Sun and it is necessary to use stellar observations in order to provide more stringent tests of the theory.

We note also that this review will concentrate primarily on heating via the dissipation of magnetic stresses in the corona. There have of course been proposed numerous other mechanisms for heating the Solar outer atmosphere, including heating by acoustic processes (*a la* Biermann 1946 and Schwarzschild 1948), by MHD body waves (Osterbrock 1961), by MHD surface waves (Ionson 1978) and by *in situ* dissipation of currents (Rosner *et al.* 1978). Most of these theoretical papers have not directly addressed the question of scalings among observable quantities by which the theories may be tested. A summary of the present overall state of heating theories and an up to date bibliography of work in these various areas may be found in Kuperus, Ionson and Spicer (1981).

2. SOLAR LOOP ATMOSPHERES

a) The T, p, L Scaling Laws

The first theoretical investigation which explicitly took into account the fundamental nature of closed loop structures as building blocks for the Solar corona was that Rosner, Tucker and Vaiana (1978); similar calculations were also performed independently by Craig, McClymont and Underwood (1978). The Rosner *et al.* study adopted the view that coronal heating as a whole must be understood by first studying the energy balance of the individual loop structures, which are to be considered as relatively isolated mini-coronae.

The work considered the energy balance relation for the loop as a whole

$$\int_V (E_H + \mathbf{f}\cdot\mathbf{v})\, d^3r = -\int_V E_R\, d^3r + L_{foot} + L_{sides}\,, \tag{2.1}$$

where the *L*'s represent losses across the boundaries of the loop, E_H is the local mechanical heating function, f is the net volume force exerted on the fluid, v the bulk plasma velocity and E_R the total radiative loss. The boundary conditions are treated by noted that cross-field effects are expected to be small, so that L_{sides} is negligible and eq 2.1 becomes one-dimensional; in addition, they assumed v=0 at the boundaries and then argued that the conductive flux at the base of the loop must be small. In that case, conduction serves only as an energy redistribution mechanism and radiation is the only remaining loss mechanism for a closed, quiescent (quasi-static) loop.

The major result of this model, for the case in which the heating function is constant in space, is a relation involving the *observable* quantities temperature, pressure and loop length:

$$T_{max} = 1.4X10^3 \, (pL)^{1/3} \tag{2.2}$$

This relation containes no free parameters and depends, in addition to the above assumptions, only on the form assumed for the radiative loss function P(T), or equivalently on equality between the radiative and conductive loses out of the "coronal" portion of the loop, i.e. the segment above the height at which the divergence of the conductive flux vanishes. As pointed out by Rosner, Tucker and Vaiana (and by Vesecky *et al.* (1979) and Galeev *et al.* (1981)), this equality holds over a surprisingly large range of coronal loop size, temperature and pressure values so that for essentially all Solar coronal loops the arguments leading to eq 2.2 hold.

The RTV paper treated loops for which the pressure scale height of the coronal plasma is large compared to the loop length, so that the pressure could be considered constant throughout the coornal portion of the loop. Extension of this work to larger loops, such as those making up the large scale structures in the corona, requires numerical simulations. these have been peformed by Serio *et al.* (1981) who used numerical computations of stationary loop atmospheres to calculate: i)loops having L larger than s_p and also, ii) loops having a local temperature minimum at the top. For the first class they found a simple extension of the previous scaling law:

$$T = 1.4X10^3 \, (p_oL)^{1/3} \exp[-.04L \, (2/s_H + 1/s_p)], \tag{2.3}$$

where p_o is now the coronal base pressure, s_H is the heating deposition scale length and s_p is the pressure scale height.

In addition to this extended scaling law, which we will adopt in the following, they also found that over a broad range of parameter values loops having $L > \sim 3 \, s_p$ do not have stable solutions unless they have a local temperature *minimum* at the top. They discussed this result within the context of prominence formation and stability, as part of the overall evolutionary history of magnetic field structures in the corona.

In the remainder of this paper, when we investigate other scaling relations involving, e.g. the magnetic field strength we will assume that scaling relations 2.2 or 2.3 are generally valid for Solar-type coronae. In the case of the TRV scaling law, which contains no adjustable parameters, we use it directly. The Serio *et al.* extension contains the pressure scale height and the heating deposition scale height as parameters. The former is obtained directly by measurement of the coronal temperature and knowledge of the stellar surface gravity and the latter quantity will be assumed to be large (usually infinite).

b) Models of Single Loops

A simple but general model of the energy balance in a single loop may be presented as shown in figure 4; the present approach follows that of Golub *et al.* 1980. The energy available for heating the coronal portion of this loop is ultimately supplied by shearing the longitudinal magnetic field in the region of the solar atmosphere for which $\beta > 1$, i.e., the photosphere or below. The energy available for heating is

$$W_m = B_j^2 V/8\pi , \qquad (2.4)$$

where B_j is the nonpotential component of the magnetic field and V is the volume over which the stresses are induced. The azimuthal field is generated from the longitudinal field by

$$\partial B_j/\partial t = B_z \,\partial v_\phi/\partial z. \qquad (2.5)$$

If we approximate the derivative by v_{ϕ_o}/D, where D is the depth of the shear zone, then since $V = \pi a^2 D$, we have

$$\partial B_j/\partial t = B_z \,\pi a^2 \, v_{\phi_o}/V. \qquad (2.6)$$

Note that the quantity a in these equations represents the cross-sectional radius of the magnetic flux tube in the high-β portion of the atmosphere and the subscript on ϕ indicates that the velocity below the photosphere is to be considered.

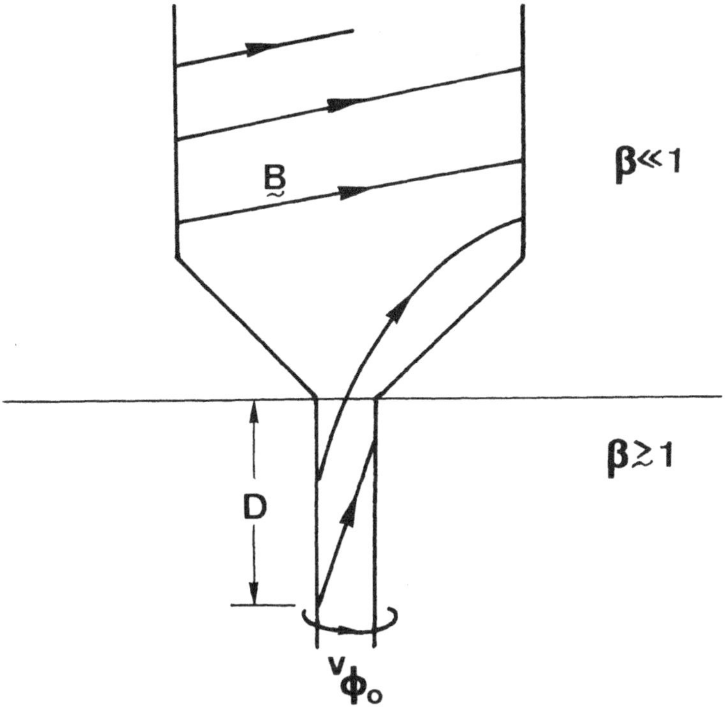

Fig. 4. Schematic representation of the loop model parameters.

The coronal extension of such a loop will have a larger cross-section due to the drop in the external gas pressure with increasing height above the photosphere. Because of this expansion, any twist which is present in the high β region will be transmitted upward and amplified, as shown by Parker (1974). As viewed from the corona, this transmition of stresses will be an effective twisting velocity at the base of the loop, which we label v_ϕ. The energy available for heating the corona is again given by eq. 2.1, but with $V = 2\pi R^2 L$, where R is the minor radius of the loop and L is the loop semi-length (i.e. distance along the loop from its base to its apex). The energy generation rate is then

$$dW_m/dt = 0.25\, B_j\, B_z\, v_\phi\, R^2. \qquad (2.7)$$

Combining this result with the scaling law of Rosner, Tucker and Vaiana (1978) and assuming that *all of the available energy goes into heating*, we find that the coronal gas pressure is related to the magnetic field strength

$$p = 2.6\, B_2^{12/7}\, L_9^{-1/7}\, (\alpha\, v_4)^{6/7}, \qquad (2.8)$$

where the subscripts indicate division by the indicated powers of ten and $\alpha \equiv B_j/B_z$.

Unfortunately, the parameter α involves the azimuthal field B_j, which is not a measureable quantity, while the effective twisting velocity v_ϕ may be treated as an adjustable normalization parameter for our purposes. In dealing with α several approaches are possible: Golub *et al.* (1980) simply assumed that all coronal loops have α constant, so that the normalization constant becomes αv_ϕ instead of just v_ϕ. Subsequently, Galeev *et al.* (1981), in a theoretical study of coronal heating via collisionless tearing modes, arrived at β ($\equiv 8\pi\, p/B^2$) = 0.3 for all loops, leading to $B_j = 8.8\, p^{1/2}$, which may be substituted directly into eq. 2.5 in order to eliminate B_j. These two approaches lead respectively to the following scalings of coronal pressure with other measureable quantities:

$$p \propto B_z^{12/7}\, L^{-1/7}\, (av_\phi)^{6/7}, \qquad \text{(constant } \alpha) \qquad (2.9a)$$

$$p \propto B^{3/2}\, L^{-1/4}\, v_\phi^{3/2}, \qquad \text{(constant } \beta) \qquad (2.9b).$$

A comparison of these two theoretical predictions with Solar data is shown in figure 5, for which we have integrated over an active region in each of the five observations. We note that the example shown here is of a single active region which was followed throughout its entire evolutionary history, from initial emergence through its blending into the large scale quiet corona.

In the present discussion we have treated v_ϕ as an adjustable parameter; its value is found to be ~ 1 km s^{-1}, which is in accord with the observed magnitude of the turbulent velocities in the Solar outer convective zone. However, this agreement can be further tested if we turn to the stellar observations, examining stars having very different values of the surface turbulent velocities. If the scaling laws can be reformulated in a suitable manner and the relevant observations performed, then these additional tests can be applied. We will look into this question in some detail in the remainder of this paper.

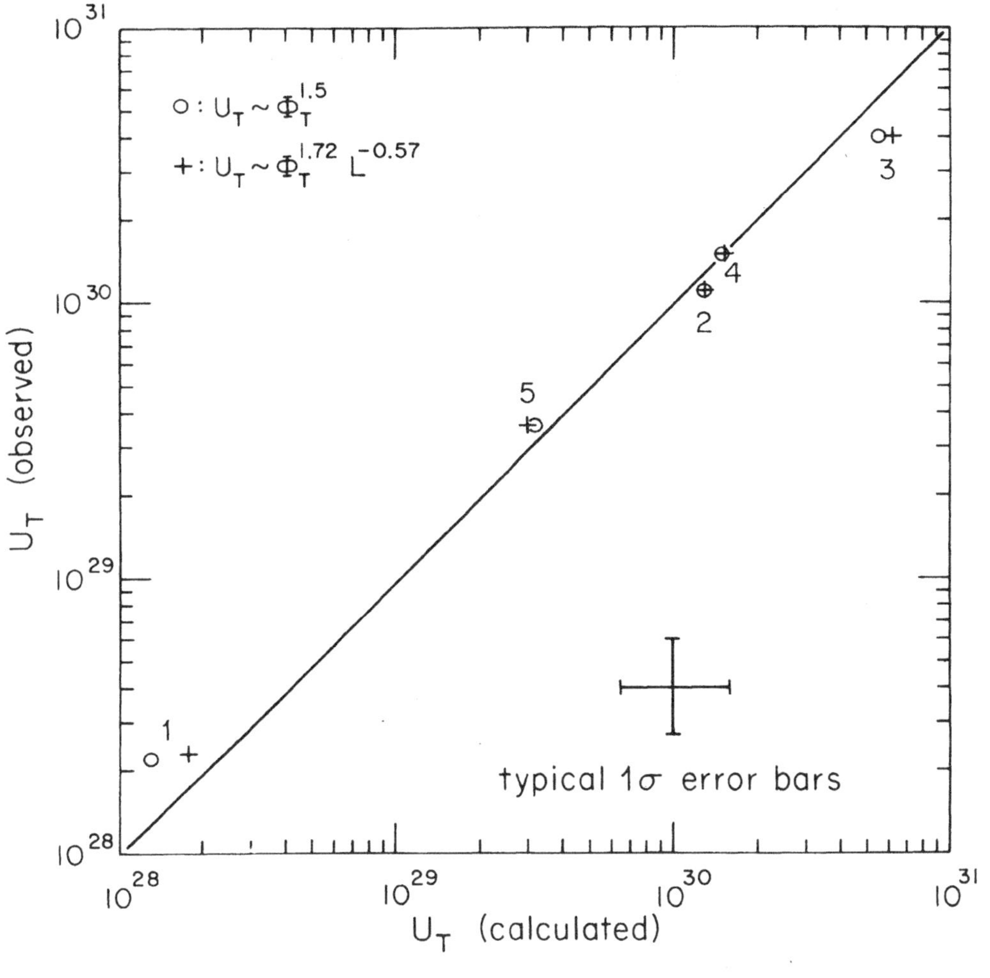

Fig. 5. Comparison of observed x-ray emission *vs.* calculated, based on the measured values of magnetic flux and region size. Data are for a single Solar active region followed throughout its entire lifetime, from initial emergence through evolution into large scale corona. Data points are based on x-ray and magnetic field (KPNO) measurements, using the predictions of equs. 2.9a,b.

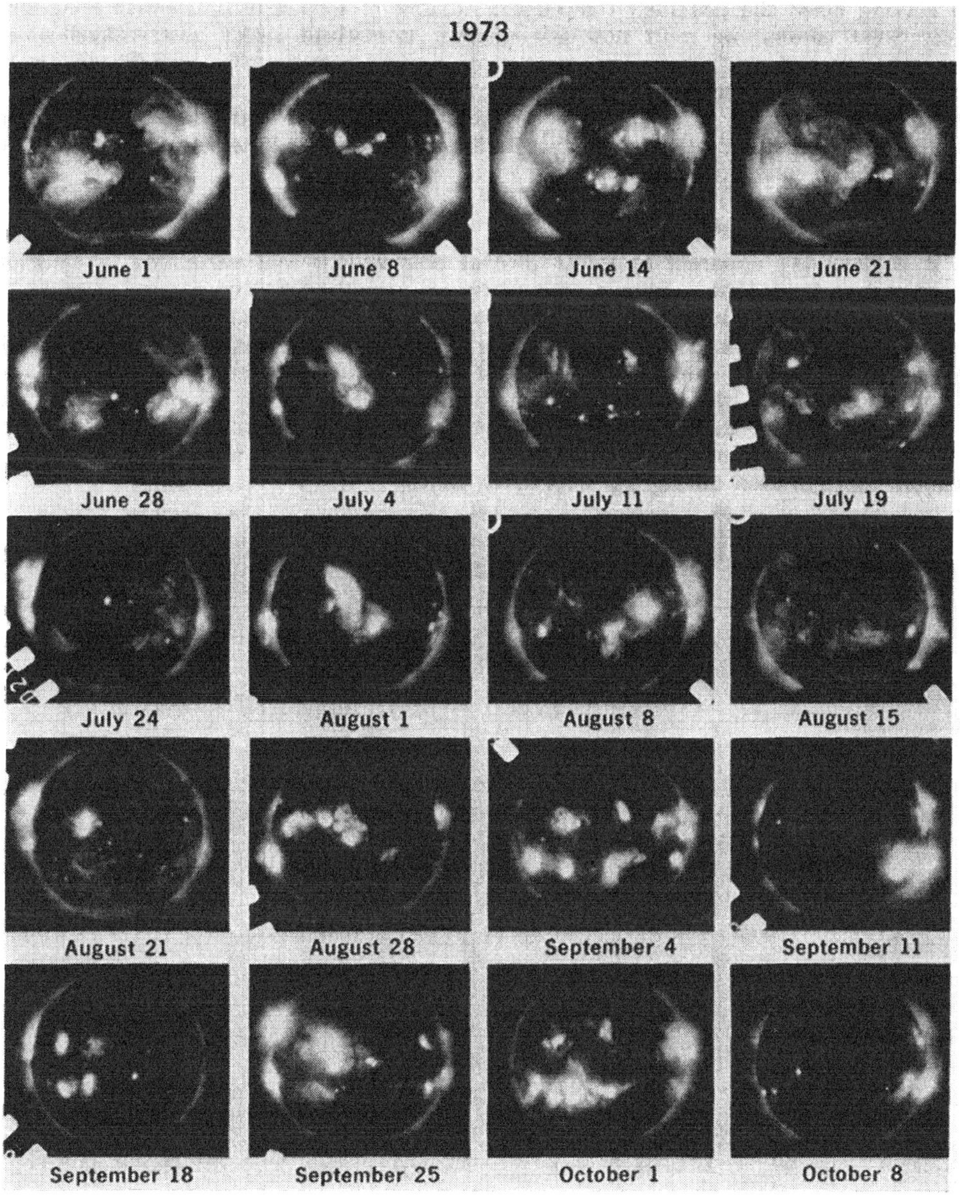

Fig. 6. Five rotations of the X-ray corona as seen from Skylab, illustrating the major changes in coronal structure and emission occurring on relatively short time scales.

3. OBSERVATIONS OF UNRESOLVED ATMOSPHERES

Having noted the extremely complicated mixture of loop structures which are found in the Solar corona, we must now ask whether unresolved stellar observations which necessarily average over the entire ensemble of structures in a stellar atmosphere can be of any real use in studying the nature of the coronal heating mechanism. We have addressed this question in some detail (Golub 1982) and find that within certain limits there is a good possibility that unresolved stellar data can answer some questions initiated within the Solar context.

a) The Sun as an Unresolved Source

It is becoming apparent in stellar coronal observations that variability in emission properties is quite common. This is not surprising, since the Solar corona is known to be variable on relatively short time scales, even when flares are excluded. This is illustrated in figure 6, which shows five Solar rotations observed from *Skylab*; in this figure, each neighboring image in a row is ~90^{0} separated from the adjacent images. Thus each vertical column shows the evolution of the same Solar longitude during the observing period. It is clear that the corona is highly variable, even from one week to the next. In particular, during the fourth and fifth rotations the corona varied from a solar maximum to a minimum configuration during less than seven days.

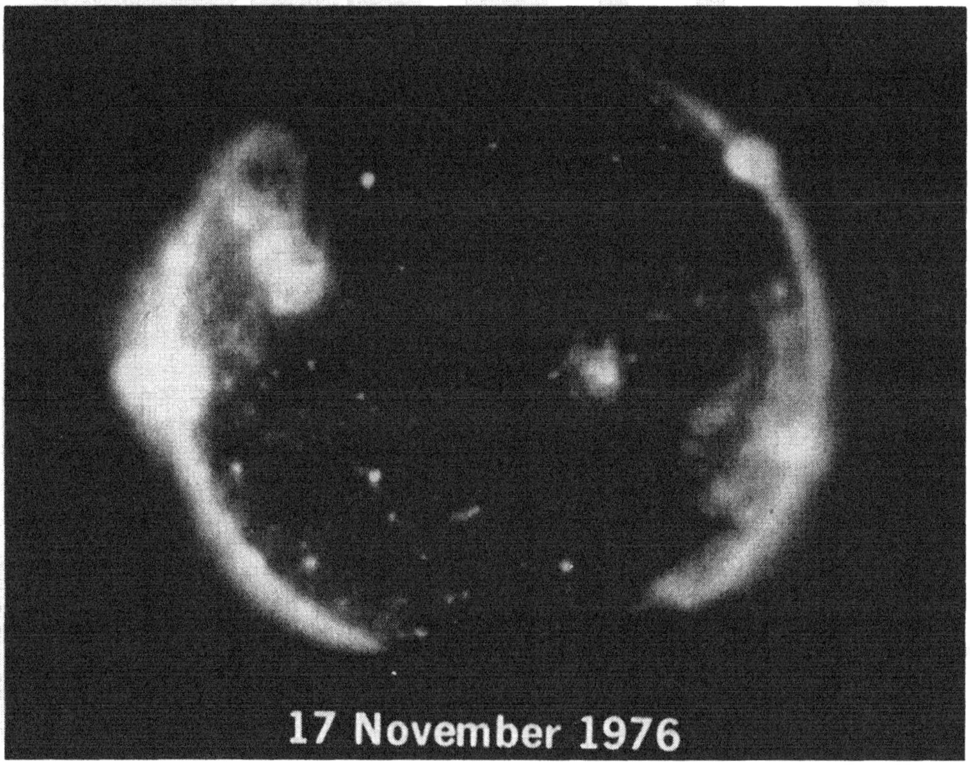

Fig. 7. The Solar x-ray corona at minimum; note that the structure is dominated by the evolved remnants of the numerous small emerging flux regions.

The variation in coronal composition during different portions of the Solar cycle can be even more extreme than those illustrated above. For example, x-ray data obtained near Solar minimum show an entirely different kind of corona (figure 7) which is dominated by large numbers of small active regions. These small regions are short-lived, do not develop sunspots and have relatively low x-ray emission measure but they are found to contain as much magnetic flux as the larger, longer-lived active regions found at times of higher Solar activity (Golub 1980).

In order to produce a quantitative representation of coronal emission in terms of the various known atmospheric components we must take into account the Solar cycle variation of each component and the evolutionary histories of the different types of emerging field regions. A first attempt at modelling this complicated mix of structures is shown in figure 8, which shows the fractional area coverage by various atmospheric components (top) and the relative percentage contribution to the total coronal emission (bottom), all as functions of the Solar cycle.

From quantitive studies of this kind we may draw the following preliminary conclusions:

The average Solar corona is dominated by a single type of structure approximately half of the time. At Solar maximum, the entire coronal emission may be characterized by the properties of large active regions, with $T \sim 10^{6.6}$ K and $EM \sim 10^{50.5}$. At Solar minimum the corona is characterized by the small emission regions and their diffused states, with $T \sim 10^{6.3}$ K and $EM \sim 10^{49}$.

b) Model Predictions and Comparison to Data

We may combine the scaling laws eq. 2.2 (2.3) and eq. 2.9 discussed above in order to eliminate one of the vairables. For the case of stellar observations, the coronal pressure is not directly observable as it is on the Sun, so that the choose the new scaling law to be of the form

$$T = 1.2 \times 10^3 \, B^{1/2} \, L^{1/4} \text{ (cgs)}, \tag{3.1}$$

where we have omitted the twisting velocity v_ϕ as a parameter and will restrict our attention to stars having turbulent velocities close to that of the Sun.

Following our discussion in §III.a we will assume that all of the coronal emission in each stellar source comes from a single type of loop structure. Since we are dealing with quiescent emission levels from these stars, we may take the loops to be as large as the pressure scale height

$$L = 5 \times 10^3 \, T \, (g/g_\odot)^{-1}. \tag{3.2}$$

The reason for this is that the activity and rapid fluctuations in emission will derive predominantly from newly emerging active region loops. These will rapidly grow in size until they become comparable in size to the pressure scale height, at which time the level of activity will have decreased substantially (see Galeev *et al.* 1981). Further growth in loop size will not change eq. 3.2, since the emission scale parameter will thereafter remain equal to the pressure scale height.

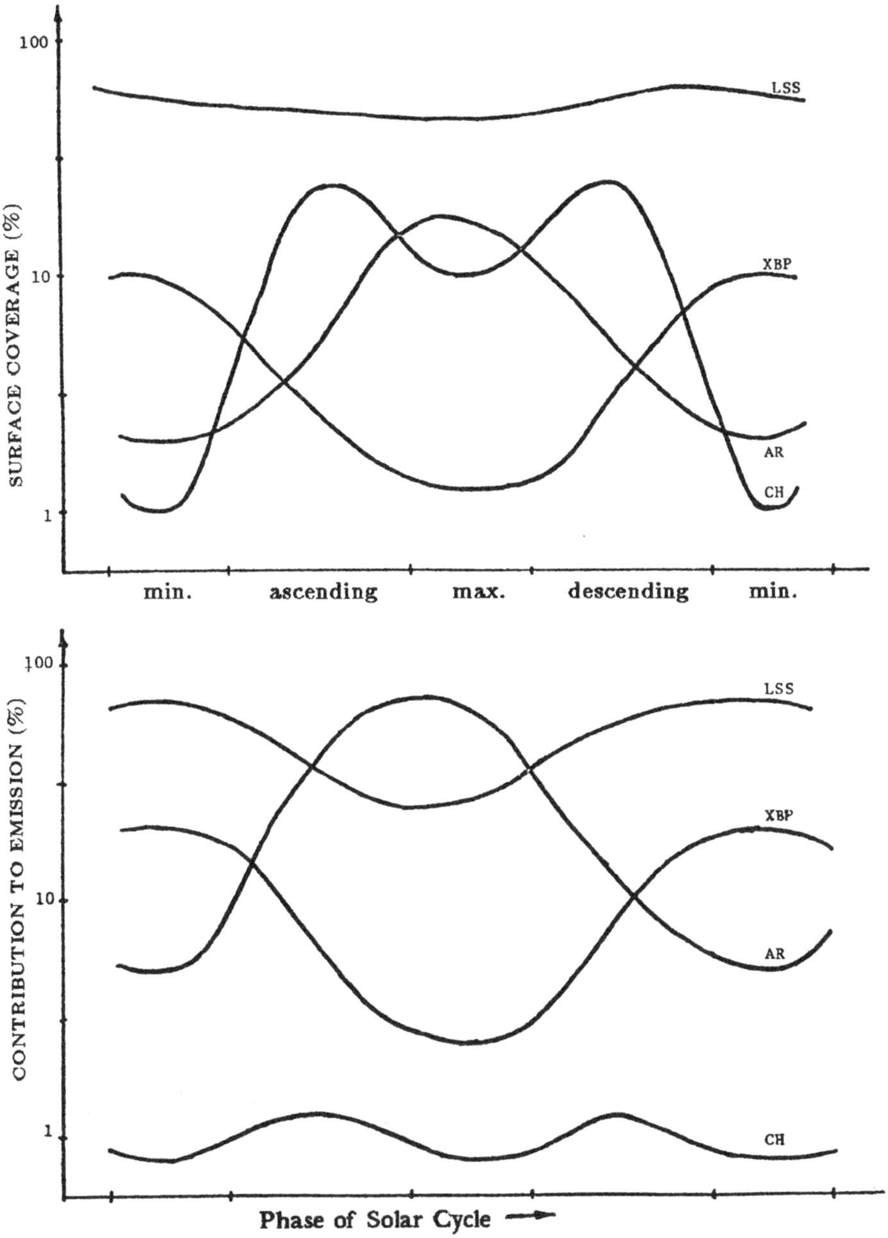

Fig. 8. Fractional area coverage factors (top) and relative contributions to coronal luminosity (bottom) for the various atmospheric components of the Solar corona, all as a function of phase in the Solar cycle.

For an atmosphere of such loops, we have shown elsewhere (Golub *et al.* 1982) that measurement of the coronal temperature and xray luminosity are sufficient to specify the atmosphere. The quantities of interest in the present discussion are the coronal filling factor f and the magnetic field strength B_{em} in the emitting regions. These are given by

$$f = [3.4 \times 10^{-9} / P(T)]\, F_x\, [T^3/(g/g_\odot)]^{-1}\,, \tag{3.3a}$$

$$B_{em} = 1.2 \times 10^{-8}\, [\, T^3/(g/g_\odot)]^{1/2}\,, \tag{3.3b}$$

where P(T) is the plasma emissivity and F_x is the average surface flux of xrays at the stellar surface; the average magnetic field on the star is given by $\langle B \rangle = f\, B_{em}$. We have calculated these quantities and list them in table 1 for some Solar-type stars. It will be quite informative in the next few years to find out whether magnetic field values measured on some of these objects agree with these predictions.

Table 1. Predicted Stellar **B** Values.

Star	F_x (erg cm^{-2} s^{-1})	T_{cor} (K)	f	B_{em} (g)	$\langle B \rangle$ (g)
Quiet Sun	3×10^4	1.8×10^6	0.15	30	5
Active Sun	1×10^5	3.5×10^6	0.07	80	6
α Cen A	1.3×10^4	2.1×10^6	0.06	90	5
α Cen B	6.1×10^4	2.1×10^6	0.18	60	11
π^1 UMa	1.2×10^6	4.0×10^6	0.22	100	20
λ And	1.5×10^6	5.0×10^6	>1	670	670

4. ACKNOWLEDGEMENTS

I would like to thank G. S. Vaiana, R. Rosner, S. Serio and J. Ionson for many helpful discussions. This work was supported in part by NASA under Grants NAGW-112 and NAGW-79 and by the Secretary's Fluid Research Fund of the Smithsonian Institution.

REFERENCES

Biermann, L.: 1946, Naturwissenschaften 33, p. 118.
Billings, D.E. and Saito, K.: 1964, Astrophys. J. 140, p. 760.
Craig, I.J.D., McClymont, A.N., and Underwood, J.H.: 1978, Astron. Astrophys. 70, p. 1.
Galeev, A.A., Rosner, R., Serio, S., and Vaiana, G.S.: 1980, Astrophys. J. 243, p. 201.
Golub, L.: 1980, Phil. Trans. Royal Soc. London A, 297, p. 595.
Golub, L.: 1982, in "Proc. 2nd Cambridge Workshop on Cool Stars, Stellar Systems, and the Sun", eds. M.S. Giampapa and L. Golub (Smithsonian Special Report No. 392), p. 39.

Golub, L., Maxson, C.W., Rosner, R., Serio, S., and Vaiana, G.S.: 1980, Astrophys. J. 238, p. 343.
Ionson, J.A.: 1978, Astrophys. J. 226, p. 650.
Kuperus, M., Ionson, J.A., and Spicer, D.S.: 1981, Ann. Rev. Astron. Astrophys. 19, p. 7.
Maxson, C.W. and Vaiana, G.S.: 1977, Astrophys. J. 215, p. 919.
McIntosh, P.S., Krieger, A.S., Nolte, J.T., and Vaiana, G.S.: 1976, Solar Phys. 49, p. 57.
Osterbrock, D.E.: 1961, Astrophys. J. 134, p. 347.
Pallavicini, R.P., Serio, S., and Vaiana, G.S.: 1977, Astrophys. J. 216, p. 108.
Parker, E.N.: 1974, Astrophys. J. 191, p. 245.
Rosner, R., Golub, L., Coppi, B., and Vaiana, G.S.: 1978, Astrophys. J. 222, p. 317.
Rosner, R., Tucker, W.H., and Vaiana, G.S.: 1978, Astrophys. J. 220, p. 643.
Schwarzschild, M.: 1948, Astrophys. J. 107, p. 1.
Serio, S., Peres, G., Vaiana, G.S., Golub, L., and Rosner, R.: 1981, Astrophys. J. 243, p. 288.
Vaiana, G.S., Krieger, A.S., and Timothy, A.F.: 1973, Solar Phys. 32, p. 81.
Vaiana, G.S., Krieger, A.S., and Van Speybroeck, L.P.: 1970, in "Solar Magnetic Fields", ed. R. Howard, IAU Symp. 43, p. 397.
Van Speybroeck, L.P., Krieger, A.S., and Vaiana, G.S.: 1970, Nature 227, p. 818.
Vesecky, J.F., Antiochos, S.K., and Underwood, J.H.: 1979, Astrophys. J. 233, p. 987.

DISCUSSION

WALTER: I have got a picky question about the temperature of λ And. Swank and White's SSS (Solid State Spectrometer) observations indicate a two-component corona, with a temperature of $\sim 8 \times 10^6$ K in the cool component, and a hot component of a few tens of 10^6 K. What about the hot component, and how would the higher temperature of the cooler component affect your estimates of f and B ?

GOLUB: First, I have neglected the hot component, since the work I presented applies to the large-scale quiet corona, which I equate with the cooler and steadier component of λ And. The temperature I used was the one that we (CFA) obtained from the IPC. If we use the 8×10^6 K value, then the agreement with the Giampapa and Worden value of B is even better than the one I quoted.

ROBERTS: Since scaling laws are increasingly being used in stellar work it is very important that their form is properly checked and understood for the Sun, where observations can be made in greatest detail. Now, as you mentioned in your review, the simplest available scaling law, due to Rosner et. al, has no free parameters (such as the heating rate). If I understand it correctly, this is because at the base of a loop the conductive flux vanishes. But it should be remembered that the L in the scaling relation is not the same as the measured L, because a small part of the loop lies hidden in the chromosphere (where, incidentally, much of the important and complex physics of loops occurs). If we avoid this problem by letting L correspond to higher levels in the loop (measuring L from a point in the corona say), then we no longer get a simple scaling relation but instead find that the explicit form of the heating enters into the determination of T_{max}. It would, of course, be nicer to stick to the simpler relation. But to do this — indeed, to use scaling laws at all — we must be certain that such relationships are well borne out by solar observations.

My question, then, is: Do we get adequate agreement with observations, especially if we compare directly T_{max} observed with T_{max} calculated (rather than comparing some integrated quantity, where errors may well be masked), and what in fact is the extent of disagreement in any one active region? What, too, are the error bars involved?

GOLUB: A full answer to this question would be quite involved. First, the loop length L as used in the scaling law is entirely insensitive to the thickness of the transition zone, since it is so thin. Of course, there is much physics there, but as a parameter in the scaling law, L is much more influenced by measurement problems such as projection effects and curvature of field lines. Concerning the agreement with data, as Dr. Rosner points out it is quite remarkable that such a simple theory agrees so well with such a wide variety of observed loop parameters. If we view the scaling as an empirical result, then it may be more acceptable to you.

ROSNER: On the whole, I agree with Robert's point. However, a more significant difficulty is that the energy equation may need to include effects such as enthalpy fluxes. Nevertheless, the $T = T(p, L)$ scaling does provide an adequate description of the data. Thus, although one might argue with details, the main point of having temperature, pressure, and size scale of emitting structures correlated in a very particular functional way is a very useful device for interpreting stellar observations, as shown by Dr. Golub.

IONSON: For solar-type stars you have set the emission scale height equal to the loop height. This seems reasonable. What is your proposal with regard to non-solar stars?

GOLUB: Since I am modelling solar-type stars and restricting attention to quiescent emission, I believe that it is reasonable to say that we are looking at the evolved, large-scale loops. The more compact, higher-pressure loops are likely to be more active, as they are on the Sun, and will represent a variability on top of a quiescent baseline emission level.

CRAM: Would you comment on the possible magnetic structure in a star like λ And, which has a coronal magnetic field about equal to its photospheric magnetic field?

GOLUB: You are asking what a star would look like if we squeezed the magnetic fields together to an extent not observed on the Sun. We have no experience with this question, and so it is difficult to guess. Perhaps we could use field extrapolation programs to do some modelling.

KUIN: I have a comment on the large filling factor in λ And and the assumption of loop lengths of the same magnitude as the scale height. If one assumes smaller loop lengths, one obtains the same results for smaller filling factors, because the loops have higher pressures. Two components may be present with different filling factors and lengths. An example of this kind of analysis is the work of Mewe et al. (1982, *Astrophys. J.* **260**, p. 233). In this case one of course has to assume something for the filling factors.

GOLUB: Certainly it is true that the X-ray measurements do not fully constrain the model atmosphere. For instance, we can define a locus of points in the f - B plane, with different points along this curve representing different loop lengths. Therefore we need an additional constraint on any one of these parameters in order to specify the other two. Since I want to estimate B, I have chosen to present an argument on what the loop length L should be. The value of f also comes out of this calculation, but was not my main concern. It only becomes important if unreasonably high values of f come out of the calculation.

RADIO WAVELENGTH OBSERVATIONS: STRUCTURE, STRENGTH AND VARIABILITY OF MAGNETIC FIELDS IN THE TRANSITION REGION AND THE SOLAR CORONA

Kenneth R. Lang
Department of Physics, Tufts University, Medford, MA 02155,USA

ABSTRACT

The 6 cm radiation of solar active regions marks the legs of dipolar loops which have their footpoints in lower-lying sunspots. The temperatures $T \approx 10^6$K and longitudinal magnetic field strengths $H_\ell \approx 600$ Gauss at heights $h \approx 4 \times 10^9$ cm above sunspot umbrae. The circularly polarized emission at 6 cm delineates the magnetic structure above sunspot penumbrae. The 20 cm radiation of solar active regions delineates the ubiquitous coronal loops previously detected at X-ray wavelengths. We infer semilengths $L \approx 5 \times 10^9$cm, maximum electron temperatures T_e(max) $\approx 3 \times 10^6$K, emission measures $\int N_e^2 d\ell \approx 10^{28}$cm^{-5} and electron densities $N_e \approx 10^9$cm^{-3} for the 20 cm bremsstrahlung. Future V.L.A. observations at 20 cm may be used to determine the magnetic field strength of coronal loops. Changes in temperature and magnetic structure before and during solar bursts are briefly discussed.

THE LEGS OF MAGNETIC DIPOLES CONNECTING SUNSPOT UMBRAE

Enhanced 6 cm emission is often most intense above sunspot umbrae. Lang and Willson (1979,1980) have shown that this sunspot-associated emission has high brightness temperatures of $T_B \approx 2 \times 10^6$K and high degrees of circular polarization of ρ_c = 60 to 80%. The 6 cm emission marks the legs of magnetic dipoles which join sunspots of opposite magnetic polarity. The highly polarized radiation ($\rho_c \approx 70$%) corresponds to the extraordinary mode of wave propagation in magnetic fields with the same polarity as the underlying sunspots. It is explained as optically thin gyroresonance emission in the low solar corona above sunspots where the temperatures $T \approx 10^6$K and the longitudinal magnetic field strength $H_\ell \approx 600$ Gauss (third harmonic at 6 cm). The observed circular polarization requires resonance at the third harmonic because the second harmonic is optically thick and gyroresonant absorption is an ineffective opacity agent at the fourth harmonic. Polarization by propagation effects requires even higher magnetic field strengths which would make the regions everywhere optically thick to gyroresonant absorption. The two most

J. O. Stenflo (ed.), Solar and Magnetic Fields: Origins and Coronal Effects, 363–367.

intense components of the 6 cm emission are displaced inwards by about 25". Assuming that the sunspots mark the feet of a circular dipole, we infer a height $h \approx 4 \times 10^9$ cm above the solar photosphere. These maps show no evidence for cool material above sunspot umbrae, but they refer to structures that are stable for hours. Cooler material with lifetimes shorter than a few hours would not be detected. There is no evidence for an abnormally thin transition region above sunspots.

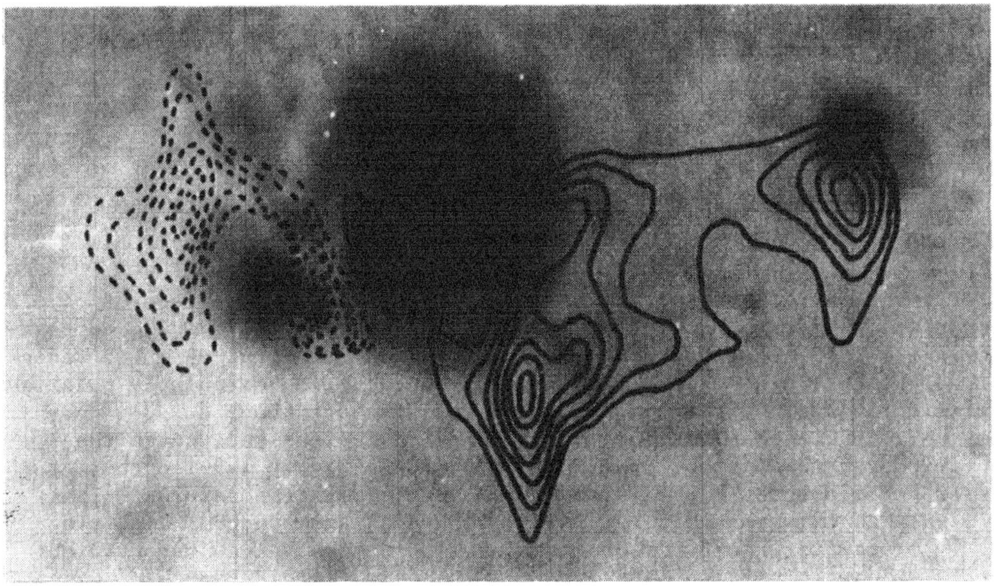

Fig. 1. A 6 cm synthesis map of circular polarization (Stokes parameter V) superimposed upon an offband Hα photograph taken on the same day. The polarized emission is concentrated above the sunspot penumbrae where the magnetic field lines are curved. (Offband Hα courtesy of Frances Tang, Big Bear Solar Observatory.)

MAGNETIC FIELDS ABOVE SUNSPOT PENUMBRAE

The intense magnetic fields of sunspot umbrae probably project radially upwards into the low solar corona with little loss of strength. This enables the third harmonic of the gyrofrequency to occur at higher, hotter levels of the solar atmosphere where the temperature gradient is small. The situation is different for the curved magnetic fields of the sunspot penumbrae. In this case, the resonance levels corresponding to the third harmonic occur in the lower-lying cooler regions of the solar atmosphere where the temperature gradients are large. In Figure 1 we show that the circular polarized emission (Stokes parameter V) at 6 cm wavelength lies above sunspot penumbrae. The high degrees of circular polarization ($\rho_c \approx 95\%$) of the peaks of the horseshoe structure requires gyroresonant emission with a longitudinal magnetic field strength $H_\ell \approx 600$ Gauss at lower temperatures $T \approx 10^5$ K above penumbrae. The

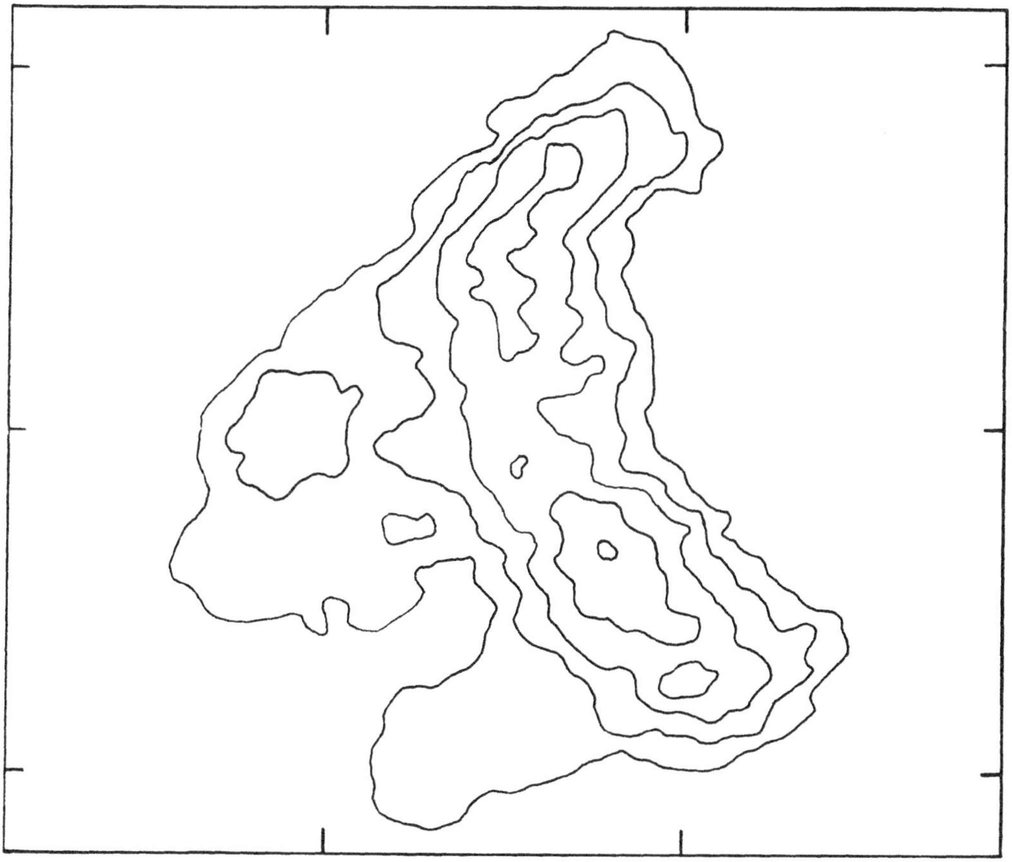

Fig. 2: A 20 cm synthesis map of total intensity delineating a coronal loop connecting two sunspots which lie just to the right of the top and bottom-most extensions of the loop. Here the contours mark levels of equal brightness temperature corresponding to 0.2, 0.4, 0.6 and 0.8 times the maximum brightness temperature of 2×10^6K. The angular scale may be inferred from the 60" spacing between fiducial marks. The solar active region is about 20° across the solar surface from the eastern limb, and for this reason we see the coronal loop bending towards the east (left).

gyroresonant interpretation is fully confirmed by the theoretical work of Gel'freikh and Lubyshev (1979) which predicted the existence of the circularly polarized horseshoes which we have observed. They are essentially a consequence of the fact that gyroresonant absorption increases with increasing angle between the line of sight and the direction of the magnetic field, as well as the fact that the magnetic fields above sunspot umbrae are intense enough to make these regions optically thick to both modes of wave propagation.

CORONAL LOOPS AT 20 CM WAVELENGTH

Lang, Willson and Rayrole (1982) have reported the existence of two looplike structures at 20 cm wavelength which join regions of opposite magnetic polarity in the underlying photosphere. We subsequently examined the observation tapes and found a total of 15 coronal loops associated with 3 active regions on 5 days. The 20 cm loops are therefore truly ubiquitous features of the solar atmosphere (see Figure 2 for an example). The maximum electron temperatures, $T_e(max)$, are characteristic of those found in the X-ray coronal loops with $T_e(max) = 2$ to 4×10^6K. The total extents of the 20 cm loops range between 4×10^9cm and 4×10^{10}cm (where 1" $\approx$ 725 km on the solar surface), and these are also comparable to the X-ray coronal loops which have semilengths L of 10^9cm to 10^{10}cm. We can use the scaling relationship of Rosner, Tucker and Vaiana (1978): $T_e(max) = 1.4 \times 10^3\ (pL)^{1/3}$ to obtain the loop pressure, p. Choosing $T_e(max) = 3 \times 10^6$K and $L = 5 \times 10^9$ cm, we obtain a pressure of $p = 1.96$ dyn cm^{-2}, which is again characteristic of the X-ray coronal loops. The inferred electron density of $N_e = 2.5 \times 10^9$ cm^{-3} can be used together with an average electron temperature of $T_e = 10^6$K to obtain the optical depth, τ, of thermal bremsstrahlung ($\tau = 2.5$), or a brightness temperature of $T_B = 0.92\ T_e$.

ACKNOWLEDGEMENTS

Radio interferometric studies at Tufts University are supported under contract F 19628-80-C-0090 with the Air Force Geophysical Laboratory and by grant INT 8006066 with the National Science Foundation. All of this work was done in collaboration with Robert F. Willson.

REFERENCES

Gel'freikh, G.B. and Lubyshev, B.I.: 1979, Soviet. Astron.A.J.23,pp.316-32

Lang, K.R. and Willson, R.F.: 1979, Nature 278, pp.24-28.

Lang, K.R. and Willson, R.F.: 1980, in I.A.U. Symposium 86, Radio Physics of the Sun, eds. M.R.Kundu and T.Gergeley (Dordrecht: Reidel), pp.109-117.

Lang, K.R. and Willson, R.F.: 1982, Astrophys. J.(Lett.)255,pp.L111-L117.

Lang, K.R., Willson, R.F. and Rayrole, J.: 1982, Astrophys. J., July.

Rosner,R., Tucker,W.H. and Vaiana,G.S.:1978, Astrophys.J. 220,pp.643-665.

DISCUSSION

KOUTCHMY: Do you see anything far from active regions, for example at the supergranulation boundaries of the quiet sun?

LANG: This is potentially measurable during quiet sun periods. We have only observed active regions, which dominate the signal when the sun is active.

SPICER: Can you give a physical interpretation of the observed polarization reversals during the flare you have discussed?

LANG: There are two possibilities. One is that we are observing sequential emission from the feet of a small coronal loop about 5 sec of arc in extent. The other is that the first burst emits a plasma cloud, and subsequent bursts propagate through this cloud. The polarization change is then a propagation effect.

LINSKY: Since the radio observations provide a means of measuring magnetic fields in the corona, and as it is possible to measure photospheric fields for the same active regions, could you comment on how the magnetic field varies with height?

LANG: For our published observations the longitudinal magnetic field strength $H_l \approx 1$ - 2 kG at the photospheric level in sunspots. Above sunspots at heights $h \approx 4 \times 10^9$ cm, we measure $H_l \approx 600$ G where the electron temperature $T_e \approx 2 \times 10^6$ K. This refers to the 6 cm emission, which marks the legs of magnetic dipoles. The 20 cm emission delineates the apex regions of coronal loops, and it can potentially measure H_l at these heights. The Russians using the Ratan 600 telescope obtain $H_l \approx 1$ kG at about 3 cm wavelength where $T_e \approx 10^5$ K. Collaborative Soviet-American observations (Ratan 600 – VLA) taken in July 1982 will lead to more detailed values of H_l as a function of height.

SPRUIT: Concerning the polarization reversals: Is not the most obvious and simple explanation that you have fields of both polarities within your field of view, which you do not resolve?

LANG: This is a good explanation. It requires polarization changes of ~ 100 % and complete reversal in the sense of polarization at the feet of magnetic dipoles separated by about 5 seconds of arc. Apparently X-ray flares have been observed from small loops of this size.

CORONAL AND CHROMOSPHERIC STRUCTURES IN AR LAC: I. DATA AND MODELS

D.M. Gibson[1], F.M. Walter[2], and G.S. Basri[3]

[1]Dept. of Physics, New Mexico Tech, Socorro, NM, USA
[2]JILA, U. of Colorado, Boulder, CO, USA
[3]SSL, U. of California at Berkeley, Berkeley, CA, USA

I. INTRODUCTION

Traditionally stellar sizes and shapes have been determined by analyzing the optical light curves of eclipsing binaries. Now it is possible to extend our knowledge to structures above the photosphere by making similar observations at X-ray, UV, and radio wavelengths. We report the first such observations, together with their geometrical and physical interpretation, which allow us to determine coronal and chromospheric properties of the stars in the RS CVn binary AR Lac. In this paper we discuss the properties of the system, Einstein observations which provide a reasonably good X-ray light curve, IUE data which enable us to tie the inferred coronal features to the surfaces, and radio observations which allow us to separate geometric and intrinsic variations by acting as a flare monitor. We also discuss the construction of simple coronal model which, when applied to the X-ray light curve, yields the sizes and shapes of the coronae in this system. In another paper (Paper II: Walter, Gibson, and Basri, 1982), the geometrical properties of the coronae are discussed and used to determine the physical conditions within.

II. PROPERTIES OF AR LAC

AR Lac is ideally suited for a study of this type. It is the nearest (40 pc; Chambliss, 1976) eclipsing RS CVn binary. It exhibits the full range of phenomena usually attributed to the class (Hall, 1976) including bright X-ray emission ($L_x \sim 10^{31}$ ergs s^{-1}; Walter et al., 1980) and strong radio flaring (Gibson and Hjellming, 1974). It has a short period, P = 1.98 days, and nearly central eclipses (i = 87°) which allow for geometrically simple models of coronal structures.

The primary star is of spectral type G2IV, with a radius of $1.55 R_\odot$ and mass $1.35\ M_\odot$, while the secondary is of spectral type K0IV, with a radius $2.81\ R_\odot$ and mass $1.35\ M_\odot$ (Chambliss, 1976). The semi-major axis of the system is $9.1\ R_\odot$. We have adopted the emphemeris of Hall (1980)

J. O. Stenflo (ed.), Solar and Magnetic Fields: Origins and Coronal Effects, 369–373.

$$\text{JD } (\phi=0) = 2439376.4973 + 1.9831956E$$

in computing our models.

III. THE X-RAY LIGHT CURVE

The X-ray observations of AR Lac began at phase 0.023 on 1980 June 13 and ended two days later at phase 0.027, giving a short overlap during primary eclipse. The total time on target was 29414 sec or 17% of the orbital period. Data were taken during egress from primary eclipse, ingress into secondary eclipse, and at both quadratures. We show these data in Figure 1 together with Einstein MPC (2-10keV) data and the times of complementary IUE and radio (Feldman, 1980) observations.

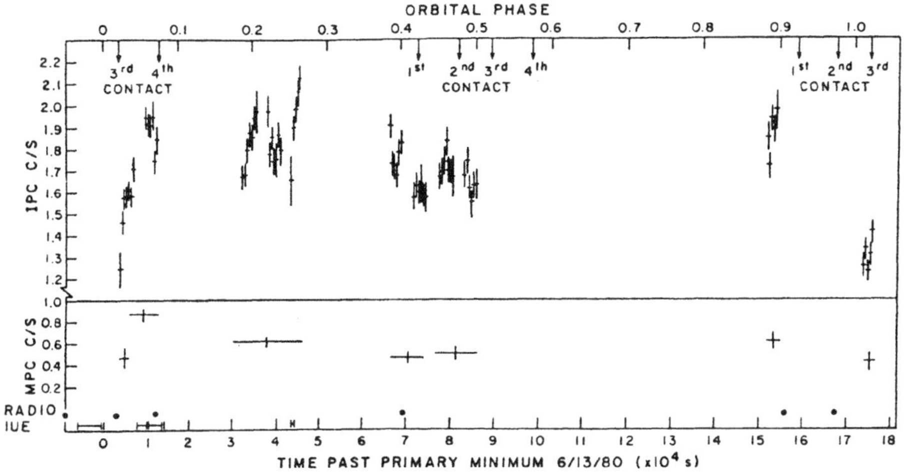

Figure 1. X-ray light curve for AR Lac. 1st and 4th contacts are points of external tangency, 2nd and 3rd contacts are points of internal tangency for the primary (G-star behind) and secondary eclipses.

There are several remarkable features in these data:

a) The out-of-eclipse observations around phases 0.2 and 0.9 show significant variations. We note this aspect of the light curve first to emphasize that although these variations complicate efforts to interpret the light curve geometrically they cannot be unexpected since bright coronae probably result from highly dynamic processes. Nevertheless, both quadratures have the same average count rate, 1.86 cts s^{-1}. We assume this represents the mean level of the system for subsequent analyses and refer eclipse levels to this value.

b) Primary eclipse appears to be total, with both observations prior to third contact giving a level of 1.26 cts s^{-1}. Thus, the G-star contributes 30% of the system's X-ray flux. Egress from primary eclipse in X-rays begins immediately after third contact (see Figure 2a) suggesting the coronal height is small ($h_G \sim 0.02R_G$). However, these data are complicated by either a flare or the uncovering of an X-ray bright active region as shown by the factor of two increase on the MPC count rate 5500 sec past primary eclipse.

c) Secondary eclipse appears to be shallow and very wide. The deepest part of the eclipse occurs near first contact where the count rate is down 14% from the quadrature values. The length of the eclipse allows one to infer the extent of the K-star's corona in the equatorial plane to be $\sim 1R_K$. The coronal height and the observed depth of the eclipse through ingress combine to constrain the size of the K-star's corona to $\lesssim 30°$ latitude.

IV. THE ULTRAVIOLET AND RADIO DATA

Six ultraviolet spectra were taken with the IUE satellite on 1980 June 13. They show the usual range of emission lines from the chromospheres and transition regions though at somewhat lower surface fluxes than other RS CVn systems. The most significant results come from the two high-dispersion long-wavelength (LWR-HI) Mg II h and k spectra which separate the two stars in velocity and were taken at phases 0.05 (when the G-star is about half uncovered) and 0.26 (at quadrature). The blue-shifted (G-star) line strengths increased by only one-third from the former to the latter exposure suggesting that the inner-leading quadrant of the G-star is three times more active than the outer-leading quadrant.

The red-shifted (K-star) line strengths, on the other hand, increased by a factor of 2.6 over the same interval. Examination of a low-resolution long-wavelength (LWR-LO) spectrum taken at phase 0.082 indicates the increased emission seen at quadrature has begun to appear. This indicates there is a bright plage region on the inner-trailing edge of the K-star at about stellar longitude 300°. It is interesting to speculate that this increased activity is related to the "flare" seen in X-rays which begins at phase 0.032. If a flaring loop were uncovered by stellar rotation, then X-rays from the top of the loop would be observable "over-the-limb" before the UV-emission from the base. The phase delay in the appearance of the UV-emission relative to the appearance of the X-ray emission could be used to estimate the height of the loop -- in this case $0.01\ R_K \lesssim R_L \lesssim 0.05R_K$.

AR Lac was not detected in radio observations by Paul Feldman (1980) using the 45-m telescope at Algonquin Radio Observatory at 2.8 cm. The individual upper limits (3σ) for each of the six 15-minute integrations (see Figure 1) was 15 mJy. Thus, the source was radio quiet even during the presumed flare during egress from primary eclipse.

V. SYNTHETIC X-RAY LIGHT CURVES

We have attempted to fit the IPC data using simple models based on the morphology of the solar corona and general characteristics of the

RS CVn binaries. Thus, our model coronae are equatorially symmetric, though they may be latitudinally confined. They are optically thin, but rather than modelling them as a sum of discrete emission regions, as is appropriate for the sun, we assume the emission arises in shells since the high X-ray luminosity suggests there should be a very large number of loops (Walter et al., 1980) on each star. To mimic the "photometric wave" in X-rays we incorporate the option of a sinusoidal coronal wave. Based on these data and studies by Swank and White (1980), we assume that the G-star has a single-temperature corona at 8×10^6K, while the K-star has a two-component corona with the inner one also of 8×10^6K and an outer one at a temperature of 3×10^7K.

Our model fits to the egress from primary eclipse and the ingress into secondary eclipse are shown in Figures 2 and 3. Egress from primary eclipse begins almost at third contact (implying the coronal depth of the G-star is small and increases rapidly until about 8000 seconds past mid-eclipse, the time of the probable flare (see Figure 2b). The increase appears too rapid to be modelled by a symmetric corona (dashed line) and is modelled more satisfactorily by a sinusoidal corona (solid line) in which the leading hemisphere of the star is brighter. This is consistent with the UV results. The high points around 10^4 sec after mid-eclipse may be due to the uncovering of progressively cooler flaring material as the K-star rotates.

Ingress into secondary eclipse is much more difficult to fit, though three aspects of the K-star's hot coronal component seem irrefutable: the corona is highly extended ($1.5\ R_K < R_{hot} < 2.0\ R_K$), is latitudinally confined ($lat_{hot} < 30°$), and is also asymmetric in that its more luminous side is above the leading hemisphere. This asymmetric extended component is shown as the dashed line in Figure 3b. Taking the inner component to the asymmetric in the opposite sense (see dotted line in Figure 3b) seems to provide a better fit to ingress into secondary eclipse (see solid line in Figure 3a) but lacking data from egress from this eclipse we feel this conclusion is much less justified.

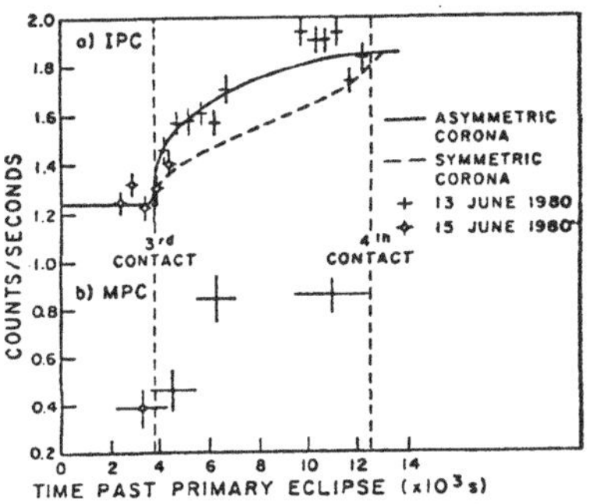

Figure 2. AR Lac's X-ray egress from primary eclipse. Though complicated by a flare, the data still require the G-star to have a brighter leading hemisphere.

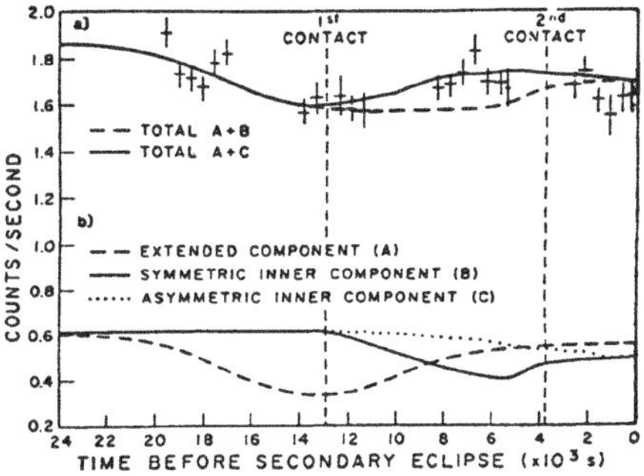

Figure 3. AR Lac's ingress into secondary eclipse. The model requires the hot outer component of the K-star's corona to be highly extended, latitudinally confined, and brighter on the leading edge.

REFERENCES

Chambliss, C.R.: 1976, Publ. Astron. Soc. Pacific 88, p. 762.
Feldman, P.A.: 1980, private communication.
Gibson, D.M. and Hjellming, R.M.: 1974, Publ. Astron. Soc. Pacific 85, p. 682.
Hall, D.S.: 1976, in "Multiple Periodic Variable Stars", ed. W.S. Fitch (Dordrecht: Reidel).
Hall, D.S.: 1980, private communication.
Swank, J.H. and White, N.E.: 1980, in "Cool Stars, Stellar Systems, and the Sun", ed. A.K. Dupree (Cambridge: SAO).
Walter, F.M., Gibson, D.M., and Basri, G.S.: 1983, in IAU Colloquium No. 71.
Walter, F.M., Cash, W., Charles, P., Bowyer, S.: 1980, Astrophys. J. 236, p. 212.

DISCUSSION

ROSNER: Do you recall at what phase(s) Swank and White measured the SSS (Solid State Spectrometer) spectra of AR Lac?

GIBSON: Yes. They observed at both primary and secondary minimum, but probably recorded a flare during the primary. Therefore, we used the secondary in setting the amount of flux at each temperature.

VILHU: Is there a photometric wave present in the optical light curve? If so, can that be explained by dark spots situated at the *same* place where your coronal belts are (at the particular time of your observations)?

GIBSON: Yes. The brighter hemisphere in X-ray emission corresponds to the longitude of the spots on the G star.

THE ROLE OF MAGNETIC FIELDS IN THE HEATING OF STELLAR ATMOSPHERES — THEORY

C. Chiuderi
Istituto di Astronomia
Università di Firenze (Italy)

ABSTRACT

The last ten years of observations have shown beyond doubt the fundamental role played by the magnetic field in the heating of stellar atmospheres. After the recognition of the extreme inhomogeneity of the solar corona, two basic new trends have appeared in the theoretical literature on the coronal heating problem. One is the adoption of a global point of view that stresses the connection of the properties of the upper layers to those of the underlying ones. In this way a general framework is provided, capable of accomodating many possible heating mechanisms that need not to be specified at this stage. The second novelty is the explicit inclusion in the theory of the inhomogeneous nature of the stellar envelopes, as a result of the presence of magnetic fields. The present status of knowledge on the subject as determined by the above evolution of the theoretical approach will be reviewed.

1. INTRODUCTION

The heating of the outer atmospheres of the Sun and the stars constitutes one of the outstanding unresolved problems of stellar physics. This problem has been on the astrophysical scene for such a long time without substantial progress that almost everybody working in the field felt discouraged one time or another. I believe that in the last few years this situation has changed: significant progress has been achieved and more is in sight. This review will try to substantiate the above optimistic statements.

Coronal heating, and in particular magnetic heating, has been the subject of a number of recent, exhaustive review papers (Heyvaerts and Schatzman 1980, Chiuderi 1981, Kuperus et al. 1981, Wentzel 1981, Priest 1982): we refer the reader to the above papers for a detailed presentatio of the general problem, a discussion of the relevant observations and an extensive bibliography. Here we shall simply outline the basic facts and concentrate mostly on the new aspects of the theoretical approach.

J. O. Stenflo (ed.), Solar and Magnetic Fields: Origins and Coronal Effects, 375–386.

The very existence of a stellar corona, namely of a hot plasma overlying a cooler one, indicates the need of a non-radiative heating mechanism. The identification of such a mechanism is the purpose of heating theories. The energetics does not pose any significant difficulty: given the very low heat capacity of a stellar corona almost any non-thermal process is capable of generating the required amount of energy. The difficult task is to describe correctly the spatial distribution of energy deposition and temperature.

The most important phenomenological constraints come from solar observations in the EUV and X-ray ranges. These have shown that the corona is a highly structured medium and that the structuring is intimately tied with the topology of the atmospheric magnetic fields. Most of the emission in the above mentioned wavelength ranges is concentrated in loop structures that form the magnetic active regions. These facts have forced the theoreticians to abandon the previously adopted homogeneous models and to investigate in detail the effects of the presence of the magnetic field both in its passive role of defining the different structures and in its active role of providing efficient heating mechanisms. There has been however a certain inertia in dropping the homogeneous assumption. Thus, the vast literature on coronal loops, generally treats them as homogeneous sub-systems without really investigating the actual consequences of the variation in space (and time) of the various physical quantities. Another shortcoming of many theoretical investigations on coronal loops is the fact that they are considered as entities separated from the underlying atmospheric layers. In spite of this, the study of coronal loops has produced a few important results and revealed the existence of simple relationships between the basic (and in principle measurable) physical parameters of the loop, such as length, pressure, temperature etc.

These relationships, or scaling laws, proved to be extremely useful to understand at least the gross features of solar coronal structures and especially in extending the knowledge thus gained to the stellar context. At the same time it become increasingly evident that no real progress toward the identification of the heating process, or processes, could be achieved without inclusion of the so far neglected aspects of the problem. The most important recent work on coronal heating, to be presented in the following, constitutes an effort to eliminate these unphysical limitations.

We shall first discuss the consequences of considering the coupling of the coronal plasma with the photospheric and sub-photospheric one. This provides a valuable general framework for the study of the heating problem. A number of important results can be obtained without specifying the actual physical mechanism at work. We shall then turn to the description of one particular mechanism, the phase-mixing of shear Alfvèn waves, that has been recently worked out. This calculation represents the quantitative counterpart of the more qualitative global description discussed in the first part. As we shall see the results fit perfectly in the general framework thus confirming its validity and strengthening our confidence in its use in other astrophysical situations.

2. A GLOBAL DESCRIPTION

It has already been mentioned that the magnetic loops are the regions where most of the coronal emission is concentrated. It is however clear that the primary source of the radiated energy is not to be found within the loops themselves. Most, if not all, heating theories place the ultimate energy source in the photospheric or sub-photospheric regions, characterized by an intense mechanical activity in the form of material motions of a more or less ordered nature, from convection to turbulence. The two regions - corona and photosphere - are distinct by their difference in the value of the plasma parameter $\beta = (8\pi p/B^2) = 2(c_S/c_A)^2$ expressing the ratio between the thermal and magnetic energy densities. In the corona $\beta < 1$ whereas in the photosphere $\beta \gtrsim 1$. The magnetic field threads the whole stellar atmosphere and therefore electrodynamically couples the regions where mechanical activity dominates to those where heating is effective. The magnetic field thus provides a potentially powerful mean of transferring energy from material motions in the $\beta > 1$ regions into thermal energy in the $\beta < 1$ regions by driving various electrodynamical processes. Notice that in a mechanically coupled stellar atmosphere (acoustic wave heating) β should remain larger than unity over the entire envelope.

The adoption of a global point of view in which the behaviour of coronal structures is causally tied to the physics of the underlying layers is clearly the only one capable of providing a self-consistent solution of the heating problem. For a long time however it has been thought that the difficulties of such an approach were so overwhelming that a better way was to separate the source regions ($\beta > 1$) from the heated regions ($\beta < 1$) and to study the latter first by introducing an unknown heating function E_H to be determined from a comparison of theories and observations. This philosophy has produced an enormous amount of possible candidate heating mechanisms - such as Alfvén waves, fast and slow magnetosonic waves, surface waves, resistive instabilities etc. - each of which has its own merits and problems. A discrimination between them is extremely difficult from an observational point of view, the observables being subject to large errors or insensitive to the differences in the heating function. On the other hand no theoretical discrimination is possible without connecting in a coherent fashion the physics of the source regions to that of the heated regions.

A first important step in the direction of a global description has been recently achieved with the introduction of the concept of the equivalent electrodynamic circuit by J.A. Ionson (Ionson 1982 a, b). He has been able to show that the electrodynamic coupling between a coronal loop and its $\beta > 1$ "driver" can be represented phenomenologically in terms of a simple LRC driven circuit whose physics is well understood. The equivalent L,R and C can be expressed in terms of the basic parameters of the magnetic loop (the field-aligned scale length ℓ_z, the cross-field scale $\ell_\perp$, the magnetic field in the loop B_0). The driving emf on the other hand is expressed in terms of the $\beta > 1$ velocity field's spectral power function. It is then possible to deduce a phenomenological form of the heating functions, E_H. Consideration of the behaviour of the equivalent electrodynamic circuit shows that the global system acts as a high-quality resonator as we are now going to sketch.

The starting point are the standard dissipative MHD equations including viscosity as well as resistivity. By linearizing the equations and considering approximately shear disturbances $(\underset{\sim}{\nabla} \times \underset{\sim}{E}) \simeq 0$ it is possible to deduce an equation that describes the local electrodynamics of the system:

$$\frac{4\pi}{c^2} \frac{\partial^2 j_z}{\partial t^2} - \eta\,[1+H(1-\beta)\,Pr_m^\perp\,] \frac{\partial^3 j_z}{\partial r^2 \partial t} - H(1-\beta)\left(\frac{4\pi c_A^2}{c^2}\right)\frac{\partial j_z^2}{\partial z^2} = H(\beta-1)\,\frac{\partial^3(v_\theta B_{ph}/c)}{\partial z\,\partial r\,\partial t} \tag{1}$$

Here the z-axis is in the direction of the ambient magnetic field, the perpendicular direction being characterized by r and x_θ. v_θ is the azimuthal component of the photospheric velocity field, the resistivity, $Pr_m^\perp = (4\pi\mu_\perp/\rho\, c^2 \eta)$ is the cross-field magnetic Prandtl number, B_{ph} the photospheric magnetic field and $H(x)$ the Heaviside step-function. c_A is the Alfvén speed.

The terms in Eq. (1) divide naturally in two types; those containing time derivatives of even order, similar to the inductive and capacitive reactances in an LRC circuit, and those containing time derivatives of odd order, analogous to the resistence and to an external source of emf that drives the currents.

Resonances occur when the inductive and capacitive reactance terms cancel each other, i.e. when

$$\frac{\partial}{\partial t} \simeq c_A \frac{\partial}{\partial z}$$

or

$$\omega \simeq \pi\, c_A/\ell_z \quad , \tag{2}$$

ℓ_z being the field-aligned scale length. Since $c_A = c_A(r)$, the condition expressed by Eq. (2) can be satisfied only at special r-locations for each ω. It is important to notice that the condition expressed by Eq. (2) needs not to be necessarily satisfied. If this happen, however, for each given ω absorption of electromagnetic energy takes place only at special r locations within a spatial bandwidth Δr that can be estimated directly from Eq. (1) expanded about the spatial resonance. The result is

$$\Delta r/a = \pi^{1/3} \, [1+H(1-\beta) \, Pr_m^{\perp}]^{1/3} \, R_m^{-1/3} \quad ,$$

R_m being the magnetic Reynolds number defined as:

$$R_m = (4\pi c_A \, a^2/\eta c^2 \ell_z) \quad .$$

Here a is the spatial scale of the Alfvén speed:

$$a = \left[\frac{d(\ell n c_A)}{dr}\right]^{-1} \quad .$$

A description of the global electrodynamics is obtained by integrating Eq. (1) over the volume of the system. This integration is greatly simplified by the assumed occurrence of spatial resonances. The result of this integration can be expressed in the form of a global electrodynamic equation, given by

$$L \frac{d^2 I}{dt^2} + R \frac{dI}{dt} + \frac{I}{C} = \frac{d\varepsilon}{dt} \qquad (3)$$

where

$$I = \int_0^{\ell_z/2} 2\pi j_z(r) \, r \, d \, r$$

$$L = 4\ell_z/\pi c^2$$

$$R = \eta(1+Pr_m^{\perp})(\ell_z/\Delta r)^2$$

$$C = (\ell_z c^2/4\pi c_A^2)$$

$$\varepsilon(t) = \ell_{\perp} v_\theta B_{ph}/c$$

and $\ell_\perp$ is the photospheric cross-field spatial scale. Eq. (3) constitutes the basic equation for the discussion of the LRC circuit analog of a coronal loop. It predicts the possibility of global structure oscillations at a characteristic frequency

$$\omega_o = (LC)^{-1/2} = \pi c_A/\ell_z$$

The precise nature of these oscillations is not important in the present philosophy: one example are the Alfvén surface waves (Ionson, 1978, Wentzel, 1979). As in any resonant LRC circuit we can define a Q-value,

$$Q = (L/CR^2)^{1/2} \quad .$$

Energy is absorbed in a limited frequency bandwidth $\Delta\omega$ around the resonant frequency ω_o with

$$\frac{\Delta\omega}{\omega_o} = 1/Q \qquad (4)$$

A direct evaluation of Q shows that for typical coronal conditions $Q \gg 1$. Thus the loop-driver system behaves as a high-quality resonator.

It is also possible to evaluate the heating function E_H in terms of suitable time-averages of the driving emf, $\varepsilon(t)$. The result (Ionson, 1982) is

$$E_H = \left(\frac{2}{\pi \ell_\perp^2 \ell_z}\right) \frac{\langle\varepsilon^2\rangle_{\omega_o}\,\omega_o}{(L/c)^{1/2}} \left(\frac{R}{R_{tot}}\right) \qquad (5)$$

where $R_{tot} = R + R_{ph}$ includes the effect of the photospheric resistance. An important feature is that the total power drain on the driver that equals $E_H(R_{tot}/R)$ is totally independent of dissipation. This is typical of all high-quality resonators, but should not be taken as an indication that the nature of the irreversibilities that determine R is not important. Their role is to fix the fraction $\Delta\omega/\omega_o$ of the available driving emf that interacts with the system in resonant conditions. The total power drain on the driver is however proportional to $\Delta\omega/R_{tot}$ which is independent of dissipation since $\Delta\omega \sim R_{tot}$. In other words, given a resonant condition, the system absorbs all the available energy within the allowed frequency bandwidth.

At this stage we have the basic missing ingredient in the phenomenological loop theories, i.e. E_H. The standard scaling laws (Rosner et al. 1978) can then be used to predict in a self-consistent manner the thermodynamic properties of a coronal loop. For instance, the following scaling laws are obtained for the maximum temperature, T, and base pressure, P:

$$T(k) = 2.10 \times 10^4 \left[T_{eff} \langle \frac{1}{2}\rho\, v_\theta^2 \rangle_{ph}\right]^{2/7}$$

$$P(\text{dyne cm}^{-2}) = (7.8 \times 10^3/\ell_z) \left[T_{eff} \langle \frac{1}{2}\rho\, v_\theta^2 \rangle_{ph}\right]^{6/7}$$

where T_{eff} is the effective blackbody temperature of the star and $\langle\frac{1}{2}\rho\, v_\theta^2\rangle_{ph}$ is the average spectral power of the photospheric velocity field evaluated at the resonance. An application of the above scaling laws to the Sun produces a remarkably good agreement with the observations.

3. A WORKED-OUT EXAMPLE

Let us now turn from the previous phenomenological discussion to quantitative study of a particular heating mechanism that explicitly takes into account the fact that the corona is a structured medium. Alfvén waves present themselves as a very natural candidate for transporting energy from the low-lying layers to the uppermost ones. The damping of Alfvén waves by frictional processes however is so small that their efficiency as a coronal heating mechanism has always been rather doubtful. The introduction of a structure perpendicular to the field lines, i.e. of a gradient of the Alfvén speed transverse to the field itself gives rise to a series of new and interesting phenomena, such as surface Alfvén waves (Wentzel 1979, Roberts 1981, Ionson 1978). Most of the past investigations, however, were concerned with the polarization which has a component of the displacement parallel to the direction of inhomogeneity, x. Quite recently (Heyvaerts and Priest 1982) a very interesting study has been performed on the propagation of shear Alfvén waves with a displacement perpendicular to x. The behaviour of these waves is amenable to a rather complete analytical investigation and leads to the possibility of a rapid damping due to phase-mixing as will be now explained.

Let us consider an equilibrium field $B_0(x)\, e_z$ and its response to motions in the y-directions at a fixed frequency ω. If the velocity is taken as $v(x,z,t)\, e_y$ the dissipative MHD (non-linear) equations give rise to the following simple propagation equation

$$\frac{\partial v}{\partial t^2} = c_A^2(x)\,\frac{\partial^2 v}{\partial z^2} + (\gamma_m + \gamma_v)\left(\frac{\partial^2}{\partial x^2} + \frac{\partial^2}{\partial z^2}\right)\frac{\partial v}{\partial t} \tag{6}$$

where γ_m and γ_v are the magnetic diffusivity and the kinematic viscosity, both assumed to be constant and small.

Without dissipation Eq. (6) implies that each magnetic surface =constant oscillates independently with a wave number

$$k_z(x) = \omega/c_A(x) \quad .$$

As the waves propagate along neighbouring surfaces at different x they become increasingly out of phase. Thus increasingly steeper x-gradients are generated up to the point where the transverse scale becomes so small that the dissipation mechanisms completely damp the wave. The basic physics is extremely simple and provides a beautiful example of the immense wealth of new possibilities that appear as soon as the unphysical assumption of perfect homogeneity is abandoned. The damping length, Λ, can be easily estimated in the long-wavelength, weak-damping and strong phase-mixing limit. The result for a purely viscous damping is $\Lambda = (6R)^{1/3}\, k_z^{-1}(x)$ where R is the Reynolds number. Numerically,

$$\Lambda(\mathrm{km}) \simeq 5\times10^3\; B^{5/3} \left(\frac{\nu_c}{\nu}\right)^{1/3} \left(\frac{\tau}{10}\right)^{2/3} \left(\frac{10^{10}}{n}\right)^{5/6} \left(\frac{2\times10^6}{T}\right)^{1/6} \left(\frac{9}{100}\right) \tag{7}$$

where B is expressed in gauss, ν_c and ν are the classical and actual viscosity coefficients, τ is the wave period in seconds, n the number density in cm^{-3}, T the temperature and $\underline{a}$ the transverse variation scale $(z\, dk_z/d_x)^{-1}$. Thus short-period waves (τ = 10 s) in weak fields damp efficiently over a distance of 5000 km. For longer periods and/or stronger fields this purely laminar mechanism does not seem to be an efficient one, even if one allows for an enhanced viscosity well above its classical value. Notice that the previous calculation refers to a field without boundaries in the z-direction and therefore does not mimic the situation in closed structures. In the latter case (coronal loops) standing waves will be formed due to the reflection from the transition regions near the footpoints of the loop. In this case then, each magnetic surface will oscillate with a frequency

$$\omega(x) = k_z \, c_A(x)$$

with k_z fixed. Phase-mixing of these standing waves will now take place in time, rather than in space. An estimate of the laminar viscous damping time, τ_d, gives

$$\tau_d = (6R)^{1/3} \, \omega^{-1}(x) \quad .$$

For the typical parameters used in the previous numerical example $\tau_D \simeq 20\,\tau$ which is short enough to be of interest. By completely solving the problem for the bounded case with a general excitation it is also possible to show that a state is produced in which the rate of dissipation exactly balances the rate at which the standing waves are produced, independently of the value of the dissipative coefficients. This very important result is in perfect agreement with the general phenomenological description, but now, we have gained a great deal in physical insight. The damping is produced by friction between close layers. Given sufficient time, phase mixing produces gradients as steep as necessary to match any dissipative efficiency, no matter how small.

Heyvaerts and Priest proceed further by observing that the damping terms depend strongly on the actual values of the dissipative coefficients. Thus an investigation is required to determine whether there are instabilities capable of creating small-scale structures in the Alfvén waves themselves that would enhance the values previously obtained for the damping lengths (or times) in the laminar case. Of particular importance are those instabilities that grow on a timescale shorter than the wave period. The results of a very elegant stability analysis for the standing waves case can be summarized as follows. Near a magnetic field node of a standing wave, where the velocity shear is maximum a Kelvin-Helmholtz instability is likely to develop, for typical coronal parameters. The induced turbulence effectively lowers the Reynolds number well below its laminar value so that damping times could be as short as a few periods. Analogously, tearing-mode instabilities are likely to develop at a velocity node, where the field shear is maximum, producing a similar shortening of the damping times.

4. CONCLUSIONS AND FUTURE PERSPECTIVES

We have tried to illustrate the basic reasons that make us believe that the coronal heating problem has possibly overcome a dead point. Looking backwards or reading past and recent reviews on the subject one realizes, as usual, that many of the ideas or schemes that now prove to be correct, or at least promising, were already there since a long time. Like in a puzzle, the pieces were there, but somehow we were unable to make them fit together.

In the last few years many people, including ourselves, pointed out that the corona was extremely inhomogeneous, but kept on making models of coronal loops that were piece-wise homogeneous with little consideration to what happened at the interfaces. Even in the cases where this class of phenomena was considered, the first examples studied were probably not the simplest or the most significant in connection with the heating problem. Similarly, the electric circuit analogy had already been used before. But those authors (see Alfvén 1977 and references therein) were mostly concerned with LR circuits and failed to recognize the importance of the capacitive properties of the coronal electrodynamic analog. Thus the possibility of exciting global oscillations was missed and the analogy of the behaviour of the coronal loops with that of a high-Q resonator was not apparent.

The framework offered by the LRC circuit analog shows that much can be gained by adopting a phenomenological description in which we do not pretend to treat the finer details and, so to speak, match the degree of theoretical finesse to what can be checked by observations. In fact the inclusion of phenomenological terms and functions in the basic equations is only meaningful if they can be directly determined by observations. In all other cases ambiguities are always present that undermine the reliability of the conclusions. Clearly, the global electrodynamical description is not the ultimate theory. It only links the corona and the photosphere. But there are still vital ingredients, the upwelling magnetic fields and the velocity of the turbulent motions, that are not introduced in a self-consistent way. To achieve this goal one should link the photosphere with the interior by producing convincing and reliable dynamo theories. We still not have them (see e.g. Gilman, 1981), but we can now wait for them because we can measure the missing quantities and use them for our theories on heating even if we do not understand exactly where they come from.

Another important piece of information is missing: the structure of coronal magnetic fields. In fact, all we know is the gross topology of coronal fields, via their influence on the thermodynamical state of the ambient plasma. Nothing is actually known on the fine structure. Since no direct measurement of coronal magnetic fields seems to be feasible in the near future, we must rely heavily on theoretical analyses. The deep connection between termodynamical and magnetic properties, whose importance can hardly be overestimated, has received only little attention so far (Einaudi et al. 1982).

Finally we would like to stress again the need of performing detailed, quantitative analyses of the type previously reviewed, along with more phenomenological studies. The comparison of the results gives strength to the simpler approach if they agree or indicates possible pitfalls or oversimplifications if they don't. In turn the more formal approach is much more effective if some general, more intuitive picture is present to provide a track.

In conclusion, something has changed in the long-standing game of understanding the heating of the stellar atmospheres. The last word has not been said, but there is a precise feeling that, after all, we are on the right track.

REFERENCES

Alfvén, H.: 1977, Rev. Geophys. Space Phys. 15, 271.
Chiuderi, C.: 1981, in R.M. Bonnet and A.K. Dupree (Eds.), Solar Phenomena in Stars and Stellar Systems, D. Reidel Publ. Co., Dordrecht, Holland, p. 269.
Einaudi, G., Torricelli-Ciamponi, G. and Chiuderi, C: 1982, "Magnetic Fields and Thermal Structure of Solar Plasmas", submitted to Solar Phys.
Gilman, P.A.: 1981, in S. Jordan (Ed.), The Sun as a Star, NASA SP-450, Chapt. 8.
Heyvaerts, J. and Schatzman, E.: 1980, in F. Moriyama and J.C. Henoux (Eds.), Proceedings of the Japan-France Seminar on Solar Physics, p. 77.
Heyvaerts, J. and Priest, E.R.: 1982, "Coronal Heating by Phase-mixed Shear Alfvén Waves", submitted to Astron. Astrophys.
Ionson, J.A.: 1978, Astrophys. J. 226, 650.
Ionson, J.A.: 1982a, Astrophys. J. 254, 318.
Ionson, J.A.: 1982b, "Electrodynamic Coupling in Magnetically Confined X-ray Plasmas of Astrophysical Origin", submitted to Astrophys. J.
Kuperus, M., Ionson, J.A. and Spicer, D.S.: 1981, Ann. Rev. Astron. Astrophys. 19, 7.
Priest, E.R.: 1982, Solar MHD, D. Reidel Publ. Co., Dordrecht, Holland, Chapt. 8.
Rosner, R., Tucker, W.H. and Vaiana, G.S.: 1978, Astrophys. J. 220, 643.
Wentzel, D.G.: 1979, Astrophys. J. 227, 319.
Wentzel, D.G.: 1981, in S. Jordan (Ed.), The Sun as a Star, NASA SP-450, Chapt. 14.

DISCUSSION

SPRUIT: I have a comment on the equivalent circuit approach. In reducing the MHD equations to this extent one sweeps some things under the carpet. I am mostly worried about the neglect of a) stratification of the atmosphere, b) inhomogeneity of the field. Both have a very strong effect on wave propagation. They dominate the behaviour of the wave between the convection zone and the base of the corona. By averaging the equations over the volume one looses the ability to keep track of these effects.

CHIUDERI: In the equivalent circuit approach we certainly loose a certain number of effects. However, from the point of view of comparing theory with observations, such an approach is likely to be more adequate to the status of present day observational knowledge than a more detailed description.

SPICER: Nowhere do I see in these calculations the effects of cooling by expansion. In fact the $\mathbf{J}\cdot\mathbf{E}$ work term given in the talk does not include the $\mathbf{J}\cdot(\mathbf{v}\times\mathbf{B})$ work of expansion. I am wondering how this affects the calculations, since there is nothing published that demonstrates cooling by expansion.

IONSON: If you take a typical speed of expansion and calculate the expansion losses within the loop, you will find that they are small.

SPICER: The relevant question is whether the expansion time is shorter than the heating time. Observationally it is impossible to tell at this time whether a loop is stationary or the top of the loop is expanding while new field is expanding up replenishing the field that has already moved up. Also the best observations of loops, due to Skylab, do not have the time or spatial resolution to support the concept of a stationary non-expanding loop.

ROBERTS: I would like to comment upon the equivalent circuit concept, and to support the remarks made by Spruit with regard to the role of stratification ($g \neq 0$). As to stratification, this is of vital importance in governing the behaviour of waves in propagating from the photosphere to the corona. As to the equivalent circuit concept based upon linear theory, this is just another way of looking at an old problem: the nature of MHD waves in a coronal field. But the linearized MHD equations may be readily investigated for a magnetically-structured atmosphere, and for this approach there is no necessity to invoke the various approximations presented in the description of circuit theory. In fact, such a treatment for coronal magnetic inhomogeneities shows the presence of oscillations with phases of the order of the Alfvén speed (see the detailed treatment in Edwin and Roberts: 1982, *Solar Phys.* **76**, pp. 239-259) — the so-called fast body magnetoacoustic modes. The circuit analogy seems to lead to the same result in that it predicts an oscillation at about the right frequency, but the detailed form of the modes is not apparent. In general, I would like to see a close comparison of the results of circuit theory with the more usual (e.g. normal mode) analysis. That would be very interesting and instructive.

CHIUDERI: I agree with the comment that the circuit analogy is not a new "theory". It simply casts the problem in a simple way and provides a general framework that includes many possible mechanisms. It shows that a certain number of features are not the consequence of a specific heating mechanism but are common to all of them. These features then cannot be used to discriminate between various mechanisms. I think I already stressed the fact that quantitative analyses must accompany this type of more intuitive approach.

CAMPOS: The model presented is based on an LRC circuit. Dr. Spruit has stated that he liked better to speak in terms of Alfvén waves including stratification effects. I would like to remark that most of your results can be proved by considering Alfvén-gravity waves propagating vertically in an isothermal atmosphere, allowing for the exponential growth in Alfvén speed. The frequencies of standing modes are given by

$$\omega_n = \frac{c_A}{L}\frac{j_n}{2},$$

where j_n is a root of the Bessel function J_0. In your formula it appears that $\omega_0 = \pi c_A/L$, so that π is replaced by $j_n/2 = 1.2, 2.0$, etc. The waveforms are given by the Bessel function J_0, and give the high gradients, etc. (cf. *Solar Phys.*, 1983).

BENZ: There are nice coronal oscillations observed after flares in metric radio emission. This radiation is probably emitted by energetic electrons trapped in high coronal loops ($\gtrsim$ 100 000 km), which are modulated by global oscillations. The damping time (about 10 - 100 periods) fits nicely with your prediction. However, the period is observed to be 1 - 2 seconds, two orders of magnitude smaller than predicted.

CHIUDERI: At this stage of the theory it is very dangerous to make direct comparisons with observations. It is not obvious that the observations are actually related to global oscillations rather than to time variations in the distribution functions.

VLA OBSERVATIONS OF QUIESCENT AND FLARE MICROWAVE EMISSION FROM LATE-TYPE STARS: A UNIQUE PROBE OF CORONAL MAGNETIC FIELDS

Dale E. Gary,[1] Jeffrey L. Linsky,[2] and George A. Dulk[1]
Joint Institute for Laboratory Astrophysics, University of Colorado, National Bureau of Standards, Boulder, CO 80309

The high sensitivity and angular resolving power of the VLA permits observations for the first time of the steady (quiescent) microwave emission from stellar coronae as well as detailed studies of stellar flares. Gary and Linsky (1981) first detected steady 6 cm emission from χ^1 Ori (G0 V) and UV Cet (M5.6e V), and Topka and Marsh (1982) subsequently detected quiescent emission from both components of EQ Peg (dM3.5e + dM4.5e).

Listed in Table 1 is a summary of sources detected in our program. χ^1 Ori was detected for two hours as a steady 1.1 mJy source at 6 cm on 1980 October 2, not detected during the previous hour of observation or on 1981 August 22, and then detected again as a 0.3 mJy source on 1981 October 16. Accordingly, the source is variable, but probably not on time scales that we associate with flaring. UV Cet was always a bright source when observed on each of four occasions, and its flux was always variable. Another dMe star system Wolf 630AB (dM4e + dM4e) also is a variable microwave source and, like the other detected systems, is a strong X-ray source. The dMe binary system YY Gem (dM1e + dM1e) is a variable 6 cm source with a flux that appears to depend on photometric phase.

The importance of these detections is that the flux levels are far larger than computed on the basis of thermal bremsstrahlung emission from the hot coronal plasma that produces the observed Einstein X-ray fluxes. Included in Table 1 are predicted values of the 6 cm flux, computed assuming optically thin bremsstrahlung emission in a 10^7 K corona consistent with the measured soft X-ray flux (cf. Gary and Linsky 1981). Since the 6 cm fluxes of the sources that have been detected all are considerably in excess of the predicted values and there is evidence for circular polarization in three of the sources,

---

[1]Department of Astro-Geophysics, Univ. of Colorado, Boulder, CO, USA.
[2]Staff Member, Quantum Physics Division, NBS.

J. O. Stenflo (ed.), Solar and Magnetic Fields: Origins and Coronal Effects, 387–390.

Table 1. VLA detections

Source	Date	Time UT	6 cm Flux (mJy) (±3 σ or 3σ upper limit)	Fractional circular polarization	Predicted bremsstrahlung flux (mJy)
χ^1 Ori	2 Oct 80	1500-1555	<0.5	---	0.14
(G0 V)		1555-1606	0.8±0.8[1]		
		1606-1613	1.2±0.8		
		1616-1625	1.4±0.7	0±0.3	
		1625-1635	1.3±0.4		
		1655-1751	1.0±0.3		
	3 Oct 80	0756-1119	<1.8	---	
	22 Aug 81	1344-1934	<0.3	---	
	16 Oct 81	1050-1252	0.3±0.2	0	
UV Cet	3 Oct 80	0358-0550	1.0-1.7±0.3[2]	0.5±0.3 RH	0.02
(M5.6 eV)	19 Jun 81	1625-1717	1.7-2.3±0.4[2]	0.3±0.2 RH	
	22 Aug 81	0715-1345	1.5-2.2±0.5[2]	0.5±0.4 RH	
	22 Aug 81	0715-1345	<10.0[3]	---	
	16 Oct 81	0420-1040	1.5-4.0±0.5[4]	0.3±0.1 RH	
YY Gem	20 Jun 81	1925-2142	<0.2	---	0.007
(dMle +					
dMle)	22 Aug 81	1942-2004	1.8±0.3		
		2007-2034	1.6±0.3		
		2037-2104	1.5±0.3	~0.1 LH	
		2107-2134	1.1±0.3		
		2137-2204	0.9±0.3		
	23 Aug 81	1340-1401	1.4±0.3		
		1404-1431	1.3±0.3	0	
		1434-1501	1.3±0.3		
	26 Sep 81	1249-1343	0.4±0.2		
	27 Sep 81	1255-1339	0.4±0.2	0	
	28 Sep 81	1248-1338	0.5±0.2		
Wolf-	22 Aug 81	2241-2400	0.5±0.2	0	0.20
630	23 Aug 81	0000-0300	<0.2	---	
AB					
(dM4e +	27 Sep 81	0027-0045	1.4±0.3		
dM4e)	28 Sep 81	0028-0041	1.5±0.3	~0.3 LH	
	29 Sep 81	0021-0039	1.2±0.2		

[1] 2 cm
[2] Variable flux.
[3] 20 cm
[4] Some flaring.

we believe that the observations can be explained by gyroresonance emission from thermal electrons trapped in coronal magnetic fields. This is the emission process thought responsible for microwave emission from magnetic flux tubes above sunspots (e.g. Pallavicini et al. 1981). Model calculations by Gary and Linsky (1981) indicate that for modest photospheric magnetic fields (1000-2000 gauss) and coronal temperatures (5-10 × 10^6 K), gyroresonance emission is optically thick and the source radius is larger than the photospheric radius for several harmonics of the gyroresonance frequency. They found that the observed 6 cm flux from χ^1 Ori could be explained by coronal fields of

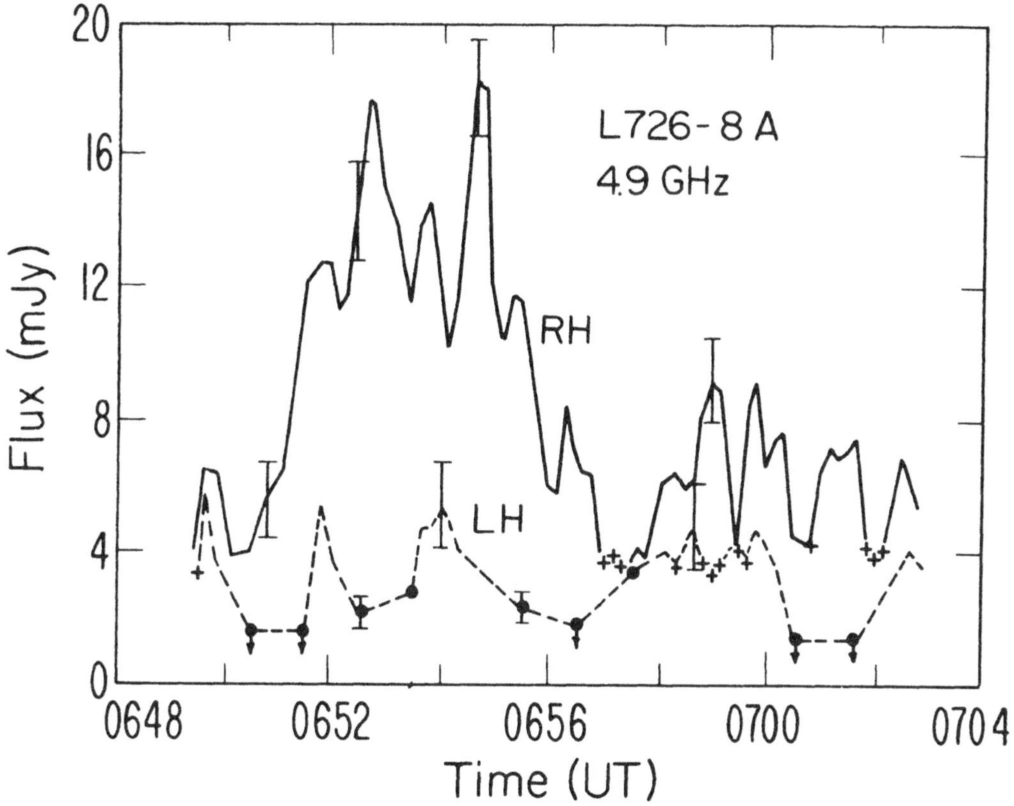

Fig. 1. 6 cm (4.9 GHz) flux of L726-8A in right-hand (RH) and left-hand (LH) circular polarization taken from VLA maps of 10 s integration time. Error bars indicate the typical 1σ rms noise level. For maps on which no source was visible, the 3σ noise level has been plotted as a '+'. In the LH flux curve, fluxes denoted by dots were taken from 1 minute integrated maps. Dots with arrows denote upper limits. From Gary et al. (1982).

at least 300 gauss and perhaps larger (if the emission is at a low harmonic). This technique provides a new and perhaps unique method for measuring coronal magnetic fields in late-type stars.

During our observing program we detected an unusual microwave flare from L726-8A (dM5.5e), whose companion is the prototype flare star UV Ceti (Gary, Linsky and Dulk 1982). The 6 cm flare had a peak flux of 10 mJy, and was almost entirely right hand circularly polarized. Furthermore, the flare exhibited a quasi-periodic variation in flux with a period of about a minute as illustrated in Fig. 1. The variability may be produced by oscillations of Alfvén or slow mode MHD waves in a magnetic loop of length L ~ 0.2 $R_{\odot}$, or perhaps by maser action. Either explanation requires large coronal magnetic fields and the observed period may turn out to be a means of measuring such fields.

This work is supported by NASA grants NSG-7287, NAGW-91, NGL-06-003-057, and NSG-9287 to the University of Colorado.

REFERENCES

Gary, D.E. and Linsky, J.L.: 1981, Astrophys. J. 250, pp. 284-292.
Gary, D.E., Linsky, J.L., and Dulk, G.A.: 1982, submitted to Astrophys. J. (Letters).
Melrose, D.B. and Dulk, G.A.: 1982, Astrophys. J. 259, in print.
Pallavicini, R., Sakurai, T., and Vaiana, G.S.: 1981, Astron. Astrophys. 98, pp. 316-327.
Topka, K. and Marsh, K.A.: 1982, Astrophys. J. 254, pp. 641-645.

DISCUSSION

BENZ: **Why do you know that the radio emission of stellar coronae and flares is gyrosynchrotron and not a plasma emission process?**

LINSKY: We have no definitive proof as yet concerning the emission mechanisms, but there is some important evidence. We believe that the emission process for the quiescent corona is gyroresonance emission as a consequence of the high percentage of circular polarization, $\approx 10^7$ K brightness temperatures, and analogy with microwave emission above sunspots. The flare emission process is less certain, but for stellar flares the brightness temperatures appear to be $\geq 10^{10}$ K, suggesting a coherent process.

ELECTRODYNAMIC COUPLING IN MAGNETICALLY CONFINED STELLAR X-RAY LOOPS

James A. Ionson
Laboratory for Astronomy and Solar Physics
NASA-Goddard Space Flight Center, Greenbelt, MD 20771

I. INTRODUCTION

The solar-coronal complex is the only astrophysical system whose X-ray structure has to some degree been spatially resolved. The most important lesson that we have learned from these relatively high-resolution observations is that the corona is highly structured, comprising a variety of closed, loop-like regions of enhanced radiative output. Furthermore, coronal radiation loops are now known to be spatially coincident with magnetic loops which confine the radiating plasma.

There is no reason to suspect that the basic X-ray morphology of the solar-coronal complex is unique, and it could in fact be fundamental in developing physical models of other as yet spatially unresolved X-ray emitting astrophysical systems. The general goals of this article are to identify and briefly discuss the primary sub-models that are involved in the global coupling between a $\beta \gtrsim 1$ mechnical energy reservoir and a contiguous site of $\beta < 1$ X-ray activity.

GLOBAL COUPLING SCENARIO

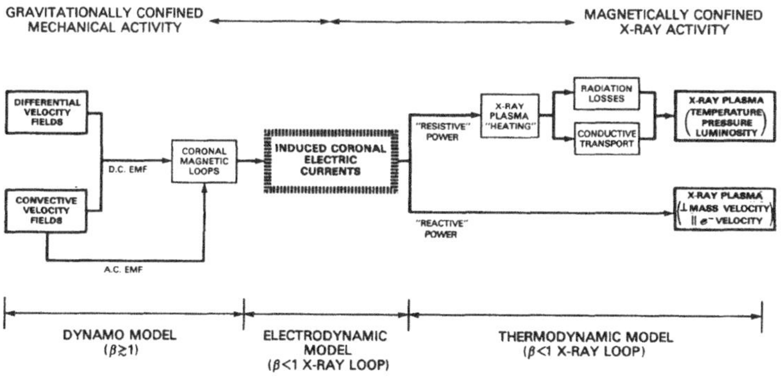

J. O. Stenflo (ed.), Solar and Magnetic Fields: Origins and Coronal Effects, 391–396.

The "Dynamo Model" establishes a quantitative connection between properties of the mechanical driver (e.g., differential and convective velocity fields) and the dimensions, field strength and number density distribution of elemental magnetic loops. As pointed out by Golub et al. (1980), the mechanical energy reservoir comprises two regions that are differentiated in terms of the velocity field's convective time and size scales. Specifically, the "surface layer" velocity field is characterized by relatively small convective time and size scales compared to the underlying "subsurface layer." As pointed out by Ionson (1982) the overall potential magnetic field structure is established by large-scale subsurface velocity fields whose characteristc time scale exceeds the magnetic loop's Alfven transit time scale (i.e., D.C. emf's). In contrast, the relatively smaller scale surface convection, whose time scale is of the order of the magnetic loop's Alfven transit time, results in A.C. emfs which drive an elemental loop into a non-equilibrium electrodynamic state characterized by the flow of electrical currents along (i.e., force-free) and across (i.e., non-force free) the coronal B-field (i.e., "Electrodynamic Model"). These induced electrical currents couple to the magnetically confined plasma within a spatial resonance shell resulting in heating as well as a confined, cross-field coronal velocity field with radiation and conduction losses determining the X-ray emitting plasma's temperature, pressure and luminosity (i.e., the "Thermodynamic Model").

This article does not attempt to present a quantitative dynamo model. This extremely complicated problem has been the subject of articles too numerous to discuss here (c.f., review by Cowling 1981), many casting doubt upon various versions of the canonically accepted α-ω dynamo. Field amplification for an α-ω dynamo is accomplished by large scale differential motion (i.e., the ω-effect) with field reversal occuring due to the diffusive effects of helical convection (i.e., the α-effect). The result of this interaction is a temporal variation in magnetic flux generation along with a corresponding spatial propagation of this flux across the $\beta \gtrsim 1$ mechanical driving zone. A major observational problem with this model is that, at least for the sun, the total magnetic flux appears to be constant throughout the activity cycle (Golub et al., 1979). This is contrary to the predictions of various α-ω dynamo models. In addition, the hydrodynamically inferred solar differential rotation is directed in a sense that is opposite to that required by kinematic α-ω dynamos (Gilman 1981).

Unfortunately, today's conventional dynamo models are the weakest link in the global coupling scenario illustrated in Figure 1. Therefore, this article will build upon the observationally rationalized assumption that elemental magnetic loops are fundmental in the $\beta \gtrsim 1$ to $\beta < 1$ coupling process. The results of this analysis will thus be presented in a form that does not explicitly depend upon magnetic properties of a particular loop. There is, of course, a strong implicit dependence upon these magnetic properties.

II. ELECTRODYNAMIC COUPLING IN MAGNETIC LOOPS

Two classes of solar coronal radiation loops have been identified. Specifically, the so called "cool loops" with temperatures of the order of 10^4 - 10^5 °K and hot "X-ray loops with temperatures of about 2.5×10^6 K. Hot and cool loops typically have the same basic dimensions (i.e., lengths < 10^{10} cm and diameters ~ $10^{8.5}$ cm). In support of Foukal's (1975, 1976, 1978) original suggestion, Raymond and Foukal (1982) have recently presented evidence that some "hot" and "cool" loops are co-spatial within the resolution limits of the observing instruments (viz., ~ $10^{8.5}$ cm). A distinct possibility is that the currently observed loops actually represent an unresolved cluster of topologically distinct radiation loops whose diameter is significantly less than $10^{8.5}$ cm. This possiblity is logically consistent with the observed existence of small diameter (< $10^{7.5}$ cm) magnetic flux elements within the photosphere which extend into the corona, confining the radiating plasma. Raymond and Foukal's observations suggest that each of these unresolved radiation loops contains both "cool" and "hot" plasma which is consistent with many contemporary coronal heating theories -- the primary point of contention being the relative volumes of cool and hot plasma.

Ionson (1982 a,b) has demonstrated that the electrodynamic coupling between a $\beta < 1$ elemental magnetic loop and its underlying $\beta \gtrsim 1$ mechanical driver can be represented by a simple yet equivalent LRC circuit analog.

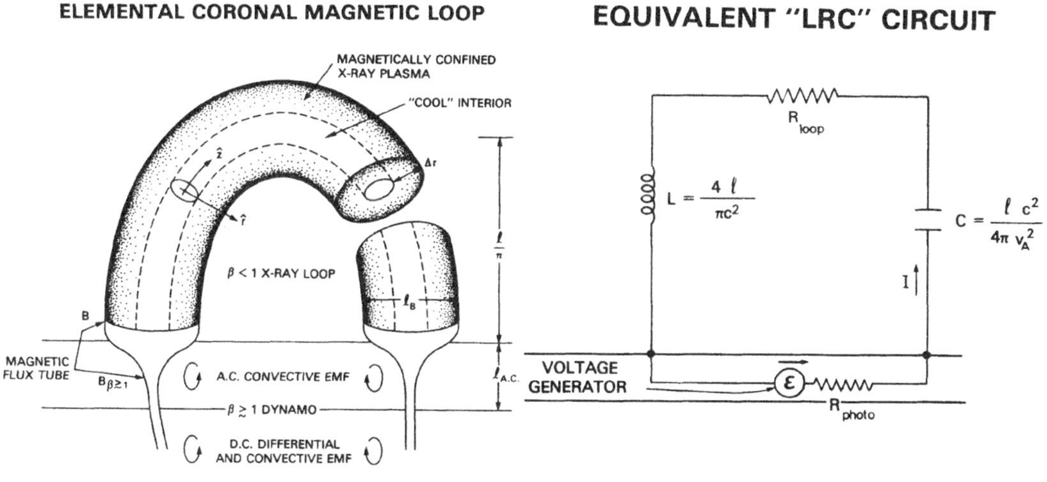

This derived analog points to the existence of global structure oscillations, such as Alfvenic surface waves, which resonantly excite internal field line oscillations within a spatial resonance shell of thickness Δr_{res}. These localized B-field oscillations result from induced currents that are driven by $\beta \gtrsim 1$, A.C., stressing emfs.

Although the width of this spatial resonance as well as the magnitude of the induced currents explicitly depend upon irreversible processes (e.g., viscosity and resistivity), the resonant form for the heating rate (i.e., the "resistive power") is virtually independent of irreversibilities. This is a classic feature high "quality" resonant systems that are driven by a broad band source of spectral power. In this case the source of spectral power is the power spectrum of the A.C. stressing velocity, $\langle v^2_{\beta \gtrsim 1} \rangle_\nu$. The essential feature of a resonant heating mechanism is that magnetic loops with different lengths and field strengths, and hence different global resonance frequencies, are heated at a rate that critically depends upon the amount of a $\beta \gtrsim 1$ spectral power at the resonance frequency ν_0 as well as the value of B.

This analog has been a powerful tool in identifying relationships that quantitatively connect the thermodynamic state of the confined $\beta < 1$ X-ray plasma (e.g., temperature, T; thermal pressure, P; cross-field flow velocity, $v_{\beta<1}$) to the confining magnetic loop characteristics (viz., field strength, B; cross-field extent, ℓ_B; inverse aspect radio, ℓ/ℓ_B; magnetic expansion factor, $B_{\beta>1}/B$) and the $\beta \gtrsim 1$ stressing velocity, $v_{\beta>1}$. These relations fall into two broad categories. Namely, (1) diffuse heating resulting from efficient transport of energy throughout the magnetic loop volume and (2) isolated heating within the electrodynamic dissipation shell. For each of these cases Joule dissipation, shear viscous dissipation and compressional dissipation have been considered. Each of these models is characterized by three thermodynamic scalings that are parameterized by the magnetic loop properties and $v_{\beta>1}$. Eliminating the magnetic field intermediary from these relations and normalizing ℓ/ℓ_B and $B_{\beta>1}/B$ with respect to solar conditions results in a single, unique velocity scaling law for each model, viz.

<u>Diffuse Heating</u> (c.g.s. units)

$$v_{\beta<1} = \begin{cases} 7.5 \times 10^{-6}\, T^{0.33}\, P^{-0.16}\, v_{\beta>1}^{2.02} & \text{Joule dissipation} \\ 9.0 \times 10^{-5}\, T^{0.66}\, P^{-0.17}\, v_{\beta\gtrsim 1}^{1.34} & \text{shear viscous dissipation} \\ 1.6 \times 10^{3}\, T^{0.51} & \text{compressional viscous dissipation} \end{cases}$$

<u>Isolated Heating</u>

$$v_{\beta<1} = \begin{cases} 1.7 \times 10^{-7}\, T^{0.99}\, P^{-0.24}\, v_{\beta\gtrsim 1}^{1.47} & \text{Joule dissipation} \\ 3.8 \times 10^{-5}\, T^{0.88}\, P^{-0.18}\, v_{\beta\gtrsim 1}^{1.12} & \text{shear viscous dissipation} \\ 85.0\, T^{0.16}\, v_{\beta\gtrsim 1}^{0.75} & \text{compressional viscous dissipation} \end{cases}$$

Although other authors have recognized the significance of temperature and pressure scaling laws, the importance of $\beta < 1$ <u>velocity</u> scaling laws which complete our description of the electrodynamic coupling process have been overlooked until now.

REFERENCES

Foukal, P.: 1975, Solar Phys. 43, p. 327.
Foukal, P.: 1976, Astrophys. J. 210, p. 575.
Foukal, P.: 1978, Astrophys. J. 223, p. 1046.
Gilman, P.A.: 1981, in "The Sun as A Star", ed. S. Jordan, (NASA Sp-450), Chapter 8.
Golub, L., et al.: 1979, Astrophys. J. Letters 229, p. L145.
Golub, L., et al.: 1980, Astrophys. J. 238, p. 343.
Ionson, J.A.: 1982a, Astrophys. J. 254, p. 318.
Ionson, J.A.: 1982b, Astrophys. J., submitted.
Raymond, J.C. and Foukal, P.: 1982, Astrophys. J. 253, p. 323.

DISCUSSION

KUPERUS: Your LRC circuit applies to coronal loops, where you have a well defined length scale (the loop length). Is it possible to extend this theory to open structures such as the coronal holes by the introduction of an equivalent length scale?

IONSON: This is an important question that we are attempting to answer.

SPRUIT: Let me respond to your statement about taking a step towards the observations. I think that even the observers would feel more comfortable in using theoretical results if these are based on solutions of equations rather than on the application of more simplified models.

IONSON: As discussed in Ionson (1982, 1983), these results are indeed based upon solutions of the equations of electrodynamics. The major point, reviewed by Dr. Chiuderi, is that these equations can be put into a *simple* form which still stresses the important physics.

UCHIDA: In your circuit model, is there any special device operating so that the energy is damped at places where the resistivity is not large, in other words, outside the photospheric level?

IONSON: As pointed out in Ionson (1982), the photospheric *resistance* is smaller than the coronal *resistance* unlike the case when one considers the resistivity.

FOING: (1) To what depths does your broad-band generator correspond in the photosphere, where the coupling is stronger? (2) How could you derive the spatial energy spectrum of subphotospheric convective motions, at the present state of velocity observations or convective theories?

IONSON: (1) The A.C. driver is found at the $\beta \approx 1$ level. (2) This model does not attempt to explain the subphotospheric energy spectrum. Rather, this spectrum is used as an observational input to the LRC coronal heating theory.

LOW: We should not think that one must regard a problem exclusively either from an MHD continuum or an LRC point of view. It would be useful if an MHD problem can be solved fully and its results be compared with the corresponding LRC model, to see to what extent the two approaches are similar and/or different. People who work with MHD cannot, with good reason, expect that a fluid system, with its impressive degrees of freedom, can be approximated by a linear circuit.

IONSON: I am surprised by this comment since the MHD-LRC connection has been demonstrated from first principles in Ionson (1982, 1983). In fact, Dr. Chiuderi has just finished reviewing this "impressive" connection during his invited review.

HEYVAERTS: Response to the comment by B.C. Low: In going from fluid picture to circuit model, an integration over loop cross section is made. The details of radial structure are lost. The oscillations are represented by a global oscillation period. In the work reported by Chiuderi (Heyvaerts and Priest, 1983), we took into account the fact that actually each magnetic surface has its own resonance frequency $\omega = \omega(x) = k_z c_A(x)$. Then, each shell of the loop resonates with the excitation at its own proper frequency and behaves individually much as described by Ionson. The net heating rate results from the integration of these individual heating rates over the loop cross section.

A THERMAL CATASTROPHE IN A RESONANTLY HEATED CORONAL LOOP

P.C.H. Martens and M. Kuperus
Sterrekundig Instituut
Zonnenburg 2
3512 NL Utrecht
The Netherlands

ABSTRACT

A theory for the thermal stability of hot coronal loops is presented, which is based on the resonant electrodynamic heating theory of Ionson (1982) and the evaporation/condensation scenario of Krall and Antiochos (1980). The theory predicts that gradual changes in the length of a loop or in its magnetic field strength can trigger catastrophic changes in the X-ray visibility of the loop, without the need for a change in the magnetic field topology.

A natural explanation is thereby given for the observations of X-ray brightenings in loops and loop evacuations with coronal rain.

1. THERMAL BALANCE OF A RESONANTLY HEATED LOOP

The electrodynamic coupling of a coronal loop to the photosphere has been described by Ionson (1982) in terms of an equivalent electric circuit with a high quality factor and resonance frequency $\nu = v_A/4L$, where v_A is the Alfvén velocity and L is the halflength of the loop. The amount of heat the loop absorbs per unit volume from the available photospheric power spectrum $S(\nu)$ is given by

$$E_H(\nu) = 9.06 \times 10^{12} \frac{S(\nu)}{L^2} \text{ erg cm}^{-3} \text{ sec}^{-1}. \qquad (1)$$

In thermal equilibrium this heat input is lost by radiation since the loops are thermally insulated. The radiative losses can be expressed in terms of the resonance frequency, the loop length and the magnetic field strength B (see Martens and Kuperus (1982) for a detailed description) and are given to a good approximation by

$$E_R \approx \frac{7.26 \times 10^2 \ B^{7/2}}{L^{15/4} \ \nu^{7/2}} \text{ erg cm}^{-3} \text{ sec}^{-1} \qquad (2)$$

provided the loops are short compared with the coronal scale height. For loops comparable to the scale height a correction is needed. Figure 1 gives the heating and the radiative losses (multiplied by L^2 for reasons of convenience (see Equation 1)) as a function of ν, with L and B as parameters for the radiative loss function.

J. O. Stenflo (ed.), Solar and Magnetic Fields: Origins and Coronal Effects, 397–400.

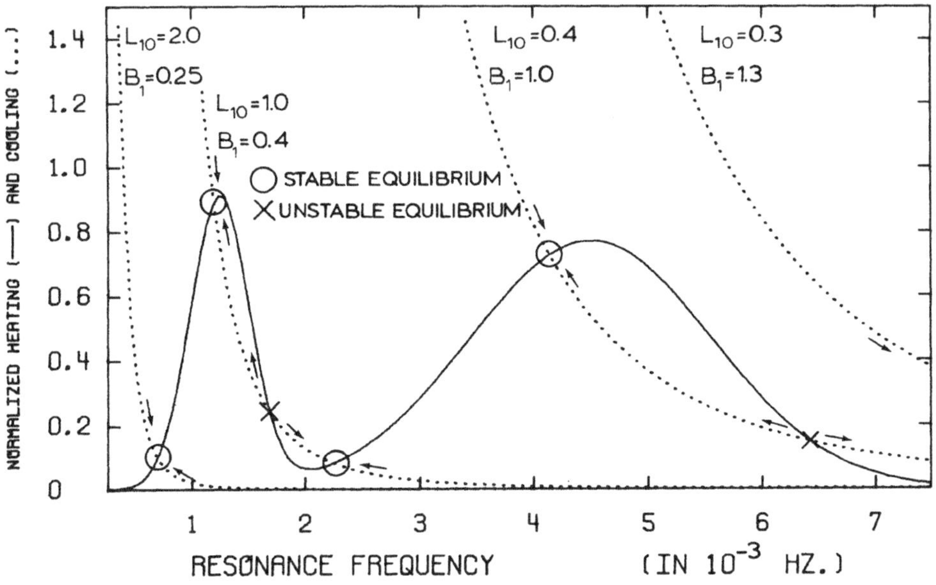

Figure 1. The heating and cooling of a coronal loop as a function of resonance frequency. The unity along the vertical axis is 10^{16}erg cm^{-1} sec^{-1}.

The heating function in Figure 1 is exactly the power spectrum that Ionson used. He found a satisfactory agreement between the heating theory and the observations of hot loops. For a given loop length and a given field strength equilibrium ($E_H = E_R$) is found when the heating and cooling curves intersect. As can be seen from Figure 1 multiple solutions are possible.

According to the evaporation/condensation scenario (Krall and Antiochos, 1980) a rise in coronal temperature results in an increased conductive flux into the transition region. The response to the increased energy input in the chromospheric upper level is an impulsive heating and subsequent expansion of the atmosphere redistributing the matter over the loop, thus increasing the average coronal density $\bar{\rho}$. As $\bar{\rho}$ increases the radiative losses increase and this continues until heating and radiative losses are balanced. The opposite ("condensation") occurs when the temperature drops below its equilibrium value. This scenario may be formalized by the expression

$$\frac{d\bar{\rho}}{dt} \sim E_H - E_R \ . \tag{3}$$

But a change in $\bar{\rho}$ is accompanied with a change in the resonance frequency ν. Hence

$$\frac{d\nu}{dt} \sim E_R(\nu) - E_H(\nu) \ . \tag{4}$$

Thus, the equilibrium at some equilibrium resonance frequency ν_0 is stable when

$$\frac{d}{d\nu}(E_R - E_H)\Big|_{\nu_0} < 0 \; . \tag{5}$$

Stable and unstable branches are indicated in Figure 1.

2. CATASTROPHIC BEHAVIOUR

Consider a loop with a slowly increasing loop length or a slowly decreasing magnetic field. Then, as can be deduced from Figure 1, the equilibrium resonance frequency slowly decreases and so do the radiative losses.

At some point in the loop development two new intersections, one stable and one unstable, will emerge. As the loop evolves further, the radiative loss curve becomes tangent to the heating curve at some point (in Figure 1 close to 2×10^{-3} Hz resp. 8.2×10^{-3} Hz). From that moment on the stable solution disappears and the loop will evolve in a thermal time scale towards the other equilibrium near 1.3×10^{-3} Hz resp. 3.2×10^{-3} Hz. The radiation losses are about eight resp. more than ten times larger than in the original situation, which must be clearly visible as a sudden X-ray brightening. Note that we propose here that a sudden brightening is a purely thermal process without the need of any magnetic reconnection. Howard and Švestka (1977) report these X-brightenings from Skylab movies.

When the length of the loop decreases or the field increases the reverse occurs. Now the loop finds a new stable equilibrium at much smaller radiative losses. This must be visible as a sudden loop evacuation and coronal rain because the cooling matter must fall down. Such observations are reported by Levine and Withbroe (1977).

It thus seems that the temporal behaviour of coronal loops can be understood as catastrophic changes in the thermal properties. The critical point in this theory is the vital importance of the photospheric power spectrum, which determines the total heating of the loop. The resonance frequency couples the heating to the power spectrum, but through the Alfvèn velocity the heating has a feedback to the resonance frequency and this determines the thermal equilibrium and its stability. Since the photospheric power spectrum, the loop length and the magnetic field strength are external parameters, any change in these may trigger the sudden loop transitions. Similar catastrophes are found numerically in an extension of the dynamical loop approach of Kuin and Martens (1982) which includes resonant heating (Martens and Kuin, 1982).

REFERENCES: Howard, R., Švestka, Z.: 1977, Solar Phys. 54, 65.
Ionson, J.A.: 1982, Astrophys. J. 254, 318.
Krall, N.A., Antiochos, S.K.: 1980, Astrophys. J. 242, 374.
Kuin, N.P.M., Martens, P.C.H.: 1982, Astron.Astrophys. 108, L1.
Levine, R.H., Withbroe, G.L.: 1977, Solar Phys. 51, 83.
Martens, P.C.H., Kuin, N.P.M.: 1982, in preparation.
Martens, P.C.H., Kuperus, M.: 1982, Astron.Astrophys. in press.

DISCUSSION

SPRUIT: The existence and number of the thermal "catastrophes" depend mostly on the assumed frequency spectrum, is that correct?

MARTENS: The existence of the catastrophes depends on the existence of peaks in the spectrum, but not specifically on those peaks that are assumed. I think that these catastrophes will occur for any narrow-band resonant heating mechanism.

MULLAN: The power spectrum of solar velocity fields has been measured by Fossat et al. (1980). It does *not* resemble the double-peaked power spectrum used by the authors. There *is* a main peak around 3 mHz, but there is *no* peak around 1 mHz.

MARTENS: We have merely used the power spectrum used by Ionson as a *demonstration* of our mechanism. Our mechanism will work for *any* power spectrum as long as the peaks have slopes that are steep enough. Only for a *flat* "shot noise" spectrum there will be no catastrophes.

CRAM: I would like to make some comments on the shape of the power spectrum used in these calculations. Unfortunately, observers cannot provide at present an accurate velocity power spectrum of the kind needed for these models, and we must assemble several different components. The short period power (3 - 5 min) is due to well-understood photospheric and chromospheric oscillations. These oscillations are temporally phase coherent for many periods; they also tend to be spatially phase coherent on a large scale (>10 Mm). The long period power (>10 min) is meant to represent the evolution of solar convective structures (granules, mesogranules, supergranules, giant granules, etc.). These structures are not spatially or temporally phase coherent. There is also some evidence for slow, quasi-periodic velocity variations in plages and near sunspots (Woods and Cram; Harvey; Lites). More work must be done to delineate the power spectrum of this phenomenon, and of the "long-period" structures in general.

KUPERUS: I would like to stress here that any power spectrum that shows a definite maximum ultimately leads to this type of catastrophic transition provided that one uses a resonance theory such as the one proposed by Ionson (1982). This very behaviour of coronal loops is actually another observational support of Ionson's LRC equivalent circuit theory.

FINITE AMPLITUDE EFFECTS OF TORSIONAL ALFVÉN WAVES IN A FLUX TUBE

Ø.Andreassen and P.Maltby
Institute of Theoretical Astrophysics, University of Oslo.

Abstract: Consider a cylindrical, homogeneous magnetic flux tube embedded in a field-free compressible plasma. We employ the method of successive approximations to solve the magnetohydrodynamic equations analytically for an axisymmetrical (m=0) torsional Alfvén mode. The m=0 mode inside the tube may excite surface-, stationary- or progressive waves in the field-free plasma. We conclude that, in contrast to the linear theory, the flux tube may lose energy by wave radiation to the surroundings.

1. INTRODUCTION

Our understanding of the mechanical energy propagation and the heating process in stellar atmospheres is still rather limited. As it appears that structured magnetic fields play a fundamental role, studies of wave propagation in magnetic flux tubes embedded in a compressible plasma are of considerable interest. Recent reviews of small-amplitude wave propagation in flux tubes are given by Roberts (1981), Wilson (1981), Spruit (1982) and Edwin and Roberts (1982).

Investigations into the nonlinear nature of the wave propagation are valuable in understanding the damping mechanisms and the heating processes. Hitherto few attempts in this direction have been done. The interesting possibility of solitons in solar magnetic flux tubes has been pointed out by Roberts and Maugeney (1982). A discussion of the nonlinear aspects of propagation of Alfvén waves has recently been given by Hollweg et al. (1982). They consider the Alfvén waves as time-dependent axisymmetrical twists and solve the equations numerically. One of their simplifying assumptions seems vital, i.e. the suppression of the particle motion in the direction perpendicular to both the average field and to the azimuthal direction. In this paper an analytical study of torsional Alfvén waves in magnetic flux tubes embedded in a field-free compressible plasma is presented. In a thorough discussion different Alfvén wave modes should be considered. We have concentrated our efforts to the axisymmetrical torsional (m=o) mode in

J. O. Stenflo (ed.), Solar and Magnetic Fields: Origins and Coronal Effects, 401–406.

order to point out differences between linear and non-linear theory. Allowing motions in three dimensions we find that the tube boundary is perturbed in the nonlinear theory. This result contrasts the linear theory of the torsional Alfvén mode where the tube boundary is unperturbed(e.g. Spruit, 1982).

2. BASIC EQUATIONS

Neglecting the effect of gravity the following set of equations will describe an adiabatic, perfectly conducting gas in magneto-hydrodynamics:

$$\frac{\partial \rho}{\partial t} + \nabla\cdot(\rho \underline{v}) = 0 \qquad \rho\frac{d\underline{v}}{dt} = -\nabla P + \underline{j}\times\underline{B}$$

$$\mu\underline{j} = \nabla\times\underline{B} \qquad \frac{\partial \underline{B}}{\partial t} = \nabla\times(\underline{v}\times\underline{B}) \tag{1}$$

$$\nabla\cdot\underline{B} = 0 \qquad \frac{d}{dt}(P\rho^{-\gamma}) = 0$$

Here $\underline{v}$ denotes the velocity, ρ the density, P the pressure, $\underline{B}$ the magnetic induction, $\underline{j}$ the current density, μ the magnetic permeability and γ the ratio of specific heats.

3. SECOND ORDER SOLUTION

We shall employ the method of successive approximations to solve equations in (1) up to second order in the field variables. Consider a cylindrical coordinate system(r, θ, z) with the unperturbed magnetic field in the z-direction. A first order symmetrical torsional Alfvén mode (m=0) is assumed and the field variables are written as follows:

$$\begin{aligned}
\underline{B} &= \underline{B}_z^{(0)} + \underline{b}_\theta^{(1)} + (\underline{b}_r^{(2)} + \underline{b}_z^{(2)}) = \underline{B}_z + \underline{b}_\theta + (\underline{b}_r + \underline{b}_z)\\
\underline{v} &= 0 + \underline{v}_\theta^{(1)} + (\underline{v}_r^{(2)} + \underline{v}_z^{(2)}) = \underline{v}_\theta + (\underline{v}_r + \underline{v}_z)\\
\rho &= \rho^{(0)} + 0 + \rho^{(2)}\\
P &= P^{(0)} + 0 + P^{(2)}
\end{aligned} \tag{2}$$

Hence, for the m=0 mode the first order perturbations are azimuthal both in the velocity and in the magnetic field. The second order perturbations occur in density, pressure and in two directions for the velocity as well as the magnetic field.

Introducing equation (2) in (1) the following expression is obtained for the second order perturbation, b_z, in the magnetic field (for details, see Andreassen, 1982)

$$\Phi \cdot b_z = \frac{B}{\rho^{(0)}} \left(\frac{\partial^2}{\partial t^2} - s^2 \frac{\partial^2}{\partial z^2}\right) \frac{1}{r} \frac{\partial}{\partial r} \left\{\frac{b_\theta^2}{\mu} - \rho^{(0)} v_\theta{}^2\right\} \tag{3a}$$

$$+ \frac{B}{\rho^{(0)}} \frac{1}{r} \frac{\partial}{\partial r} \left\{r \frac{\partial}{\partial r} \left[\frac{\partial^2}{\partial t^2} \left(\frac{b_\theta^2}{2\mu}\right)\right]\right\}$$

where the operator,

$$\Phi = \frac{\partial^4}{\partial t^4} - (s^2 + a^2) \left[\frac{\partial^4}{\partial t^2 \partial z^2} + \frac{\partial^2}{\partial t^2} \frac{1}{r} \frac{\partial}{\partial r} \left(r \frac{\partial}{\partial r}\right)\right]$$

$$+ s^2 a^2 \left[\frac{\partial^4}{\partial z^4} + \frac{\partial^2}{\partial z^2} \frac{1}{r} \frac{\partial}{\partial r} \left(r \frac{\partial}{\partial r}\right)\right] \tag{3b}$$

and s and a are sound speed and Alfvén speed, respectively. It appears from the right hand side of (3a) that the second order magnetic field perturbation b_z is caused by a combination of the magnetic tension, the centripetal force and the magnetic pressure.

Let the m=0, first order solution be written as

$$b^{(1)} = b_\theta = \tilde{b}_\theta \, f(r) \exp[i\omega(t \mp z/a)], \tag{4a}$$

$$v^{(1)} = v_\theta = \mp \frac{\tilde{b}_\theta}{B} a \, f(r) \exp[i\omega(t \mp z/a)], \tag{4b}$$

where f(r) is a function of the distance from the centre of the flux tube and the minus sign corresponds to propagation in the positive z-direction. We note that the magnetic tension and the centripetal force (see 3a) do balance in this mode.

Equation (3) will have solutions with twice the original frequency,

$$b_z^{(2)}(r,z,t) = b_z = \tilde{b}_z(r) \exp\left[2i\omega(t \mp z/a)\right]. \tag{5}$$

Introducing this solution in (3), taking (4) into account we find

$$\frac{1}{r} \frac{\partial}{\partial r} \left[r \frac{\partial b_z(r)}{\partial r}\right] = - \frac{\tilde{b}_\theta^2}{2B_z} \frac{1}{r} \frac{\partial}{\partial r} \left[r \frac{\partial f^2}{\partial r}\right]. \tag{6}$$

Integration of (6) gives

$$\tilde{b}_z(r) = -\frac{\tilde{b}_\theta^2}{2B_z}\left[f^2 + c\right], \qquad c = -2B_z b_z(0)/\tilde{b}_\theta^2 \tag{7}$$

Here c is a constant of integration and we have set the term $c_1 \int dr/r = 0$, i.e. only finite values of $b_z^{(2)}$ at r=0 are accepted. The second order solutions may be summarized as follows:

$$\begin{aligned}
b_z &= -\frac{\tilde{b}_\theta^2}{2B_z}\left[f^2 + c\right] \exp[2i\omega(t \mp z/a)],\\
b_r &= \mp \frac{i\omega}{a}\frac{\tilde{b}_\theta^2}{B_z}\left[\frac{1}{r}\int_0^r r\, f^2\, dr + cr/2\right] \exp[2i\omega(t \mp z/a)],\\
v_z &= \mp \frac{\tilde{b}_\theta^2}{2\mu\, \rho^{(0)} a}\left[f^2 + \frac{c\, s^2}{s^2 - a^2}\right] \exp[2i\omega(t \mp z/a)],\\
v_r &= i\omega \frac{\tilde{b}_\theta^2}{B_z^2}\left[\frac{1}{r}\int_0^r r\, f^2 dr + cr/2\right] \exp[2i\omega(t \mp z/a)],\\
p^{(2)}(r,z,t) &= \frac{c\, s^2}{s^2 - a^2}\frac{\tilde{b}_\theta^2}{2\mu} \exp[2i\omega(t \mp z/a)].
\end{aligned} \tag{8}$$

We note that the perturbations in pressure and density are independent of r, whereas the perturbations in velocity and magnetic induction vary with r and z. The magnitude of the perturbations depend on the boundary conditions.

4. BOUNDARY CONDITIONS

In the field-free medium ($r > r_o$) the wave solution may be written:

$$\begin{aligned}
v_{r,e}^{(2)}(r,z,t) &= v_{r,e} = \tilde{v}_{r,e}\, R_o'(\nu r) \exp[2i(\omega t \mp kz)],\\
p_e^{(2)}(r,z,t) &= p_e = -\frac{2i\omega\rho_e^{(0)}}{\nu} R_o(\nu r) \exp[2i(\omega t \mp kz)],
\end{aligned} \tag{9}$$

where $R_o'(\nu r) = \partial R_o(\nu r)/\partial(\nu r)$ and R_o is a cylindrical function (Bessel, modified Bessel or Hankel function).

We intend to fit the outside solution to an inside solution with m=0 and $\omega/k = a$. The boundary conditions at $r = r_o$ are:

$$v_{r,i} = v_{r,i}^{(2)} = v_{r,e}^{(2)} = v_{r,e} \tag{10}$$

$$p_i^{(2)} + \frac{\underline{B} \cdot \underline{b}^{(2)}}{\mu} + \frac{b^{(1)2}}{2\mu} = p_e^{(2)}. \tag{11}$$

Or alternatively,

$$i\omega \frac{\tilde{b}_\theta^{\,2}}{B_z^2} \left[\frac{1}{r_o} \int_0^{r_o} r\, f^2\, dr + c r_o/2\right] = \tilde{v}_{r,e}\, R_o'(\nu r_o), \tag{12}$$

$$c\, \frac{\tilde{b}_\theta^{\,2}}{2\mu} \frac{a^2}{s_i^{\,2} - a^2} = p_e^{(2)}\, R_o(\nu r_o). \tag{13}$$

Noting that $\tilde{v}_{r,e} = -\nu p_e^{(2)}/(2i\omega\rho_e^{(0)})$ these boundary conditions may be combined. Taking the dispersion relation for the external medium into account we find

$$\frac{1}{r_o c} \int_0^{r_o} r\, f^2\, dr = -\frac{r_o}{2} - \frac{R_o'(\nu r_o)}{4\nu R_o(\nu r_o)} \frac{a^2}{s_i^2 + \frac{1}{2}\gamma a^2} \frac{s_e^2 - a^2}{s_i^2 - a^2}\,. \tag{14}$$

For a given $f(r)$ and ω the constant c is determined. Hence, given $\tilde{b}_\theta$, $f(r)$ and ω the second order perturbation b_z is determined from equations (8) and (14). The perturbation in velocity and pressure in the external medium follows from equations (12) and (13). We find that $p_e^{(2)}$ is finite for $s_i \to a$, provided $s_e \neq a$. The waves in the external medium may occur as surface waves if $a < s_e$. The flux tube may lose energy by wave radiation if $a > s_e$, the progressive waves will travel in the direction $\cos\phi = s_e/a$ with respect to the z-direction. Hence, the sound radiation from the flux tube is in this case very similar to the sound radiation from an infinitely long cylinder excited in a single mode (e.g. Skudrzyk, 1971).

5. DISCUSSION

The present results supplement the numerical study of nonlinear Alfvén waves by Hollweg et al. (1982), where $v_r^{(2)}$ was set equal to zero, but the effect of gravity was taken into account. Our result implies that a weakly nonlinear torsional (m=0) Alfvén mode may generate sound waves in the field-free surroundings. The generated waves will have a frequency twice that of the torsional Alfvén wave. In contrast to the results of the linear theory (e.g. Spruit, 1982) we find that the flux tube may lose energy by wave radiation similar to sound radiation from an infinitely long cylinder excited in a single mode.

REFERENCES

Andreassen, Ø.: 1982, Inst.Theor.Astrophys. Report (in prep.).
Edwin, P.M. and Roberts, B.: 1982, Solar Phys. 76, 239.
Hollweg, J.V., Jackson, S. and Galloway, D.: 1982, Solar Phys. 75, 35.
Roberts, B.: 1981, "The Physics of Sunspots" eds. L.E. Cram and J.H. Thomas, Sacramento Peak Observatory.
Roberts, B. and Maugeney, A.: 1982, Mon. Not. R. astr. Soc. 198, 7p.
Skudrzyk, E.: 1971, The Foundations of Acoustics, Springer-Verlag.
Spruit, H.C.: 1982, Solar Phys. 75, 3.
Wilson, P.R.: 1981, Astrophys. J. 251, 756.

DISCUSSION

IONSON: Since nonlinear effects are important to order $v_{wave}/v_{Alfvén}$, I would expect that they are unimportant, since $v_{wave} \approx 20$ km s^{-1} if one associates it with anomalous coronal line broadening.

MALTBY: The numerical study by Hollweg et al. (1982, *Solar Phys.* **75**, p. 35) shows that the nonlinearity of Alfvén waves may be important already at chromospheric heights.

MOUSCHOVIAS: If I understood your equation for b_r correctly, you always generate an r component of the magnetic field if the wave has z and θ components. We know, however, that a torsional Alfvén wave can propagate in the $\hat{z}$ direction without generating an r component of the magnetic field, even in a compressible medium. Why do you get the result I referred to? In particular, if you do not put in a b_r initially, will you generate one?

MALTBY: The results you mentioned are correct for a homogeneous medium. However, for a fluxtube with finite diameter a second-order perturbation in b_r will be generated, the magnitude of which depends on the boundary conditions.

ROBERTS: A point of clarification. Are the boundary conditions applied on the undisturbed boundary of the fluxtube or on the (first-order) disturbed boundary?

MALTBY: In our case the tube boundary is not perturbed in the first-order calculations. Hence, the boundary conditions could be applied to the undisturbed tube boundary.

CAMPOS: The Alfvén waves in an atmosphere without dissipation grow initially exponentially, by a factor of four per scale height, and later grow linearly. If there is resistive dissipation there are opposing effects, but the velocity is ultimately bounded. Since the Alfvén speed grows exponentially with height, the Alfvén number v/c_A is small at high altitudes. So I would expect non-linear effects to be small in the corona for plane, incompressible Alfvén waves. The conclusion could be modified by magnetosonic waves, which propagate a compression.

OBSERVATIONS OF ALFVÉN WAVES SIMULTANEOUSLY IN Hα AND K

Ronald G. Giovanelli
CSIRO Division of Applied Physics, Sydney, Australia 2070

Jacques Beckers
MMTO, University of Arizona, Tucson, AZ 85721

Abstract. Simultaneous spectroheliograms of a quiet region at solar disk centre in Hα + 0.29 Å, Hα - 0.29 Å, K + 0.18 Å and K - 0.18 Å show much similarity in the asymmetries in the two lines. The fibrils are identical geometrically. Both lines show patterns of line-of-sight motions propagating along the fibrils. Close to the network, the velocity of propagation is of the order of 12 km s^{-1} towards or away from the network; further away the patterns propagate away from the network at velocities of the order of 75 km s^{-1}. The latter are interpreted as Alfvén waves, the former as due most likely to variations in longitudinal velocities along the fibrils.

1. INTRODUCTION

There is a long history of observations of Alfvén waves in Hα fibrils (McMath 1939, Sawyer, 1974). Henry Smith prepared (or at least had) a film at Sacramento Peak Observatory around 1958, demonstrating the phenomenon, though its identification as such is more recent (Giovanelli, 1975).

In 1975 we made further observations of Alfvén waves in chromospheric fibrils. The waves are detected by observing the propagation of transverse velocities along fibrils, which are aligned with the magnetic field. The transverse motions are observed via line asymmetries.

Although profile asymmetries in unblended lines are due only to Doppler shifts, differential motions along the line of sight make the interpretation of asymmetric profiles difficult (e.g. Beckers, 1962). Far more confidence can be placed in the results if those from two or more lines formed differently are in agreement. For observations of chromospheric fibrils, there are only two independent lines in which high-enough spatial resolution can be obtained, Hα and K. Therefore we undertook a study involving time-sequences of observations of a region near the centre of the disk using these lines, obtaining spectroheliograms simultaneously at Hα + 0.29 Å, Hα - 0.29 Å, K_{2R} and K_{2V}.

J. O. Stenflo (ed.), Solar and Magnetic Fields: Origins and Coronal Effects, 407–411.

2. OBSERVATIONS AND ANALYSIS

The observations were obtained with the Sacramento Peak tower telescope, spectroheliograms being obtained in the 9th order in Hα and the 15th in K with a 300 lines mm grating and a dispersion of 6.9 and 11.5 mm/Å at Hα and K respectively. Observations were obtained at two wavelengths in each line, pairs of fixed slits being placed in the plane of the spectrum to allow two symmetrically-spaced monochromatic slit images to be transmitted in each line. The orders were isolated by an appropriate red filter for Hα and by the use of a red-insensitive emulsion for K. Spectroheliograms were obtained, as usual, by scanning the solar image over the entrance slit and driving synchronously two cameras mounted on the same rigid frame, one for the pair of Hα spectroheliograms and one for the K. To avoid overlap, the width of the spectroheliograms had to be limited to less than the separation between the closer of the two pairs of slits.

With this arrangement it is possible to get fairly-well-balanced densities between the two spectroheliograms in the one line, but residual differences in the widths of the fixed slits do result inevitably in small, harmless, residual density differences. To ensure adequate densities a compromise was needed. The slit widths were as great as we dared use (exit slits 0.25 mm, corresponding to 36 mÅ at Hα and 22 mÅ at K, entrance slit 0.30 mm); and a new frame commenced every 20 s, resulting in good exposures in Hα and the brighter K features, but the big intensity variations across the solar surface in K_2 left the fainter features underexposed.

The wavelengths used in Hα were also based on a compromise giving an acceptably large separation, and hence width of spectroheliograms, but close enough to line centre to relate to the higher levels of the chromosphere. The fibrils are still evident at Hα $\pm$0.29 Å, as are the chromospheric granules; the main differences between images at line centre and at these wavelengths are due to Doppler shifts and to a relative reduction in intensity of the bright network features. In consequence we chose Hα $\pm$ 0.29 Å for this study. In K we used K $\pm$ 0.18 Å, which lie fairly close to the K_2 maxima.

A 42 min sequence was obtained on 15 May 1975, with seeing of good quality throughout. Auxiliary observations included white-light and Hα line centre (0.5 A pass-band) slit-jaw images simultaneous with the start of each scan.

The analysis has been qualitative. The velocities are shown by photographic subtractions of the spectroheliograms in one wing from those in the other, both for Hα and K. These were then placed side-by-side on the one frame, and a time-lapse film produced showing the two together. From this we can see whether similar motions are revealed in the two lines.

We have also used blink comparison of the original frames, subtractions, and photographic additions of the images in opposite wings which enable the individual fine-structures to be seen better without the disturbing influence of Doppler shifts.

The region studied is not far from disk centre, and appears to be typical of any quiet part of the disk. The system of fibrils at A (Figure 1a, b) is visible further from the network in Hα ± 0.29 Å than in K ± 0.18 Å. Blink comparison shows that the same individuals are present in the two lines, but the relative intensities vary irregularly from fibril to fibril and along their lengths. With occasional exceptions (e.g. B), fibrils brighter than average in Hα are brighter too in K; and similarly for structures darker than average. The shorter apparent lengths of the dark K fibrils may possible be due merely to inadequate exposure; for the bright fibrils, it seems more likely that they are due mainly to lower opacity in K than in Hα at the wavelengths used. Whatever the reason, there is a zone where fibrils obscure lower structures, and this is more extensive in Hα than in K.

Figure 1 also shows side-by-side the subtractions in Hα (1c) and K (1d). Where the fibrils appear in both lines, the asymmetries have identical shapes and signs, differing only in magnitude between the two lines.

In cine projection we have been able to distinguish two types of behaviour. (i) Very close to the network (C, Figure 1), there may be a slow propagation of asymmetries at velocities of the order of 12 km s^{-1} either towards or away from the network. (ii) Further away, there is a much faster propagation mainly away from the network at velocities of the order of 75 km s^{-1}. Both these types of behaviour occur in both lines, though the contrast in K is much lower in (ii), where the fibrils are also much fainter than in Hα. Giovanelli (1975) has already described the phenomenon (ii) in Hα, interpreting it as a propagation of Alfvén waves mainly outwards from the network. We can now confirm that it is observed in K. The same phenomenon is thus found in the only two types of line in which fibrils are visible.

It is not successful to present still photographs of velocitygrams showing the propagation of Alfvén waves in fibrils; this can be done adequately only with cinefilms (a film of this type was shown at the Symposium).

Fig. 1. (a, b) Spectroheliograms derived by adding (a) the Hα ± 0.29 Å and (b) the K_{2R} and K_{2V} images. (c, d) Corresponding subtractions in (c) Hα and (d) K; features darker than average are receding.

4. DISCUSSION

Where the fast waves occur, the fibrils and magnetic field are mainly horizontal, and the observed displacements are transverse. We have no doubt that the waves involved are of the Alfvén type. However these observations add nothing quantitatively to our knowledge of the amplitudes involved or the power carried, which remain as described by Giovanelli (1975).

The slower (12 km s^{-1}) propagation of asymmetries, either towards or away from the network when very close to it, seems to be explained most readily by changes in the well-known longitudinal oscillations inside magnetic elements and tubes of force. Since the fibrils are nearly vertical close to the network, motions along them have strong line-of-sight components when observed near disk-centre. On the contrary, transverse motions have only small line-of-sight components, so that the displacement velocities in an Alfvén wave very close to the network footpoint are of reduced amplitude.

ACKNOWLEDGEMENTS

The authors acknowledge gratefully the help of the staff of the Sacramento Peak Observatory in the acquisition of the observations, and of the photographic laboratory of the then CSIRO National Measurement Laboratory (now Division of Applied Physics) in their analysis.

REFERENCES

Beckers, J.M.: 1962, Aust. J. Phys. 15, p. 327.
Giovanelli, R.G.: 1975, Solar Phys. 44, p. 299.
McMath, R.: 1939, Mon. Not. R. Astr. Soc. 99, p. 559.
Sawyer, C.: 1974, Solar Phys. 35, p. 63.

DISCUSSION

SPRUIT: What is the basis for your conclusion that transversal tube waves cannot be excited by granulation?

GIOVANELLI: This comes from considering the effects of hydrodynamic forces of granule motions in displacing fluxtubes. In turn this depends on the depth of a granule. I have estimated the corresponding increase in magnetic energy in such a tube, supposed to be vertical where the granule horizontal motion reverses. This gives the equilibrium displacement δ at the surface for a given granule and tube model. δ depends on whether the internal gas pressure is small in comparison with the magnetic pressure, as expected. If this is so, Spruit's 1974 convection zone model and a typical tube diameter at the surface imply that δ is $\lesssim$ 25 km even for a depth of 2000 km to the level where the horizontal motion reverses. For bigger surface diameters, δ is even less. Unless the granule depth is much greater or the model data are inappropriate, granulation would seem unable to generate the transverse velocities required. The situation could be different if the tube diameter is effectively constant with depth, but this seems unlikely.

MOUSCHOVIAS: (1) Do you observe any mass transported in the direction in which the waves propagate? (2) Can you put any limits on the mass transport?

GIOVANELLI: (1) I do not know of any investigation aimed specifically at measuring gas flow along effectively horizontal fibrils. (2) Not from observations. I would expect any systematic flow towards or away from the network to be small in the fibrils. On the other hand, I would expect large-amplitude longitudinal waves to be present.

IONSON: Have you determined the spectral power function for these "Alfvén waves"?

GIOVANELLI: No. I suspect that it would be extremely difficult.

ROBERTS: I was very interested in your comment that the observed waves could be connected with the magnetic canopy. Could it be that you are seeing surface waves propagating along the canopy, i.e. along an interface between a magnetic field and a field-free atmosphere? At such an interface, a surface wave with phase speed of the order of the sound speed is possible. But this is probably too slow to agree with the speeds you quote.

GIOVANELLI: The Hα fibrils are higher than the canopy base, and it seems unlikely that the waves observed are purely surface waves. The observed speeds, typically 70 km s^{-1}, are much greater than the sound speed, which is unlikely to reach 10 km s^{-1}.

HEATING OF INTENSE MAGNETIC FLUX TUBES BY MAGNETOHYDRODYNAMIC WAVES

Z. Musielak and E. Bielicz
Institute of Theoretical Physics and Astrophysics
University of Gdansk, Poland

ABSTRACT

Theoretical models of intense magnetic flux tubes embedded in the photospheres of late-type stars have been calculated. Magnetohydrodynamic waves generated in the convective zone are radiatively damped in the lower part of the flux tube and then dissipated. We discuss how the presence of flux tubes in stellar photospheres influences the temperature minimum and the chromospheric activity. We suggest a possible interpretation of the Wilson-Bappu effect in stars with strong chromospheric activity.

1. INTRODUCTION

It is generally accepted that most of the features of solar activity are connected with the existence of magnetic fields. In solar active regions the magnetic field is concentrated in discrete elements which appear as bright small elements with field strengths above 1000 G. These elements may be described by a set of magnetostatic flux tube models (Zwaan, 1977). The intense magnetic flux tubes have been considered recently by Webb and Roberts (1980a, b) and Spruit (1981a, b). In all these studies the propagation of a longitudinal and a transverse wave mode along the flux tube was discussed.

The aim of the present paper is to use theoretical models of intense magnetic flux tubes to show how the temperature minimum region and the lower part of the chromosphere are formed. Besides, we try to explain the existence of strong chromospheric activity in some late-type stars.

2. ENERGY BALANCE IN MAGNETIC FLUX TUBES

The magnetostatic models of flux tubes describe the magnetic elements as structures in equilibrium with non-magnetic surroundings. We will assume that the temperature inside the tube is equal to the exter-

J. O. Stenflo (ed.), Solar and Magnetic Fields: Origins and Coronal Effects, 413–416.

nal temperature at each depth. Further, we choose at the bottom of the flux tube a value of the magnetic field which makes the gas pressure in the tube equal to 30% of the total pressure.

Radiative damping of magnetohydrodynamic waves in the flux tube is treated in the way proposed by Webb and Roberts (1980a, b). However, to calculate the Alfvén and slow mode dissipation we have used a modified version of the mechanism of Pikelner and Lifshitz (1964) (cf. Musielak and Sikorski, 1980). The energy dissipated at each layer is balanced by radiative losses (Musielak, 1982).

3. RESULTS AND DISCUSSION

Our calculations show that magnetohydrodynamic waves can propagate within the magnetic flux tubes. These waves carry a large amount of energy, which causes the temperature minimum to be higher than in the surroundings and the temperature gradient to be steeper in the chromosphere. The later the spectral type, the steeper the temperature gradient (Fig. 1). This leads to increased chromospheric activity above the flux tubes in these stars. There are observational and theoretical suggestions that a large number of flux tubes can be produced in late-type stars (Linsky, 1980; Zwaan, 1980), which increases the activity area factor. The mechanisms mentioned above can explain the existence of late-type stars with particularly strong chromospheric activity.

The effect of appearance of a great number of magnetic flux tubes in active stars cause the "average" T_{min} and "average" magnetic field to be higher for active than for non-active stars (Bielicz and Musielak, 1982).

Our calculations of models of flux tubes embedded in the photospheres of late-type stars (EQ Vir) with strong chromospheric activity, suggest an increase of $\pi F_M^{\,o}$ (mechanical energy flux generated in the convection zone) compared to the corresponding quantity given by Bohn (1981). It is generally accepted that an additional mechanical flux can be produced when a magnetic field is present (e.g. Ulmschneider and Stein, 1982). However, we conclude that the generation of magnetohydrodynamic waves in the convective zone in the presence of strong magnetic fields may be less effective than has been assumed (Musielak and Bielicz, 1982).

Knowing the theoretical values of T_{min} for flux tubes embedded in photospheres of main sequence stars and giants, we can explain the Wilson-Bappu effect corrected by Glebocki and Stawikowski (1978) for stars with strong chromospheric activity. The intensity of the K reversal (I_K) depends mainly on the activity level of the chromosphere model, i.e., on the relative area of the active region in comparison with the whole surface of the star. Accordingly, I_K is proportional to the number

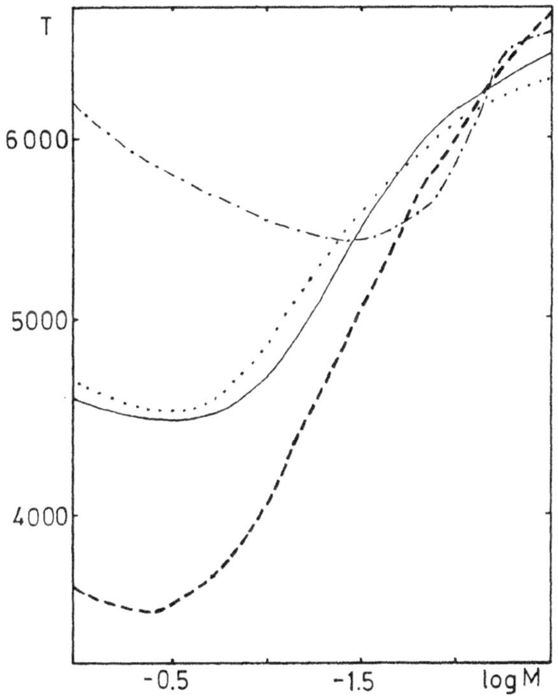

Fig. 1. Temperature as a function of the logarithm of the mass column density for flux tubes embedded in the photospheres of γVirN (dashed-dotted line), Sun (dotted line), and EQ Vir (dashed line). The models were obtained for a value of $\pi F_M{}^o = 1.0 \times 10^8$ erg $cm^{-2}s^{-1}$ and for the magnetic fields 600 G, 2000 G, and 2300 G for γVirN, Sun, and EQ Vir, respectively. The solid line gives the solar empirical model for very bright network elements.

of flux tubes. Further, according to the method proposed by Ulmschneider et al. (1979) to tentatively explain the Wilson-Bappu effect, we can give a quantitative interpretation of the width W_o as an "average" property of the discrete magnetic flux tubes.

REFERENCES

Bielicz, E. and Musielak, Z.: 1982, Acta Astr. 32 (in press).
Bohn, H.U.: 1981, Ph.D. thesis, University of Würzburg.
Glebocki, R. and Stawikowski, A.: 1978, Astron. Astrophys. 68, pp. 69-74.
Linsky, J.L.: 1980, "Solar Phenomena in Stars and Stellar Systems" ed. R.M. Bonnet and A.K. Dupree (Reidel) pp. 99-123.
Musielak, Z.: 1982, Astron. Astrophys. 105, pp. 23-36.

Musielak, Z. and Bielicz, E.: 1982, Acta Astr. 32 (in press).
Musielak, Z. and Sikorski, J.: 1980, Acta Astr. 30, pp. 167-182.
Pikelner, S.B. and Lifshitz, M.A.: 1964, Astron. Zhurnal 41, pp. 1007-1021.
Spruit, H.C.: 1981a, Astron. Astrophys. 98, pp. 155-160.
Spruit, H.C.: 1981b, Astron. Astrophys. 102, pp. 129-133.
Ulmschneider, P.: 1980, "Solar Phenomena in Stars and Stellar Systems" ed. R.M. Bonnet and A.K. Dupree (Reidel) pp. 239-263.
Ulmschneider, P., Schmitz, F., and Hammer, R.: 1979, Astron. Astrophys. 74, pp. 229-234.
Ulmschneider, P. and Stein, R.F.: 1982, preprint.
Webb, A.R. and Roberts, B.: 1980a, Solar Phys. 68, pp. 71-85.
Webb, A.R. and Roberts, B.: 1980b, Solar Phys. 68, pp. 87-102.
Zwaan, C.: 1977, Memorie della Societa Astron. Italiana 48, pp. 525-557.
Zwaan, C.: 1981, "Solar Phenomena in Stars and Stellar Systems" ed. R.M. Bonnet and A.K. Dupree (Reidel) pp. 463-486.

V. STELLAR WINDS AND SPINDOWN IN LATE – TYPE STARS

STELLAR WINDS AND SPINDOWN - OBSERVATIONS

L. Hartmann
Harvard-Smithsonian Center for Astrophysics
60 Garden Street
Cambridge, MA 02138, U.S.A.

1. INTRODUCTION

Stars with masses $\lesssim 1\ M_{\odot}$ are observed to rotate more slowly as they age. The angular momentum loss is undoubtedly caused by the coupling of the stellar magnetic field to the escaping wind (Schatzman 1962). Chromospheric and coronal radiative losses depend upon rotation (Wilson 1966a,b; Kraft 1967; Skumanich 1972; Hall 1976; Bopp 1980; Walter and Bowyer 1981; Walter 1981; Vaiana et al. 1981). It is therefore likely that both magnetic fields (Skumanich 1972) and the mechanical energy fluxes required to drive mass loss also depend upon rotation as well. This complicated feedback between magnetic fields, winds, and rotation must control the variation of solar-type activity over much of the HR diagram, and may have very important effects on pre-main sequence evolution.

Unfortunately, we have almost no direct observations of magnetic fields in late-type stars. In addition, there are very few stars for which we can observe both winds and the effects of spindown. The only exceptions are the pre-main sequence T Tauri stars, whose positions in the HR diagram and interior structures are not well known. In general the observational situation is distressingly unconstrained, although there are several new clues which constrain spindown and the connection between magnetic fields and mass loss.

2. STELLAR WINDS

A. Evolved Stars

It is much easier to detect cool stellar winds than hot, solar-type flows. The reason is that in the absence of spatial resolution, mass loss must observed from doppler shifts in line profiles, and the spectral signatures of hot winds mostly occur at inaccessible wavelengths. So any discussion of stellar mass loss must begin with the study of cool winds and circumstellar shells.

J. O. Stenflo (ed.), Solar and Magnetic Fields: Origins and Coronal Effects, 419–438.

Mass loss is detected in the spectra of evolved, late-type giants in the form of circumstellar lines of low-excitation ions like Mg II, Ca II, Na I, and K I. As indicated in Figure 1, cool winds are observed only in the upper-right hand corner of the HR diagram. Circumstellar shell lines are generally not observed in stars with log g >1.5. Stars of higher surface gravity exhibit transition-region and coronal emission patterns similar to the Sun's (Linsky and Haisch 1979; Dupree and Hartmann 1980; Vaiana et al. 1981; Hartmann, Dupree, and Raymond 1982), and so presumably have solar-type winds.

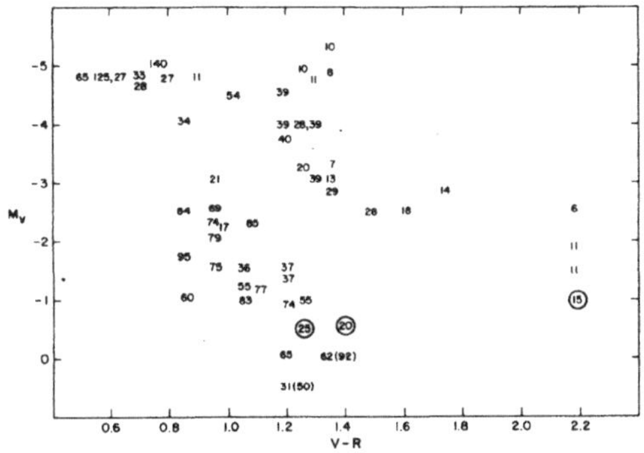

Figure 1. Domain of cool stellar winds in the HR diagram, with terminal velocities in km s^{-1}. Adapted from Reimers (1977).

Recent studies suggest that there may be a closer relation between cold flows and solar-type winds than previously thought. One line of evidence for such a connection comes from the observations of Ca II and Mg II line profiles by Stencel and Mullan (1980). They showed that red giant stars often exhibit an asymmetry in which the blue emission peaks of these chromospheric lines are weaker than the red peaks. This asymmetry is seen with increasing frequency in stars of lower gravity. Although the line profiles cannot be interpreted in a unique way, the sense of this asymmetry is consistent with what is expected from differential expansion due to mass loss.

Progressing towards the upper righthand corner of the HR diagram, the change in asymmetry is first seen in Mg II, then Ca II, followed by the appearance of true circumstellar shell lines (Figure 2). This behavior is exactly what one would expect if mass loss rates are smoothly increasing, rather than a discontinuous jump due to the turn-on of a new mass loss mechanism. Mass loss can be seen first in the Mg II resonance lines, since they have an order of magnitude more opacity per gram than the Ca II lines.

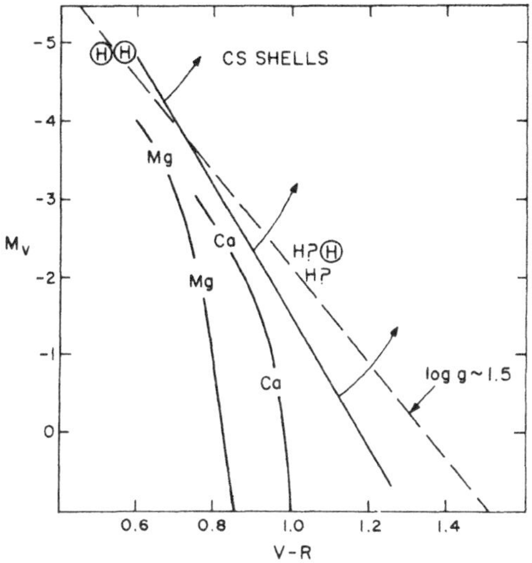

Figure 2. Hybrid stars (H symbols) are observed near the CS boundary in the HR diagram. Stars to the right of the Mg II and Ca boundaries tend to exhibit asymmetries indicative of mass loss in the respective resonance lines.

Another reason for suspecting a connection between hot and cool winds is the observation of stars with both hot outer atmospheres and cool winds (Hartmann, Dupree, and Raymond 1980,1981). These objects, called hybrid atmosphere stars, show C IV and N V emission and have UV line ratios very similar to the Sun's. They also exhibit circumstellar absorption features in Mg II and Ca II, demonstrating that solar-type activity is not incompatible with cool, low-velocity winds. The hybrid stars have positions in the HR diagram near the boundary of CS lines and the limit of transition-region emission (Figure 2). This property also seems to suggest a continuous progression in atmospheric properties, in agreement with the asymmetry data.

A more subtle reason for suspecting a connection between cool winds and solar- type flows concerns the energy balance of the extended atmospheres of red giants. Several investigations have led to the conclusion that "extended chromospheres" surround late-type stars in which transition-region and coronal emission is absent (cf. Goldberg 1979). The extended chromospheres are regions of temperatures $\sim 10^4$ K with scale heights far in excess of what would be expected from thermal pressure balance.

Stencel et al. (1981) used the IUE satellite to observe the density sensitive C II lines near 2325 A in a number of evolved stars. The line ratios indicate electron densities of order 10^8 cm^{-3} in the line-forming region for K giants like α Boo and α Tau. Combining the density with the observed flux, the emitting region is calculated to have a path length comparable to the stellar radius.

Examination of spectral features, such as the Hα and Ca II infrared triplet lines, also support the concept of extended chromospheres in low- temperature winds. For example, these lines exhibit blue-shifted cores in supergiant stars like α Ori (Figure 3), suggesting that at least part of the chromosphere is being formed in an expanding region. The photosphere of α Ori undergoes small, irregular pulsations, which the Hα core does not share (Goldberg 1979). This means that Hα is formed in a wind region physically distinct from the photosphere, probably extended over several stellar radii. Occultation techniques and speckle interferometry directly demonstrate the existence of the extended Hα zone (White, Kreidl, and Goldberg 1982; Goldberg et al. 1982).

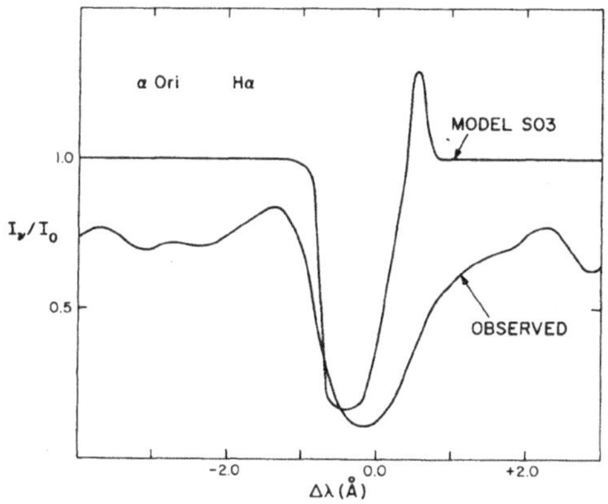

Figure 3. Observed Hα profile for α Ori (Goldberg, private communication). A wave-driven wind model has been calculated for comparison purposes (Hartmann, MacGregor, and Avrett 1982) (see section 2B).

The ionization of the extended chromosphere and the emission in excited- state lines such as Hα require some source of mechanical heating. It is difficult to make an accurate estimate of the required energy flux. The best present values are $\sim 10^4$-10^5 erg cm^{-2} s^{-1} emanating from the surface of the star (Hartmann and MacGregor 1980; Hartmann, MacGregor, and Avrett 1982). In my view this result may also indicate a connection between cool winds and solar activity, because the energy fluxes required by the solar wind are also of this order of magnitude (Withbroe and Noyes 1977). There is also evidence that energy and/or momentum is being deposited in the solar wind out to several $R_\odot$ (Munro and Jackson 1977).

If there are related mechanisms operating in both hot and cold stellar winds, why are the temperatures so different? Theoretical calculations show that the high densities and long flow times characteristic of the winds of low-gravity stars promote radiative cooling (Hartmann and MacGregor 1980; Hartmann, MacGregor, and Avrett 1982). In contrast, the solar wind does not radiate efficiently (Withbroe and Noyes 1977), so that heat energy being supplied to the gas goes into potential or kinetic energy.

It has been suggested that some features of the solar wind, particularly high-speed streams, cannot be explained by thermal acceleration alone, but require some direct deposition of momentum (Belcher 1971; Hollweg 1973; Jacques 1978). This suggestion leads to the idea of a natural sequence of winds which connects solar-type activity with circumstellar shells. In this view the winds of high-gravity stars are propelled by both momentum deposition and heating. In winds from low-gravity stars, the energy dissipated into heat is quickly radiated away before it can play a major role in wind acceleration. The mass loss observed is then driven primarily by the momentum deposition mechanism. This picture qualitatively accounts for the gradual increase in Ca II and Mg II asymmetries, the existence of hybrid stars, and the occurrence of extended chromospheres.

What is the momentum deposition mechanism? Alfven waves have been observed in the solar wind (Belcher and Davis 1971), and theoretical calculations show that wave fluxes $\sim 10^5$ erg cm^{-2} s^{-1} can account for high-speed streams (Hollweg 1973; Jacques 1978). Although magnetic fields or Alfvenic fluctuations on stars are not directly observable, it is possible that line broadening due to wave motions might be detected. There are indications of large "turbulent" motions in the envelopes of low-gravity stars. For instance, it is difficult to obtain the very large widths seen in the $H\alpha$ profile of a star like α Ori (cf. Figure 3) without invoking some large (supersonic) turbulent velocity.

A remarkable and unexpected example of the large motions seen in low-gravity atmospheres is the discovery of broad transition-region lines in some stars. The optically-thin Si III 1892A and C III 1909A emission lines in the hybrid atmosphere star α TrA (K4II) exhibit velocity widths of 100 km s^{-1} (Hartmann, Dupree, and Raymond 1981). Such supersonic motions clearly indicate that hydrostatic equilibrium cannot be assumed. Somewhat smaller broadening velocities have been observed in the G2II star β Dra and in an active RS Cvn star (Ayres et al. 1982). These observational results make the idea that propagating waves are being excited more plausible.

B. Wave-driven Winds

In order to see what the physics of a wind driven by Alfven waves is like, consider the simplest possible situation, in which the flow is spherically symmetric and the magnetic field lines are all radial. With these assumptions, the introduction of pure Alfven modes results in an equation of motion of the form

$$v\frac{dv}{dr} + \frac{1}{\rho}\frac{d}{dr}\left(P + \frac{\langle\delta B^2\rangle}{8\pi}\right) = -\frac{GM}{r^2} \tag{1}$$

(Hartmann and MacGregor 1980), where v, ρ, and P are the gas velocity, density, and pressure, respectively, and δB is the magnetic field fluctuation. The structure of eq. (1) suggests that the effect of the turbulent magnetic pressure is analogous to the role played by the gas pressure in a thermally-driven wind.

The steady flow solution passes through a critical point, which determines the mass loss rate for given initial values of the magnetic field, wave flux, and gas density. The analogy with thermal-pressure driven winds makes the critical point conditions transparent. For example, the numerator = 0 condition at the critical point of a thermally driven (isothermal) wind is

$$\rho a^2 = \frac{GM\rho}{2r}$$

where a is the isothermal sound speed. The corresponding relation for a low-temperature wind driven by undamped Alfven waves is

$$\delta B^2/16\pi \sim GM\rho/2r \quad . \tag{2}$$

Similarly, the denominator = 0 equation for the thermally-driven wind, $\rho v^2 = \rho a^2$, becomes

$$\rho v^2 \sim \delta B^2/32\pi \tag{3}$$

for wave-driven winds.

Combining (2) and (3) with the energy flux dependence for undamped waves (which turns out to be $F_w \propto r^{-2}$ for low Alfvenic mach number), and assuming that the critical point distance scales with the stellar radius, one can explain the detailed wind calculations with the approximate scaling law

$$\dot{M} \sim 10^{-25}\ M^{-1.5}\ R^{3.5}\ (F_5/B_o)^2\ M_\odot\ yr^{-1} \quad , \tag{4}$$

where M and R are the stellar mass and radius in solar units, F_5 is the initial wave flux measured in units of 10^5 erg cm^{-2} s^{-1}, and B_o is the magnetic field strength in gauss. The difficulty in evaluating the applicability of eq. (4) is that mass loss rates are poorly known (Goldberg 1979) and magnetic field strengths are completely unknown. However, it is suggestive that the order of magnitude mass loss estimates (Bernat 1977; Reimers 1977; Hagen 1978) can be reproduced by solar-type surface conditions, with $F_5 \sim 1$ and $B_o \sim$ gauss.

The wave fluxes possible are obviously constrained by the magnetic field strengths. One can show that plausible limits on the initial density leads to the approximate scaling (Hartmann, MacGregor and Avrett 1982)

$$\dot{M} \sim 10^{-14} M^{-0.5} R^{2.5} B_o^{2} M_\odot yr^{-1}. \quad (5)$$

If Alfven waves propagate without dissipation, high wind velocities result (Hartmann and MacGregor 1980). This is fine for producing high-speed streams, but terrible for matching the low terminal velocities observed for cool winds (Figure 1). Hartmann and MacGregor (1980) showed that the Alfven waves must dissipate on length scales of a stellar radius in order to drive cool mass loss with acceptably small terminal velocities. Although there are several possible ways in which the waves can damp, there is no clear way at present to predict the detailed manner of dissipation. However, if one parameterizes the dissipation, it is clear that substantial gas heating must occur. If the radiative cooling rate is known, the balance of heating and cooling permit the calculation of the wind temperature structure, which can be compared with observation.

Radiative transfer calculations show that the wave heating predicted by this parameterized dissipation results in extended chromospheres which are in qualitative agreement with the previously mentioned empirical studies (Hartmann, MacGregor, and Avrett 1982). Specifically, the wave dissipation explains: (1) the C II density diagnostic line ratios observed in K giants; (2) free-free radio emission from α Ori (Altenhoff, Oster, and Wendker 1979; Bowers and Kundu 1979; Newell and Hjellming 1982); and (3) extended Hα emission surrounding α Ori (Figure 2). The quantitative agreement of theory with observation is reasonable, considering the simplifications introduced to make the calculations tractable.

Since most of the wave energy introduced into the cool wind is radiated away, the wave flux needed to produce a given mass loss rate can be empirically determined from an estimate of the radiative losses of the extended chromosphere. Present values for these losses are obviously consistent with the (poorly-determined) observed mass loss rates. In addition, Hartmann, MacGregor, and Avrett (1982) have shown that the limit of the cool wind region in the HR diagram can be explained by assuming $B_o \sim 3$ gauss and $F_5 \sim 3$.

These considerations show that there are many attractive features of wave- driven winds which suggest application to cool low-velocity flows. Although there are many observational uncertainties, and the simplifications of present theory may not be very realistic, the convergence of theory and observation to emphasize the importance of extended chromospheres is encouraging. We may learn a great deal about momentum deposition in the solar wind, which is difficult to distinguish from heating (cf. Munro and Jackson 1977), by studying the cool winds of late-type stars.

C. Pre-Main Sequence Stars

At first glance, the Hα profile of a T Tauri star displayed in Figure 4 has little in common with the Hα line of α Ori (Figure 3). Despite the obvious differences, there are two underlying aspects in common. The asymmetry of the Hα emission in DF Tau is one indication that mass loss is being driven off at temperatures which are too low for thermal acceleration to be important, as is the case for α Ori. A second point of contact is that there is evidence for large turbulent velocities in both T Tauri variables and evolved stars. P Cygni-type profiles in which the blue-shifted absorption reversal is at a smaller velocity than the emission component generally require some sources of line broadening larger than the local expansion velocities. Indeed, many of the T Tauri stars show much more symmetric line profiles, with hardly a hint of mass loss (De Campli 1981; Hartmann 1982).

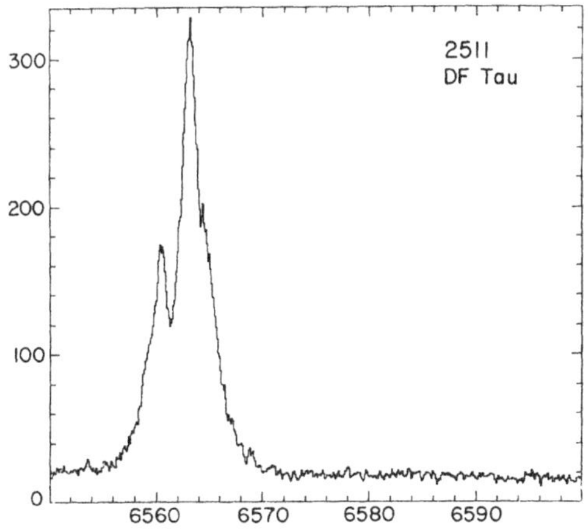

Figure 4. Hα line profile of the T Tauri star DF Tau (Hartmann 1982).

Recognition of the need for a momentum source which drives T Tauri winds led De Campli (1981) to investigate Alfven wave-driven wind models. He concluded that the mass loss rates must be $\lesssim 10^{-8}$ $M_\odot$ yr^{-1} if one makes the plausible assumption that wave energy fluxes are less than the stellar luminosity. Although this mass loss limit was in conflict with previous estimates, De Campli showed that the introduction of large turbulent velocities could bring the observational mass loss estimates into better agreement with the theory.

Subsequently Hartmann, Edwards, and Avrett (1982) showed that the very existence of the waves driving the flow demanded large "turbulent" velocities. The critical point conditions include the intuitively reasonable constraint (cf. eq. (2)) that the wave force is comparable to gravity at this point. Since $\delta B/B = \delta v/A$ for an Alfven wave, where A is the Alfven speed, eq. (2) is equivalent to

$$\delta v_c^2 \sim GM/r_c \tag{6}$$

Flow speeds are typically less than the escape velocity within the critical point. Therfore wave motions generally dominate expansion in the dense regions of the wind, where most of the emission is expected to originate. The wave motions have the effect of producing much larger equivalent widths for the optically-thick Balmer lines at a given mass loss rate than would be calculated in the absence of wave broadening.

The terminal velocities of T Tauri stars are not well known, and so there is no requirement for wave damping as there is for giant and supergiant winds. However, there are very large emission-line fluxes for which some source of heating must be supplied. Adopting the same parameterization of wave dissipation for T Tauri stars as for evolved stars, Hartmann, Edwards, and Avrett (1982) showed that the strongest observed emission can be accounted for with mass loss rates $\sim 10^{-8}$ $M_\odot$ yr^{-1}, and wave fluxes $\sim 10^{-1}$ of the stellar luminosity, requiring $B_o \sim$ 200 gauss.

D. Main Sequence Stars

Unfortunately, we have really no observational information concerning the winds of late-type main-sequence stars. The only thing one can do at present is to make crude estimates based on comparison with the Sun.

Alternatively, one might consider the effects of momentum deposition by waves. It is interesting to note that eq. (7) reduces to the solar mass loss rate for field strengths $\sim$ gauss; this is consistent with suggestions that Alfven waves provide significant acceleration of the solar wind (cf. Hollweg 1973).

3. STELLAR SPINDOWN AND WINDS

A. Main Sequence

Little progress on determining stellar rotational velocities was made for many years after Kraft's (1967) work. Although improvements in spectroscopic observational techniques have led to more accurate measurements of low rotational velocities (Smith 1979; Soderblom 1980), a real breakthrough came from the realization that the rotational modulation of Ca II H and K emission can provide direct measurements of the rotational periods (Vaughan et al. 1981).

Although the objects observed in this survey are all field stars, I have attempted to place them in age groups according to their Ca II emission (S index). In particular, I have identified objects with emission levels that would place them within observed limits for the Hyades and for old field stars (cf. Wilson 1970; Vaughan and Preston 1980). One cannot be sure that objects with strong emission are all of Hyades age, but comparison with v sin i's measured for some Hyades members by Soderblom (1980) and Baliunas (1982) shows good agreement.

The combined data are shown in Figure 5, in which I have placed the new observations in the context of previous studies of earlier spectral types. The data give an impression of a slight decline in v_{rot} with decreasing effective temperature for stars of Hyades S index. Similarly, there is an indication of a modest decline in v_{rot} for G-K dwarfs of advanced age, although again it is difficult to be precise about this since the ages of the weak-emission field stars are not well known. In any event, the data suggest that rotational braking timescales are roughly similar in G-K dwarfs.

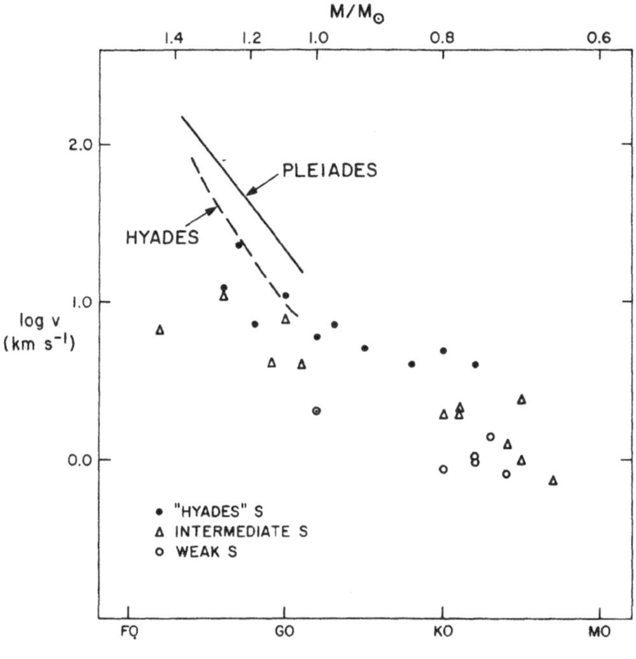

Figure 5. Main sequence rotational periods vs. spectral type.

The main sequence stars studied here clearly fall below the total angular momentum law $J \propto M^{0.57}$ suggested for earlier spectral types (Kraft 1970). For stars of Hyades emission, the data suggest $J \propto M^4$, although the exponent must be considered rather uncertain due to small number statistics and a small baseline in stellar mass M. This exponent is similar to the value derived by Kraft (1967) for the range $1.2 \lesssim M/M_\odot \sim 1.4$.

What can we infer about stellar winds and their braking effects from Figure 5? If the Skumanich law in the form $v_{rot} = v_o t^{-1/2}$ strictly held everywhere, the rotational velocities of all very old stars should approach the same value. The observational evidence seems to suggest that K dwarfs may have the same braking timescales as G dwarfs, but at lower equatorial velocities. If further studies bear out this result, the implication would be that angular momentum loss rates at a given equatorial velocity or rotational period remain relatively constant while the stellar moment of inertia decreases with decreasing mass ($\sim m^3$). What this means in terms of the dependence of $\dot{M}$ and B on rotation and spectral type is unclear, given the absence of any observational limits on mass loss rates.

Durney and Stenflo (1972) pointed out that the Skumanich law can be understood on the simple Weber-Davis model for stellar spindown if mass loss rates are constant, the mean magnetic field B_o scales as v_{rot}, and the Alfven radius r_A scales as B_o. It is amusing to point out that the Durney-Stenflo result also holds for the case in which the mass loss rate varies as B_o^2 (cf. eq. (7)). In this case the Alfven radius will remain relatively constant if wind velocity structure does not change dramatically; the stars would tend to remain more in the slow-rotator domain rather than becoming fast magnetic rotators (Belcher and MacGregor 1976).

There are other indirect means of studying main-sequence stellar spindown, involving close binaries. It has been suggested that angular momentum loss from a stellar wind can assist in forming contact binaries of the W UMa type by bringing the component stars together from an initially detached state (Huang, 1966; Okamoto and Sato 1970; Van't Veer 1979; Vilhu 1982). Similarly, it has been suggested that magnetic braking of the dwarf component of cataclysmic variables is important in driving mass transfer (Eggleton 1976; Verbunt and Zwaan 1981). The aforementioned authors have shown that a Skumanich-type braking law leads to the correct order-of-magnitude angular momentum loss required. In principle it might be possible to estimate spindown properties by these studies, but it seems more likely that observations of the braking of single stars will be more definitive, and more useful in analyses in the opposite direction, i.e. using observed angular momentum loss rates to predict cataclysmic variable accretion rates in more detail. Clearly it is of importance to extend our knowledge of stellar rotation to smaller masses (later spectral types) for this particular purpose.

B. Origin of the Sharp Break at F5

It has been clear for some time that the rapid break in main-sequence rotation near F5 is the result of a decrease in the efficiency of stellar wind braking in early spectral types. Since chromospheric emission disappears in this region, and convective envelopes shrink rapidly in proceeding to hotter stars, the natural conclusion is that dynamo-generated magnetic activity also disappears, and so no magnetic braking is expected (Schatzman 1962; Wilson 1966a,b; Kraft 1967). However, it is difficult to detect Ca II and Mg II emission at reasonable levels in early spectral types due to the increase in photospheric background. Therefore the disappearance of chromospheric emission cannot be regarded as a definitive measure of magnetic activity.

In this regard the observations of X-rays from A and F stars were somewhat of a surprise (Vaiana et al. 1981). Although it is possible that late-type companions account for some of the observed X-ray emission, high-resolution images of Sirius and Vega make it likely that these stars are actual (albeit weak) sources of X-ray emission. We have no direct evidence for magnetic fields in these stars, but is plausible to suppose that they may result from some analog of solar coronal emission.

Skumanich (1981) pointed out that the presence of X-ray emission might indicate significant magnetic fields, despite the observed lack of spindown. He suggested that one way out of this difficulty would be to assume a general change in magnetic topology, in which all the field lines close near the surface and therefore do not couple to the wind. However, such an extreme change in field structure may not be necessary. A and F stars appear to deviate from the general relationship between rotation and X-ray luminosity, in the sense that they have relatively weak emission for their rapid rates of spin (Pallavicini et al. 1981). Noyes (this volume) has discussed evidence from the Mt. Wilson survey that chromospheric emission in F stars is weaker than in later spectral types of comparable rotational periods. This suggests that the magnetic and mass loss activity of these stars may be more representative of more slowly rotating stars.

The median X-ray luminosity of the A-F stars is $\sim 10^{29}$ erg s^{-1}, comparable to the mean luminosity of G dwarfs in the Hyades (Stern et al. 1981). Suppose this means that the magnetic field strengths and mass loss rates, and therefore angular momentum loss rates, are the same for the A-F field stars and the Hyades G dwarfs. The latter have a braking time of order 10^9 years. From this it is clear that the rapidly rotating early F dwarfs would suffer very little spindown with the same angular momentum loss rate over their main-sequence lifetimes $\sim 10^9$ yrs. It appears that a change in the dependence of magnetic field and energy generation rates on rotation, rather than a complete cessation of mass loss and magnetic activity, can be compatible with the observations of both X-ray emission and rapid rotation in the A-F stars.

C. Rotation of Evolved Stars

In general there are fewer data concerning the rotation of red giants than dwarfs, because of (generally) smaller equatorial velocities. In addition, the internal structural changes involved in evolution off the main sequence make the connection between surface rotation and angular momentum loss more problematical.

Gray (1981,1982) has suggested that there is a sharp reduction in rotation near G5 III, and that these stars slow down from about 25 km s^{-1} to about 5 km s^{-1} in only a few spectral subclasses. Gray (1982; see also Gray and Endal 1982) find that redistribution of angular momentum within the envelope is insufficient to account for the reduction in equatorial velocity. Instead, Gray (1982) suggests that rapid braking occurs at this point, due to the increase of depth of the convective zone. This generates a very strong dynamo-generated magnetic field, which in turn couples to the stellar wind, and reduces the stellar rotation very rapidly.

There are problems with this scenario. Certainly stars from types F5 III to G5 III have chromospheric emission, which on the basis of solar studies suggests magnetic fields are already present. Thus, a subtle change in stellar structure must result in a large increase in the magnetic field. On the basis of the simple Weber-Davis (1967) formulation, one can show that the braking timescale τ is approximately given by

$$\tau \sim \frac{\alpha M v_A}{B_o^2 R^2} \tag{7}$$

where M and R are the stellar mass and radius, respectively, B_o is the surface magnetic field strength, and v_A is the wind velocity at the Alfven radius r_A. The factor α is the moment of inertia constant, $\sim 10^{-1}$.

Evolutionary calculations (cf. Iben 1965) suggest that the first crossing of the Hertzsprung gap is very rapid, and that the evolution of a giant through a few subclasses occurs in a few x 10^5 years or so. The rotational braking time must therefore be of this order. Substituting the values $M \sim 3\ M_\odot$, $R \sim 10\ R_\odot$, $v_A \sim 200$ km s^{-1}, and $\tau \sim 3 \times 10^5$ yr into Eq. (8), one finds $B_o \sim 50G$. Comparison with the Sun suggests that such surface averaged field strengths would result in (unobserved) strong chromospheric and coronal emission in G5 III stars.

The evolutionary calculations also show that stars with $M > 2.25\ M_\odot$ have blue loops in the HR diagram, and that they cross part of the Hertzsprung gap a second time after helium ignition (see also Gray and Endal 1982). This second crossing takes much longer than the first crossing, and so stars in this evolutionary state can be expected to dominate the distributions of giants in certain regions of the HR diagram. Alschuler (1975) suggested that G5 III and earlier stars are first-crossing stars of intermediate mass and age, while G8 III and later giants are mostly second-crossing stars, and backed up this contention with statistical analyses and improved evolutionary tracks. Gray and Endal (1982) suggest that the majority of G5 III stars must be second-crossing objects. If this picture is correct, then one should not be surprised to see a break in v sin i near G5 III, for this spectral type is near the division between objects of significantly different ages and internal structures.

If we do not need to invoke sharply peaked braking at G5 III, then we have little indicated need for wind-related spindown; Endal and Sofia (1979) have shown that the general decay of v sin i can be explained without mass loss, due to the effects of convection in transporting angular momentum and in changing the moment of inertia.

Chromospheric surface fluxes decline with decreasing effective temperature (Basri and Linsky 1979; Middelkoop 1982), which is likely to be correlated with spindown. If there is a sharp change in rotation at $\sim$ G5 III, irrespective of its explanation, one might expect this to be reflected in stellar activity. There is some evidence for this effect in Ca II emission (Middelkoop 1982), and transition-region fluxes generally become undetectable at K0 III (Linsky and Haisch 1979; Simon, Linsky, and Stencel 1982).

D. Spindown of Pre-Main Sequence Stars

Although we have more observational information about pre-main sequence winds than for solar-type stars, obtaining rotational velocities for these stars is very difficult. Not only are they relatively faint, but their chromospheric activity tends to wash out strong photospheric absorption lines or even turn them into emission lines. In addition, it is very difficult to place these stars on the HR diagram. Even with reasonable estimates of log g and T_{eff}, there are still substantial uncertainties in the internal structure of these stars, and so there is some question as to the accuracy of the evolutionary tracks to be used. The importance of this point can be appreciated from Figure 6. One can see that the assignment of the correct mass depends crucially on the adoption of the correct evolutionary tracks. This in turn is vital to understanding spindown, since the rotation of main sequence stars is a very sensitive function of mass in the 1-2 $M_{\odot}$ range.

Vogel and Kuhi (1981) have studied stars with ages $\sim 10^7$ years in several regions of the sky. It appears that the rotation of these pre-main sequence stars do not follow the Skumanich law for V_{rot} vs. age, in the sense that V_{rot} is smaller than the projection of V_{rot} backwards in time from the Hyades-Pleiades relation. To put it another way, the T Tauri stars seem to be slow rotators. However, if the change in the moment of inertia as a star evolves is taken into account, and the Skumanich law is changed to an angular momentum loss rate, then the agreement is improved (Vogel and Kuhi 1981). This result is not a very accurate test, given all of the uncertainties involved.

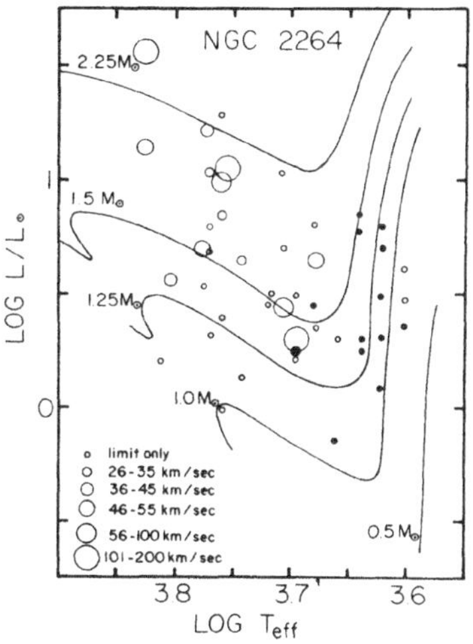

Figure 6. V sin i's of pre-main sequence stars in NGC 2264, with approximate evolutionary tracks indicated (Vogel and Kuhi 1981).

If T Tauri mass loss rates are less than 10^{-8} $M_{\odot}$ yr^{-1}, then clearly there must be a magnetic field to brake the stars as observed. Although the exact braking timescales are uncertain, they must be of order 10^6 yr. Then, if we use the simple Weber-Davis estimate (7), and assume $M \sim M_{\odot}$, $R \sim 3R_{\odot}$, $v_A \sim 200$ km s^{-1}, and $\tau = 10^6$ years, $B_o \sim$ 170G. This field strength is of the order necessary to have the possibility of an Alfven-wave driven wind, and so suggests that we may be on the right track in interpreting T Tauri mass loss.

Another result of importance found by Vogel and Kuhi (1981) is that the fast rotators seem to generally occur among the high-mass pre-main sequence stars. In other words, there is some evidence that the break in rotation seen on the main sequence near F5 is already present in the T Tauri phase. This indication, if true, is extremely important, for it seems to suggest a qualitative difference between the formation processes of low and high mass stars.

One aspect of Figure 6 I find intriguing is the suggestion that there is a wide range of v sin i's present in the more massive stars. This seems qualitatively consistent with the results of Wolff, Edwards and Preston (1982), who found a similarly broad distribution of rotation in B-A stars and suggested that this distribution has a protostellar or pre-main sequence origin. These data indicate that protostars may form with widely varying angular momentum. The operation of wind braking with a Skumanich-type law, i.e. one in which the braking time decreases for increasing rotation, will tend to bring the older stars on the lower main sequence into a narrower angular momentum distribution than was initially present.

In recent years, observational studies have begun to address mass loss from protostars. Radio frequency observations of regions of star formation indicate that surprisingly large mass outflows are associated with very early stages of stellar evolution (Genzel et al. 1981; Bally and Lada 1982; see also Hartmann and MacGregor (1982) for further references). Space does not permit an adequate description of the problems associated with these outflows, and the field is evolving so rapidly that it is premature to attempt to review it. However, it seems quite possible that many of the concepts of magnetic wind braking developed to explain the angular momentum loss of main sequence stars may also be extended to describe protostellar spindown.

REFERENCES

Alschuler, W.: 1975, Astrophys. J. 195, p. 649.
Altenhoff, W. J., Oster, L., and Wendker, H. J.: 1979, Astron. Astrophys. 73, p. L21.
Ayres, T. R., Linsky, J. L., Basri, G. S., Landsman, W., Henry, R. C., Moos, H. W., and Stencel, R. E.: 1982, Astrophys. J. 256, p. 550.
Baliunas, S. L.: 1982, private communication.
Bally, J. and Lada, C.: 1982, preprint.
Basri, G. S. and Linsky, J. L.: 1979, Astrophys. J. 234, p. 1023.
Belcher, J. W.: 1971, Astrophys. J. 168, p. 509.
Belcher, J. W. and Davis, R. L.: 1971, J. Geophys. Res. 76, p. 3534.
Belcher, J. W. and MacGregor, K. B.: 1976, Astrophys. J. 210, p. 498.
Bernat, A. P.: 1977, Astrophys. J. 213, p. 756.
Bopp, B. W.: 1980, Highlights of Astronomy 5, p. 847.
Bowers, P. F. and Kundu, M. R.: 1979, Astron. J. 84, p. 791.
De Campli, W.: 1981, Astrophys. J. 244, p. 124.
Dupree, A. K. and Hartmann, L.: 1980, in IAU Colloq. 51, ed. D. F. Gray and J. L. Linsky, Springer, Heidelberg, p. 279.
Durney, B. R. and Stenflo, J. O.: 1972, Astrophys. Space Sci. 15, p. 307.
Eggleton, P. P.: 1976, IAU Symposium 73, eds. P. Eggleton, S. Mitton, and J. Whelan, Reidel, Dordrecht, p. 209.
Endal, A. and Sofia, S.: 1979, Astrophys. J. 232, p. 531.
Genzel, R., Reid, M. D., Moran, J. M., and Downes, D.: 1981, Astrophys. J. 244, p. 844.
Goldberg, L.: 1979, Quarterly Journ. Roy. Astron. Soc. 20, p. 361.
Goldberg, L., Hege, E. K., Hubbard, E. N., Strittmatter, P. A., and Cocke, W. J.: 1982, in "Second Cambridge Workshop on Cool Stars, Stellar Systems, and the Sun", ed. M. S. Giampapa and L. Golub,

SAO Special Report 392, p. 131.
Gray, D. F.: 1981, Astrophys. J. 251, p. 155.
Gray, D. F.: 1982, Astrophys. J., in press.
Gray, D. F., and Endal, A. S.: 1982, Astrophys. J. 254, p. 162.
Hall, D. S.: 1976, in IAU Colloq. 29, ed. W. S. Fitch, Reidel, Dordrecht, p. 287.
Hagen, W.: 1978, Astrophys. J. Supp. 38, p. 1.
Hartmann, L.: 1982, Astrophys. J. Suppl. 48, p. 109.
Hartmann, L., Dupree, A. K., and Raymond, J. C.: 1980, Astrophys. J. (Letters) 236, p. L143.
Hartmann, L., Dupree, A. K., and Raymond, J. C.: 1981, Astrophys. J. 246, p. 193.
Hartmann, L., Dupree, A. K., and Raymond, J. C.: 1982, Astrophys. J. 252, p. 214.
Hartmann, L., Edwards, S., and Avrett, E. H.: 1982, Astrophys. J., in press.
Hartmann, L. and MacGregor, K. B.: 1980, Astrophys. J. 242, p. 260.
Hartmann, L. and MacGregor, K. B.: 1982, Astrophys. J., in press.
Hartmann, L., MacGregor, K. B., and Avrett, E. A.: 1982, in preparation.
Hollweg, J. V.: 1973, Astrophys. J. 181, p. 547.
Huang, S.-S.: 1966, Ann. d'astrophys. 29, p. 331.
Iben, I.: 1965, Astrophys. J. 142, p. 1447.
Jacques, S. A.: 1978, Astrophys. J. 226, p. 632.
Kraft, R. P.: 1967, Astrophys. J. 150, p. 551.
Kraft, R. P.: 1970, in "Spectroscopic Astrophysics", ed. G. H. Herbig, Univ. of California, Berkeley, p. 385.
Linsky, J. L. and Haisch, B. M.: 1979, Astrophys. J. (Letters) 229, p. L27.
Middelkoop, F.: 1982, Astron. Astrophys., in press.
Munro, R. H. and Jackson, B. V.: 1977, Astrophys. J. 213, p. 874.
Newell, R. T. and Hjellming, R. M.: 1982, Astrophys. J. (Letters), in press.
Okamoto, I. and Sato, K.: 1970, Pub. Astron. Soc. Japan 22, p. 317.
Pallavicini, R., Golub, L., Rosner, R., Vaiana, G. S., Ayres, T., and Linsky, J. L.: 1981, Astrophys. J. 248, p. 279.
Reimers, D.: 1977, Astron. Astrophys. 57, p. 395.
Schatzman, E.: 1962, Ann. d'astrophys. 25, p. 18.
Simon, T., Linsky, J. L., and Stencel, R. E.: 1982, Astrophys. J., in press.
Skumanich, A.: 1972, Astrophys. J. 171, p. 565.
Skumanich, A. and Eddy, J. A.: 1981, in "Solar Phenomena in Stars and Stellar Systems", ed. R. M. Bonnet and A. K. Dupree, Reidel, Dordrecht, p. 368.
Smith, M. A.: 1979, Pub. Astron. Soc. Pacific 91, p. 737.
Soderblom, D. S.: 1980, Ph.D. Thesis, Univ. of Calif.-Santa Cruz.
Stencel, R. E., Linsky, J. L., Brown, A., Jordan, C., Carpenter, K. G., Wing, R. F., and Czyzak, S.: 1981, Monthly Notices Roy. Astron. Soc. 196, 47P.
Stencel, R. E. and Mullan, D. J.: 1981, Astrophys. J. 238, p. 221.
Stern, R. A., Zolcinski, M.-C., Antiochos, S. K., and Underwood, J.

H.: 1981, Astrophys. J. 249, p. 647.
Vaiana, G. S. et al.: 1981, Astrophys. J. 245, p. 163.
Van't Veer, F.: 1979, Astron. Astrophys. 80, p. 287.
Vaughan, A. H., Baliunas, S. L., Middelkoop, F., Hartmann, L., Mihalas, D., Noyes, R. W., and Preston, G. W.: 1981, Astrophys. J. 250, p. 276.
Vaughan, A. H. and Preston, G. W.: 1980, Pub. Astron. Soc. Pacific 92, p. 385.
Verbunt, F. and Zwaan, C.: 1981, Astron. Astrophys. 100, p. L7.
Vilhu, O.: 1982, Astron. Astrophys., in press.
Vogel, S. N. and Kuhi, L. V.: 1981, Astrophys. J. 245, p. 960.
Walter, F. M. and Bowyer, S.: 1981, Astrophys. J. 245, p. 671.
Walter, F. M.: 1981, Astrophys. J. 245, p. 677.
Weber, E. J. and Davis, L. Jr.: 1967, Astrophys. J. 148, p. 217.
White, N. M., Kreidl, T. J., and Goldberg, L.: 1982, Astrophys. J. 254, p. 670.
Wilson, O. C.: 1966a, Astrophys. J. 144, p. 695.
Wilson, O. C.: 1966b, Science 151, p. 1487.
Wilson, O. C.: 1970, Astrophys. J. 160, p. 225.
Withbroe, G. L. and Noyes, R. W.: 1977, Ann. Rev. Astron. Astrophys. 15, p. 363.
Wolff, S. C., Edwards, S., and Preston, G. W.: 1982, Astrophys. J. 252, p. 322.

DISCUSSION

NUSSBAUMER: I have a short comment on your mass loss formula. I have recently published (1982, *Astron. Astrophys.* **110**, p. L1) a short note on the possibility that *randomly emerging magnetic flux* may be at least partially responsible for stellar winds. I introduced factors into the formula to account for the fact that magnetic fields do not cover the total surface of the star, and that closed fields could rather inhibit than produce mass loss. The formula can explain the total solar mass loss with $B = 100$ G, and if it is assumed that 0.1% of the solar surface is covered by such fields, and that 0.1% of the emerging flux is efficient in producing mass loss. For the sun the area coverage factor can be determined observationally; to determine the second (efficiency) factor is a more difficult task. I think, however, that it is important not to neglect these efficiency factors.

HARTMANN: I agree that some area factors must be involved. However, nobody knows what they are for stars. Furthermore, the observed mass loss rates are uncertain by an order of magnitude or more. So for reasons of simplicity and ignorance I have left these factors out.

MULLAN: Mass loss in cool giants can be driven by unstable emerging fluxtubes. This appears to be as viable a mechanism as wave driving. In particular, emerging flux can account for macroturbulent broadening, episodic mass loss, highly variable chromospheric/coronal emission, etc. As regards the break in rotational velocity at $\sim$F5, I have already suggested that the coincidence of main sequence lifetime with the age of the galaxy at ~F5 is important (cf. 1972, *Astrophys. Letters*). There need not be any break in convective efficiency at F5. It may occur much later (G-K). In the sun the major correlation between wind flux and surface features occurs when one considers X-ray bright points. These are known to be regions of emerging flux. This is a strong indication that emerging magnetic flux plays an important role in driving mass loss in the sun.

HARTMANN: The wind can open field lines up, so to some extent there is a question of which is the chicken or the egg, the open field lines or the mass loss. In addition, to the extent that emerging field lines exert an outward force on the gas, they might have properties in a time-averaged sense which look like some of the steady wave-driven wind properties.

GIAMPAPA: (1) Let me just comment that the X-ray emission observed in A stars is likely to be due to dMe stellar companions, although there is A star X-ray emission observed for some apparently single A stars. (2) Would you care to comment on Dr. R. Stencel's hypothesis that magnetic fields contribute significantly to the hydrostatic balance in the 10^4 K "extended chromosphere" of α.Ori?

HARTMANN: The out-flow of α Ori is fairly slow. So there is an extended region where the flow velocity is considerably less than the sound speed, and this region could be nearly in a hydrostatic equilibrium, with the gravitational acceleration being balanced by the turbulent magnetic pressure gradient.

UCHIDA: In talking about Alfvén waves I presume that you have in mind the stochastic generation of Alfvén waves in convective turbulence distributed all over the stellar surface. How then do you explain the time variations seen e.g. in T Tau stars? Such large time variations should be due to fluctuations coming from a smaller number of sources.

HARTMANN: The calculations are spherical and steady-state. These conditions have been assumed, not because we feel that they are correct, but because they simplify the problem. Undoubtedly magnetic fields are structured and variable if the sun is any guide, so we should expect wave-driven winds to be variable as well.

SHEELEY: You have used the term "extended chromospheres" several times. Have you tried to imagine what an extended chromosphere might look like if you took a spectroheliogram of it? Would it consist of a "hayfield" of very long spicules much like your lawn when you return from an extended vacation? Or would it consist of many large prominences? Or what?

HARTMANN: I have not thought much about it. There are speckle observations of α Ori by Goldberg and coworkers in Hα, which show irregular structure. In addition, α Ori exhibits irregular variations in polarization. These data indicate an irregularly moving surface, perhaps similar to solar spicules, but moving much further in a gravitational field $\sim 10^{-4}$ of the solar field.

CRAM: There is another class of giant stars whose "extended" chromospheres may be observed: eclipsing binaries like 31 Cyg and 32 Cyg. There is evidence for time-dependent structures of fairly large size in the chromospheres of these stars, and most observers of these objects have drawn an analogy to solar prominences, rather than spicules. The idea that these observed structures may be analogous to magnetically supported solar prominences is perhaps the only evidence we have for the existence of "important" magnetic fields in the atmospheres of giant stars.

GIAMPAPA: Just a comment on Lawrence Cram's remark: A radio flare has been observed by Boice et al. (1981, *Astrophys. J. Lett.* **245**, pp. L71-L74) on the M giant star α Ceti.

SHIELDS: The slightly blueshifted absorption line in the Hα emission profile of DF Tauri resembles some ultraviolet emission line profiles of quasars. How does one understand the existence of high velocity emitting gas when lower velocity absorbing gas is able to escape from the star's gravitational field?

HARTMANN: The blueshifted absorption occurs at a distance of a few stellar radii, where the escape velocity is lower than it is near the surface, while the broad emission originates near the surface, deeper in the gravitational potential well. In addition, the emission component is broadened further by optical depth effects.

ROSNER: Did I understand correctly that $v_{\parallel} \ll v_{\perp}$ at the base of the wind models? I am surprised by that because, in the presence of magnetic fields, I would have expected any anisotropy to obey just the opposite inequality, $v_{\perp} \lesssim v_{\parallel}$, because the base coincides with the region of wave generation, i.e. the region where $\beta \gtrsim 1$.

HARTMANN: We have not considered the wave generation zone, and have assumed that only transverse modes are propagating. This simplification is not correct. What is important is that δv can be much larger than the average or background net expansion of the wind. We assumed transverse modes, because it appears that compressive longitudinal modes dissipate fairly rapidly as they propagate outwards.

KOUTCHMY: Coming back to the comment by Dr. Mullan, I would like to say that we completely agree that large mass flows coming from coronal holes are produced by very small-scale events occuring in regions cospatial with X-ray bright points and elements of the chromospheric network. Dr. Brueckner's group at NRL and we in Paris have pointed out the importance of the so-called superenergetic events occuring very suddenly above enhanced network elements. They are named super-spicules, macro-spicules, or coronal spikes, and they are clearly the source of the filamentary structure of the wind flow.

MESTEL: This is by way of a philosophical comment. In magnetic braking one is trying to eat one's cake and have it. The angular momentum transport is equivalent to effective corotation being maintained out to the Alfvén surface. Within this surface, by definition the wind energy is below the magnetic energy, so that the magnetic stresses control the rotation of the outflowing gas; but equally the magnetic field will either inhibit or channel the wind. In particular, one expects that if the total magnetic flux emerging from the star is imagined being increased — e.g. due to a higher rotation yielding a more efficient dynamo — then there will be a correspondingly larger zone of closed field lines. In some computations done some sixteen years ago, I assumed the field structure to be dipolar and curl-free out to the Alfvénic surface, and radial beyond. Increasing the stellar magnetic flux did not then increase the angular momentum loss; the larger radius of the Alfvénic surface was compensated by the reduced area of the wind zone, so that we got the same rate of braking, but for a smaller mass loss. My model was certainly too conservative, but the results illustrate that it is dangerous to assume an enormous scaling-up of the braking rate with increase in stellar flux; the problem is a good deal more subtle.

ANGULAR MOMENTUM LOSS AND CHROMOSPHERIC ACTIVITY IN LATE-TYPE DWARFS

David R. Soderblom
Harvard-Smithsonian Center for Astrophysics
Cambridge, Massachusetts USA

A chromospheric color-magnitude diagram has been constructed from the Vaughan-Preston survey of Ca II emission among solar neighborhood stars. The relative flux in the H and K lines declines with mass on the lower main-sequence as a consequence of the decline of the ZAMS rotational velocity with mass. The features of this C-M diagram are discussed, and some evidence for a gap in the distribution of Ca II emission with age is examined.

We know qualitatively that chromospheric emission (CE) is a manifestation of stellar magnetic fields, and that it diminishes with age for cool main-sequence stars. Skumanich (1972) fit a $t^{-1/2}$ law (t=age) to the available data for solar-type stars (near G2V), but the survey of Vaughan and Preston (1980, hereafter VP) shows an apparent gap in the distribution of Ca II emission vs. (B-V). This gap would imply a discontinuity in the relation between CE and age for stars bluer than (B-V)=1.0.

VP showed S, the equivalent width of the Ca II emission cores, as a function of (B-V). S is sensitive to the stellar temperature because the continuum in the ultraviolet weakens for the cool stars; this accounts for the steep slope of VP's log S vs. (B-V) diagram. Middelkoop (1982) has recently derived a correction for this effect so that the true emission flux (except for an arbitrary scale factor) can be obtained. His method gives values that agree with an independent derivation for solar-type stars (Soderblom 1983b). For the range over which Middelkoop's formulae apply ($0.45 \lesssim$(B-V)$\lesssim 1.50$), I have computed R_{HK}=(Ca II flux)/σT_e^4, i.e., the fraction of the star's luminosity that appears as emission in the H and K lines.

Figure 1 shows this chromospheric color-magnitude diagram. After correcting for the changing continuum, the CE shows a trend opposite to that seen in VP's plot of log S vs. (B-V). In other words, the cooler stars produce relatively less CE (as a fraction of their luminosity) than the hotter stars. This trend can be attributed to the decline of the zero-age-main-sequence (ZAMS) rotational velocity with mass, which in turn is a consequence of the greater time spent in the pre-main-sequence

J. O. Stenflo (ed.), Solar and Magnetic Fields: Origins and Coronal Effects, 439–444.

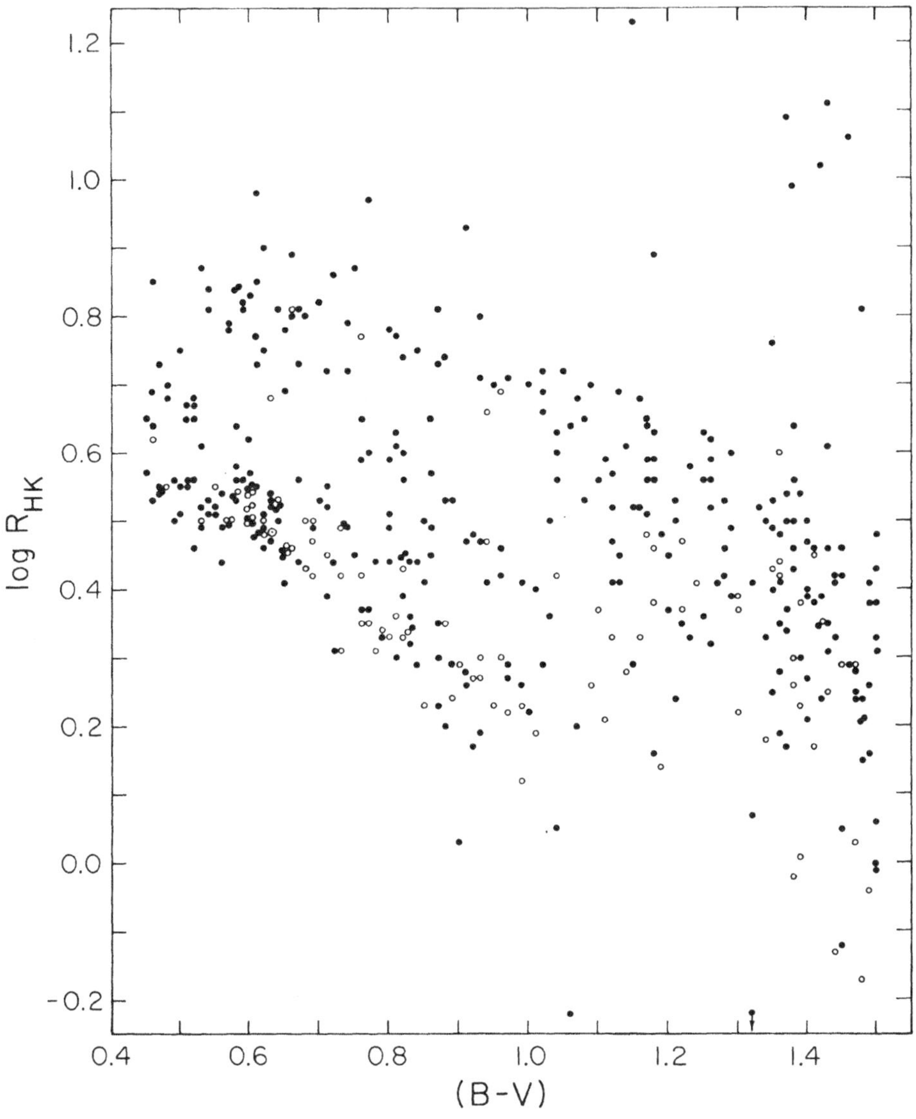

Figure 1. A chromospheric color-magnitude diagram. R_{HK} is the fraction of the stellar luminosity appearing in the emission reversals of the Ca II H and K lines, i.e., the equivalent width S of Vaughan and Preston (1980) transformed with Middelkoop's (1982) formulae. The solid points are disk population stars, while the open circles are old disk and halo objects (either $|W+10|>30$ km s^{-1} or $(U^2+V^2)^{1/2}>65$ km s^{-1}.

contraction stage, where most of the initial angular momentum is lost. The youngest, most rapidly rotating stars at (B-V)=0.60 have $v \sin i$'s = 10-12 km s^{-1} (Soderblom 1982). This drops to about 6 km s^{-1} at (B-V) = 0.80 (Soderblom 1981) and goes below 4 km s^{-1} for the M dwarfs (Soderblom *et al.* 1982). The rotation periods found by Baliunas *et al.* (1983) and the tight correlation between R_{HK} and period (R.W. Noyes, this conference) confirm this steady decline of rotation with mass on the lower main-sequence. The weakest chromospheric emitters among the M dwarfs must be rotating very slowly indeed.

In Figure 1, VP's sample has been divided into two kinematic groups: the young disk stars (filled circles) and the old disk/halo population objects (open circles). For (B-V)<1.0, the concentration of halo stars near the lower edge of the distribution underscores VP's conclusion that kinematically old stars have weak CE. Among the chromospherically active stars at the upper edge of the distribution are objects kinematically associated with the Pleiades and Hyades clusters, in addition to the Ursa Major group stars that VP found. Members of old groups such as Eggen's (1969) Wolf 630 and 61 Cygni groups lie along the lower edge with the halo objects. If one postulates that CE$\propto t^{-1/2}$ (Skumanich 1972) and if a constant stellar birthrate is assumed, the vertical distribution of points is reasonable given: 1) the size of the sample, 2) the attrition of old stars in the solar neighborhood due to their large space velocities, and 3) the photospheric contribution to R_{HK} which condenses the distribution for the oldest stars. The vertical spread in R_{HK} implies a ratio of ages of oldest to youngest stars of about 100.

As VP cautioned, the authenticity of a gap in this diagram is problematical. Stars above and below the "gap" may indeed differ in some fundamental way, but there may be systematic effects responsible for this hiatus as well. For example, there is not so much a gap as a concentration of stars near the upper edge of the distribution. If the CE saturates at large magnetic field strengths, such an effect would arise. Also, the stars in the survey were mostly observed once or twice. Significant variability is seen, particularly for the strong emitters, as well as cyclic modulation over periods of years (Wilson 1978). If the mean CE averaged over, say, 20 years were plotted, would the gap disappear? The sample size also affects our perception of the gap. Figure 1 contains about 40 more stars than VP's diagram did because (B-V)'s were now available. By coincidence, about 6 of these new points tend to fill in the gap near (B-V)=0.80. VP's planned extension of the survey to southern declinations is a necessary augmentation of the sample. One last effect on this diagram is the lower average metal abundance of old stars. The conditions of formation of the Ca II H and K lines are too complex to easily predict the result of this effect.

There is some circumstantial evidence relevant to this "gap". The tight correlation between R_{HK} and rotation period (Noyes, this conference) implies that any discontinuity in the relation between CE and age should be matched by a corresponding discontinuity in the rotation-age data, but no such lacuna is seen (Soderblom 1983b). Similarly, the relation between lithium depletion and age is smooth for solar-type stars (Soderblom 1983b).

Whether or not a gap truly exists in the distribution for (B-V)<1.0, the distribution for cooler stars seems significantly different. Except for a few BY Draconis and flare stars in the upper right corner, there is no apparent gap and the vertical spread is less. The halo stars still exhibit lower R_{HK}'s overall. The Pleiades, Ursa Major and Hyades group stars lie along an extension of their loci among the hotter stars. The Wolf 630 and 61 Cygni group stars cooler than (B-V)=1.0 are displaced upward by 0.3 in log R_{HK} relative to the hotter stars in those groups. This shear in the distribution at (B-V)=1.0 is also seen in members of visual binaries whose components span that color.

This jog in the isochrones at (B-V)=1.0 could be rationalized if in fact a gap runs across the entire diagram. For (B-V)>1.0, the old stars would not have evolved across the gap because their ZAMS rotational velocities are so low that angular momentum loss requires most of the stellar lifetime. (The main-sequence lifetime approaches the age of the Galaxy for (B-V)>1.0). Before accepting this explanation, however, the data for these cool stars should be carefully examined. First, (B-V) is a poor photometric index for M stars, but (V-R)'s are not available for most of the sample. A plot of log S vs. (R-I) (Soderblom 1983a) is ambiguous because (R-I) is inappropriate for F and G stars. For (B-V)> 1.0, VP's continuum bands do not behave as expected (see their Fig. 3), suggesting that the S's may be muddled. Observations of the flux in the continuum bands of these stars should clarify the situation.

REFERENCES

Baliunas, S.L., Vaughan, A.H., Middelkoop, F., Hartmann, L.W., Mihalis, D., Noyes, R.W., Preston, G.W., Frazer, J., and Lanning, H.: 1983, Astrophys. J., in press.

Eggen, O.C.: 1969, Pub. Astron. Soc. Pacific 81, 553.

Middelkoop, F.: 1982, Astron. Astrophys. 107, 31.

Skumanich, A.: 1972, Astrophys. J. 171, 565.

Soderblom, D.R.: 1981, in M. Giampapa and L. Golub (eds.), *Second Cambridge Workshop on Cool Stars, Stellar Systems and the Sun*, Cambridge, Mass., Smithsonian Astrophys. Obs., Special Report 393, p. 197.

Soderblom, D.R.: 1982, Astrophys. J. 262, 275.

Soderblom, D.R.: 1983a, in P. Byrne and M. Rodonò, *Activity in Red Dwarf Stars* (IAU Coll. 71) D. Reidel, Dordrecht.

Soderblom, D.R.: 1983b, Astrophys. J., in press.

Soderblom, D.R., Vogt, S.S., and Penrod, G.D.: 1982, Bull. Am. Astron. Soc. 13, 873.

Vaughan, A.H. and Preston, G.W.: 1980, Pub. Astron. Soc. Pacific 92, 385

Wilson, O.C.: 1978, Astrophys. J. 226, 379.

DISCUSSION

WALTER: Can the apparent saturation at the top of the gap be, in part, attributed to peculiarities in the Pleiades relative to the Hyades? If, as I understood you to say, the v_{rot} for the Pleiades is lower than expected from the Skumanich relation, and Ca II fluxes scale as v, and not as $t^{-1/2}$, then the inferred saturation may be due only to the low $\langle v_{rot}\rangle$ for the Pleiades. How important is the scatter from cluster to cluster about the Skumanich relation? There is some evidence for differing initial velocities in pre-main-sequence (PMS) stars: M. Smith finds that the G PMS stars in the Orion Ic association rotate very slowly, whereas the G PMS stars in Taurus, which are of similar ages (and presumably similar masses), rotate far more rapidly.

SODERBLOM: My observations of main-sequence $1M_{\odot}$ stars from the age of the Ursa Major group and older show no greater dispersion about the $t^{-1/2}$ law than can be ascribed to a random distribution of rotation axes, observational error, and a modest intrinsic dispersion about the mean. In other words, nothing peculiar is seen. If you consider high mass stars which do not lose angular momentum on the main sequence, such stars in the Pleiades were seen by Abt to rotate faster than comparable field stars. Abt attributed this property to a relative lack of binaries in the Pleiades. This suggests that the Pleiades stars have greater than average angular momenta, which is opposite to what is needed to agree with your explanation.

MARCY: There is support for your contention that the Ca II H and K emission saturates in the young stars. The X-ray, ultraviolet, and magnetic field observations of the more active stars suggest the necessity of rather high photospheric fluxtube densities to explain the high values of the activity diagnostics. Such high densities of diverging fluxtubes would result, perhaps, in considerable field-line overlap in the lower chromosphere, and hence in saturation of the H and K emission.

SODERBLOM: I forgot to mention that solar observations support this saturation hypothesis. Skumanich, Smythe, and Frazier (1975, *Astrophys. J.* **200**, p. 747) looked at Ca II emission strength as a function of the local magnetic field strength. For weak fields the relation was linear, but for strong fields saturation occured. Their data seemed to fit something like Ca II emission $\sim B^{1/2}$, where B is the magnetic field.

FOING: How sensitive is the lithium depletion to the depth of the convective zone? Does it give a quantitative measurement of the distance between the Li burning shell and the subconvective layer, whatever the overshooting process could be?

SODERBLOM: The lithium depletion is *very* sensitive to the convective zone depth. Both myself and Duncan have examined the Li abundances of solar-type stars to derive Li abundance vs. age relations. In stars as young as the Hyades, the Li is already depleted in the K dwarfs. The transition from late F to K is fairly smooth, providing some information on the depth of the convective zones in these stars. I personally do not know how to interpret the data, however. There is evidence that the convective zone depth is sensitive to [Fe/H]. Recently the Spites observed a few G subdwarfs that are definitely Population II. Despite their age, these stars have strong Li lines, which is completely contrary to the simplest models of Li depletion. Their observations can be explained if the convective zones of low metal-abundance stars are thinner due to the lower opacity in the outer layers.

The Li data provides one of the few observational constraints on the convective properties of late-type dwarfs. Careful theoretical interpretation of the data is needed.

VAUGHAN: (1) In your statistical models have you tried yet to take the saturation in the Ca emission into account (for example $S \sim exp(-P_{rot}/P_0)$)? (2) Considering the space motions of the stars involved, could Pleiades "escapees" have wandered into the solar neighbourhood yet? In the case of the Ursa Major group the velocity dispersion is exceedingly small.

SODERBLOM: (1) No, I have not tried any other set than the $t^{-1/2}$ law. There are two reasons: First, the simulations were done in the last couple of weeks, so we tried the most obvious case first. Second, we tried to minimize the number of assumptions needed to simulate the chromospheric color-magnitude diagram to avoid any systematic effects. For example, we deliberately avoided any attempt at accounting for short and long time scale variability for that reason. (2) I am not sure. That question occured to me, but I have not addressed it. The Pleiades, Ursa Major, and Hyades group members all have space motions essentially identical to the cluster velocity, considering the uncertainties. Eggen has applied liberal criteria to determine group membership — he only requires that the stars corotate in the Galaxy. My criteria are stricter: all three space velocities should agree to 3 or 4 km s^{-1}. The spectroscopic properties of these group members make their membership very plausible. You pointed out some group members (Ursa Major and 61 Cygni groups) in your original paper.

MAGNETIC TOPOLOGY AND THE HEATING OF EXTENDED CHROMOSPHERES

Robert E. Stencel
Joint Institute for Laboratory Astrophysics, University of Colorado and National Bureau of Standards, Boulder, CO 80309

ABSTRACT

New data indicate that red giants are surrounded by geometrically thick chromospheres of several stellar radii extent. Such chromospheres occur among stars which apparently lack coronae. Maintenance of this extended warm region may require non-compressional wave heating of a magnetic character, and this may provide a crucial clue to the mechanism of rapid mass loss from red giant stars.

INTRODUCTION

Red giant and supergiant stars are known from their optical spectra to possess cool circumstellar (CS) envelopes and winds. At issue is the nature of the interface between the warm photosphere and the cool CS shell. New spectroscopic evidence from optical and non-optical regimes, as well as direct imagery, is providing new insights into this question.

SPECTROSCOPIC EVIDENCE

Ultraviolet observations with the IUE satellite have made possible the exploitation of the mid-UV lines of C II near 2325 Å and 1335 Å as density and temperature diagnostics (Stencel et al. 1981). Combined with estimates for the ionization equilibrium and abundances, one can deduce column densities and lower limits to the physical thickness of the emitting region. For Arcturus, we deduced the existence of a 5000-10,000 K chromosphere of at least one stellar radii extent. The same technique yields geometrically thick chromospheres (5-10 stellar radii) for other non-coronal red giants, but thin chromospheres (few percent or less of a stellar radius) for G giants and the Sun (Stencel and Carpenter 1982).

The He I 10830 Å line is variably in emission and absorption in the spectra of non-coronal red giants (Zirin 1982). In contrast, the coronal-type, yellow giants show non-variable 10830 Å absorption, as is the case in the solar spectrum. However, the low upper limits on soft X-ray emission from red giants call into question any solar-like line formation mechanism of recombination following ionization of He by

J. O. Stenflo (ed.), Solar and Magnetic Fields: Origins and Coronal Effects, 445–448.

X-rays. Alternatively, line formation in shocks over extended regions may explain the variable 10830 Å emission of the red giant stars.

Chromospheric velocity fields can, in principle, be derived from the profiles of the emission cores of the resonance doublets of Ca II, Mg II and similar features. Statistics on the asymmetries of these doubly reversed emission cores argue for a boundary in the H-R diagram between solar-like asymmetries (among yellow giants) and outflow asymmetries, among red giants (Stencel 1978; Stencel and Mullan 1981). These boundaries correlate with X-ray and transition region line statistics (Simon et al. 1982) and CS features (Reimers 1977). However, stars like Arcturus, which lie near these boundaries show a strong outflow asymmetry in Mg II, but a solar-like (and slightly variable) asymmetry in Ca II. These "discrepant asymmetry" stars may provide the crucial clue to the evolution of atmospheres between coronal and non-coronal states (Mullan and Stencel 1982). Extended chromospheres can help in the interpretation of this phenomenon.

Wischnewski and Wendker (1981) and Newell and Hjellming (1982) have discussed the implications of the recently established two component radio emission of the M supergiant Betelgeuse, in terms of its atmospheric structure. They demonstrate that the star probably possesses an extended chromosphere in substantial agreement with the C II results. Comprehensive models such as that by Jura and Morris (1981) should be re-explored with this new information.

Additional spectroscopic evidence which suggests the existence of extended chromospheres includes: (a) atmospheric eclipsing binaries (Stencel et al. 1982); (b) cool star ionization equilibria (Ramsey 1981); and (c) fluorescent emission in the mid-UV (Carpenter et al. 1982) and near Ca II H and K (Stencel 1977). For details concerning these factors, see Stencel (1982).

DIRECT EVIDENCE

Goldberg et al. (1982) have discovered that the apparent diameter of the M supergiant Betelgeuse increases by a factor of 5 when viewed in narrow band H-alpha speckle interferometry, compared to broadband continuum diameters of 0.05 arcseconds. Goldberg and colleagues also report changes in the intensity distribution over a three-month timescale which are correlated with changes in polarization measurements independently obtained by Hayes (1980). Further results are likely to be fundamentally important.

Similar efforts involving narrow band occultation observations have also indicated extended chromospheric emission among red giants and supergiants: White et al. (1981) found a two stellar radii chromosphere around the M supergiant 119 Tau; Radick and Africano (1981) reported the suggestion of an increased angular diameter of Aldebaran (K5 III) in intermediate band occultation studies.

CHROMOSPHERIC HEATING

Dimensional arguments suggest that extended chromospheres are orders of magnitude larger than their isothermal pressure scale heights. The average chromospheric densities implied by the C II diagnostics for Arcturus and Betelgeuse are 10^8 and 10^7 cm^{-3}, respectively, and the isothermal pressure scale heights in 10^4 K chromospheres are 10^{10} and 10^{12} cm, respectively. These are 200-500 times smaller than the dimensions implied by five stellar radii chromospheres for these stars. It appears that hydrostatic pressure alone is incapable of supporting these extended chromospheres.

Another source of pressure in red giant atmospheres is due to turbulent and expansion velocities. The average pressure implied by the gravitational potential of material of 10^8 cm^{-3} density suspended at 5 stellar radii corresponds to approximately 50 km s^{-1} rms turbulent velocities. Although this velocity corresponds to transient CS absorption features seen in Arcturus (Reimers 1977), it appears too large for velocities normally associated with Betelgeuse (Bernat 1981), whether chromospheric turbulent or expansion. A plausible alternative to hypersonic velocities involves the support inherent in a modest magnetic field energy density. Assuming an r^2 divergence, a surface field of less than 10 gauss would suffice. Magnetic fields could provide forms of non-compressional wave heating which the extended emission distribution appears to require.

At present, two hypotheses appear capable of explaining several of the observed phenomena associated with extended chromospheres among non-coronal red giants and supergiants. Mullan (1982) has proposed a magnetic topology argument, with chromospheric heating and extent maintained by continually emerging, star-sized, discrete flux loops, which are unable to find stable, closed configurations (as they can in coronae). Hartmann and MacGregor (1980) have proposed an Alfvén wave heating theory which predicts extended chromospheres among red giants. Whereas the Alfvén wave theory predicts temperature and velocity distributions which could explain several observations, including perhaps the discrepant asymmetries, the unstable loop model matches the observed variability which is a major factor in the outer atmospheres of red giants.

CONCLUSIONS

Extended chromospheres may provide the missing link in the evolution of the outer atmospheres of cool stars between coronal and non-coronal states, as well as important clues to the nature of the mass loss process. Much work remains to be done to quantitatively improve the analysis of the spectroscopic clues listed above, including spectral synthesis work which includes spherical symmetry and velocity fields in chromospheres. Additional discussion of this point may be found in Stencel (1982). The enhanced spatial resolution offered by narrow band speckle interferometry ("speckle spectroscopy") ought to be pursued with vigor.

I am happy to acknowledge useful discussions with Reiner Hammer, and to thank NASA for support in this research, and the Symposium organizers for an invitation to participate in this meeting.

REFERENCES

Bernat, A.: 1981, Astrophys. J. 246, p. 184.
Carpenter, K., Wing, R., and Stencel, R.: 1982, Bull. A.A.S. 14, p. 614.
Goldberg, L., Hege, E., Hubbard, E., Strittmatter, P., and Cocke, J.: 1982, in "Second Cambridge Workshop on Cool Stars, Stellar Systems and the Sun," ed. M. S. Giampapa and L. Golub, S.A.O. Special Report 392, p. 131.
Hartmann, L. and MacGregor, K.: 1980, Astrophys. J. 242, p. 260.
Hayes, D.: 1980, Astrophys. J., 241, p. L165.
Jura, M. and Morris, M.: 1981, Astrophys. J. 251, p. 181.
Mullan, D.: 1982, Astron. & Astrophys. 108, p. 279.
Mullan, D. and Stencel, R.: 1982, in "Advances in Ultraviolet Astronomy: Four Years of IUE Research," eds. Y. Kondo, J. Mead and R. Chapman, NASA Conference Publ. No. 2238, p. 235.
Newell, R. and Hjellming, R.: 1982, Astrophys. J. (submitted).
Radick, R. and Africano, J.: 1981, Astron. J. (in press).
Ramsey, L.: 1981, Astrophys. J. 245, p. 984.
Reimers, D.: 1977, Astron. & Astrophys. 57, p. 395.
Simon, T., Linsky, J., and Stencel, R.: 1982, Astrophys. J. 257, p. 225.
Stencel, R.: 1977, Astrophys. J. 215, p. 176.
Stencel, R.: 1978, Astrophys. J. 223, p. L37.
Stencel, R.: 1982, in "Second Cambridge Workshop on Cool Stars, Stellar Systems and the Sun," ed. M. S. Giampapa and L. Golub, S.A.O. Special Report 392, p. 137.
Stencel, R. and Carpenter, K.: 1982, in "Advances in Ultraviolet Astronomy: Four Years of IUE Research," eds. Y. Kondo, J. Mead and R. Chapman, NASA Conference Publ. No. 2238, p. 243.
Stencel, R., Chapman, R., Kondo, Y., and Wing, R.: 1982, in "Advances in Ultraviolet Astronomy: Four Years of IUE Research," eds. Y. Kondo, J. Mead and R. Chapman, NASA Conference Publ. No. 2238, p. 492.
Stencel, R., Linsky, J., Jordan, C., Brown, A., Carpenter, K., Wing, R., and Czyzak, S.: 1981, Mon. Not. R. A. S. 196, p. 47p.
Stencel, R. and Mullan, D.: 1981, Astrophys. J. 240, p. 718.
White, N., Kriedel, T., and Goldberg, L.: 1981, Astrophys. J. 254, p. 670.
Wischnewski, E. and Wendker, H.: 1981, Astron. & Astrophys. 96, p. 102.
Zirin, H.: 1982, Astrophys. J. (submitted).

STELLAR WINDS AND SPINDOWN IN SOLAR TYPE STARS

I.W. Roxburgh
Department of Applied Mathematics
Queen Mary College, University of London
Mile End Road, London E1 4NS

ABSTRACT

The angular momentum loss produced by stellar winds is reviewed and a simple model of angular momentum loss with multipole fields is presented in which the field is potential when the flow speed is less than the Alfvén speed, and radial when greater than the Alfvén speed. The simpler the magnetic geometry, the larger is the angular momentum loss rate. This result is used to explain the rotational discontinuity across the Vaughan-Preston gap as being due to a sudden increase in angular momentum loss when the dynamo field switches from a quadropole to a dipole geometry.

The evolution of the internal rotation of stars as a result of surface angular momentum loss is considered. In the absence of a magnetic field, differential rotation can drive instabilities which then transport angular momentum out from the interior down the angular velocity gradient. Other instabilities such as that caused by the build up of ^{3}He can also transport angular momentum outwards. If angular momentum is transported by such weak turbulence, it also makes the star more homogeneous than standard evolutionary models and lowers the predicted value of the solar neutrino flux.

.The recent results on rotational splitting of solar oscillations are considered: these suggest that the inside of the sun is spinning faster than the surface and are compatible with models in which angular momentum is transported by mild turbulence. But data is scarce — and in such circumstances the speculations of the theorist must be viewed with caution!

1. INTRODUCTION

Mass loss in stellar winds from the surface of rotating stars will carry away angular momentum and thus lead to a decrease in the stars' angular momentum, but the details of mass and angular momentum loss are uncertain; so too are the dynamics of rotating stars. In particular the mechanism by which angular momentum is transported from the interior to the surface layers is still a matter for debate.

J. O. Stenflo (ed.), Solar and Magnetic Fields: Origins and Coronal Effects, 449–460.

A magnetic field can cause the material in the stellar wind to corotate with the star. The actual angular momentum loss therefore depends on the detailed magnetic structure of the open wind region. If the inside of a star contains a magnetic field, this can be very effective in causing nearly uniform rotation, and angular momentum is transported outwards by magnetic forces. On the other hand, if there is only a negligible internal magnetic field, the spindown of the surface layers by a stellar wind produces differential rotation which can be unstable - the resulting instabilities then transport angular momentum. If such instabilities exist they can lead to diffusion of chemical inhomogeneities and thus affect the evolution of the star.

The problems to be solved are difficult: this paper necessarily can only cover a few topics and even then only superficially.

2. ANGULAR MOMENTUM LOSS

The simplest model of a stellar mass loss is Parker's (1958) spherically symmetric isothermal solar wind. For a given temperature T and density ρ_o at the base of a corona of radius r_o, the wind velocity V is given by

$$\frac{V^2}{V_s^2} - 2\ln\frac{V}{V_2} = 2\ln\left(\frac{r}{r_c}\right) + 4r_c\left(\frac{1}{r} - \frac{1}{r_c}\right) + 1 , \qquad (1)$$

where V_s is the isothermal sound speed, $r_c = GM/2V_s^2$ is the critical point where $V = V_s$ and M the mass of the star. More sophisticated models have been calculated including energy transport in the one fluid and multicomponent plasma approximations, but for illustrative purposes, the simple isothermal model will suffice. The mass loss in such a wind is

$$\frac{dM}{dt} = -4\pi r_o^2 V_o \rho_o . \qquad (2)$$

The velocity V_o at $r = r_o$ is given by the isothermal wind equation with $r = r_o$.

A naive model of angular momentum loss in such a wind was given by Weber and Davis (1967) taking a spherically symmetric flow along radial magnetic field lines. The essential features of their model was that the wind corotates with the star out to that distance where the wind speed equals the Alfvén speed $V_A = (B^2/4\pi\rho)^{1/2}$, which is larger than the radius of the star. If the magnetic field is radial, then $B \propto r^{-2}$ and the angular momentum loss is

$$\frac{dH}{dt} = -\frac{8\pi}{3}\rho_A V_A r_A^4 \Omega = -\frac{2}{3}\frac{B_o^2 \Omega}{V_A} , \qquad (3)$$

where B_o is the strength of the field at the base of the corona.

Now the strength of the field B_o itself depends on Ω through the dynamo mechanism that is believed to operate in stellar convective zones. If the dynamo is such that $B_o \propto \Omega^2$, then with V_A approximately constant, one can reproduce (c.f.Durney 1972) the Skumanich (1972) relation

$$\Omega \propto t^{-1/2} . \tag{4}$$

Such a model is clearly incorrect; the magnetic field on the sun is far from radial in the inner regions. Moreover the heating of the corona and hence the wind speed may themselves depend on the magnetic field and hence on the angular velocity.

3. MAGNETIC FIELD MODELLING

The classical model of the solar coronal magnetic field is the hairy ball model of Altschuler and Newkirk (1969) and Schatten(1970). This model takes the measured value of the line of sight component of the magnetic field, assumes the field is current free (so curl $\underline{B} = 0$) out to some spherical source surface r_s and radial thereafter. This is a very coarse approximation to the physics. In my opinion, the apparent fit between observations and modelling is just a consequence of conservation of magnetic flux ($\nabla . \underline{B} = 0$) . The detailed structure of the field cannot be given by a model that combines detailed data on the surface field with a simple global approximation of a spherical source surface.

Durney and Pneuman (1975) took the hairy ball model and solved the one fluid solar wind model along field lines. Not surprisingly they found that the Alfvenic surface where $4 \pi \rho V^2 = B^2$ was nothing like the spherical source surface.

Several authors have sought to find self-consistent models of a magnetic solar wind but for a much simpler magnetic structure. Pneuman and Kopp (1971) solved the problem of a steady isothermal wind with a dipole base field, and Endler (1971) solved the same problem dynamically. More recently Robertson (1983) and Steinolfson et al (1982) have examined and extended this to one fluid conducting and polytropic models. All these models require considerable effort and expense and still are far removed from the actual physical conditions in stellar coronae.

Approximate methods for dipole fields were developed by Mestel (1968) and by Rowse and Roxburgh (1981). In these models the magnetic field is assumed to be current free (curl $\underline{B} = 0$) within a surface S_A, and radial outside S_A, but the position of the surface is determined by the condition that the wind velocity which is along the field lines equals the Alfven speed. Rowse and Roxburgh (1981) included growing terms in the potential field and were able to make the potential field radial at S_A. These models satisfactorily reproduce the essential features of the full MHD solutions.

During the course of the meeting I extended these models to higher multipole fields in order to seek an explanation of the discontinuity of rotation speed across the Vaughan-Preston gap (see Vaughan's article in these proceedings). I therefore now give some details of the method.

4. APPROXIMATE MODELS OF ANGULAR MOMENTUM LOSS FOR MULTIPOLE FIELDS

I assume that when the velocity V is less than the Alfvénic velocity, V_A, the field is essentially current free and the flow follows the field lines, when $V > V_A$ the flow and the field lines are radial. With this approximation the tangential component of the field is given by

$$B_\theta = \frac{B_o}{(n+1)\,r^{n+2}}\left\{1 - \frac{[(n+1)\,r^{2n+1} + n]}{[(n+1)\,r_A^{2n+1} + n]}\right\} P_n^1\,(\cos\theta) \tag{5}$$

and the radial component follows from $\nabla \cdot \underline{B} = 0$. Note that this field goes radial, $B_\theta = 0$ at $r = r_A$. The field strength $|B|$ varies with latitude but is given approximately by

$$B^2 = \frac{B_o^2}{r^{2n+4}} \tag{6}$$

where r is the distance in units of the radius of the star. The fraction of the star that has open field lines is $f = r_A^{-n}$, and the area of a flux tube increases as r^{n+2}.

If we now assume that the wind corotates out to the Alfvénic radius — which is a good approximation for calculating angular momentum loss (c.f. Mestel 1968) then this loss rate is

$$\frac{dH}{dt} = \frac{8\pi}{3}\,\rho_o\,V_o\,\Omega\,r_A^{2-n} \tag{7}$$

where r_A is given by the condition that the flow speed equals the Alfvénic speed

$$r_A^{n+2}\,V_A = \frac{B_o^2}{4\pi\rho_o V_o} \tag{8}$$

It is worth pointing out that a strong field B_o does not necessarily increase the angular momentum loss rate (c.f. Mestel 1968). Whilst r_A increases with B_o, the fraction of the star with open field lines decreases as B_o increases. From equations (7) and (8) we see that (with V_A slowly varying) the angular momentum loss rate is proportional to B_o^a, where $a = (2-n)/(2+n)$. For radial ($n = 0$) and dipole ($n = 1$) fields, the loss rate does increase with B_o, for a quadropole field $n = 2$, it is independent of B_o, whilst for $n > 2$ the loss rate decreases as B_o increases.

The flow of the wind along these field lines is easy to calculate: the equations for an isothermal flow can be integrated to give

$$\frac{V^2}{V_s^2} = 2 \ln \frac{V}{V_s} = 2(n+2) \ln \frac{r}{r_c} + 4 X_c \left(\frac{1}{r} - \frac{1}{r_c}\right) + 1 \tag{9}$$

where $X_c = GM/2 V_s^2 r_o$, $r_c = 2 X_c/(n+2)$ and the factor $(n+2)$ comes from the variation of the area of a flux tube with distance r. For given base conditions ρ_o, B_o, T_o these equations can be solved to determine the angular momentum loss (Roxburgh 1983).

5. EFFECT OF CHANGING MAGNETIC FIELD GEOMETRY ON ANGULAR MOMENTUM LOSS

As a star's angular velocity decreases we may expect the dynamo generated field to go through a series of bifurcations to simpler magnetic geometries, with consequential sudden changes in the angular momentum loss rates. For a change from quadropole ($n = 2$) to dipole ($n = 1$) the ratio of angular momentum loss rates is

$$\frac{\dot{H}_1}{\dot{H}_2} = \left(\frac{V_{01}}{V_{02}}\right) r_{A1} \tag{10}$$

Note that for a quadropole field the angular momentum loss rate is independent of the field strength. Now V_{01}, V_{02} are given by the wind equation and depend only on n, not on B_o. On the other hand, r_{A1} does depend on B_o and increases as B_o increases. Hence provided B_o is large enough, the ratio $\dot{H}_1/\dot{H}_2 > 1$ and the change from quadropole to dipole magnetic field produces a sudden increase in angular momentum loss rate.

The response of a star to a sudden increase in angular momentum loss depends on the coupling between the interior and the surface convective zone. If this coupling is weak, the convective zone (with its small moment of inertia) would spin down on a short time scale and the rotation of the inner region would readjust on a longer time scale.

It is therefore tempting to use this as an explanation of the jump in rotational velocity across the Vaughan-Preston gap (see Vaughan, this volume). Provided the spindown time of the convective zone is short compared to the age of the star, we would expect to see such a jump as a consequence of a change in the magnetic field geometry. Indeed, the model could be refined to calculate the spindown time of the convective zone and hence to predict the change of finding stars crossing the gap.

6. THE INTERIOR OF SOLAR TYPE STARS

Models of the internal structure of solar type stars are based on the assumptions that energy is generated by hydrogen burning nuclear reactions (principally the proton-proton chain in the sun). This energy is transported outwards by radiation, but near the surface the lowering of specific heats due to ionisation and the temperature variation of the opacity, combine to produce a convectively unstable outer layer. It is further assumed that rotation, magnetic fields, oscillations, and mass loss, can be safely neglected. Such models can reproduce broad features of the observed main sequence.

Evolutionary sequences are then evaluated by calculating the change in chemical composition due to the nuclear reactions with the assumption that mixing only takes place in convectively unstable regions. Such models eventually develop a helium core surrounded by a hydrogen burning shell and become red giants.

I should however emphasise that we have no reliable theory of turbulent convection. Standard models use a local mixing length theory, which excludes the effect of overshooting into the stable layers (c.f. Roxburgh 1978) and the mixing length, ℓ is treated as a free parameter - adjusted to obtain agreement between models and the observed solar radius. Again, for the sun we do not know the helium abundance, Y, this is again taken as a free parameter and is adjusted so that models based on these assumptions give the observed solar luminosity for a model with an age of $t_o \simeq 4.7.10^9$ yrs. Clearly other assumptions could be made to fit the observations!

The standard solar model has its problems - the predicted neutrino flux is some three times greater than the measured upper limit (Davis 1978) and as was first pointed out by Gough and his coworkers (Dilke and Gough 1972, Christensen-Dalsgaard et al 1974), this standard model predicts a build up of the ^{3}He away from the centre that after some 3.10^8 years is large enough to excite a global over-stability. These problems have not been resolved, and must inevitably call into question the simple spherical standard model.

One exciting possibility is that the observations of solar oscillations will give an empirical model of the internal density distribution which can then be compared with the models. I suspect we are in for a few surprises when this is done!

The standard model has been questioned by several authors, myself included (c.f. Rood 1978, Roxburgh 1976, 1978). One school of thought argues that rapid internal rotation or strong magnetic fields can so distort the star as to change its structure. Another school argues that there must be some internal mixing caused by the ^3He instability and/or rotation and magnetic fields that changes the evolutionary models (Dilke and Gough 1972, Roxburgh 1976, Maeder and Schatzman 1981). A mixed, or partially mixed model gives a lower neutrino flux, essentially because mixing brings hydrogen into the central regions and with a higher hydrogen abundance, a lower temperature is required to produce a given amount of energy.

7. INTERNAL ROTATION OF THE SUN

The first indications that the inside of the Sun is rotating more rapidly than the surface layers come from the rotational splitting of the oscillation modes as observed by Claverie et al (1982) and Hill (1982). The interpretation of these observations is a matter of debate, but suggest that the average internal rotation is some 2 - 10 times the surface value. A detailed discussion is given in Gough (1982), where he uses the splitting of 7 lines observed by Hill to estimate the internal distribution of angular velocity. A word of caution is required:before such an analysis can be carried through, it is necessary to identify the particular normal mode that is being observed - this is by no means a simple matter.

The analysis by Gough (1982) shows that the observed modes fall into two classes, those where the main effect is from central regions , $0.1 < r/R < 0.3$, and those where the main effect is from regions near the surface. This permits one to conclude that the central regions are spinning faster, but makes it difficult to determine the variation of angular velocity with radius $\Omega = \Omega(r)$, the details depending on the method used to invert from the data to $\Omega(r)$. All models used by Gough give an angular velocity at $r = 0.1 R_\odot$ of between 6 and 10 times the surface value, and would be consistent with an even larger rotation in the very central regions, but were unable to distinguish between a steady inward increase in Ω, and a model with Ω constant $\simeq 6 \Omega_o$ in a large central core $r < 0.7 R_\odot$, and a sudden discontinuous drop to near the surface value in the outer regions. The various models gave a solar quadropole moment, J_2, between 10^{-6} and 5.10^{-6}

The situation will become clearer as and when more data is available.

8. ROTATIONAL INSTABILITIES

There are potentially a range of instabilities that could be excited in the stable regions of rotating stars. In particular, if the angular momentum per unit mass decreases outwards over a surface of constant entropy, the star would be dynamically unstable (Høiland 1941). Shear instabilities have been considered by Zahn (1975), who concludes that horizontal shear, i.e. along surfaces of constant entropy, is readily excited but has a long growth time unless there is some other turbulence already present. The question of shear instabilities in the radial direction is much more difficult as if it is present, it is a finite amplitude instability.

As was pointed out by Goldreich and Schubert (1967) and Fricke (1967, 1968), departures from cylindrical rotation laws drive an instability in which the stabilising effect of stratification is overcome by thermal diffusion. For a spherically symmetric angular velocity distribution $\Omega(r)$, instability sets in when

$$\left(\frac{d\Omega}{dr}\right)^2 > \frac{4\nu}{\kappa} \frac{N^2}{r^2} \tag{11}$$

where ν is the viscosity, κ the thermal diffusivity and N the Brunt Vaisala frequency. In solar conditions $\nu/\kappa \simeq 10^{-6}$ and $N \sim 10^{-3}$ $\sec^{-1}$, and this condition gives instability if $d\Omega/dr$ is greater than about $5\Omega_\odot/R_\odot$. This leads to a very nice picture that as the sun spins down, the interior becomes unstable: the resulting turbulence then diffuses angular momentum outwards and the resulting model has an angular velocity gradient similar to that deduced from rotational splitting of solar oscillations.

This picture is not correct since even a very small build up of chemical inhomogeneity can suppress the instability; the actual condition then becomes (Biermann 1968, Roxburgh 1975)

$$\left(\frac{d\Omega}{dr}\right)^2 > -\frac{4g}{r^2}\frac{1}{\mu}\frac{d\mu}{dr} \tag{12}$$

which permits very large angular velocity gradients to be stable.

However, the baroclynic instability is then effective, thermal diffusion destabilising what would otherwise be a stable state (Knobloch and Spruit 1982). This instability is not suppressed by a radial distribution of chemical composition $\mu = \mu(r)$ and the condition for instability becomes

$$\left(\frac{d\Omega}{dr}\right)^2 > \frac{8\nu}{\kappa} \frac{N^2}{r^2} \tag{13}$$

Even this analysis is not correct; whilst a radial distribution of chemical composition does not suppress the baroclinic instability, a particular non spherical distribution can. Thus the instability can redistribute matter until the system is stable again (Roxburgh 1982). However, as the chemical composition changes through nuclear reactions, the non-spherical distribution required to suppress instability changes leading to a continual slow turbulent diffusion (Roxburgh 1982) which will again transport angular momentum outwards.

The detailed analysis of rotational instabilities is still incomplete. Very little has been done on non axisymmetric perturbations, but it does seem that in the absence of a magnetic field, there will be some weak turbulence in the solar interior, but how effective it is remains to be determined.

9. ROTATIONAL SPINDOWN AND STELLAR STRUCTURE

We can now try to put together the previous discussion to try to build a plausible model (or models!) of the evolution of stars as they spin down, but such models are highly speculative.

Initially, stars are rapidly rotating, mass and angular momentum loss slow down the surface layers and produce differential rotation beneath the outer convective zone. This differential rotation is unstable, leading to mild turbulent diffusion which transports angular momentum from the interior to the convective zone. (cf Dicke 1972)

If the star has an oscillatory dynamo operating in the convective zone, or possibly just below the convective zone in the overshoot region, any central magnetic field is uncoupled from the dynamo field. If the star has an internal magnetic field, which is stable against 'floating' instabilities, this will keep the core uniformly rotating and there must then be a shear zone some distance beneath the convective zone, the gradient of angular velocity being determined by the efficiency of the rotationally driven turbulence.

If there is no internal magnetic field, we may expect a monotonic inwards increase in the angular velocity and some mild turbulence in the central regions which diffuses both angular momentum and chemical composition, leading to a greater degree of homogeneity in the central regions than predicted by the standard solar model, and hence a lowering of the neutrino flux. Attempts to model the effect of spin down with somewhat different assumptions have been made by Endal and Sofia (1981) and Schatzman and Maeder (1981).

There still remains the problem of the ^{3}He instability. One possibility is that this instability produces mild turbulence that keeps it close to marginal stability - again leading to diffusion of angular momentum and chemical composition. If this were so, we might expect to see the unstable mode - perhaps we do in the 160 min period global oscillation. The ^{3}He instability might be suppressed by rotationally driven turbulence diffusing ^{3}He and keeping the central regions stable. A magnetic field may also suppress this instability.

If some mixing is caused by rotational spin down, it will not suppress evolution to the red giant phase. The effect of mixing will decrease as the star slows down so that the main effect will be to increase the main sequence life time of such stars. This could lead to significant changes in estimates of the ages of globular clusters.

10. CONCLUSIONS

At the end of his book on the Internal Constitution of the Stars, Sir Arthur Eddington (1926) concluded: "It is reasonable to hope that in the not too distant future we shall be competent to understand so simple a thing as a star". We have not yet reached that position. Our knowledge is still very limited and where observational constraints are few there is much room for theoretical speculation. Let us hope that improved data, particularly of solar oscillations, will soon be available to help us reach that goal Eddington thought was close at hand, but which seems still to be remote.

REFERENCES

Altschuler, M.D. and Newkirk, G.: 1969, Solar Phys. 9, p. 131.
Biermann, P.: 1968, Diplomarbeit, Göttingen.
Claverie, A., Isaak, G.R., McLeod, C.P., and van der Ray, H.B.: 1981, Nature 293, p. 443.
Christensen-Dalsgaard, J., Dilke, F.W.W., and Gough, D.O.: 1974, Monthly Notices Roy. Astron. Soc. 169, p. 429.
Davis, R.: 1978, Proceedings Conf. on Solar Neutrinos, Brookhaven Nat. Lab. BNL 50879, Vol. I, p. 1.
Dicke, R.H.: 1972, Astrophys. J. 171, p. 331.
Dilke, F.W.W. and Gough, D.O.: 1972, Nature 240, p. 262.
Durney, B.R.: 1972, in "Solar Wind", eds. C.P. Sonnett, P.J. Coleman, and J.M. Wilcox, NASA SP 308, p. 282.
Durney, B.R. and Pneuman, G.W.: 1975, Solar Phys. 40, p. 461.
Eddington, A.S.: 1926, "Internal Constitution of the Stars, Cambridge University Press, p. 393.
Endal, A. and Sofia, S.: 1981, Astrophys. J. 243, p. 625.
Endler, F.: 1971, Mitt. Astron. Ges. 30, p. 136.
Fricke, K.: 1968, Z. Astrophys. 68, p. 317.
Goldreich, P. and Schubert, G.: 1967, Astrophys. J. 150, p. 571.
Gough, D.O.: 1982, Nature 298, p. 334.
Hill, H.: 1982, Irish Astron. J. (in press).
Høiland, E.: 1941, Avhandl. utgilt Norske Vidensk. Akad. Oslo, I Mat.-naturv. kl. No. 11.
Knobloch, E. and Spruit, H.C.: 1982, Astron. Astrophys. (in press).
Mestel, L.: 1968, Monthly Notices Roy. Astron. Soc. 138, p. 359.
Parker, E.N.: 1958, Astrophys. J. 128, p. 664.
Pneuman, G.W. and Kopp, R.A.: 1971, Solar Phys. 18, p. 258.
Robertson, B.R.: 1983, Solar Phys. (in press).

Rood, R.T.: 1978, Proceedings Conf. on Solar Neutrinos, Brookhaven Nat. Lab. BNL 50879, Vol. I, p. 175.
Rowse, D.P. and Roxburgh, I.W.: 1981, Solar Phys. 74, p. 165.
Roxburgh, I.W.: 1975, Mem. Soc. Roy. Sci. Liège, 6e Série 8, p. 69.
Roxburgh, I.W.: 1976, in "Basic Mechanisms of Solar Activity", eds. V. Bumba and J. Kleczek, Dordrecht: D. Reidel Publ. Co., p. 453.
Roxburgh, I.W.: 1978, Proceedings Conf. on Solar Neutrinos, Brookhaven Nat. Lab. BNL 50879, Vol. I, p. 207.
Roxburgh, I.W.: 1978, in "Pleins Feux sur la Physique Solaire", Centre Nat. de la Recherche Scientifique, No. 282, p. 21.
Roxburgh, I.W.: 1978, Astron. Astrophys. 65, p. 281.
Roxburgh, I.W.: 1982, (to be published).
Roxburgh, I.W.: 1983, (to be published).
Schatten, K.H.: 1970, Solar Phys. 15, p. 499.
Schatzman, E. and Maeder, A.: 1981, Astron. Astrophys. 96, p. 1.
Skumanich, A.: 1972, Astrophys. J. 171, p. 565.
Weber, E.J. and Davis, L., Jr.: 1967, Astrophys. J. 148, p. 217.
Zahn, J.P.: 1975, Mem. Soc. Roy. Sci. Liège, 6e Série 8, p. 31.

DISCUSSION

STIX: The shear layer near the bottom of the convection zone is just where the dynamo seems to need it (because of the magnetic buoyancy argument). On the other hand, the 20-fold angular velocity at the Sun's center that you showed on an earlier viewgraph is in conflict with the oblateness measurements, isn't it?

ROXBURGH: I consider the shear layer below the convective zone as a possible site for dynamo action. The diffusive rotation model is really only an illustration, and one would obtain different solutions by changing either or both the magnitude and radial variation of the diffusion coefficient.

MOUSCHOVIAS: The solar wind emanates primarily from coronal holes, which are found more frequently at higher latitudes. Higher latitudes, therefore, are magnetically braked faster than lower latitudes. Does a quantitative comparison of time scales (of magnetic braking and of redistribution of angular momentum on the solar surface) allow an explanation of the observed differential rotation in the above manner?

ROXBURGH: The diffusion time scale for the convective zone is of the order of years — very much shorter than the wind spindown time scale, so this effect should not be important in determining the differential rotation.

HARTMANN: Would you expect a difference in the way that stars with shallow convective zones would spin down as compared with the spindown of more fully convective stars?

ROXBURGH: The easy answer is yes! But in what way? First I would expect the differential rotation and the dynamo and the mass loss to all vary with convective zone depth. Even if all these were unchanged, the response of the star would be different as the diffusion coefficient, which is probably high in convective zones and probably smaller in radiative zones, determines how the star responds to a given rate of mass loss. I would anticipate that the deeper the convective zone the longer would be the surface rotation spindown time. But it is not reasonable to assume that all other things are equal!

KÖMLE: If you change the magnetic-field structure in your model from a dipole to a quadrupole — what would the effect on the solar wind bulk velocity be?

ROXBURGH: It is increased. The quadrupole field diverges more rapidly than the dipole, and this leads to higher velocities.

KOUTCHMY: (1) My first question deals with the problem of calculating the angular momentum loss using the concept of a source surface. According to this concept, coronal structures outside the source surface should show the characteristic spiral pattern. This is clearly not observed, but the large-scale coronal structures are rather showing the rigid rotation properties up to distances of at least 12 - 15 $R_\odot$. Does this mean that the angular momentum loss of the sun is larger than what we think it is, and is this kind of observations relevant to the problem of angular momentum loss? (2) My second question is: Did you compute the solar-cycle related angular momentum loss due to coronal transient events?

ROXBURGH: (1) I believe it is misleading to refer to the source surface model for comparison with observations. At best it can only be an average of the real sun. The extent of individual corotating structures are determined by the actual properties of the magnetic field and corona — strong fields could cause corotation out to $15R_\odot$. (2) The answer to the second question is no!

ENDAL: A difficult problem that I encountered in my calculations (with S. Sofia) of solar spindown and internal rotation is that the composition gradient in the core prevents sufficient angular momentum removal as needed to satisfy the oblateness constraints. If the recent analysis of Knobloch and Spruit is correct, this may indicate the way out of this dilemma, and this would be a very important result.

ROXBURGH: I agree — indeed the model I calculated in 1971 for marginal stability of the Goldreich-Schubert-Fricke instability had similar problems. The triple diffusive instability does offer the possibility of increased diffusion — although it depends on the non-linear saturation of the instability.

TARBELL: Bill Press of Harvard has suggested another possible problem with the standard solar model. He suggests that gravity waves generated in the convection zone propagate into the core without too severe damping. The enhanced transverse gradients in temperature lead to enhanced radiative energy transport, simulating a reduced opacity. Are you aware of any calculated evolutionary sequences considering this effect? Do you think it is important?

ROXBURGH: In fact Henry Hill suggested this some years ago (see also Roxburgh, 1975). I doubt if this could be important on a thermal diffusion time for the sun — but perhaps in other stars with larger amplitude oscillations. But there is a need for a detailed study of the effect of oscillations on internal structure and angular momentum transport.

ROSNER: In response to Tarbell's question, I understand that W. Press and G. Rybicki have recently evaluated the diffusive properties of the modes in question. They find that as far as radiative transport is concerned, these modes probably do not resolve the neutrino problem. However, these modes probably do have important consequences for mixing, and hence may play an important role in the development of instabilities such as the triple-diffusive instability of Knobloch and Spruit (1983).

ROXBURGH: I would be pleased to receive a copy of that work.

THE REMARKABLE ROTATIONAL BRAKING OF G5 GIANTS

David F. Gray
University of Western Ontario, London, Ontario, Canada

ABSTRACT
The strong rotational braking seen in the G5 III stage of evolution may be the key to understanding how stellar dynamos work.

We are all familiar with the leisurely spin-down seen in cool main-sequence stars like our Sun. The time scales here are ~ 10^9 years and the accepted cause is the loss of high angular momentum mass in the form of stellar winds interacting with the stellar magnetic field. The magnetic field is believed to result from the interaction of envelope convection with the rotation of the star through a dynamo mechanism. Our understanding of how a dynamo actually operates, how that operation depends on the driving forces of rotation and convection, what kind of stochastic and secular time variations are to be expected, remains fragmentary even though many inventive minds have contributed. One reason for slow progress is simply that the Sun is almost the only example of a stellar dynamo we have had. But nature has given us another, much more powerful dynamo in the G5 giants, it just took us a little longer to discover it.

Figure 1 shows the observational results: v sin i vs. spectral type for luminosity class III stars. Most of these v sin i values were obtained by Fourier analysis of the line broadening (details can be read in Gray 1981, 1982b). The crosses in Fig. 1 represent means for groups by spectral type but increased by the statistical projection factor of $4/\pi$. The inset graph shows these same means along with means for earlier spectral types according to Fukuda (1981). The observed fact, then, is that the rotation drops abruptly at G5 III from ~ 25 km s^{-1} to ~ 5 km s^{-1}.

Note here how the giant spectral sequence tells us the time order of events in contrast to the main-sequence where we get an ordering by mass. Several people have used evolutionary models to study the expected changes in rotation as a star moves off the main-sequence. I have chosen to use the comprehensive and up-to-date calculations of

J. O. Stenflo (ed.), Solar and Magnetic Fields: Origins and Coronal Effects, 461–466.

Endal and Sofia (1979). Starting with a main-sequence rotation of ~ 150 km s^{-1}, simple evolutionary expansion results in a sixfold increase in the moment of inertia and satisfactorily reproduces the observed 25 km s^{-1} rotation as a star enters G5 III. This is shown by the dashed lines in Fig. 1. The theoretical calculations did not anticipate the remarkable brake that occurs at this point. However, if I scale the dashed line down arbitrarily by 5.7 times, it agrees very well with the observations (crosses in Fig. 1). In other words, from G5 III, after the braking, to K2 III, the internal moment of inertia once again controls the rotational decline. The braking not only started abruptly at G5 III, it also ends abruptly at G5 III.

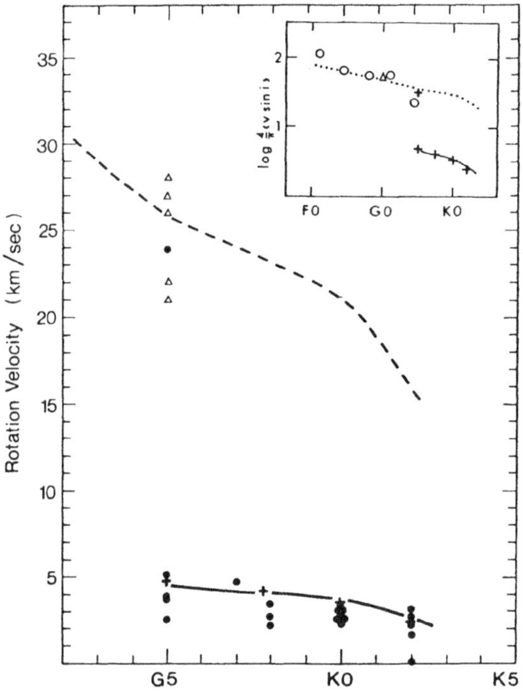

Fig. 1. Observed rotation velocities, v sin i, are shown by the dots and triangles. Means for spectral type groups which have been increased by 4/π are shown by crosses. Model calculations are shown by the dashed and solid lines. The inset graph shows 4/π times the mean v sin i's (on a log scale) over a wider spectral interval (from Gray 1982b).

Considering that the moment of inertia for a G5 giant is an order of magnitude larger than for a solar-type star, and the braking time scale ratio is ~ 10^4 yrs/10^9 yrs, the brake at G5 III is ~ 10^6 times stronger than the solar variety. Admittedly this is a rough estimate, but it brings home the point that this dynamo brake is unequivocally vigorous compared to what we are accustomed to considering for the Sun.

Let me jump now directly to the dynamo explanation and refer you to the cited references for some considerations of alternative hypotheses. First, we are led to a dynamo explanation because it is at G5 III where significant envelope convection is added to the rotating star, giving the star the driving forces it needs to generate a dynamo field and the resulting magnetic brake. Second, as angular momentum is removed from the star, the rotation eventually becomes too small to drive the dynamo. This happens at a surface rotation rate of ~ 5 km s^{-1} for G5 III stars, but we would guess it might be a function of mass.

We are now in a position to estimate the rate of dissipation of angular momentum and the size of the dynamo magnetic field. Further, we can ask about what actually happened when the dynamo stopped and what the envelope parameters (thickness, run of velocities, densities, etc. with depth) were at the time of cessation. These data will help us understand how a dynamo works.

Now let us turn back to main-sequence stars. When stars destined to be on the cool half of the main-sequence are in their adolescent stages, they have those two special ingredients of rotation and convection. We saw the remarkable reaction in G5 giants when these were brought together. Who then could deny the possibility of a similar magnetic brake functioning in young stars and/or pre-main-sequence stars? The important feature I wish to consider here is the abrupt turn-off of the brake. Adolescent stars would be spun down until the rotation was insufficient for dynamo action---at least the virile type of dynamo responsible for the brake we are considering. Therefore, we would expect young stars to appear on the main-sequence with a rotation rate reflecting the dynamo turn-off characteristics appropriate to their mass. That in turn implies an upper bound to the rotation rate along the main-sequence. Figure 2 shows that the observed main-sequence rates do indeed have an upper bound, although this observed one corresponds to a somewhat later time ~ $10^{8.5}$ yrs. Under the dynamo-magnetic-brake hypothesis, the upper bound becomes a map proportional to the dynamo turn-off

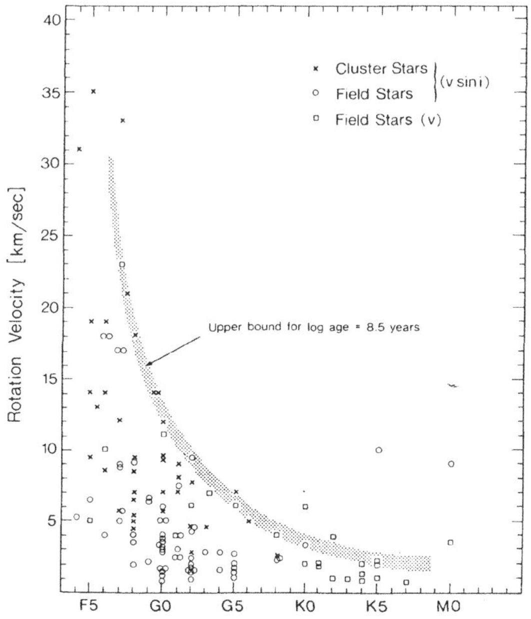

Fig. 2. Rotation rates for main-sequence stars are shown as a function of spectral type. The subjectively drawn upper bound is indicated (from Gray 1982a).

rotation rates. We are faced with the delightful prospect of being able to identify the internal structure, as a function of mass, when the strong dynamo mechanism ceases operating.

Certainly there is more to the rotational history of a main-sequence star, and much of it has been known for a decade or longer. Specifically points in Fig. 2 will lie below the upper bound because of (1) the sin i projection factor and (2) the slow spin-down going ~ $\text{age}^{-\frac{1}{2}}$, resulting from the puny solar-type dynamo brake.

It is also of interest to notice that the quantity L = MRv, a psuedo-angular momentum, (M=mass, R-radius, v=surface rotational velocity) varies as M^7 for points along the upper bound of Fig. 2. The relation

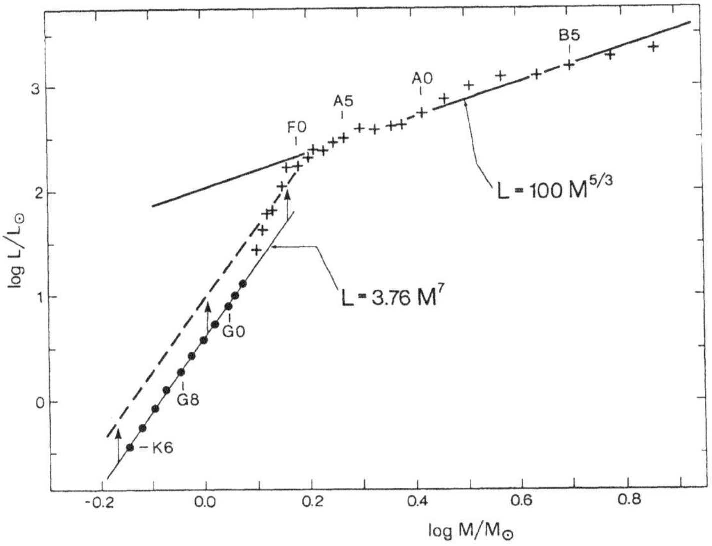

Fig. 3. The psuedo-angular momentum, L = MRv, is shown as a function of mass. The points defining the M^7 relation correspond to the uppper bound shown in Fig. 2. The dashed line corresponds to the upper bound at a younger age ~ 10^8 yrs.

to the upper main-sequence is shown in Fig. 3. A "de-aging" of the M^7 portion by 0.3-0.4 dex leads to one smooth break in slope near F0 or a mass of ~ $1.5M_\odot$. This may imply that the actual brake-turn-off velocities are 2 to 2.5 times larger than those shown in the observed upper bound of Fig. 2.

Studies of stellar braking may give us the observational guidance we need to firm up our understanding of dynamos.

REFERENCES

Endal, A.S. and Sofia, S.: 1979, Astrophys. J. 232, p. 531.
Fukuda, I.: 1981, preprint.
Gray, D.F.: 1981, Astrophys. J. 251, p. 155.
Gray, D.F.: 1982a, Astrophys. J. 261, Oct. 15 issue.
Gray, D.F.: 1982b, Astrophys. J. 262, Nov. 15 issue.

DISCUSSION

SCHÜSSLER: Have these stars also been observed at X-ray and Ca II wavelengths and—if yes — do these observations show the same sort of behaviour?

GRAY: The X-ray data for class III giants show the same discontinuity at G5 III. But the X-ray flux scale has to be compressed by a factor of two relative to the rotational velocity scale, implying that X-ray flux varies with the square of the rotational velocity. The Ca H and K line emission seems to vary proportional to the rotational velocity in dwarfs. Other chromospheric and transition-region lines appear to be correlated with rotation, but show no simple functional dependence. We will have to look into this further.

SIMON: (1) Comment in response to Schüssler's question: Ultraviolet observations of G-K giants do show an apparent break in transition-region properties near G5 III, but the observational situation is very complex. In particular, transition-region lines among early G giants show a wide range in strength (Simon, Linsky, and Stencel: 1982, *Astrophys. J.* **257**, p. 225). (2) I am puzzled by your remark that the existence of a rotational brake in the Hertzsprung gap was previously unknown, since R. Kraft published a classic paper on this question more than 15 years ago in the *Astrophysical Journal*. I would like to know how your results differ from Kraft's. Moreover, since Kraft, and later Alschuler, showed that 75-90% of G5-K2 giants are likely to be second crossers, travelling from right to left across the H-R diagram, how can we be sure that the apparent drop in surface velocity occurs *at* G5 rather than elsewhere in the H-R diagram, say in the red giant region?

GRAY: Kraft measured a few Hyades giants (K0 III) and found upper limits of 5 or 6 km s^{-1}. From this he concluded that angular momentum loss must have occurred. Using these upper limits, Endal and Sofia (1979, *Astrophys. J.* **232**, p. 531) reversed this conclusion on the basis of their ability to calculate fairly realistic models, which met the 5–6 km s^{-1} upper bound. My measured values are half of Kraft's upper limits and reinstate the original conclusion of mass loss being needed. To my knowledge, Kraft made no measurements of G5 - K0 giants and so did not see the discontinuity. Similarly, if my memory is correct, Alschuler's work (1975, *Astrophys. J.*) extended only as late as G5 III and was really mainly concerned with F stars. So he did not see the discontinuity either.

The question of first vs. second vs. third crossing is unresolved. Most evolutionary calculations in this mass range seem to show loops extending to ~G8 but not to G5. Similarly, color-magnitude diagrams of globular clusters seem to support the G8 position for the blue end of the loop. Even if the K0 III and K2 III stars I have measured are on the second or third crossing, it does not destroy the picture I have presented, but of course you are free to develop your own interpretation of the G5 III discontinuity.

WALTER: I grant you that the rapidly rotating G giants are likely to have evolved from rapidly rotating upper main sequence stars. However, by the time that you reach K0, this is no longer necessarily true: many may be low-mass stars like Arcturus. If so, the K giant population, or some proportion thereof, began as slow rotators. Could your alleged break be due to selection effects, primarily sampling low-mass K0 giants, as well as second crossing stars?

GRAY: It seems to me very unlikely that all of the G5 - K2 giants of my sample should be low-mass stars. If some low-mass stars were intermixed with others having evolved horizontally across the H-R diagram, then I would expect to see a wide spread of rotational velocities. Instead, the G5 III to K2 III observations are indicative of a rather homogeneous group. Further, the evolutionary tracks that I have seen remain basically horizontal until at least K0 III. Certainly by K5 III more vertical-like tracks are computed, but that is well to the cool side of my stars.

SODERBLOM: Let me comment on this question: Although I am skeptical of your evolutionary interpretation, I remind you that four of your stars are the Hyades giants, which rotate just as slowly as the other K0 III's. These are definitely not low mass stars, in fact they have about $3M_\odot$.

MOUSCHOVIAS: The quantity $\bar{\rho}_*^{2/3} R_*/R_\odot$ is close to unity for stars earlier than F5. Does any observer here know whether this quantity is significantly smaller than unity for later-type stars? (If it is, the break in the rotational velocity with spectral type can be understood very simply.)

GRAY: I don't know.

ENDAL: In reply to Ted Simon's remark about older observations, I would note the following: The Endal and Sofia (1971) paper showed that Kraft's observations (upper limits) could be explained without angular momentum loss. Gray's results, by contrast, definitely require angular momentum loss.

GRAY: Yes, my observations show the rotation to be about half of Kraft's upper limit.

SODERBLOM: I wish to take issue with your claim that the distribution of velocities that is present before the braking is completely eliminated, resulting in a single rotational velocity for each spectral type, G5 and later. Your data appear to me to show the same proportionate spread after G5 as before. Given these modest $v \sin i$'s and observational error, how can you rule out a similar Maxwell-Boltzmann distribution among the slow rotators?

GRAY: Let us go back and look at Figure 1. The individual measurements are shown by the solid dots. If I take means for the four obvious spectral groups, and then increase this mean by $4/\pi$ to statistically account for the average $\sin i$ value, I get the crosses you see. Notice how these crosses fall near the top of each set of dots, at least for the G5, G8, and K0 groups. This is what you would expect for a $\sin i$ distribution, and what you would not expect for a Maxwell-Boltzmann distribution. I have done a more complete analysis (1982, *Astrophys. J.* **262**, pp. 682 – 699), from which I conclude that a $\sin i$ distribution is the better representation of the observations. With statistics on about two dozen points, the conclusions of this sort are obviously not firm.

NORDLUND: If you read off the upper envelope on your slide with the main sequence stars, you get approximately the same 5 km s^{-1} as for the giants. Do you consider this significant?

GRAY: I suspect that the main sequence velocities of the brake turn-off need to be scaled up so that they correspond to an earlier age (as I indicated in my talk). In such a case, the 5 km s^{-1} correspondence between giants and dwarfs is coincidental.

EXPULSION OF MAGNETIZED PLASMAS FROM CORONAE

B. C. Low
High Altitude Observatory
National Center for Atmospheric Research[1]
Boulder, Colorado, USA.

Abstract. A theory of the white light transient phenomena in the solar corona is given, based on recent analytic treatments of magnetohydrodynamics.

I. INTRODUCTION

When the solar corona is observed in electron-scattered white light, large-scale structures are found to undergo reconfiguration on average about twice a day, often involving an outward ejection of matter estimated to be of the order of 10^{15} gm. This new awareness of an active corona came largely as the result of data obtained in the past decade with the use of space-born coronagraphs (MacQueen 1980). Some progress has been made in treating analytically the magnetohydrodynamics (MHD) of stellar atmospheres. We present a physical picture, suggested by the theoretical MHD results obtained, to explain how and why magnetized plasma is sporadically expelled out of the gravitational bound of the Sun. Conceivably, the same physical process takes place in coronae other than that of the Sun.

II. THE ERUPTION OF CORONAL MAGNETIC FIELDS

An essential effect of solar activity is the introduction of fresh magnetic flux into the corona from below. The magnetic field at first evolves slowly, driven by gradual changes of boundary conditions due to the transport of magnetic flux along and across the coronal base. Slow changes also result from fluctuation in the ambient heating and energy transport taking place in the corona everywhere. On the very large scale, say 10^5km, the electrical conductivity may be regarded to be infinite. Consequently, free energy can be stored in the form of electric currents as a magnetic field is gradually

J. O. Stenflo (ed.), Solar and Magnetic Fields: Origins and Coronal Effects, 467–471.

distorted during its evolution. In the low corona, pressure gradients and gravity have important roles to play. One situation in which this is the case, is the formation of prominence filaments supported by Lorentz force against gravity. It is well-known from eclipse pictures that these filaments are encased in cavities of low density (Newkirk 1967). Such cavities do not collapse presumably because of an internal magnetic field (Low 1980). In the stratified atmosphere, the cavity is buoyant but the "magnetic" buoyancy (Parker 1955) is countered by the tension force of the field lines anchored to the photosphere. As the magnetic footpoints evolve apart, the tension force eventually fails to hold the cavity in equilibrium. Then, the cavity rises indefinitely as the prominence erupts (Low 1981, Wolfson 1982). The point to stress is that the corona is an active system capable of storing free energy to erupt. The eruption can take place in association with a prominence as we just described. It can also take place, more impulsively, in association with a flare in the low corona (Munro et al. 1979). In both types of events, the flare, prominence and coronal re-configuration need not have a simple cause-effect relationship with one another but are all consequences of a global loss of equilibrium.

Preliminary results on the coronal transients in its early development below $2.2R_{\odot}$, observed with HAO's ground-based coronagraph at Mauna Loa, showed that two main classes of events can be discerned (MacQueen and Fisher 1982). Those that are flare-associated, moving at speeds of 500 km/s or more and characterized by bright loops and those that are associated with eruptive prominences, slowly moving at speeds typically below 200 km/s and characterized by a moving void. It seems natural to identify the latter "dark" transients with a prominence cavity that has broken away, and the case can be supported with a comparison between theoretical models and observation (Fisher, Garcia and Seagraves 1981 and Low, Munro and Fisher 1982). A dark transient may start to move before the associated prominence erupts. This, as well as its characteristic low speed, is consistent with a spontaneous loss of equilibrium. Since the motion is driven internally by the hydromagnetic system, drawing upon the stored free energy, the transition is not impulsive and the speed of the plasma can, at most, be accelerated to the order of the signal speeds at which the fluid system communicates between its different parts. The sound speed at the coronal temperature of 1.5×10^6 K is about 120 km/sec. For a plasma beta in the neighborhood of unity, applicable to the prominence environment, the hydromagnetic signal speeds are perhaps not more than a factor of two larger. Many of the dark transients are, in fact, observed to have speeds of less than 100 km/sec. The dark transients develop bright loops as they travel above 2.2 $R_\odot$, the typical field of view of orbiting coronagraphs. Except for their low speeds, the appearances of these transients in this field of view are similar to the flare-associated ones that developed bright loops during their early stages.

For the fully developed coronal transients, the time-dependent MHD equations provide the lowest order description. For a $\gamma=4/3$ polytrope, the atmosphere can expand radially outward in a self-similar flow (Low 1982a, b, 1983). Explicit solutions can be constructed. An essential feature is that solar gravity is included and the expelled plasma can travel outward at speeds less than the local escape speed. From these self-similar solutions, two implications are noteworthy.

For a polytropic magnetostatic atmosphere in a $1/r^2$ gravity, the total energy is given by $E=-(3\gamma-4)U$, where U is the internal energy and various singularities at r=0 have been appropriately re-normalized. The polytrope $\gamma=4/3$ is the marginal case between gravitationally bound (E<0) and unbound (E>0) states. That the self-similar MHD solutions exist for $\gamma=4/3$ is a manifestation of the fact that the static atmosphere for $\gamma\leq 4/3$ is gravitationally unstable. It follows that the outward expansion of a coronal transient could be interpreted in general terms to be the result of a hydromagnetic system that has gained sufficient energy to become gravitationally unbound. The transition to an unbound state may be gradual, such as through the introduction of new magnetic flux and ambient heating, or, in the case of a flare-related event, through the enhanced heating as part of the flare build-up. The transition may be impulsive such as through a sudden input of energy released by a flare. The flare with its rapid rise time is effectively an external driver, being able to impart speeds much in excess of the characteristic speeds of the coronal medium above it. Transient speeds of the order of 1000 km/sec are sometimes observed.

We interpret the coronal transient to be the outflow of a gravitationally unbound, pre-existing hydromagnetic structure in the corona. The observed white-light density structure is a part of this moving pre-existing structure. This contrasts with the physical picture of some numerical MHD models (e.g., Dryer _et al_. 1979), in which the observed transient density structure is interpreted to be a wave-like response in the ambient atmosphere due to an impulsive input at the coronal base. In our picture, the ambient atmosphere is swept upward and sideways by a large-scale outflow.

III. CONCLUSION

If there is no magnetic field, the corona expands outward in a spherically symmetric solar wind, the impossibility of a gravitationally bound state being the consequence of heating and the absence of adequate confining pressures in the interstellar medium (Parker 1963). Magnetic fields anchored at their footpoints can trap pockets of coronal matter, first in gravitationally bound local systems and then through subsequent heating and evolution, transit into an outward moving unbound state. This, we believe, is the origin of coronal transients.

ACKNOWLEDGMENT

I thank Dick Fisher, Rainer Illing and Bob MacQueen for stimulating discussions.

REFERENCES

Dryer, M., Wu, S.T., Steinolfson, R.S., and Wilson, R.M.: 1979, Astrophys. J. 277, p. 1059.
Fisher, R., Garcia, C.J., and Seagraves, P.: 1981, Astrophys. J. Letters 246, p. L161.
Low, B.C.: 1980, Solar Phys. 65, p. 147.
Low, B.C.: 1981, Astrophys. J. 251, p. 352.
Low, B.C.: 1982a, Astrophys. J. 254, p. 796.
Low, B.C.: 1982b, Astrophys. J., in press.
Low, B.C.: 1983, Astrophys. J., submitted.
Low, B.C., Munro, R.H., and Fisher, R.R.: 1982, Astrophys. J. 254, 335.
MacQueen, R.M.: 1980, Phil. Trans. Roy. Soc. London A297, p. 605.
MacQueen, R.M. and Fisher, R.R.: 1982, private communication.
Munro, R.H., Gosling, J.T., Hildner, E., MacQueen, R.M., Poland, A.I., and Ross, C.L.: 1979, Solar Phys. 61, p. 201.
Newkirk, Jr., G.: 1967, Ann. Rev. Astron. Astrophys. 5, p. 213
Parker, E.N.: 1955, Astrophys. J. 121, p. 491.
Parker, E.N.: 1963, "Interplanetary Dynamical Processes" (Wiley).
Wolfson, R.L.T.: 1982, Astrophys. J. 255, p. 774.

Note:

[1]The National Center for Atmospheric Research is sponsored by the National Science Foundation.

DISCUSSION

SPRUIT: **Can your self-similar solutions have the form of observed loops?**

LOW: Yes, and that is easy to see. In the self-similar space, explicit time dependence has been removed, and one has a problem very much like that of magnetostatic equilibrium. Loop solutions occur naturally. Transformed into real space-time, the loop solutions become expanding loop structures.

CAMPOS: Your self-similarity solution depends on the ratio of specific heats, $\gamma = 4/3$. For a fully ionized gas, with monoatomic molecules, $\gamma = 5/3$, and you say that the difference is due to heating. The value $\gamma = 4/3$, however, is the classical ratio of specific heats for a polyatomic gas, with three-dimensional molecules, and 6 degrees of freedom. Why do you need a value of γ for polyatomic molecules when the coronal gas is monoatomic? Is it a coincidence that heating accounts exactly for the difference between $\gamma = 4/3$ and $\gamma = 5/3$?

LOW: In my analysis I do not regard γ to refer directly to the degrees of freedom of the constituent gas. The polytropic relation is a mathematical device to close the equations of motion. The interesting properties of the $\gamma = 4/3$ polytrope, namely, those properties arising from a marginally bound state, are applied to the corona in the very crude approximation that a $\gamma < 5/3$ polytrope simulates a $\gamma = 5/3$ gas subject to heating.

MARTENS: Do I understand that by taking $\gamma = 4/3$, you simulate or solve for the coronal heating problem?

LOW: One does not solve the coronal heating problem by taking $\gamma < 5/3$. The idea is that a $\gamma < 5/3$ for an expanding gas implies a pressure going down with density more slowly than the 5/3 power, and in this sense simulates some heating during the expansion.

PLASMA FLOW AND SOLAR WIND ACCELERATION IN NON-RADIAL CORONAL MAGNETIC FIELDS

H. Biernat, N. Kömle and H. Rucker
Institut für Weltraumforschung, Österreichische Akademie der Wissenschaften, Observatorium Lustbühel, Graz, Austria

In the vicinity of the Sun - especially above coronal holes - the magnetic field lines show strong non-radial divergence and considerable curvature (see e.g. Kopp and Holzer, 1976; Munro and Jackson, 1977; Ripken, 1977). In the following we study the influence of these characteristics on the expansion velocity of the solar wind.

1. BASIC CONCEPT OF THE SOLAR WIND FLOW IN A PRESCRIBED GEOMETRY

The steady state conservation equations for mass, momentum, energy and magnetic flux for an ideally conducting one-fluid plasma moving along a magnetic field line are given by the following equations (see e.g. Suess, 1979; Biernat et al., 1982):

$$nvA = F_1 \tag{1}$$

$$mv \frac{dv}{ds} = - \frac{1}{n} \frac{d}{ds} (2nkT) - \frac{GmM}{r^2} \frac{dr}{ds} \tag{2}$$

$$Anv \left[\frac{mv^2}{2} + \frac{\gamma}{\gamma-1} 2kT + Q - \frac{GmM}{r} \right] = F_2 \tag{3}$$

$$AB = F_3 \tag{4}$$

Here n is the number density of the protons, v the bulk velocity of the plasma, A the cross section of the flow tube; s the coordinate distance along the flow tube measured from the solar surface, T the temperature, r the radial distance from the solar centre and B the magnetic induction; Q is an energy addition function and γ the ratio of specific heats. F_1, F_2 and F_3 are constants of integration.

With the aid of (1) and (4) the Eq.(3) can be written as

$$\frac{1}{v} \frac{dv}{ds} \left[v^2 - \frac{c^2}{\gamma} \right] = \left[- \frac{GM}{r^2} \frac{dr}{ds} - \frac{1}{\gamma} \left(\frac{c^2}{B} \frac{dB}{ds} + \frac{dc^2}{ds} \right) \right] \tag{5}$$

J. O. Stenflo (ed.), Solar and Magnetic Fields: Origins and Coronal Effects, 473–477.

where c is the adiabatic sound speed $c = (\gamma 2kT/m)^{1/2}$. We set $\rho = r/R$ and $\sigma = s/R$, where R is the solar radius. Now we construct ρ, $d\rho/d\sigma$, B, $dB/d\sigma$ as given analytic functions of σ and T as an analytic function of ρ. For this purpose we use a function of the form

$$f(x) = \frac{1}{2} \{a + b + (a - b) \sin\left[\pi \left(\frac{1}{2} \pm \frac{x - p}{q - p}\right)\right]\} \qquad (6)$$

which describes a sinusoidal increase (positive sign) or decrease (negative sign) of f from a to b within the interval $p \le x \le q$.

Fig. 1 shows three field lines of different curvature. Field line (I) is radial, i.e. $d\rho/d\sigma = 1$ over the whole range. The field lines (II) and (III) are obtained by a decrease of $d\rho/d\sigma$ ($d\rho/d\sigma = f(\sigma)$) from 1 at the solar surface $\sigma = 0$ to 0.766 and 0.5, respectively, at $\sigma = 1.5$. $d\rho/d\sigma$ is not changed up to $\sigma = 1.6$ and increases from there again according to (6) until the value 1 is reached at $\sigma = 3$.

The divergence of the flow tubes (Kopp and Holzer, 1976) is given by the magnetic field strength along the field line. B is assumed to decrease from 2Γ at the coronal base according to

$$B(\sigma) \sim (1 + \sigma)^{-f(\sigma)} \qquad (7)$$

By setting f = 2 over the whole range (case (A)) a divergence is obtained as suggested by the simple radial case, where $A \sim r^2$.In order to study the effect of a stronger flow tube divergence, we let $f(\sigma)$ increase between $\sigma = 0$ and $\sigma = 1.5$ and decrease between $\sigma = 1.6$ and $\sigma = 3$ according to Eq.(6). At $\sigma = 0$ we set f = 2. In the cases (B) and (C) we vary f up to 3 and 4, respectively. Within the interval $1.6 \le \sigma \le 3$ the value of f again decreases to 2.

We now determine the temperature as a function of the radial distance from the Sun. We investigate two cases: (i) the isothermal case with $T = 2 \times 10^6$ K and (ii) the case with a temperature function fitted to observational data. At the coronal base we assume a low temperature (10^6 K) as might be expected for a coronal hole. From 1.5 up to 4 solar radii we take the values found by Withbroe et al. (1982) from EUV-obser-

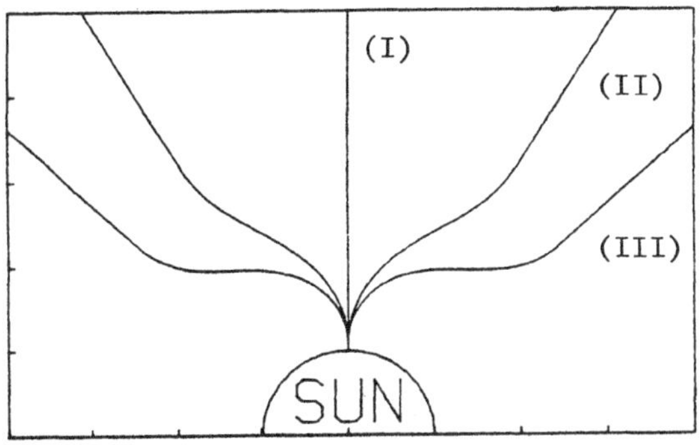

Fig. 1: Magnetic field lines of different curvature, along which the velocity profiles shown in Fig. 2 are calculated.

vations over a polar coronal hole. From 75 solar radii to 1 AU data of the HELIOS spacecraft as given by Cuperman et al. (1978) can be used. The exact values are given in Table 1.

Tab. 1: Temperature values as a function of the distance from the Sun

ρ	1.0	1.5	4.0	75.0	215.0 ($\hat{=}$ 1 AU)
T(K)	10^6	2.6×10^6	1.2×10^6	4.69×10^5	2.08×10^5

The increase of the temperature near the Sun is described by $T(\rho) = 10^6 \times (1 + 1.6 \sin\{\pi (\rho - 1)\})$. The temperature decrease beyond $\rho = 1.5$ is supposed to be

$$T(\rho) \sim \rho^{-f(\rho)} \tag{8}$$

where $f(\rho)$ in each temperature interval of Table 1 is given by (6).

Finally F_1 and F_2 are determined from observations given by Feldman et al. (1976) for typical high speed streams. They find at 1 AU for the average particle flux density $\langle nv \rangle = 3.3 \times 10^8$ cm^{-2} s^{-1} and for the energy flux density $\langle E \rangle = 2.4$ erg cm^{-2} s^{-1}. The constants F_1 and F_2 are obtained by multiplying with the cross section A of the flow tube at 1 AU. The constant F_3 is obtained by using (4) at the solar surface.

Starting from the critical points (which are obtained by setting the two brackets in (5) equal to zero), Eq.(5) is integrated numerically.

2. RESULTS

For strong curvature and/or divergence multiple critical points occur (see also Kopp and Holzer, 1976). For all cases investigated here the solution which describes an expanding supersonic solar wind runs through the innermost critical point. Different velocity profiles are shown for the isothermal case in Figures 2a and 2aa and for the observed temperature in Figures 2b and 2bb. Figures 2a and 2b show a solar wind expansion along field lines with different curvature (I), (II) and (III) in the case (A). Figures 2aa and 2bb illustrate an expansion along the radial field line (I), but with different flux tube divergence (A), (B) and (C). A comparison of Figures 2a and 2b shows that an increase of the curvature leads either to a decrease or an increase of the asymptotic velocity, depending on the temperature profile. In the isothermal case a slight decrease occurs; in the case of the observed temperature a somewhat stronger increase of the final velocity v is found. From Figures 2aa and 2bb it is seen that increasing divergence in all cases increases the asymptotic velocity of the solar wind.

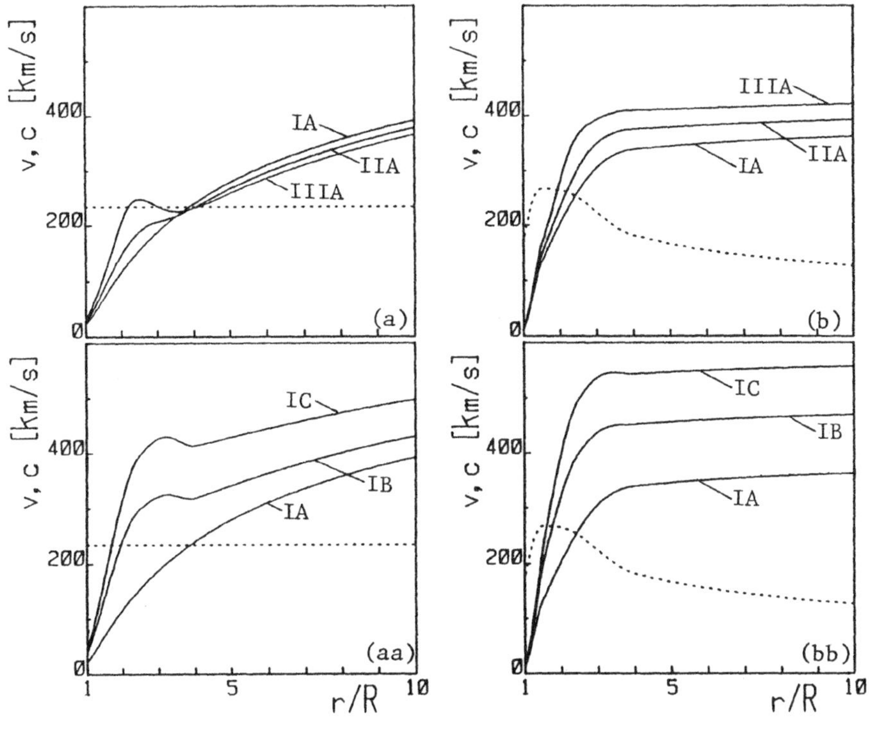

Fig. 2: Bulk velocity and adiabatic sound speed (dashed line) versus the distance from the solar centre in the isothermal case, (a) and (aa), and in the case of the observed temperature, (b) and (bb). IA denotes expansion along field line (I) with tube divergence (A), etc.

Acknowledgement: The authors wish to thank H. Lichtenegger for his help in preparing this paper.

REFERENCES

Biernat, H., Kömle, N., Rucker, H.: 1982, Kleinheub. Ber. 25, pp. 269-277
Cuperman, S., Levush, B., Dryer, M., Rosenbauer, H., Schwenn, R.: 1978, Astrophys. J. 226, pp. 1120-1128.
Feldman, W.C., Asbridge, J.R., Bame, S.J., Gosling, J.T.: 1976, J. Geophys. Res. 81, pp. 5054-5060.
Kopp, R.A., Holzer, T.E.: 1976, Solar Phys. 49, pp. 43-56.
Munro, R.H., Jackson, B.V.: 1977, Astrophys. J. 213, pp. 874-886.
Ripken, H.W.: 1977, Thesis, Univ.Bonn.
Suess, S.T.: 1979, Space Sci. Rev. 23, pp. 159-200.
Withbroe, G.L., Kohl, J.L., Weiser, H., Noci, G., Munro, R.H.: 1982, to appear in Astrophys. J.

DISCUSSION

ROXBURGH: The isothermal model equations can be integrated in closed form — and the solutions immediately obtained — for any area factor A(r) (cf. Rowse and Roxburgh: 1981, *Solar Phys.* **74**, p. 165).

CAMPOS: Have you plotted the variation of the various energy terms (radial and tangential kinetic energy, potential energy, etc.) along the magnetic field lines? I think this would illustrate why the asymptotic velocity increases with curvature for one temperature profile and decreases for another.

KÖMLE: We have not done this yet, but I think this is a good suggestion.

BENZ: What radial dependence of the temperature have you assumed for your second case?

KÖMLE: We assumed a low temperature (10^6 K) at the base of the corona, a temperature maximum of 2.6×10^6 K at 1.5 solar radii, and from there a decrease fitted to EUV measurements given by Withbroe et al. in a recent paper.

KUIN: The main difference between your work and previous models is the study of the influence of the curvature on the physical properties along a field line, while in previous work only the influence of the expansion has been considered. Can you point out what the separate effect of the curvature is?

KÖMLE: The effect of the curvature depends on the assumed temperature profile. In the isothermal case the asymptotic bulk velocity decreases as the curvature of the field line is increased. In the case of the more realistic temperature profile, increasing curvature leads to an increase of the asymptotic solar wind speed.

MAGNETIC BRAKING AND ANGULAR MOMENTA OF PROTOSTARS

Telemachos Ch. Mouschovias
University of Illinois at Urbana-Champaign
Departments of Physics and Astronomy

ABSTRACT

A contracting, typical, isothermal interstellar cloud or fragment, magnetically linked to the surrounding medium, loses angular momentum so efficiently that it remains in nearly synchronous galactocentric orbit as long as the field is frozen in the matter. At a high enough density n_{dec}, the field decouples from the matter due to ambipolar diffusion. Thereafter angular momentum is essentially conserved and, consequently, there is a one-to-one correspondence between n_{dec} and the angular momenta of binary and single stars that can ultimately form. Although the calculated rotational velocities (typically, 100 km/sec) are not yet refined enough to account for peculiarities among different spectral classes which may be due to phenomena following protostar formation, they are encouragingly consistent with observations.

1. INTRODUCTION

Under the premise that stellar angular momentum is a remnant of the original angular momentum of the parent cloud, it has been shown that the entire range of periods of binary stars from 10 hours to 100 years (see Abt and Levy 1976) can be accounted for by a single mechanism (Mouschovias 1977; 1978a). The angular momentum of a stellar system (single or multiple) is simply that which is present at the time the magnetic field decouples from the matter and can no longer brake the rotation of a cloud or fragment by transporting angular momentum to the surrounding medium. [Contrary to many statements in the literature, the angular momentum of a parent cloud is much too large to have been stored in orbital motions of individual stars in a stellar cluster (see reviews Mouschovias 1978b; 1981). In this paper we show that the observed angular velocities of <u>single</u> stars (Vogel and Kuhi 1981; Wolff, Edwards and Preston 1982) can be accounted for by the same mechanism.

The observed temperature, ion and neutral densities, mass and magnetic field strength of diffuse H I clouds imply that the field is dy-

J. O. Stenflo (ed.), Solar and Magnetic Fields: Origins and Coronal Effects, 479–485.

namically important. In fact, clouds should form mainly by material motions along field lines, with lateral contraction setting in only when the density increases sufficiently for gravitational forces to come into play. The density of protons ($n_{p,0}$) at which this occurs is approximately given by the expression

$$n_{p,0} = 2.1 \times 10^{2} \frac{(B_{ext}/1\ \mu G)^{3/2}}{(M/1\ M_{\odot})^{1/2}} \quad cm^{-3}, \tag{1}$$

where B_{ext} is the "external" (background) field, and M the cloud mass. For the typical values $M = 10^{3}\ M_{\odot}$ and $B_{ext} = 3\ \mu G$, we find that $n_{p,0} = 34.6\ cm^{-3}$. The corresponding cloud radius is

$$R_0 = 1.9\left(\frac{M/1\ M_{\odot}}{n_{p,0}}\right)^{1/3} \quad pc, \tag{2}$$

which yields $R_0 = 5.8$ pc for the above example. (The constants on the right-hand sides of equations [1] and [2] are slightly different in the case of a disk or cylinder of radius R_0 and half height $Z_0 \approx R_0$. This is unimportant in what follows because the ultimate goal is to obtain the angular momenta of single stars, after several orders of magnitude of angular momentum has been transported away by magnetic braking). Once gravity comes into play, balance of gravitational and thermal-pressure forces is nearly established along field lines, and the cloud contracts only as rapidly as magnetic forces allow it to contract laterally. During such contraction the field strength in a cloud's core (B_c) increases with mass density (ρ_c) as (Mouschovias 1976a,b; also recent review 1982)

$$(B_c/B_0) = (\rho_c/\rho_0)^{1/2}, \tag{3}$$

where the quantities B_0 and ρ_0 refer to values at which gravitational forces become important.

In § 2 below we obtain an expression for the characteristic time of magnetic braking of a cloud whose angular momentum and magnetic field vectors are aligned; the "fanning out" of field lines in the immediate neighborhood of a cloud (or fragment) is now taken into consideration. The density at which the field decouples from the matter due to ambipolar diffusion (Mestel and Spitzer 1956) is calculated in § 3 by equating the time scales for dynamical contraction and ambipolar diffusion. It is also shown that, up to such a density, the time scale for the magnetic braking of a fragment is smaller than the other two time scales. The implied angular velocities of single stars are obtained and compared with observations in § 4.

2. MAGNETIC BRAKING REVISITED

The characteristic time for magnetic braking of an aligned rotator (disk or cylinder) of density $\rho_{c\ell}$ and half thickness $Z_{c\ell}$ is given by $\tau_{\parallel} = (\rho_{c\ell}/\rho_{ext})(Z_{c\ell}/v_{A,ext})$, where $v_{A,ext}$ is the Alfvén speed in the external medium (Mouschovias and Paleologou 1980a,b). This is an exact expression for a rigid rotator, and an excellent approximation (except for short-lived transient effects) for a differentially rotating disk. It applies to the case in which the magnetic field strength in the external medium is comparable to that of the interior field. In a more realistic geometry, in which the field strength decreases from its value in the cloud, through a transition region, to that of the background field, the time scale $\tau_{\parallel}$ acquires a correction factor whose value depends on the thickness of the transition region. If the transition region is extended, so that the moment of inertia of the medium therein (assumed to have density ρ_{ext}) is larger than that of the cloud, then $\tau_{\parallel}$ is reduced from its above value approximately by the factor 2/15. If, on the other hand, the transition region is characterized by a relatively small moment of inertia, it is the external medium beyond the transition region which accepts the cloud's angular momentum. In such a case, $\tau_{\parallel}$ is given by (Mouschovias 1983)

$$\tau_{\parallel} = \frac{\rho_{c\ell}}{\rho_{ext}} \frac{Z_{c\ell}}{v_{A,ext}} \left(\frac{R_{c\ell}}{R_0}\right)^4 , \tag{4}$$

where R_0 is the original radius of the disk-cloud, when its density was $\rho_0 > \rho_{ext}$ (e.g., see eq. [1]), due to its formation by motions along field lines, and its magnetic field was equal to that of the external medium. Since ρ_{ext} and $v_{A,ext}$ refer to the unperturbed medium away from the cloud, it is clear that $\tau_{\parallel}$ decreases upon contraction (but not indefinitely, unlike the perpendicular-rotator case studied by Mouschovias and Paleologou 1979). Thus, even for an aligned rotator, magnetic braking is a more efficient process than previously realized. Our previous conclusion that the angular momentum problem can be resolved, at least for binary stars, in the early, diffuse stages of cloud contraction is further strengthened.

Equation (4) is, equivalently, written as

$$\tau_{\parallel} = \frac{\sigma_0}{2\rho_{ext} v_{A,ext}} \left(\frac{R_{c\ell}}{R_0}\right)^2 , \tag{5}$$

where $\sigma_0 \equiv M_{c\ell}/\pi R_0^2$ is the original column density of the cloud. It is clear that an upper limit on $\tau_{\parallel}$ for the cloud as a whole can be obtained by setting $R_{c\ell} = R_0$. Typically, one finds that $\tau_{\parallel} \lesssim 10^6$ yr. For a fragment, however, with mass of order $M_{\odot}$, the "external" medium is the surrounding cloud. This, as we shall see below, results in a significant decrease in $\tau_{\parallel}$ even if the factor in parentheses on the right-hand side of equation (5) is ignored. In any case, a cloud contracts at a nearly constant angular velocity until the magnetic field decouples from the matter.

3. AMBIPOLAR DIFFUSION: THE DENSITY AT DECOUPLING

The shortest time scale on which a cloud (or fragment) can contract is the free-fall time, τ_{ff}. Nakano and Tademaru (1972) obtained a density of order 10^9 cm^{-3}, at which the field decouples from the matter, by equating τ_{ff} and the characteristic time for ambipolar diffusion, τ_D. However, the presence of magnetic fields dilutes gravity (Mestel 1965). This is represented roughly as a reduction of the universal gravitational constant G by the factor $(1 - M/|\Omega|)$, where M and Ω are, respectively, the magnetic and gravitational energies of the cloud. Since we are concerned with clouds having $M \sim |\Omega|$, the dilution of gravity can be very significant. For such clouds, detailed, numerical, collapse calculations now under way show that the dynamical contraction time typically exceeds τ_{ff} by a factor of 5 - 30. We therefore take $\tau_{dyn} = a\tau_{ff}$, with $a > 1$, in what follows. We thus have that

$$\tau_{dyn} = 4.36 \times 10^7 \; n_p^{-1/2} \; a \quad \text{yr}, \tag{6}$$

where n_p is the number density of protons in the cloud --a 10% helium abundance is accounted for.

The characteristic time for ambipolar diffusion is defined as $\tau_B = R/v_D$, where v_D is the relative drift speed between ions and neutrals, determined by equating the driving magnetic force on the ions and the retarding collisional force on the ions due to neutrals. One finds that (see Spitzer 1978)

$$\tau_B = 8\pi \, R_c^{\,2} \, n_i \, n_c \, m_n \, \langle \sigma w \rangle_{in} \, B_c^{-2} . \tag{7}$$

The quantites n_i and n_c are the ion and neutral densities, respectively; m_n is the mass of a neutral particle (which we take to be H_2) and is assumed to be much smaller than the ion mass (N_a^+ or HCO^+, in our case); $\langle \sigma w \rangle_{in}$ is the average ion-neutral collisional rate and is approximately equal to 1.7×10^{-9} cm^3/sec in our example. We have introduced the subscript c because we shall apply equation (7) only to the cloud's core, not to the cloud as a whole, because it is neither necessary nor likely that ambipolar diffusion decouple the field from the matter over the entire cloud for stars to form (see Mouschovias 1978b). We now use equation (3) to eliminate B_c from equation (7), we express R_c in terms of the mass and density of the core, and we evaluate the constants to find that

$$\tau_B = 1.2 \times 10^{10} \; \frac{n_{p,0}}{(B_0/3 \; \mu G)^2} \; \frac{n_i}{n_{p,c}^{\;2/3}} \left(\frac{M_c}{M_\odot} \right)^{2/3} \quad \text{yr}; \tag{8}$$

where the quantities $n_{p,0}$ and $n_{p,c}$ are, respectively, the density of protons in the original cloud when lateral contraction began and the present density of protons in the core. In the density range $2 \times 10^3 \lesssim n_{p,c} \lesssim 2 \times 10^7$ cm^{-3}, we may relate the ion density to the neutral density by $n_i \approx 9.5 \times 10^{-6}\, n_{p,c}^{1/2}$, which is consistent with Nakano's (1979) Figure 1. To obtain the proton density at which the field essentially decouples from the matter, we set $\tau_{dyn} = \tau_B$. The result is

$$n_{p,dec} = 2.0 \times 10^6 \left(\frac{a}{10}\right)^3 \left(\frac{30\ \mathrm{cm}^{-3}}{n_{p,0}}\right)^3 \left(\frac{B_0}{3\ \mu G}\right)^6 \left(\frac{M_\odot}{M_c}\right)^2 \quad \mathrm{cm}^{-3}. \qquad (9)$$

The sensitivity of $n_{p,dec}$ on the quantities on the right-hand side of equation (9) is only apparent because these quantities cannot be specified independently of one another. Using equation (1) for the cloud as a whole to eliminate $n_{p,0}$ from equation (9), we find that $n_{p,dec} \propto B_0^{3/2}\, a^3\, M_{c\ell}^{3/2}\, M_c^{-2}$. In addition, the larger $M_{c\ell}$ is at a fixed B_0, the smaller the dilution of gravity by magnetic forces, so that the quantity a bears some inverse relation to $M_{c\ell}$ which will be determined by the detailed numerical calculations. Further discussion of these points is presented elsewhere (Mouschovias 1983).

The crucial question still remains, whether magnetic braking remains efficient up to the density $n_{p,dec}$. To check this, we calculate $\tau_\parallel$ (actually, its upper limit) and compare it with τ_{dyn} ($= \tau_B$). A core mass of order 1 $M_\odot$ at the typical density given by equation (9) is contained within a radius $R_c = 4.7 \times 10^{16}$ cm, which we take to be the half thickness of the disk, Z_c, in equation (4), which is applied to the core, not the cloud as a whole. With $B \propto \rho^{1/2}$, the Alfvén speed in the "external" medium, which is now the cloud surrounding the core, remains constant and somewhat larger than 1 km/sec. This gives a strict upper limit on $\tau_\parallel$ of 1.5×10^4 yr. At the same density and mass for the fragment, equation (8) yields τ_B ($= \tau_{dyn}$) $= 3.1 \times 10^5$ yr, which is larger than the upper limit on $\tau_\parallel$. It therefore follows that the angular velocity of a fragment at decoupling can be taken to be comparable with that of galactic rotation, i.e., 10^{-15} rad/sec. (A more detailed calculation shows that the angular velocity of a fragment begins to increase somewhat above 10^{-15} rad/sec before decoupling, but that does not affect our present conclusions much.)

4. BEYOND DECOUPLING: ROTATIONAL VELOCITIES OF SINGLE STARS

At densities in excess of $n_{p,dec}$, the angular momentum of a collapsing blob is essentially conserved, so that the angular velocity of a protostar (ω_*) that will ultimately form is

$$\omega_* = \omega_{dec}\, (\rho_*/\rho_{dec})^{2/3}; \qquad (10)$$

where ω_{dec} and ρ_{dec} are, respectively, the angular velocity and matter density of the blob at decoupling, and ρ_* is the mean density of the protostar. Since ω_{dec} is maintained at a value near 10^{-15} rad/sec by magnetic braking, equation (10) implies a linear velocity at the protostellar equator given by

$$v_{*,eq} = 252\ \rho_*^{2/3} \left(\frac{R_*}{R_\odot}\right) \left(\frac{10}{a}\right)^2 \left(\frac{n_{p,0}}{30\ cm^{-3}}\right)^2 \left(\frac{3\ \mu G}{B_0}\right)^4 \left(\frac{M_c}{M_\odot}\right)^{4/3} \frac{km}{sec}. \quad (11)$$

As explained below equation (9), the sensitive dependence of $v_{*,eq}$ on the quantities a, $n_{p,0}$ and B_0 is <u>only apparent</u>. For main-sequence stars of spectral type between O5 and F5, the product $\rho_*^{2/3}$ $(R_*/R_\odot)$ varies by less than a factor of 2 (e.g., see Allen 1974); its extreme values are 0.82 $g^{2/3}$ cm^{-2} for O5 stars and 1.33 $g^{2/3}$ cm^{-2} for F0 stars. It therefore follows that, since protostars lose typically half of their angular momentum (compared to a few orders of magnitude already lost) between the onset of the Hayashi phase and their arrival on the main sequence, typical equatorial stellar $\langle v_{*,eq} \sin(i)\rangle$ should be around 100 km/sec. This is surprisingly (but encouragingly) close --given that three independent physical processes (collapse, magnetic braking and ambipolar diffusion) worked together to give this prediction-- to the observed rotational velocities for such stars, which vary from about 25 to 160 km/sec (Vogel and Kuhi 1981; Wolff <u>et al.</u> 1982).

REFERENCES

Abt, H.A. and Levy, S.G.: 1976, Astrophys. J. Suppl. 30, p. 273.
Allen, C.W.: 1974, "Astrophysical Quantities" (Athlone Press, London).
Mestel, L.: 1965, Quart. J. Roy. Astron. Soc. 6, p. 265.
Mestel, L. and Spitzer, L., Jr.: 1956, Monthly Not. Roy. Astron. Soc. 116, p. 503.
Mouschovias, T.Ch.: 1976a, Astrophys. J. 206, p. 753.
Mouschovias, T.Ch.: 1976b, Astrophys. J. 207, p. 141.
Mouschovias, T.Ch.: 1977, Astrophys. J. 211, p. 147.
Mouschovias, T.Ch.: 1978a, Astrophys. J. 224, p. 283.
Mouschovias, T.Ch.: 1978b, in "Protostars and Planets", ed. T. Gehrels (University of Arizona Press, Tucson), pp. 209-242.
Mouschovias, T.Ch.: 1981, in "Fundamental Problems in the Theory of Stellar Evolution", eds. D. Sugimoto, D.Q. Lamb, and D.N. Schramm (Reidel, Boston), pp. 27-59.
Mouschovias, T.Ch.: 1982, Advances in Space Research, in press.
Mouschovias, T.Ch.: 1983, Astrophys. J., to be submitted.
Mouschovias, T.Ch. and Paleologou, E.V.: 1979, Astrophys. J. 230, 204.
Mouschovias, T.Ch. and Paleologou, E.V.: 1980a, Moon Planets 22, p. 31.
Mouschovias, T.Ch. and Paleologou, E.V.: 1980b, Astrophys. J. 237, p. 877.
Nakano, T.: 1979, Publ. Astron. Soc. Japan 31, p. 697.
Nakano, T. and Tademaru, E.: 1972, Astrophys. J. 173, p. 87.
Spitzer, L., Jr.: 1978, "Physical Processes in the Interstellar Medium" (Wiley-Interscience, New York).
Vogel, S.N. and Kuhi, L.V.: 1981, Astrophys. J. 245, p. 960.
Wolff, S.C., Edwards, S., and Preston, G.W.: 1982, Astrophys. J. 252, p. 322.

DISCUSSION

MESTEL: I would not quarrel with your picture as something that *can* happen, if the parameters — i.e. the initial conditions — are appropriate. A crucial point is that the mass of your cloud is extremely close to the critical mass for the cloud's flux, so that the magnetically diluted contraction time is 20 times the free-fall time. This certainly enables the magnetic stresses to maintain corotation with the background. But if the cloud is only a little more massive, then centrifugal forces will instead be maintained close to gravity (with the contraction of the cloud being *determined* by the rate of braking). Subsequent loss of flux by ambipolar diffusion will leave a rapidly rotating cloud of only moderate density. On the other hand, if the cloud mass is below the critical mass, it will indeed be kept in corotation with the background, but it can contract only through slow flux leakage. My point is that one will want to satisfy oneself that the choice of parameters leading to a line of evolution is not implausibly restrictive.

A further physical question is that of the topology of the field lines. Once detachment of much of the cloud field from the background field has occured, then angular momentum transport will be much removed. But perhaps this rather subtle question should be left for the next Symposium on Star Formation.

MOUSCHOVIAS: As I am sure Dr. Mestel knows, I agree with his first point, but not with the implication that I have unduly restricted the range of the input parameters. I have let observations suggest a "typical" case. Of course, in our detailed calculations we perform a much more complete parameter study. In the ten minutes available to me, I can describe only the typical case and even that, only in an incomplete way. In any case, the equations in my paper show explicitly the dependence of the solution on the relevant parameters.

On the magnetic detachment of the cloud from the background, our calculations suggest that it takes place at a much later stage of contraction, long after the angular momentum problem is resolved. More calculations, aiming specifically at resolving this issue, are needed.

CLOSED AND OPEN MAGNETIC FIELDS IN STELLAR ATMOSPHERES: EFFECTS ON MASS LOSS FROM COOL GIANT STARS

D. J. Mullan
Bartol Research Foundation of the Franklin Institute
University of Delaware, Newark, Delaware 19711, USA

ABSTRACT

We propose that the onset of rapid mass loss among cool giants, and the absence of hot coronal material from their atmospheres, is associated with a transition in the large-scale magnetic topology of the atmosphere from closed to open. According to this view, field loops in the atmospheres of giants of spectral class K and later cannot find equilibrium, but are in a state of dynamical evolution throughout their lifetime in the atmosphere.

INTRODUCTION

Mass loss from the highly ionized atmosphere of a star is permitted only in regions where the magnetic field is not rooted in the star in the form of a closed loop. Thus, the field can be either open to space (as in coronal holes on the sun), or bubbles of magnetic flux may become disconnected from the stellar surface and transport material away from the star in discrete events. In both cases, since magnetic fields are created in late-type stars by dynamo action in the form of bipolar pairs, the mass loss process requires that a mechanism must be at work in the stellar atmosphere to convert originally closed loops of flux into either open form or else disconnected bubbles.

In the solar corona, loops cannot always find a means of becoming open or disconnected. In such cases, the loops enter a stage of evolution which, although basically magnetohydrodynamic in character (due to the jostling of footpoints by convective flows), can nevertheless be described with fair accuracy as a sequence of quasi-steady equilibria. During this phase, the closed loop traps material, prevents mass loss, and the trapped material emits X-rays copiously.

Mass loss among cool giant stars sets in across a "velocity dividing line" in the HR diagram (Stencel and Mullan, 1980) which coincides rather well with a "temperature dividing line", along which hot coronal material disappears from the atmosphere (Linsky and Haisch, 1979). In the context

J. O. Stenflo (ed.), Solar and Magnetic Fields: Origins and Coronal Effects, 487–491.

of the above discussion, the coincidence of these two dividing lines in the HR diagram can be interpreted as follows: when a star evolves across either dividing line, closed loops can no longer exist in quasi-steady state, but must immediately enter a phase of dynamical evolution which ultimately leads to opening of the field lines, or disconnection from the star. In either case, enhanced mass loss is permitted, and the disappearance if closed loops removes the prime source of coronal X-rays from the atmosphere. Hence, the atmosphere no longer contains the hot material which is so characteristic of the closed loops in the solar corona.

The purpose of this paper is to report on a test of this hypothesis.

CLOSED MAGNETIC FIELDS IN STELLAR WIND

Pneuman (1968) has analyzed the case of an isothermal wind flowing around and above a region where the field is closed at low altitudes, but open higher up. This magnetic configuration gives rise to helmet streamers in the solar corona. Pneuman found that the outflow above the helmet passed through a sonic point at a radial distance r_s which was related to the radial distance of the last closed field line, r_c, by a simply relation: $r_s = 2\ r_c$.

Subsequently, Pneuman and Kopp (1971) considered the case of a steady solar wind flow imbedded in a global dipole magnetic field. They found that the field lines remained closed near the equatorial plane (at least with the initial conditions which they selected). And as before, the ratio of r_s to r_c which emerged from their calculations turned out to be close to the value of 2.

To make these estimates more precise, I have recently collaborated with R.S. Steinolfson in a time-dependent, two-dimensional MHD calculation of the interaction between the solar wind and a global dipole field. Rather than assuming isothermal flow, as was assumed in the previous studies, the flow was assumed to be polytropic, with index 1.05, for the initial, non-magnetic wind. A value was chosen for the coronal base temperature, T_b. At time t=0, the global dipole is superposed on the spherical wind solution, and the time-dependent evolution is followed until a steady state is reached. In the case of a solar model, some (1-2) thousand seconds of real time must elapse before steady state is reached. The dipole strength is parametrized by a beta value, which is the ratio of gas pressure to magnetic pressure at the base of the corona in the equatorial plane. The calculations showed that the steady state results were insensitive to β for all $\beta \lesssim 1$. In what follows we report on calculations which assumed β=0.5.

We found that as T_b increases, the size of the closed field region near the equator shrinks, until finally, for a critical value of T_b, no closed field lines exist in steady state. For T_b less than critical, we evaluated the ratio of r_s/r_c in the equatorial plane. In the case of the solar corona, the value turned out to be remarkably constant, varying by only ±6% around a mean value of 2.55 as T_b varied by 220%.

We calculated the flow for a giant star also, with solar mass, but with radius equal to 10 times solar. Here, the critical T_b value was found to be less than solar by a factor of 10 (i.e. 4×10^5K, rather than 4×10^6 K in the solar case). This is the scaling which would be expected if the corona were assumed to be strictly isothermal. And also in the case of the giant model, r_s/r_c turned out to be insensitive to T_b. The value remained constant (within ±16%) at 2.25, as T_b varied by 220%.

These calculations suggest that the radial distance of the last closed field line can be characterized rather well by a knowledge of the radial distance to the sonic point.

DISAPPEARANCE OF CLOSED FIELD LINES IN STELLAR CORONAE

This leads us to consider what happens in a star where the sonic point approaches closer than 2.2-2.6 stellar radii. In such a case, the last closed field line extends to a radial distance of less than one stellar radius, i.e. lies completely inside the star. Thus, in such a star, if a closed field line were to appear in the stellar atmosphere (by emerging from the photosphere due to buoyancy, for example), it could not find a steady state in the presence of the thermal wind from that star. In the absence of steady state, such a field line enters a phase of dynamical evolution. The final state of such a field lines is expected to be either open entirely, or else a state of disconnection from the stellar surface. In such a star, no long-lived steady closed loops persist. Thus, there are no analogs to the bright "building-blocks" which characterize the solar corona.

We now ask: where in the HR diagram does the sonic point radius approach closer to the star than 2.2-2.6 stellar radii? To answer this, we use the same technique which we used several years ago in estimating the "supersonic transition locus" (Mullan, 1978), i.e. a semi-empirical technique in which we combine the coronal base pressures derived by Kelch et al. (1978) from model chromospheric fits to spectral line data, with the minimum flux coronal concept. In the present case, however, we are not relying on the requirement that mass loss become rapid because the sonic point sinks into chromospheric material. Now, the sonic point remains well out in the corona. We have used evolutionary tracks for stars of 1.5, 3, and 5 solar masses to determine where the last closed field line disappears. The location of the relevant points in the HR diagram have already been published for the case $r_s/r_c = 2.0$ (Mullan, 1982). With the revised values of r_s/r_c reported here, the locations of the relevant points in the HR diagram shift slightly, but not by more than the size of the symbols which were used in Mullan (1982). Thus, the major point to emerge from that paper remains: the location of the transition from closed field lines to absence of closed field lines occurs in good coincidence with the empirical velocity dividing line.

We consider this support for our conjecture that the velocity and temperature dividing line(s) in the HR diagram are to be identified with a magnetic topology transition locus (MTTL).

ARGUMENT FROM MAGNETOSTATIC EQUILIBRIUM

Independent support for the MTTL concept is provided by Low's (1981) work on magnetic equilibrium in stratified atmospheres. Equilibrium requires that the parameter C ($=BL/4H(\Delta p)^{0.5}$) satisfy $C<1$. (Here, B is the field strength, H is the scale height, L is the separation between loop footpoints, and Δp is the pressure differential between magnetic and non-magnetic gas.) To optimize stability, we set $\Delta p=p_o$, the coronal base pressure, determined by Kelch et al. (1978). (In active chromosphere stars, where p_o may be larger, B is also probably larger in proportion to $p^{0.5}$ (cf. Mullan, 1975). Hence, C may be roughly equal in active and inactive stars.) Suppose $L \approx 0.1R$ (star). Then equilibrium requires that the atmospheric temperature T exceed $(2-4)\times 10^5$ B along the VDL. If solar coronal values of B are typical ($B \simeq 1G$), then stars above the TDL (with $T<<10^5$) contain loops which are <u>not</u> in equilibrium, whereas stars below TDL (where $T>>10^5$K), can have closed loops in equilibrium.

CONCLUSION

Above MTTL, magnetic flux loops emerging into a stellar atmosphere are always unstable. Mass loss can be rapid, and X-rays will be weak because of the lack of closed field loops. Unstable loops can be used to interpret a wide variety of characteristics of late-type stellar chromospheres/coronae, including episodic mass loss, extended atmospheres, variability of emissions from chromosphere and coronae, hybrid atmospheres, enrichment of certain isotopes in the atmosphere, and supersonic macroturbulence (cf. Mullan and Cram, 1982; Mullan and Stencel, 1982).

REFERENCES

Kelch, W.L. et al.: 1978, Astrophys. J. 220, p. 962.
Linsky, J.L. and Haisch, B.M.: 1979, Astrophys. J. 229, p. L27.
Low, B.C.: 1981, Astrophys. J. 251, p. 352.
Mullan, D.J.: 1975, Astron. Astrophys. 40, p. 41.
Mullan, D.J.: 1978, Astrophys. J. 226, p. 151.
Mullan, D.J.: 1982, Astron. Astrophys. 108, p. 279.
Mullan, D.J. and Cram, L.E.: 1982, Astron. Astrophys. 108, p. 251.
Mullan, D.J. and Stencel, R.E.: 1982, in "Four Years of IUE Research", (in press), (Washington: NASA).
Pneuman, G.W.: 1968, Solar Phys. 3, p. 578.
Pneuman, G.W. and Kopp, R.A.: 1971, Solar Phys. 18, p. 258.
Stencel, R.E. and Mullan, D.J.: 1980, Astrophys. J. 240, p. 718.

DISCUSSION

KUIN: What is the influence of the temperature distribution that you assume in the helmet streamer on the position of the sonic point? I am afraid that this would have a large influence on your results. In a different structure like a polar coronal hole one also observes that the sonic point lies much closer to the star than one would expect from the Parker solar wind.

MULLAN: I do not *assume* a temperature distribution. It is calculated at each point in the grid. The temperature is naturally found to be higher on closed field lines than in the open region. The calculation actually includes a coronal-type structure: around the poles, the field lines are open, and diverge faster than radial in steady state. Hence, the sonic point certainly lies closer to the surface in that part of our calculated flow. In fact, the polar sonic point approaches the surface (to $r \lesssim 1.2 r_\odot$) at essentially the same value of T_b as the "last closed field line" disappears. Our steady state is definitely very different from a Parker solar wind.

GRAY: If one looks at those stars of luminosity class III showing hot temperatures vs. those showing cool temperatures in transition-region lines, you find that they fall on opposite sides of the rotational discontinuity. Would you comment on the possibility of rotation being the driving mechanism for the things you have discussed?

MULLAN: I have discussed magnetic effects without specifying where the fields originate. Presumably rotation contributes to field creation in some way. Until this contribution is better known, I cannot comment on how important the rotational discontinuity is. However, it is clear that onset of rapid mass loss across the velocity dividing line may be important in braking the G-K giants. Thus the connection between dividing lines in the H-R diagram is still obscure in the sense of isolating causes from effects.

ZWAAN: How do you explain that synchronized K-type binaries do show high-temperature lines, or, in other words, how does the atmosphere know that there is a companion?

WALTER: Let me comment on this question: I don't think that there is a problem with the RS CVn giants. As I recall, at K0 the various dividing lines are at about luminosity class II, and do not get to luminosity class III until about K5. Most, if not all the RS CVn giants lie below the lines and probably have closed fields.

STELLAR WINDS AND SPINDOWN IN RED GIANTS

A. S. Endal
Applied Research Corporation
8201 Corporate Drive, Landover, MD 20785 USA

ABSTRACT

Recent observations of rotation rates in G and K giants show a sudden drop in the rotation rate at spectral type G5. The observed behavior requires a strong external brake, such as applied by a magnetized stellar wind. Since spectral type G5 coincides with the onset of deep envelope convection in giants, a dynamo mechanism is suggested as the controlling factor. Stellar evolution models have been used to estimate the time scale for the rotational braking, the angular momentum loss rate, and the required magnetic field strengths.

1. INTRODUCTION

In an earlier paper in this session, D.F. Gray described some very interesting observational results on the rotation of G and K giants. I will present an interpretation of these results, based on first-crossing (FC) models of intermediate mass ($3M_\odot$) stars. The mass was chosen on the basis of the estimated luminosities of these stars; the results are not very sensitive to this parameter. The question of whether FC models are appropriate will be deferred to a later section.

2. PREDICTIONS FOR STARS WITH CONSTANT ANGULAR MOMENTUM

Figure 1 shows the observed rotational velocities (v sin i) for giants of spectral types F8-K2. The lines indicate predicted velocities for a $3M_\odot$ star evolving from the ZAMS with an initial velocity of 180 km/s. This initial value corresponds to the observed $\langle v \rangle = (4/\pi) \langle v \sin i \rangle$ for late-B main sequence stars. The solid line shows the predictions for convective envelopes in rigid-body rotation (Case 2 of Gray and Endal, 1982) and the dashed line shows predictions for uniform specific angular momentum in convective regions (Case 3). The models were computed by the techniques described by Endal and Sofia (1976, 1978).

J. O. Stenflo (ed.), Solar and Magnetic Fields: Origins and Coronal Effects, 493–498.

For spectral types F8-G2 (and for earlier spectral types - see Endal and Sofia, 1979) the observed <v> agrees with the predictions. At G5, there appears to be two groups of stars. One group is consistent with the predicted velocities of 20-30 km/s, while the second group shows values of 5 km/s or less. Gray (1982) interprets this as evidence for a strong rotational brake setting in near G5. This is consistent with the uniformly slow rotation of the giants of later spectral types.

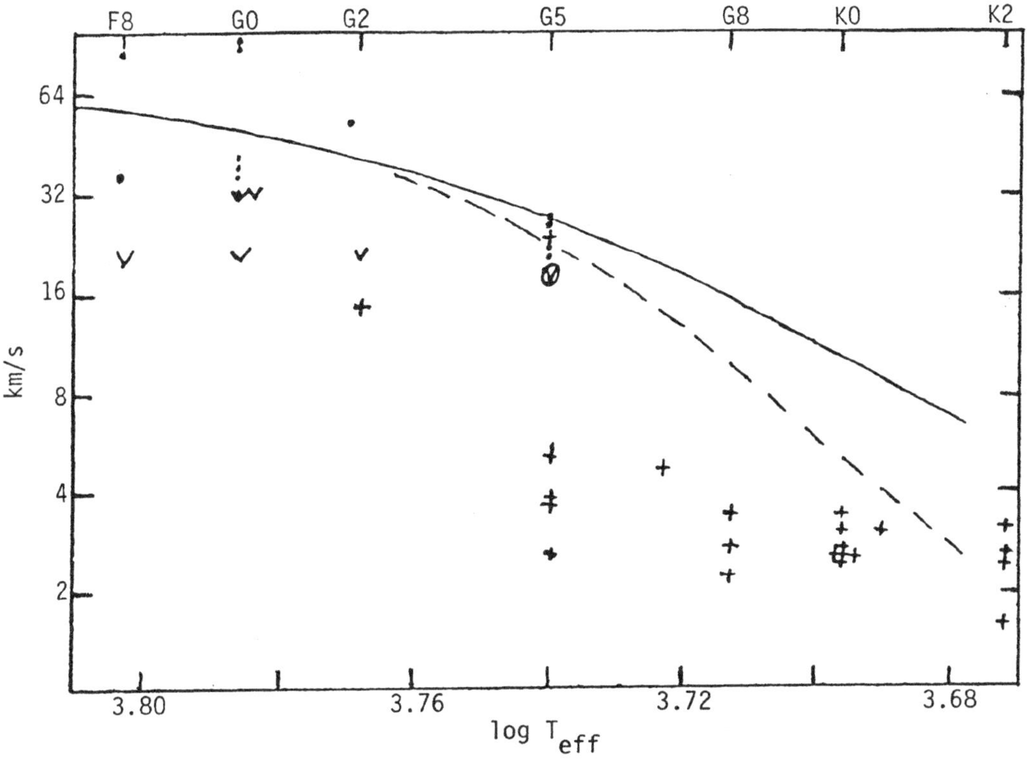

Fig. 1 - Observed and predicted velocities for F8-K2 giants. Heavy dots and carets (upper limits) show observations by Alschuler (1975). The circled caret at G5 represents 12 stars. Crosses show measurements listed by Gray (1982).

3. ANGULAR MOMENTUM LOSSES

Using the information from the stellar evolution models, we can calculate the rates of angular momentum removal required to bring the predictions into agreement with the observations. The results of such a calculation are given in Table 1. I have assumed, in these calculations, that the braking affects only the convective envelope. This is a reasonable (though not entirely safe) assumption. The last two lines will be explained in Section 4.

Table 1. Spindown of $3M_\odot$ Giants*

Sp. Type:	G5	G8	K0	K2
Observed <v>	4.8	4.1	3.5	2.5
Predicted <v>	27.4 (23.1)	15.3 (9.6)	10.9 (5.3)	5.6 (2.3)
$\dot{J}$ (dyn-cm)	$>4.5\times10^{36}$ ($>4.3\times10^{36}$)	1.6×10^{37} (1.2×10^{37})	6.5×10^{36} (2.4×10^{35})	none (none)
$f^{-1/2}B_s$ (gauss)	>16.7 (>16.4)	67.2 (57.8)	34.9 (6.7)	-- --

*Values in parentheses refer to Case 3 rotation.

Several points should be made with regard to the values given in Table 1. First, the growth of the convective envelope during the evolution from G5 to K2 plays a very important role in the calculations. The deepening of the envelope is illustrated in Figure 2. This brings up additional angular momentum from the unbraked core and requires that the braking continue throughout the later spectral types. This could be avoided by assuming that the initial braking affects almost all of the star but this seems unlikely since the convective envelope at G5 contains only ∿1% of the total mass. Second, the time scale for the braking at G5 is not known. I have used the time required to evolve from G2 to G5 as an upper limit. Considering the sharp distinction between rapid and slow rotators at G5, the real time scale is probably much shorter. Third, the values at K2 should not be taken seriously. This group of stars is very close to the red giant branch and, therefore, contaminated by low-mass stars with very different histories.

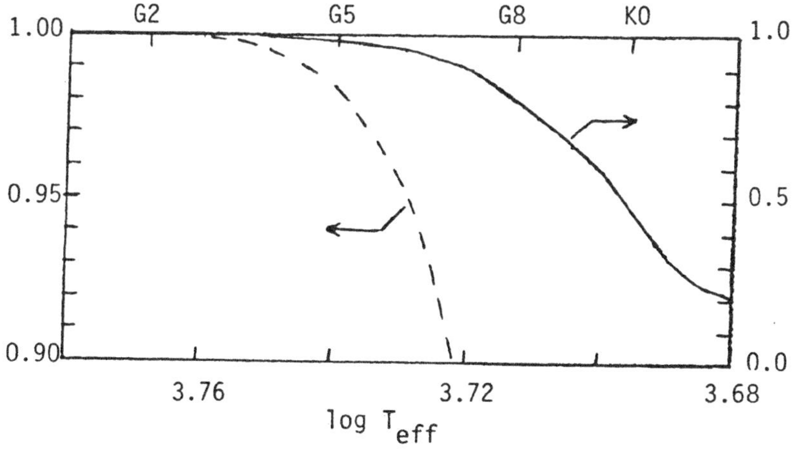

Fig. 2 - The depth of the convective envelope (M_r/M at the bottom of the envelope) for a $3M_\odot$ star. Arrows indicate whether values should be read from the left or right axis.

The rate at which the solar wind carries away angular momentum is $\sim 7 \times 10^{30}$ dyn-cm (Brandt and Heise, 1970). The $\dot{J}$ values in Table 1 are roughly a million times larger. The total loss during the evolution from G5 to K0 is 65% (Case 2) or 35% (Case 3) of the initial angular momentum.

4. MAGNETIC FIELDS

The large $\dot{J}$ values in Table 1 require a very efficient braking mechanism, such as provided by a large-scale magnetic field embedded in a stellar wind. This hypothesis is supported by the correspondence between the start of the observed braking and the onset of deep convection (see Figure 2). Via a dynamo, this provides a means for generating the required magnetic field. In order to estimate the magnitude of this field, I have used a simple model of a magnetized stellar wind. The derivation given here will necessarily be sketchy; a more detailed derivation and justifications for the assumptions will be published elsewhere.

Following Weber and Davis (1967), the rate at which a magnetized wind carries away stellar angular momentum may be written as

$$\dot{J} = (2/3)\,\dot{M}\,(\Omega_s r_A^{\,2}) = (2/3)(4\pi r_A^{\,2}\rho_A w_A)(\Omega_s r_A^{\,2}), \tag{1}$$

where $\dot{M}$ is the mass-loss rate, Ω_s is the angular velocity of the stellar surface, r is radius, ρ is density, and w is the radial wind speed. A subscript A means the quantity is to be evaluated at the Alfvén point, though $\dot{M}$ could be evaluated at any distance. At the Alfvén point, the kinetic energy density of the wind equals the energy density of the radial magnetic field:

$$\text{kinetic energy} = (1/2)\rho_A w_A^{\,2} = (1/2)\rho_A w_A f(GM/R)^{1/2}, \tag{2}$$

$$\text{magnetic energy} = (1/8\pi)B_A^{\,2} = (1/8\pi)B_s^{\,2}(R/r_A)^4. \tag{3}$$

In equation (2), I have assumed that w_A is some fraction f of the plasma escape velocity and equation (3) assumes that the radial field scales back to the stellar surface ($r=R$) as $1/r^2$. Equating the energies at the Alfvén point, solving for $\rho_A w_A$, and substituting into equation (1), the result can be written as

$$f^{-1}B_s^{\,2} = (3/2)(GM/R)^{1/2}R^{-4}(\dot{J}/\Omega_s). \tag{4}$$

Note that all of the quantities on the right side of equation (4) can be determined from either the stellar models or the observations. The results are given in the last two lines of Table 1. Since $f^{-1/2} \sim 1$, this gives an estimate of the required magnetic field which couples the wind to the stellar surface. Note that this estimate refers only to that component of the field which extends to the Alfvén point. Localized surface fields may be much larger.

5. WARNINGS AND CONCLUSIONS

The interpretation presented above is based solely on FC models. An alternate interpretation would identify the rapid rotators of spectral types G5 and earlier as FC stars, while the slow rotators from G5 to K2 would be considered second-crossing (SC) stars. This would still require rotational braking, but the time scales would be much longer and $\dot{J}$ would be correspondingly smaller. I estimate that this would reduce the required magnetic fields by a factor of 5 (Case 2) or 10 (Case 3).

Theoretical models (Iben, 1965) suggest that the SC loop does not extend as early as G5 in intermediate mass stars. This conclusion is supported by the "clump giants" commonly observed at spectral types G8 and later in intermediate-age clusters (Cannon, 1970). Giants of spectral type G5 are relatively rare, in agreement with the short time scale of the FC evolution. Thus, the available information favors the FC interpretation for the G5 giants but may require SC models to interpret the G8 and later giants.

I thank David Gray for communicating his results prior to publication and for useful correspondence regarding their interpretation.

REFERENCES

Alschuler, W.M.: 1975, *Astrophys. J.*, 195, pp. 649-660.
Brandt, J.C., and Heise, J.: 1970, *Astrophys. J.*, 159, pp.1057-1066.
Cannon, R.D.: 1970, *Mon. Not. Roy. Astron. Soc.*, 150, pp.111-135.
Endal, A.S., and Sofia, S.: 1976, *Astrophys. J.*, 210, pp.184-198.
Endal, A.S., and Sofia, S.: 1978, *Astrophys. J.*, 220, pp.279-290.
Endal, A.S., and Sofia, S.: 1979, *Astrophys. J.*, 232, pp.531-540.
Gray, D.F.: 1982, preprint.
Gray, D.F., and Endal, A.S.: 1982, *Astrophys. J.*, 254, pp.162-167.
Iben, I., Jr.: 1965, *Astrophys. J.*, 142, pp. 1447-1467.
Weber, E.J., and Davis, L., Jr.: 1967, *Astrophys. J.*, 148, pp.217-227.

DISCUSSION

ROXBURGH: It is dangerous to use a radial field — radial fields do not exist — only drawn out non-radial fields (e.g. dipole - multipole). In this case the angular momentum loss does not vary as B_0^2. Indeed for a dipole field the variation is like $B_0^{2/3}$, for a quadrupole field it is independent of B_0.

ENDAL: I agree that the radial-field assumption is an oversimplification. However, I do not think that we know enough about the multipole structure of non-solar fields to justify a more complex treatment.

WALTER: If you look at the figure showing the data, and scale the curve down to fit the G stars (to account for $\sin i$), the fit looks reasonable. In the K stars, the *average* $v \sin i$ may not be physically meaningful, because you are likely to be mixing evolving A stars with evolving (non-rotating) stars of approximately solar mass plus (non-rotating) second-crossing stars, which will decrease the average. The curve is not clearly discrepant — does the break really exist?

ENDAL: More data are certainly needed to be absolutely sure about the break in the rotation rate. I have interpreted the available data as they exist. On the other hand, the Hyades giants (at K0) certainly came from A stars. Their rotational velocities are consistent with those of the other stars in the sample.

VI. SUMMARY AND FUTURE DIRECTIONS

OBSERVATIONAL NEEDS FOR PROGRESS IN SOLAR/STELLAR MAGNETIC ACTIVITY: OVERVIEW

Robert W. Noyes
Harvard-Smithsonian Center for Astrophysics
Cambridge, MA 02138, U.S.A.

Recent observational and theoretical findings have clarified the physical mechanisms which underlie magnetic activity production in stars, and point the way naturally to a number of new or more crisply defined questions, whose answers can lead to major progress in the near future. Concerning observational programs, a guiding principle has been evident throughout this symposium: We should rely heavily on the Sun for understanding the detailed physics of magnetic activity and its generation, while at the same time we study analogous stellar phenomena for comparison with the Sun, and for new insights and extension to different regions. I list below some broad observational areas in which conditions seem ripe for important progress in understanding solar and stellar magnetic activity, leaving to other summarizers the discussion of particular observational programs.

1. SOLAR AND STELLAR INTERIORS

Solar "seismology" has given tantalizing hints of surprises in store, such as the possible existence of a rapidly rotating interior. This type of investigation holds great promise both for pinning down fundamental properties like the depth d_c of the convection zone and of the depth dependence of angular velocity $\Omega(r)$. Both are basic to understanding magnetic field generation.

Although the analogous stellar problem is vastly more difficult than that of solar seismology, the detection of stellar oscillations in the near future seems likely. Empirical determination of d_c or $\Omega(r)$ for another star may lie well in the future, but initial detection of the analogue of the spectrally unresolved "5-minute" solar oscillation will start us on the way.

On the Sun, the search for "giant cells", i.e. a large-scale circulation pattern, is continuing at improved levels of velocity sensitivity and precision, and one may hope for a definitive resolution of this important problem in the near future. It is imortant to continue efforts to determine convection-related properties at the surfaces of late-type stars, by analysis of line profile broadening and asymmetries. In addition, such studies can yield information on the surface differential rotation of relatively rapidly rotating stars (see below).

The above observations should not be confined solely to main sequence stars. Effects may be larger and more observable in giants, and the insights gained from the different gravity and internal structure of giants should provide further clues to the overall physical mechanisms involved.

J. O. Stenflo (ed.), Solar and Magnetic Fields: Origins and Coronal Effects, 501–502.

2. STELLAR ROTATION, DIFFERENTIAL ROTATION, AND ACTIVITY

The relation between rotation rate and activity needs to be extended to a large number of lower main sequence stars of varying degrees of chromospheric and coronal activity and age. This is required to build a convincing statistical picture of the relation between activity level, (i.e. chromospheric emission, coronal emission, surface magnetic fields, or other indicators) and rotation rate, spectral type, and age. Of particular interest is extending the data base to both earlier and later spectral types, as well as to very old and very young stars. In the latter connection, independent determination of stellar age is very important.

In the light of recent results on the rotation and differential rotation of the solar surface layers, and their variation with time, it is clear that further studies along this line stand to give us critically important information about the solar dynamo. In a complementary fashion, the vastly cruder information on differential rotation seen on some stars (both from line profile measurements as mentioned above and from rotationally-modulated light curves) represents potential key extensions of the data to different physical regimes. Observations spanning a stellar activity cycle could reveal cycle-dependent changes of mean rotation rate or differential rotation, if these are significantly larger than the Sun's.

3. ACTIVITY CYCLES

Needless to say, programs to study activity cycles, similar to the seminal long-term series of observations begun by Olin Wilson, should continue and be extended. Important questions which can be addressed include: where in the H-R diagram do cycles occur? How do their amplitude, period, cycle shape, or other attributes vary with rotation, age, and basic stellar properties? Are stellar analogues of the Maunder Minimum revealed by the statistics for cycles of other stars? It would be useful to study how other properties vary over stellar cycles, in addition to the Ca II emission by which they were originally discovered. Examples include: coronal emission levels, surface manetic fields, and spot areas, as well as the rotation and differential rotation parameters mentioned above.

4. SURFACE STRUCTURE OF MAGNETIC ACTIVITY

The realization that magnetic flux on the Sun is concentrated in flux bundles with field strengths of a kilogauss or more has significantly changed our understanding of surface activity, and of the underlying field generation mechanisms. Several times during the Symposium the need has been emphasized to resolve these flux knots completely, in order to delineate better their interaction with the convective motions below, and with heating of the structured atmosphere above. This objective is a key application of new high resolution techniques under development for both space- and ground-based observations.

On main sequence stars, significant spatial resolution does not seem directly attainable in the near future. Nevertheless, much information is potentially available in rotationally-modulated light curves, concerning: longitude and latitude distribution of activity, size and brightness (or darkness) of chromospheric plages (or spots), strengths and area coverage of photospheric magnetic fields, lifetime of active regions or spots, and growth and decay properties of surface magnetic activity. The rich store of information available for such data is just beginning to be tapped.

All in all, it appears that research into solar and stellar magnetic activity continues to be an area driven heavily by observations. It is abundantly clear from present observational opportunities that this situation will continue for some time to come. It is creative observers, always with due regard to the theoretical implications of their work, who will lead and in some cases define the new attacks on one of astronomy's most fascinating research areas.

OBSERVATIONAL REQUIREMENTS: GROUND-BASED

Cornelis Zwaan
Observatory, Zonnenburg 2
3512 NL Utrecht, The Netherlands

Vigorous pursuit of the investigation of magnetic activity in stars with convective envelopes requires a diversity of observational studies and a variety of instrumentation. I leave out purely solar studies of magnetic structure and activity; it goes without saying that any well designed solar investigation contributes to better understandiug of general astrophysical problems. Hence I restrict myself to observational programs pointed at stars (sections 1, 2, 3) and to solar studies necessary to interpret stellar data (section 4).

1. MAGNETIC DATA

In Marcy's review two possible ways of obtaining data on magnetic structure in late-type stars have been indicated:

(a) Determination of mean field strengths $<B>$ and magnetic filling factors from line profile analysis; this yields the most direct information on magnetic structure in stellar atmospheres;
(b) Measurements of linear polarization, which is presumably due to a difference in saturation between the π and the σ components in the spectral window. If this interpretation of the instrinsic linear polarization is correct, then it may yield the mean direction and an average magnitude of the transverse field component across the stellar disk.

2A. EXTENDED SERIES OF MEASUREMENTS

Long series of measurements of chromospheric emission flux are needed to follow the variability for many stars, for instance to determine rotation periods, or to search for activity cycles. These investigations require (i) *efficient line photometers* similar to the Mt. Wilson Ca II H and K photometer, and (ii) *dedicated telescopes* with apertures $D \gtrsim 1$ m. It is of prime importance that the program with the Mt. Wilson 60 inch telescope plus the H and K photometer be contintued for at least several more years. Moreover, similar instruments should be made available; at least one is needed in the southern hemisphere.

2B. CHROMOSPHERIC EMISSION FLUX MEASUREMENTS OF FAINT STARS

In order to compare the activity of stars of different ages, masses and chemical compositions, measurements are needed of stars in clusters and in the Magellanic Clouds. Thus very efficient photometers and large telescopes ($D \gtrsim 3.0$ m) are required.

J. O. Stenflo (ed.), Solar and Magnetic Fields: Origins and Coronal Effects, 503–504.

Programs under 2A and 2B require many man-years of steady labor in well defined programs in order to solve basic astrophysical problems. We should convince our scientific community, and eventually our funding institutions, that in the field of solar-stellar astrophysics cleverly designed "routine" programs are essential. These programs are very rewarding on time scales ranging from weeks to years and even decades.

The need for a large body of observational data will keep many observers busy. Before long we may need a standing working group coordinating and stimulating these observational efforts.

3. MEASUREMENTS OF GLOBAL VELOCITY FIELDS

The discovery of the stellar rotation rate as an important parameter in stellar magnetism calls for measurements of rotation rates, both from the rotational modulation in the chromospheric emission flux (see 2A), and from line profile analysis yielding $v \sin i$. Since synchronized spectroscopic binaries are extremely important in the study of stellar activity, we need programs for the detection of spectroscopic binaries and for the determination of the v_{rad} curves.

4. SOLAR-STELLAR STUDIES

The proper interpretation of chromospheric and coronal indicators of stellar activity requires special studies; some examples are given below.

(1) *Calibration* of the *"chromospheric magnetometers"*. For the Sun, magnetic flux densities and chromospheric emission fluxes should be compared for a variety of magnetic regions in various positions on the solar disk. For active stars simultaneous spectra for magnetic field determinations (see 1) and Ca II H and K line-core fluxes are needed.
(2) *"Hyperfine" fluxtube structure*: the basic structure of magnetic fluxtubes, their evolution and their role in chromospheric and coronal heating, will be studied by means of the *S*olar *O*ptical *T*elescope. SOT is not just another toy for a bunch of solar physicists; it is an instrument for fundamental astrophysical studies.
(3) *Relations between soft X-ray–, transition-region–, and chromospheric emission fluxes* for a variety of magnetic regions at various positions on the solar disk. These studies may rely on ATM data, to be supplemented at a later stage by SOT.

OUTLOOK FOR GROUND-BASED AND SPACE OBSERVATIONS

G.S. Vaiana
Osservatorio Astronomico di Palermo (Italy) and
Harvard-Smithsonian Center for Astrophysics, Cambridge, MA (USA)

1. INTRODUCTION

The presentations given at this Symposium have made clear that we are witnessing an exciting period in the study of solar activity, and in the ancillary study of magnetic field dynamics in solar-type stars: the new space observations obtained within the past four years, together with the renaissance in ground-based observations, have occasioned a burgeoning in our phenomenological understanding of stellar activity in its manifold facets, and have encouraged a substantially greater interest in the problem of magnetically-coupled activity in stars by the theoreticians. This excitement must not distract us from the realization that in this branch of our science, much of the impetus towards new research has come, and very likely will continue to come, from observations; with this fact in mind, I would like to address the possible evolution of our experimental science in response to the new observational and theoretical results.

As we look to the future, two key facts ought to be kept in mind. First is that scientific progress in our science has basically resulted from instrumental developments which had their genesis some 10 to 20 years ago. Thus, the recent results in stellar optical and UV astronomy have been based on detector technology which predates current photon-counting techniques; the design of the EINSTEIN Observatory dates back to the 1960's. This historical lesson must be considered as we think about the future because it is very likely today's technology that will delimit the observational possibilities over timescales of the next decade or so. Second, we must ask ourselves whether the pace of technological development has remained steady; can we see on the technological horizon any new, promising techniques that may promise a real breakthrough in observational capabilities (for example, CCD technology has brought about in extragalactic optical astronomy), or have we reached a period of quiescence in which we are just likely to do more of the same (with some incremental improvements).

2. POSSIBILITIES FOR NEW OBSERVATIONAL CHALLENGES

In order to provide an organizational scheme for the discussion, I shall phrase the future observational goals in terms of specific scientific goals which I have heard articulated at this Symposium.

(i) *Structure of stellar interiors and dynamo activity.* A continuing problem in understanding stellar magnetic activity revolves about the nature of stellar interiors: how deep is the outer convection zone as a function of stellar parameters on the main sequence;

J. O. Stenflo (ed.), Solar and Magnetic Fields: Origins and Coronal Effects, 505–507.

what is the rotational state of stellar interiors; how does compressibility (departure from Boussinesq behavior and removal of the assumptions of mixing length theory) affect the flow structure of the convection zone; what is the nature of magnetic field generation in the interior of stars; how does the rate of magnetic flux generation depend upon the physical properties of the underlying star? Some of these problems may well be addressed by observational tools now under development. For example, the field of stellar seismology is just getting under way, having gotten off to a strong start with the recent successes of global mode detection for the Sun; in particular, the ACRIM results for bolometric solar fluctuations (which show evidence for 5-minute oscillations!) could well find application in the stellar domain, with possible trade-offs between sensitivity and long-term stability. Such measurements of the interior structure of stars will be complemented by direct measurements of surface magnetic fields, using recently developed techniques such as those used by Robinson, Worden, and Harvey (1980), and will be particularly powerful if one is able to go to the IR (as Zeeman splitting varies with wavelength), and will be complemented by other magnetically-sensitive diagnostics, such as Ca II and X-ray emission for solar-type stars. These may allow us to determine the variation of the rate of magnetic flux production as a function of stellar parameters: mass, radius, composition, age, rotation rate.

(ii) *The nature of photospheric surface activity.* The Sun has taught us that the essence of surface activity resides in its inhomogeneous nature: by and large what we refer to as "activity" occurs in association with observationally well defined photospheric features: the active regions, composed of the sun spots and ancillary plage regions. Perhaps the most crucial question related to the activity problem (which only stellar observations can address) is whether stellar surface magnetic fields are invariably inhomogeneous (intermittent): is the near equipartition between gas and magnetic fields at the solar surface an accidental attribute of the Sun, or is such behavior generic to solar-type (surface-convecting) stars? If generic (as we suspect it is), does the surface structuring depend upon the stellar state (in particular, its rotational state)? Such questions may also be susceptible of observational test in the foreseeable future: the newly-developed technique of "Doppler imaging" (Vogt 1983), which requires the combination of very high spectral resolution and sensitivity, will allow determination of the spatial structure of spectrally-anomalous surface features (such as "star spots"); while the further, continuing refinement of speckle imaging (which also places a premium on sensitivity) may allow direct spatial resolution of stellar surfaces of dwarf stars (at least in a gross sense).

(iii) *Nature of chromospheric and coronal activity.* It is by now one of the common-places of solar and stellar astronomy that chromospheric and coronal activity are largely governed by the dynamics of magnetic fields which emerge from the interior of stars. From this perspective, stellar surface activity can be used as a probe for stellar magnetic activity, and hence, indirectly, as a probe for dynamo action in stellar interiors. One of the prerequisites for such studies is however a better understanding of chromospheric and coronal activity in and of themselves. Specific problems to be considered include the gross characteristics of emission from stellar outer atmospheres, such as the mean luminosity and temperature, as functions of stellar parameters (as will be addressed by the coming generation of major space-borne observatories such as the Space Telescope, AXAF, EUVE, EXOSAT, and ROSAT); the physical (= geometrical) structure of stellar outer atmospheres (a problem which will only be accessible to study by the planned VLBA, which will allow us to go to the milliarcsecond resolution level at roughly VLA sensitivities); detailed atmospheric diagnostics, including more refined temperature, density, and velocity measurements (as will be accessible to study by means of the high spectral resolution spectrometers on board ST and AXAF).

SUMMARY

As we look to the future, three points ought to be kept in mind. First, it should not be surprising that the available instrumental capabilities are not particularly well-matched to the observational requirements: after all, much of the recent space-based observational work was based on the serendipitous discovery of the remarkable breadth of stellar activity; and certainly very few of the space instruments which fly today (or have recently flown) were designed with stellar activity observations in mind. Thus, in defining the capabilities of new instrumentation, we now have the not inconsiderable advantage that we roughly know what sensitivities and spectral resolution to aim for, and what appropriate design of observational programs is, in order to answer current observational questions.

Second, I must emphasize that many of the imposed instrumental requirements for the new observations do not demand major development of new technology: by comparison with the more esoteric branches of astrophysics, the instrumental constraints set by stellar activity observations are, in most cases, relatively modest, and can be easily met by available technology. It is also in this context that the issue of simultaneous multi-bandpass observations must be placed: it is most certainly the case that observations of stellar activity (especially those which aim at diagnostic information) find observations made in near temporal coincidence in the radio, optical, UV, and X-ray of great usefulness. Such observations can be accomplished by either a multiplicity of space platforms, or a single larger "bus" platform, either one in conjunction with ground-based observatories. Which is optimal is not obvious, but certainly an important ingredient in any calculation must be the relative probability of actually bringing either possibility to fruition: this will depend upon a number of trade-offs, including the relative ease of individual instrument optimization, the relative difficulty of insuring coordinated observations, and the relative total system reliability; my own assessment is that these trade-offs on the whole favor the notion of a succession of separate platforms, each optimized in instrumental and programmatic capability for the peculiarities of the particular instrumental passband.

Finally, the third point: the most critical issue in defining the capabilities of new observational tools is accessibility: stellar observations are peculiar in demanding large blocks of time, on a fairly regular and extended basis, demands which are generally difficult to deal with in the context of instruments which are to be shared with galactic and extragalactic astronomers. This crucial requirement implies one or more instruments dedicated to stellar observations (e.g., Explorer–class stellar satellites), and hence imposes a very special obligation on the solar-stellar activity community, especially in light of present fiscal contingencies: I suspect that we, as a community of scientists with common research interests, must be prepared to define a commonality of scientific goals to the point of achieving a consensus on what instrumentation ought to be built in the near future; this has been one of the crucial elements in the successes of extragalactic research within the past decade, and we must be prepared to do likewise.

THEORETICAL PROBLEMS

M. Stix
Kiepenheuer Institut für Sonnenphysik
Schöneckstrasse 6, D-7800 Freiburg i.Br.
Federal Republic of Germany

The purpose of theoretical work is twofold: we want to understand physical mechanisms — and we want to predict numbers which can be compared with observations. In the field of solar and stellar magnetism much of the work still belongs to the first category.

A typical example is the migration of magnetic fields in dynamo models. It has been known for a number of years that such migration occurs in $\alpha\omega$-dynamos along surfaces of isorotation. We now learn from calculations presented by Gilman that this is more general: without the explicit assumption of an α-effect the field migrates in the same way. And we obtain this result from a model which does not (yet) describe details of the solar cycle.

Another example is the attractor model described by Weiss. Although strongly simplified, this model retains some of the physical essentials of the dynamo: it produces cycles, Maunder minima and transition to chaotic behaviour. Obviously the MHD equations which govern stellar dynamos satisfy the requirements of order and non-linearity which characterize the attractor.

Or take the equivalent circuit as an analog of the coronal heating process. As explained earlier during this symposium by Chiuderi and Ionson, an ordinary and linear differential equation can demonstrate how a coronal loop is heated by waves with the loop's global resonance frequency (and it appears that in this case even quantitative predictions are possible if only the power spectrum of the driving perturbation is known).

A fourth example is the braking of stellar rotation. A star's angular momentum loss can essentially be determined from a stellar wind co-rotating out to it's Alfvénic point. This morning Roxburgh told us that such simple pictures may also illustrate the star's accelerated rotational braking when a change of magnetic multipole structure occurs in the corona: a possible explanation of the much-discussed Vaughan–Preston gap.

Detailed modelling does however depend on a knowledge of *what happens on the small scale*. This is evident for all examples mentioned above: the global circulation and dynamo model — when applied to a stellar convection zone — depends on turbulent diffusivities, and we have heard repeatedly from Schüssler and Frisch that such concepts are very much in question in the case of large Reynolds numbers. We must therefore continue to study small-scale structures such as fluxtubes in detail. We must also continue to study the detailed balance of forces in a fluxtube surrounded by a helical flow in order to get a better estimate of the α-effect which so critically enters the attractor model. So far not even the order of magnitude of α can safely be predicted. As far as coronal heating is concerned, we must continue to study MHD waves in detail in order to understand the

J. O. Stenflo (ed.), Solar and Magnetic Fields: Origins and Coronal Effects, 509–510.

dissipation mechanism, even if the total amount of heating depends little on such details. The wave properties must also be understood before we can safely employ the turbulent (wave) pressure as a driver for the stellar wind.

It is the small scale processes which determine the form and coefficients of our global models, and I am convinced that this is the area which finally will control the speed of theoretical progress.

DISCUSSION

STENFLO (CHAIRMAN): After you have now heard the four "provocative" summaries, let me ask the audience if you find that there is something fundamentally missing in the presentations of Noyes and Stix concerning the scientific needs. In particular I would like the questions to focus on the interface observation – theory, because the close interaction between observation and theory seems to be fundamentally important for progress in the "solar-stellar connection".

FRISCH: One (obvious) tool that was not mentioned in the flash-reviews of our needs is the *computer*. From the view-point of theoreticians it appears useful to distinguish (as done by M. Stix) numerical experimentation from numerical modelling. In the former case we are mostly interested in understanding some mechanisms involving only a few physical processes (e.g. flux expulsion); in the latter case the description must be sufficiently comprehensive to allow actual prediction of numbers. I wish to stress that even the fastest existing "supercomputers" are hardly able to cope with realistic modelling of phenomena, which are actually three-dimensional (as is generally the case with magnetic fields present). Until we have access to substantially more powerful machines ($\geq 10^8$ words of core, $\geq 10^{10}$ FLOPS), it might be wise to put the emphasis on numerical experimentation more than on modelling; in particular we want to be very careful with semi-realistic modelling which (i) will not add much to our knowledge, and (ii) can be extremely expensive.

TITLE: One should be aware that the observations have the same problems as the computers. On SOT our experiment will be collecting data from 8, perhaps 10, 800 $\times$ 800 arrays at 12 bits per pixel, hopefully once every second. This will contain part of the information that the theoreticians have been asking for, but some clever people have to figure out how to get this information out of the numbers. We want to understand things on a scale of 1/10 arcsec, which means that we have to take several samples over 1/10 arcsec over a range of heights and over an area that is typically tens of arcsec. People are going to have to think very hard how to describe the data that they get.

IONSON: I think that we should guard ourselves against mathematical overkill. This is a common trap that theorists fall into when doing purely computational studies. It is not efficient to construct theoretical models that are orders of magnitude more detailed than observational input.

MOUSCHOVIAS: I have two comments. (1) This symposium has convinced me that the stellar-solar connection has been made. I think, however, that it is equally important to make the interstellar-stellar connection. The results of cloud collapse calculations can yield the distribution of angular velocity and magnetic field strength within a protostar. These are quantities which are essential as input to calculations such as those of P. Gilman. More effort should be put in this direction. (2) On the interaction between theory and observation: I find it ironic that, for a week now, the complaint has been expressed that theorists ignore observations, and, at the same time, when a theorist gets up and restricts the range of his input parameters as suggested by observations, another theorist complains of such an approach. What is needed is less (rushed) criticism and an effort to understand what each research endeavour can and cannot accomplish.

J. O. Stenflo (ed.), Solar and Magnetic Fields: Origins and Coronal Effects, 511–513.

GIOVANELLI: I agree with the two previous speakers. But it is also important to remember the development of theory over the past 10 years or so. The theorists have been concerned with the interactions of fields and motions — building up from roughly uniform weak fields. The mathematical solutions to these problems are elegant and informative. But the observers have been showing for 15 years or more that the fields are concentrated into tiny strong fluxtubes at the surface, and these should be concentrated to much greater strengths when carried down by convection. The problem then is very different from the original one — for example it is hard to find processes that will bring the fluxtube fields down towards equipartition values. The theoretical problem of the interaction of fields and motions becomes different. Of course it is also of the greatest importance to produce theories and mechanisms which conform to the modern discoveries such as the observed meridional flows, torsional oscillations, etc., rather than merely feeding in boundary conditions, which are purely fictional.

STENFLO: Let us now shift the discussion to the instrumental problems, to consider the summaries given by Zwaan and Vaiana. They presented the separate needs for ground-based and space observations, but I would like you to also consider the important question of coordinated observations in different wavelength ranges and with different telescopes.

MARCY: I would like to suggest that high-resolution infrared spectroscopy should be useful for measuring mass-loss rates and Zeeman broadening in cool stars.

LIVINGSTON: Much of the observational work today is at, or near, the state-of-the-art in terms of sensitivity or resolution. For this reason it is important that observers check up on one another in a critical way. It is dangerous to depend on single observers, or groups, to supply critical data. Duplication in this sense is a healthy thing.

DUPREE: I would like to add to Dr. Vaiana's list a mission that was mentioned in the US "Field Report", a far ultraviolet explorer satellite, which would primarily aim for the 900 - 1200 Å region, but, as talks are developing now, may well go from 100 to 1500 Å. This would provide us simultaneous coverage of plasma diagnostics over a very large temperature range, without having to try to schedule different satellites like Einstein and IUE to look at the same star at the same time.

WALTER: I think it should be pointed out that the Space Telescope does have an instrument which is basically stellar — the HRS (although it may be monopolized by the hot star observers). It has the high spectral resolution (up to about 2×10^5) and temporal resolution ($\sim$200 ms) which IUE lacks. With such resolution, chromospheric and transition-region flows and flares will certainly be observable. The stars, of course, will not be spatially resolved, but Doppler imaging techniques in the UV lines can be used to map out large-scale structures in stellar chromospheres. In addition, high time resolution observations of eclipsing binaries, as in the crude observations reported here by Gibson, Walter, and Basri, can also be used to determine chromospheric and coronal structures and their absolute sizes. Often a picture is worth thousands of words.

ROXBURGH: One of the top priorities must be to measure solar oscillations with high precision — this is perhaps the most promising development in recent years and will permit us to determine the internal structure of the sun. We will need a satellite such as DISCO to achieve the precision needed.

GIBSON: One needs to be aware that radio observations may provide the best available data on structures and physical parameters in solar and stellar coronae. Even better data may be available with the simultaneous 2 cm/6 cm system, which will become available at the VLA next year, as well as the extention of its wavelength capability to 327 MHz. Also VLBI will become available for observations of stars, and will be quite capable of resolving their coronal structures.

STENFLO: In conclusion, may I ask for your brief opinions on future international collaboration in the area of the "solar-stellar connection", in particular for bringing theoreticians and observers together? In considering future meetings as a follow-up of this one, we have several options: from regular small-scale meetings of workshop type to larger but infrequently occurring symposia. What are your feelings about the optimum size of a meeting, and how frequently should it be organized?

KUPERUS: There are already so many meetings going on, so one should be careful when organizing new ones. It is not good for communication to have too many meetings.

GIOVANELLI: May I just mention that there is this week another solar meeting in Boulder

SHEELEY: ... and a third one in Ireland.

DUPREE: I should mention that we have in Cambridge an ongoing series of Workshops, called "Cool Stars, Stellar Systems, and the Sun", the next one to be held in October 1983. Usually we have 100 - 150 solar and stellar astronomers, including a fair European representation. This seems to be a successfull way of integrating the communities.

SODERBLOM: Speaking for myself, I have to say that for the meetings I have ever been to, on the whole the quality of the meeting and the results obtained have been inversely proportional to the number of people attending, although some critical mass may be needed. One should not try to agglomerate the meetings. The value of a meeting like this is simply that it brings people with common interests together.

CAMPOS: About the question of size of a meeting there is an intermediate optimum: large enough to have a broad range of views, and small enough so that everyone can discuss with everyone else. I think that this meeting is of the right size.

About cooperation between theoreticians and observers, I think that this is what we all have been doing this past week.

RELATIONSHIPS BETWEEN SOLAR AND STELLAR MAGNETIC FIELDS

L.E. Cram
CSIRO Division of Applied Physics, Sydney, Australia 2070

ABSTRACT

Observational and theoretical studies of solar and stellar magnetic fields are reviewed. The discussion draws together several themes that emerged during IAU Symposium 102 on "Solar and Stellar Magnetic Fields: Origins and Coronal Effects," and identifies a number of areas in which research over the next few years is likely to be fruitful.

1. INTRODUCTION

In July, 1908, George Ellery Hale, working at Mt. Wilson, demonstrated that the observed splitting of some of the spectral lines in sunspots is due to the Zeeman effect. This discovery represented a turning point in the study of solar physics, since it showed for the first time that magnetic fields played a role in solar activity.

Before Hale's discovery there had been many studies of the interrelations between the various manifestations of solar activity, but there was no unifying concept and theories were thus highly fanciful. Since the time of Hale's work it has become increasingly clear that magnetic fields play a dominant role in many solar phenomena, and much of contemporary solar physics is devoted to the observational and theoretical study of magnetic fields and their consequences.

One of Hale's reasons for founding the Mt. Wilson solar observatory was "the investigation of the Sun as a typical star". However, in the decades that have passed since the discovery of solar magnetic fields, the promise of a fruitful synergy between solar and stellar physics has never been fully realized. There has been a tendency for stellar studies to concentrate on abundances and evolution, while solar studies have tended to concentrate on activity and "meteorology." It was known that some stars possess magnetic fields, but these were thought to be somehow different from those of the Sun. On the other hand, while it was suspected that solar-like phenomena did occur on other stars, there was

J. O. Stenflo (ed.), Solar and Magnetic Fields: Origins and Coronal Effects, 515–531.

no way to study them, and consequently they were viewed as being unimportant. Recent observational studies of stellar chromospheres, particularly those undertaken by Olin Wilson at Mt. Wilson, have demonstrated the widespread occurrence in many stars of phenomena that were previously viewed as being of interest only to solar physicists. It is clear that a new phase of solar-stellar research has begun, in which the main unifying theme is the physics of magnetic fields in stars. This Symposium provided a timely opportunity to discuss what is known of this subject at present, and to define some of the important questions that remain to be answered.

This summary does not attempt to provide a detailed commentary on the previous contributions: they speak for themselves. Rather, it seeks to provide an account of the background research that has led to the present picture that we have of magnetic effects in the stars, with a special attempt to describe the logical (and not so logical) connections between key concepts. I believe that much can be learned by carefully following the development of "patterns of thought" in a discipline that is evolving as rapidly as the topic of this meeting.

The paper first reviews basic observational data for both the Sun and the stars. Then it discusses the two central theoretical problems that have been exposed during this Symposium: (i) the influence of magnetic fields on atmospheric structure, and (ii) the origins of solar and stellar magnetic fields. The paper concludes with an assessment of some promising directions for research over the next few years.

2. OBSERVATIONS OF SOLAR ACTIVITY AND MAGNETIC FIELDS

Prior to Hale's work at Mt. Wilson, there had already been a great deal of work done to characterize the nature of solar activity. Early work on sunspots had revealed their changeability, and the motion of sunspots across the solar disk had been used to determine the solar rotation rate and to demonstrate the presence of differential rotation. The existence of a solar cycle was recognized, and the systematic migration of the birthplaces of sunspots towards the equator during the progress of the cycle had been studied. Solar flares had been seen on a few occasions, and it was suspected that there was a connection between solar activity and terrestrial phenomena such as aurorae and magnetic disturbances. There was intense interest in expeditions to study the chromosphere and corona during solar eclipses.

In the latter part of the 19th century, advances in instrumental design led to a growing number of studies of the detailed structure of the solar atmosphere. The manufacture of superb high resolution gratings, combined with the application of photography to astronomical observations, led to remarkable advances in the quality and spectral purity of images of the Sun. These advances were particularly evident after the invention of the spectroheliograph, which allowed detailed studies to be made of the chromosphere, both on the disk and above the limb. It was apparantly the observation of "vortices" in H-alpha with this instrument that led Hale to suspect that electrical or magnetic forces could be important in sunspots, and this prompted his Zeeman effect studies.

For three decades following Hale's work, observational studies of solar magnetic fields were devoted to discovering properties of the gross relationships between magnetic structure and activity. The polarity laws of active regions were discovered, and the basic 22-year cycle was identified. A great deal of effort was expended in attempts to measure the global (dipole) field of the Sun, apparantly in the belief that such a field would help to explain coronal structure (polar plumes) and possibly to establish a connection between solar and terrestrial magnetism. With hindsight, it can now be seen that that this effort was probably not focussed on a critical problem, and even today the significance of observations of solar magnetic fields made on a global scale is not clear.

It is worth noting that the main thrust of solar reseach in this period was not towards the study of magnetism and activity, but rather towards the study of gross photospheric structure and abundances (e.g. Minnaert 1953). The development of quantum mechanics, and a growing interest in the interpretation of the Hertzsprung-Russell diagram in terms of stellar evolution, seem to have been responsible for this particular focus of solar physics. A similar emphasis on photospheric structure and abundances has occupied stellar physics until very recently. One might suspect that this emphasis will be redirected over the next few years, into areas that lie closer to the central themes of contemporary solar physics. Instrumental capabilities naturally played an important part in determining the course of research in these areas. For example, the limited sensitivity and inconvenience of photographic techniques for measuring magnetic fields, combined with the poor resolving power of the solar telescopes of the time, retarded work on activity and magnetic fields in the first half of this century. Only now are the instruments becoming available to study the related phenomena in stars.

A new era of research on solar magnetic fields and activity was born in the 1940s, when narrow-band filters were applied to the study of evolutionary aspects of activity, and photoelectric methods were applied to the measurement of polarization in spectral lines. The seminal discovery in this period was the recognition of the extremely high temperature of the corona and chromosphere. This discovery revolutionised concepts regarding the basic physics of astronomical gases, and can be identified as the beginning of "high-energy astrophysics." Another important advance in this period was the introduction of "patrol" observations of solar activity. By the mid-1950s, the photoelectric magnetographs at Mt. Wilson had been developed into quite sensitive instruments, and they were used in studies of the detailed correlations between plages and magnetic fields, and of the large scale organisation of the solar magnetic field into bipolar and unipolar regions (e.g. Kiepenheuer 1953; Babcock and Babcock 1955; Howard 1959). The comprehensive review by de Jager (1959) summarizes the results of this work, and vividly illustrates the depth of knowledge 25 years ago.

The advent of space-borne instrumentation opened a new vista on solar activity. Gradually, this new way of exploring the outer parts of the solar atmosphere reinforced the belief that magnetic fields are the prime driving force behind most of the phenomena seen in the solar atmosphere. **Skylab** observations were particularly influential in triggering awareness of the importance of magnetic fields (see Vaiana and Rosner 1978). A singularly important observation from this era was the detection and study of coronal holes, since these provided new insight into the connections between solar magnetism and the solar wind (Zirker 1977). I must confess to the feeling that there has been an overemphasis in recent years on the magnetic aspects of solar atmospheric structure and heating. This has been prompted in particular by an almost universal chorus among space observers of the theme that the corona is composed of only two basic structures, loops and open-field regions. It is not clear to me that this kind of description of the corona focusses on a fundamentally important property (after all, it is hard to imagine a topology that contains other kinds of structures and still remains quasi-stationary), and such a description is probably so simplistic that it hinders the development of theoretical ideas.

A more illuminating description of coronal structure should add to the present emphasis on **global topology** (open versus closed) a characterisation of coronal structure using another parameter that measures the **local importance** of magnetic fields (a suitable parameter might be the ratio of magnetic to gas pressure, or a measure of the magnetic free energy). Such a parameter describing the local importance of the coronal magnetic field is not presently available because there is no way to measure accurately the coronal magnetic field. It is reasonable to expect that theoretical ideas regarding the role of magnetic fields in the corona will change when methods for measuring the field in situ become available and are widely applied. It will be interesting to discover whether there exist large volumes of the corona in which the magnetic field plays only a secondary role, just as it appears that magnetic fields are not particularly important in most parts of the photosphere.

Many important observational discoveries made over the past 20 years have resulted from major improvements in the spatial resolution of ground-based telescopes, combined with improvements in the stability and sensitivity of focal-plane instrumentation. As the sensitivity and spatial resolution of magnetographs has improved, the intimate correlation between photospheric magnetic fields and chromospheric fine structure has become ever clearer. A study of special significance in this connection was undertaken by Skumanich, Smythe, and Frazier (1975), who showed that there is a **linear** correlation (with some scatter) between the K-line intensity and the "magnetic field strength" on scales as small as 2.4x2.4 arc sec. This result represents a key link between solar and stellar observations, and it is important to note the caveat that the magnetic quantity measured in this study is an "equivalent field representing the total flux" and is **not** an actual field strength.

The discovery of supergranular velocity cells and their relation to magnetic fine structure and the chromospheric network provided the first evidence of a connection between convection and magnetic structure. The importance of this connection has been reinforced by studies of the filigree and other fine-scale structure discussed by Tarbell at this meeting. However, it must be emphasized that many of our concepts regarding the physics of the fine-scale magnetic structure are based on inferences and educated guesses about processes that occur on scales too small to be studied with contemporary instrumentation. For example, with only a few exceptions, the information that we have regarding the fine-scale structure of the solar magnetic fields is based on indirect methods that exploit low-noise observations made with magnetographs that do not resolve the magnetic structure (Stenflo 1973). It is inevitable that new telescopes (such as SOT) will show that the picture we have today is severely "blurred" by the observing methods we have been obliged to use.

If progress is to be made in quantitative magnetic studies, futher work must be done in the area of spectroscopic diagnostics. There is an extremely complex relationship between the structure of three-dimensional solar magnetic fields and the polarisation signature they impress on spectral lines. Methods used today to unravel this relationship are crude, and lead to major errors in interpretation. For example, several discussions at this meeting have confused the terms "magnetic flux" and "magnetic field strength." The main reason for this probably lies in the inability of modern magnetographs to provide unambiguous measures of one or other of these quantities. Clarification of the relations between these two properties would represent a valuable contribution to the understanding of solar magnetism, and it is disappointing to note how little is being done in this area.

The development of extremely sensitive and stable magnetographs has led to new discoveries and clarifications of the nature of the large-scale structure of solar magnetic fields. In particular, the torsional oscillation discussed in this volume by Howard and LaBonte is evidently a manifestation of a fundamentally important process in the solar cycle. Observations of this kind, concentrated on the large-scale, slowly evolving aspects of solar magnetism and activity, require a dedicated, synoptic observing program. The astronomical community must support programs of this kind, despite pressures for rapid results and limited efforts. A related kind of research program involves the investigation of the evolution of solar activity over very long time scales. In this field, Eddy's work on the Maunder Minimum has provided a strong stimulus to attempts to determine the variation of solar activity over many cycles. The remarkable historical survey by Xu and Jiang (1982) which recovered 32 previously "lost" sunspot sightings during the years 1603-1665, combined with their arguments concerning the destruction of records during the Boxer rebellion, throws open the entire question of the reality of the Maunder Minimum. It is clear that painstaking work will be required to characterize the nature of solar variability on time scales longer than those covered by reliable historical records.

3. OBSERVATIONS OF STELLAR ACTIVITY AND MAGNETIC FIELDS

The observational search for stellar analogues of solar magnetic activity is extremely challenging. Direct observations of stellar magnetic fields are hard to make, and until very recently most detections referred to A and F stars. At the present Symposium, the physics of such "magnetic A stars" has not been discussed (see Cameron 1967 for a review), apparantly because it is generally felt that there are fundamental differences between the magnetic fields in these objects and those in solar-type stars. However, it might be unwise to view the magnetic A stars as a distinctly different class of objects. The basic difference between the magnetic fields of these stars and those of "solar-type" stars is that the magnetic A stars possess a marked global magnetic field, while the solar-type stars are characterised by strongly inhomogeneous fields. However, it is quite possible that the magnetic A stars do have an inhomogeneous field that is hidden from current observing methods by the global field, and on the other hand it is well known that the Sun exhibits large unipolar magnetic structures reflecting organization on a global scale. There are some intriguing similarities between the two kinds of stellar magnetism (including irregular variability, changes in photospheric structure, and possibly coronal effects), and a closer examination of the magnetic A stars might provide some important insight into the possible forms of stellar magnetic activity. If the evidence for a rapidly rotating magnetic solar core continues to accumulate, the connection between solar magnetism and the magnetic A stars will assume even greater importance.

The magnetic fields of magnetic A stars can be studied directly by analysing the circular polarisation of Zeeman-sensitive lines, since the global fields impose a strong signature on the line profiles. The magnetic fields of solar-type stars are much harder to study, because the mixture of almost equal areas of positive and negative polarity in relatively small active regions implies that the Zeeman splitting of magnetically sensitive lines cannot be detected by an analysis of circular polarization. Recently, however, the measuring techniques reviewed by Marcy in these proceedings have been applied to late-type stars, and some direct detections of magnetic fields have been reported. While the importance of such direct detections of stellar magnetic fields must not be underestimated, the method will not yield much reliable quantitative data without a breakthrough in instrumental sensitivity and (possibly) in basic technique. It is important that the observers who are applying this method constantly remind the theoreticians who use their data that the measurements do not provide reliable quantitative information.

Because of the limited possibilites for direct measurements of magnetic fields on cool stars, almost all of the information that we have regarding stellar magnetic activity on such stars is based on observations of "activity indicators," such as those reviewed by Zwaan in this volume. The principal indicator in use at present is the H and K flux parameter, which was used by Olin Wilson at Mt. Wilson in his pioneering studies of stellar chromospheres. In an early paper on this technique, Wilson (1968) justifed his research on the grounds that "it is a reasonable supposition that if analogous cycles could be detected in other stars with different values of the fundamental stellar parameters, the results would be of considerable value in sharpening the theoretical attack on the whole problem". Without the dedication and foresight displayed by Wilson, some of the critical data that is used today would not yet be available.

The Mt. Wilson H and K flux program, described in this volume by Vaughan, has made important discoveries in several areas, including (i) the delineation of the star-to-star variation in H and K fluxes across the cool part of the H-R diagram, (ii) the detection of stellar activity cycles, and (iii) the determination of stellar rotation rates and the gross structure of stellar active regions. Some aspects of these discoveries will be discussed below.

A number of important effects combine to produce the observed star-to-star differences in H and K fluxes. Some of the differences are undoubtedly due to variations in the degree of chromospheric activity. It is important that extensive, long-term studies be made in an effort to characterise the maximum amplitude and time scale of variations of this kind. Another component of the star-to-star variation is due to the fact that the Mt. Wilson H and K photometer measures a quantity that is not a "pure" indicator of the degree of chromospheric activity. In the hotter stars, for example, there will be an important "photospheric" contribution to the line fluxes: the precise calibration of this factor will require a dialogue between theoreticians who construct radiative equilibrium models of stellar photospheres, and theoreticians who carry out non-LTE diagnostics of the H and K lines. In the cooler stars, the calibration of the H and K index in terms of the adjacent continuum bands becomes uncertain, since the continuum becomes heavily blanketed by spectral lines. In intrinsically luminous stars, the H and K emission width exceeds the pass-band of the Mt. Wilson spectrophotometer, and the measured fluxes are thus less reliable indicators of the degree of chromospheric activity.

Despite these diagnostic difficulties, observations of stellar H and K fluxes exhibit a number of intriguing tendencies. Among the main sequence stars, there is a wide range in H and K flux for a given spectral type. It is observed that the emission strength depends on the age of the star, or its rotation rate, or both (Wilson 1963). It would be a particularly valuable contribution to theoretical work in this field if observational desiderata were available to separate the age and rotation factors. This is not easy, since indicators of stellar age are not very accurate. Moreover, there is a clear correlation between age and rotation period, and it may be impossible in practice to separate cleanly their respective influences on the H and K line fluxes. Among the giant stars, H and K emission is widespread, but more work is required to quantify the star-to-star variation in terms of stellar type, class, and evolutionary status. The study of H and K fluxes in binary systems is also instructive, since the stars presumably have identical ages. The relation between H and K strengths and evolutionary stage can thus be studied, although one must always be suspicious of the possibility that the chromosphere of the star in question is somehow influenced by the companion star.

Wilson's (1978) report on the nature of stellar activity cycles provides a fascinating first glimpse into the range and possible forms of variability of stellar chromospheres. The connection between these stellar cycles and the solar cycle is immediately clear, but we should be cautious in reading too much into the very limited information we have on the activity of the stars and of the Sun. For example, it is quite possible that over relatively short (in cosmic terms) time scales, the range of cycle morphology exhibited by the stars in Wilson's sample is duplicated by variations in the form of the solar cycle (one has only to review the evidence of palaeomagnetic variations to be struck by the absence of stable patterns in magnetic cycles). It is not known whether or not there are episodes of intensified activity in the past history of the Sun that could be related to a changed cycle morphology such as that seen in the stars above the "Vaughan gap." Furthermore, there is a distinct possibility (discussed below) that there is not a linear, or even a one-to-one relation between the measured H and K flux and the presumably more fundamental parameter -- the total magnetic flux. Attempts to reproduce quantitatively the forms of the cycles seen in the K-line fluxes may fail unless the possibility of a non-linear and perhaps multi-valued relation between K-line flux and magnetic flux (or field) is included in the theory.

The detection of rotational modulation of the H and K fluxes in a number of cool main sequence stars provides a reliable new method for measuring stellar rotation periods. This will provide important data for studies of the rotation-activity connection. Furthermore, detailed studies of flux changes made with adequate time resolution will reveal important properties of the physical organization of active regions on the stellar surface. This kind of study will prove to be particularly rewarding if it finds that different kinds of stellar active regions (depending on size, magnetic configuration, and degree of maturity) have different spectrophotometric signatures. If these can be measured, and calibrated by solar observations, we will be able to learn a great deal about the influence of magnetic fields on the structure of stellar active regions. Noyes has discussed these possibilities further in his contribution to these proceedings.

In fact, observations of other signatures of stellar activity already provide some insight into the structure of stellar active regions. At the present time, the main signatures that have been exploited (in addition to H and K) are (i) starspots and related variations of broad-band fluxes; (ii) He 10830 and to some extent H-alpha , both observed from the ground; (iii) the UV spectrum from Ly-alpha to the Mg II resonance lines, observed mainly with IUE; and (iv) the X-ray spectrum from 0.2 to 3.0 keV, observed mainly with the Einstein satellite. Other instruments and other parts of the electromagnetic spectrum have also been used.

Starspots have long been recognised by the presence of rather regular modulation of the light-curve of BY Draconis and RS CVn stars. Observations of stars with spots provide important data on the size, life-time, and physical properties of the dark regions. New techniques reviewed by Vogt (1981) can reveal details of the shape and position of spots on the stellar surface.

Ground-based observations of He 10830 show a remarkable variety of line profiles, and some perplexing time variations (e.g. Zirin 1975; O'Brien and Lambert 1979). Although controversy surrounds the interpretation of He absorption or emission, it seems probable that in main sequence stars, at least, the He spectrum is photo-excited by soft X rays from a corona. In this case, the He observations can be used to infer the presence of a corona, and to study the coronal evolution by proxy. In luminous stars it is not clear at present how the He absorption is formed, although it is evident that it requires temperatures in excess of radiative equilibrium values. Unfortunately, He 10830 lies in a heavily blended part of the spectrum, but despite this the greater flux at this wavelength favours He over H and K as a chromospheric-coronal diagnostic in cool stars. H-alpha has provided a good deal of the information we have on the solar chromosphere, and it is likely that a concerted study of this line in other cool stars would give valuable information on stellar chromospheres.

The IUE satellite has revealed a number of important properties of chromospheres of late-type stars. As emphasized by Linsky in this volume, a vital clue to the structure of stellar active regions is provided by the studies that show that the ratio of two line fluxes (say, Mg II to C IV) depends on the absolute level of activity measured, for example, by the surface flux in one of the lines. This result shows that active regions exhibit collective effects, and cannot be regarded as composed of a conglomeration of "elementary" active structures with a universal spectral signature. This result is not surprising, since solar studies show that the "activity" in network regions, in plages, and in sunspots, is not characterized only by a linear superposition of elementary structures. This discovery has the unfortunate consequence that it complicates the procedures that might be used to unravel the various factors that control the strength of activity indicators in flux spectra. These factors include (i) the area covered by the active regions, and (ii) the "intensity" of the active regions. Moreover, it strongly implies that there is not necessarily a simple correlation between the H and K flux and quantitative measures (such as flux or field strength) of the magnetic field.

The detection of X-ray emission in a large sample of late-type stars by the Einstein satellite represents an important step in the study of stellar activity. The results and implications of this work have been reviewed in this volume by Vaiana. There is a widely held view that stellar X-ray emission provides very firm evidence not only for the existence of a stellar corona, but also for a **magnetic** origin of the emission. Presumably, this view has been reinforced by the growing acceptance of the idea that magnetic processes determine the energy balance of the solar corona. However, while such an agument is appealing in the context of main sequence stars, I find it less compelling in intrinsically luminous stars. Perhaps such stars have magnetic fields that can excite stellar coronae, or play a role in mass loss, but one would be more comfortable if the theoretical edifice being erected on this hypothesis were based on firmer foundations. A direct detection of magnetic fields in such stars would be extremely convincing, but this is a remote possibility. It seems more probable that observations of flare-like events, or time dependent phenomena that could be related to active region changes, will provide the kind of evidence that we seek. Observations in X rays or at radio frequencies would be useful in this context, since they are more sensitive to transient activity than are the usual chromospheric indicators.

This brief review of the observational aspects of stellar activity would be incomplete without a mention of the "transition locus" in the H-R diagram between solar-type stars with a hot corona and low-mass winds, and luminous stars with cooler outer atmospheres and high-mass winds. The exact nature of the dividing line between these two classes of stars is not yet clear, and in fact, there are "hybrid" stars that fall between them. The analogy between this division of stellar types and an apparently similar difference between solar active regions and coronal holes suggests an explanation of the dichotomy in terms of differences in stellar magnetic field configuration. Several theoretical papers at this meeting have explored this question. Again, however, one would feel more secure if there were firmer evidence for the existence of significant magnetic fields in the atmospheres of late-type giant stars.

4. THEORIES OF MAGNETIC EFFECTS IN STELLAR ATMOSPHERES

Magnetic fields influence the structure of stellar atmospheres in many ways. One crude classification of these influences divides them into those that change the mass distribution in the atmosphere (pressure and velocity) and those that change the energy distribution (temperature). Of course, these two aspects cannot be clearly separated, but it is nevertheless convenient to discuss their physics from these two points of view.

4.1 Mass Balance

Magnetic fields in the solar photosphere appear almost always to exist in a form of magnetostatic equilibrium, as described in the contribution by Spruit in this volume. Essentially, the gas pressure inside the almost vertical magnetic structures is reduced, and the horizontal pressure of the surrounding non-magnetic region is partially balanced by magnetic pressure. An important question regarding this configuration is why the gas pressure inside the magnetic structure is lowered. The answer is not known, although a plausible suggestion involves the inhibition of heat flux into the magnetic region, with a consequent lowering of the temperature and decrease of the pressure scale height. However, there are several arguments suggesting that such a configuration is unstable. If the degree of inhibition depends on the size of the magnetic structure, it would be possible to explain the observation of the existence of an ordered sequence of photospheric magnetic structures, characterized by one dominant parameter, the total flux in the structure (Frazier 1978).

In the solar chromosphere the filamentary structure seen in H-alpha, Ly-alpha, and other lines strongly suggests the existence of a filamentary magnetic field. It is tempting to postulate that the chromospheric filaments (or fibrils) are individually connected to a discrete magnetic "knot" in the photosphere, but this is probably not the case. Attempts to discover such a connection have been unsuccessful (e.g. Dunn and Zirker 1973), and it is likely that the filamentary appearance of the chromosphere is due to the fact that the chromospheric magnetic field guides the velocity patterns, without necessarily being composed of discrete tubes.

The existence of loops in the solar corona points to magnetic control of the distribution of matter. Presumably, much of the corona lies in a configuration close to magnetostatic equilibrium, so that the sum of gas pressure and magnetic pressure is approximately constant. In this case, since the loops are regions of high pressure (high temperature and possibly high density), they will be regions with relatively low magnetic field. This conclusion is perhaps counter-intuitive, and it is important to make accurate measurements of coronal magnetic fields, both inside and outside loops. It is also of considerable importance to study the velocity field in coronal loops, to test whether dynamical forces can be important in controlling the mass distribution. Magnetic fields are obviously involved in the difference between closed regions and coronal holes, but it is not yet clear whether the main effect of the magnetic field in a hole is to force the outflow to diverge, or to modify the local energy budget through changes in the heating mechanism or heat transfer processes. As emphasized by Hartmann in this volume, this is an exciting area of research, since there is a growing body of evidence that suggests that magnetic fields play an essential role in the winds from a large variety of stars.

It seems reasonable to conjecture that processes akin to the solar phenomena described above are operative in stellar atmospheres, and that similar magnetic distortions of the mass distribution do occur. If this is the case, we can infer that photospheric field strengths will increase down the main sequence, in parallel with the increase in photospheric pressure, while any fields in the photospheres of low gravity stars will be much weaker than those on the Sun. The latter fact suggests that convective motions (which are more vigorous in luminous stars) will have a very strong influence on any magnetic structures existing in giant stars. Note that while it might be possible to make some reasonable suggestions about the field strengths encountered in stars other than the Sun, it is impossible to make corresponding statements about the total fluxes that might be found in magnetic elements on such stars. It is thus not possible at this time to discuss important questions such as the relative importance of sunspot-like and plage-like structures in other stars.

Since we have such a limited understanding of the factors influencing the magnetic field configuration of the solar chromosphere and corona, it is hard to make plausible suggestions regarding field configurations in other stars. For main sequence stars, there is some evidence that the more active objects have plage-like chromospheric components that essentially cover the entire star, and inferences based on scaling laws for magnetic loops (see the contribution by Golub in this volume) suggest a similar blanketing by coronal loops in very active stars. It seems clear that once the surface coverage by magnetic fields becomes larger than, say, 50%, the entire stellar atmosphere will have to adjust to this fact, and extrapolation from the Sun may then give a false picture. In stars with low surface gravity, the pressure scale height can be as large as the stellar radius, and in this case a magnetostatic structure will be enormously extended.

4.2 Energy Balance

As discussed above, the inhibition of energy transport in the deeper layers of a photospheric magnetic element seems to be an important factor in intensifying the field strength. Spectroscopic studies of sunspots and pores show that the photospheric velocity field corresponding to granulation in the quiet photosphere is indeed suppressed, in accord with the theory, although it is not clear that motions on unresolved spatial scales are similarly inhibited. In fact, there are many unsolved problems regarding the nature of the processes that modify energy transport in and near sunspots (Wilson 1981).

While it appears that magnetic fields act to inhibit the flow of energy in the deepest parts of the photosphere, this tendency yields rapidly to the extra heating evident in the magnetic parts of the upper photosphere (bright points, network) and in the chromosphere. The connection between magnetic fields and chromospheric heating is of vital concern to this Symposium, since it forms the basis of the supposition that the H and K fluxes of stars provide a quantitative estimate of some parameter associated with the magnetic field. Given the importance of this connection it is sobering to note that despite a great deal of work we have no real understanding of the mechanisms responsible for the extra heating that occurs in magnetic regions of the solar chromosphere.

A view that remains popular is that the magnetic regions are heated by "waves" that are generated by turbulent buffeting of flux tubes in the convection zone, and which dissipate their energy in the chromosphere by shock formation. A similar idea has been studied by several workers as a model for heating the non-magnetic chromosphere, but there are many unsatisfactory assumptions and approximations in these models. It seems likely that the heating results from the dissipation of electric currents generated locally by interactions between magnetic elements and waves of various kinds. A satisfactory theoretical study of this process would require the developement of a detailed multi-fluid model for the partially ionized chromospheric plasma. This model should include the possibility of drift and diffusion of several of the dominant components of the plasma (e.g. electrons, neutral hydrogen, protons, etc.), and should describe the electron-ion components as an ambipolar fluid. This fluid model could be used to describe the behaviour of the plasma in the neighbourhood of magnetic structures, including a consistent description of the external "exciting" processes. Such an ambitious program may lie many years in the future, so that the important question of the connection between magnetic fields and H and K emission will remain unanswered.

As reviewed by Chiuderi in this volume, there has been a major conceptual breakthrough in the related problem of coronal heating, provided by the "electrodynamic coupling" model of Ionson (1982). This is not an entirely new approach to problems in cosmic electrodynamics, since "circuit models" have been used widely in magnetospheric studies for several decades (e.g. Alfven 1979). The application of this approach to coronal heating is new, however, and the explicit derivation of the equivalent circuit is more firmly based on physical principles than the picturesque models that are sometimes used in these studies.

Since several heated debates regarding this model arose, both inside and outside the Symposium, it is appropriate to make a few comments at this point. First, it may be noted that the existing theories of coronal heating were unfruitful. These theories focussed on specific processes that might transport and dissipate energy in the corona, but they could not be tested because either (i) the theories were too abstract to be applicable to the solar atmosphere (which is highly inhomogeneous on all scales), or (ii) the crucial tests could not be made because of observational constraints (e.g. the coronal magnetic field cannot be measured accurately). In addition to these objections, the existing theories have generally assumed that the problem could be divided into one part dealing with the propagation and dissipation of energy, and another dealing with the generation of the necessary Poynting flux. This approach could never explain the coupling between coronal heating and the primary energy source in the convection zone.

It is important to judge Ionson's model in the spirit of its objectives, and not in terms of the older, failed approaches. For example, by focussing on the global electrodynamic coupling, Ionson cannot answer questions regarding the details of specific generation, propagation, or dissipation modes. This is seen as a shortcoming by many commentators, but future work within the spirit of the global model will be able to make specific predictions that may prove to be more readily tested than those made by previous workers. Finally, Ionson's model provides a close analogy between the electrodynamic behaviour of the solar atmosphere and the physics of electrical circuits, and thus provides a new way of looking at several old problems. It is quite enlightening to reverse the analogy, and to seek parallels in the solar atmosphere for several processes that are familiar in electrical circuit theory (the reader might speculate on the parallels between open transmission lines and coronal holes, or electrical discharges and solar flares, or analogies to feedback processes).

Most of the theoretical ideas discussed above in the solar context may be readily modified to produce speculations regarding corresponding processes in other stars. However, it seems that so little is **understood** (as opposed to believed!) about the role of magnetic fields in the mass and energy balance of the solar atmosphere that such speculations are premature. It is evident that while studies of the stars offer us new insights into the possible range of behaviour of stellar magnetic fields, it is only by studies of the detailed phenomena occurring on the Sun that we will develop an understanding of the fundamental processes at work. I share strongly the view expressed by Linsky, that we must study the Sun much more if we are to understand the stars.

5. THEORIES OF THE ORIGINS OF STELLAR MAGNETIC FIELDS

The structure and evolution of magnetic fields in stellar atmospheres are controlled by processes occurring beneath the visible layers of the atmosphere. We have only rudimentary ideas regarding the physical conditions in these sub-atmospheric layers, and consequently the theoretical investigation of the behaviour of magnetic fields in stellar interiors is severely hampered. It is convenient to separate the theoretical problem into two parts, one addressing the question of the factors that determine the large-scale, long-term behaviour of the magnetic field, and the other dealing with the processes responsible for the small-scale structure. Some workers believe that these two aspects cannot be separated, and that both are tied directly to the existence of discrete magnetic elements (flux ropes of various sizes) in the stellar interior.

Most attempts to explain the large-scale, long-term aspects of solar (and stellar) magnetism are based on the theory of the turbulent dynamo, reviewed in this volume by Schuessler. The essential features of dynamo theory are (i) the generation of toroidal magnetic fields from poloidal fields by differential rotation, (ii) the expulsion of this field from the solar interior under the influence of convective motions and buoyancy, with a Coriolis-induced helical motion leading to the generation of a poloidal field with reversed polarity, (iii) dispersal of the magnetic flux by eddy diffusion, concentrating the residual poloidal field near the poles, and (iv) repetition of this process to yield the full 22-year cycle.

Each part of this explanation contains questionable theoretical arguments. For example, it is by no means clear what causes solar differential rotation, so that it is difficult to estimate the distribution of rotation rate within the interior, and hence to estimate the rate of shearing of the magnetic fields. The role of buoyancy is poorly understood, and a critical problem of dynamo theory concerns the mechanisms that are responsible for holding the magnetic field down for a sufficiently long time against the strong buoyancy forces. The concept of turbulent diffusion of magnetic fields also contains a number of questionable features. A comparison of the monographs by Piddington (1981) and Parker (1979) provides complementary pictures of many of the important issues.

It is important to realize that there are a number of hotly debated questions that touch at the heart of attempts to explain the gross behavior of stellar magnetic fields. One of these concerns the role played by turbulent diffusion. Another concerns the question of modelling the effects of turbulent convection. In the model discussed by Gilman, for example, the behaviour of the magnetic field is explained in terms of the properties of large-scale "global convection" flow patterns, and smaller scales of turbulence are described by parameters that are isotropic. Another approach to the problem (e.g. Durney and Spruit 1979) identifies the primary factor in the theory as anisotropic turbulence. Some aspects of these alternatives might be tested when very accurate synoptic observations of large-scale solar velocity fields are available, but this is not certain.

Another important question concerns the properties of the solar core, in particular whether it is rotating rapidly, and whether it posesses a strong, primordial magnetic field. A related problem concerns the behavior of the overshooting region between the radiative core and the convective envelope. A number of theoreticians believe that this region plays a central role in the physics of solar magnetic fields. A cloud hangs over all of these questions, since the persistent failure of attempts to account for the solar neutrino deficit may indicate that there are fundamental processes occurring in the solar core about which we presently know nothing.

Given these problems with solar dynamo theory, it is not surprising to find that applications of the theory to other stars can not be made with confidence. One might expect that stars with properties similar to those of the Sun would behave as the Sun does, but this expectation would not be met if stellar magnetism has a long "memory" of the history of the star. Given the rudimentary state of the theory, the kind of taxonomical approach discussed in this volume by Rosner is particularly appropriate. We need more information on the **empirical** connections between rotation, activity, age, and so on, to assist the development of reliable models. It is especially difficult to extend the solar analogy to stars with internal structures and evolutionary histories that may differ greatly from those of the Sun, without much more insight into the factors controlling the operation of the solar magnetic generator.

The existence of discrete magnetic structures of various sizes in the solar atmosphere presumably results from the interaction between convection and magnetic fields in the interior, but the mechanisms responsible for this intermittency of magnetic flux are unknown. However, the question of the relationship between the global aspects of magnetic fields in the Sun, and the relatively fine-scale structure seen in its atmosphere is an important one, for two reasons. First, it is possible that an essential part of the theory of magnetic generation is the existence of discrete magnetic structures: an example of this would be the "flux-rope dynamo" mentioned by Schuessler. Second, the structural aspects of atmospheric fields clearly play a central role in controlling the gross properties of the atmosphere. Since a key link in the rotation-activity connection is the problem of the loss of angular momentum and spin-down, and since these processes depend on the structure of atmospheric winds and magnetic fields, there is an intimate connection between the evolution of stellar magnetism on a long time scale and the detailed atmospheric structure of the star. Aspects of this link have been discussed by Roxburgh in this volume. However, it is premature to attempt to formulate a detailed explanation of the interconnections between interior and exterior processes insofar as they influence the long-term evolution of stellar magnetic fields.

6. CONCLUSION

This review has identified a number of important areas where productive research should be possible. Solar observations should be afforded high priority, since they offer a unique opportunity to study the relevant processes in action on astrophysical scales. It is important to study both the global solar properties and the fine scale structure. An area where much more work should be done involves the measurement of vector magnetic fields, and the development of adequate diagnostic methods to interpret such measurements.

Stellar observations, particularly the Mt. Wilson H and K flux program, have rekindled interest in Hale's vision of the solar-stellar connection. There is a multitude of observational problems that should be pursued in connection with this observing technique. Supplementary spectrophotometric data should be obtained in lines such as H-alpha and He 10830. Attempts to refine the measurement of magnetic fields in late-type stars should be encouraged. The wealth of knowledge added by satellites such as IUE and Einstein will require several years to be thoroughly digested, but thought must be given now to the next generation of even more powerful instruments.

Theoretical work in the area of solar and stellar magnetism will continue to be very challenging, since it will be necessary to treat complex interactions between physical processes that are not understood well at present. The problem of atmospheric heating will be solved only when present approaches to the problem are revised to provide a much more realistic description of the atmospheric plasma (see Alfven 1979). Basic assumptions (such as ideal MHD) are simply not adequate to account for the important processes. Ionson's model might provide a vital ingredient in this revision. Dynamo theory is in a rudimentary state, with increasingly complicated models being constructed on the basis of assumptions that are clearly suspect. Unfortunately, this Symposium did not reveal a new conceptual approach to the dynamo problem, and progress is likely to be slow. It seems inevitable that the explosion of new observational data on stars of diverse characteristics will generate new theoretical ideas regarding the generation of stellar magnetic fields, and the effects of these fields on atmospheric structure.

REFERENCES

Alfven, H.: 1979, J. Phys. 40, p.C7-1.
Babcock, H.W., and Babcock, H.D.: 1955, Astrophys. J. 121, p.349.
Cameron, R.C.: 1967, "The Magnetic and Related Stars" (Baltimore: Mono Book Corporation)
Dunn, R.B., and Zirker, J.B.: 1973, Solar Phys. 33, p.281.
Durney, B.R., and Spruit, H.: 1979, Astrophys. J. 234, p.1067.
Frazier, E.N.: 1978, Astron. Astrophys. 64, p.351.
Howard, R.: 1959, Astrophys. J. 130, p.193.
Ionson, J.: 1982, Astrophys. J. 254, p.318.
Jager, C. de: 1959, Handbuch der Physik 52, p.80.
Kiepenheuer, K.O.: 1953, Astrophys. J. 117, p.447.
Minnaert, M.: 1953, in "The Sun" (ed. G.P. Kuiper), (Chicago: University of Chicago Press), p. 88.
O'Brien, G. and Lambert, D.L.: 1979, Astrophys. J. 229, p.L33.
Parker, E.N.: 1979, "Cosmical Magnetic Fields" (Oxford: Clarendon Press).
Piddington, J.: 1981, "Cosmic Electrodynamics" (Malabar: Krieger).
Skumanich, A., Smythe, C., and Frazier, E.N.: 1975, Astrophys. J. 200, p.747.
Stenflo, J.O.: 1973, Solar Phys. 32, p.41.
Vaiana, G.S, and Rosner, R.: 1978, Ann. Rev. Astron. Astrophys. 16, p.393.
Wilson, O.C.: 1963, Astrophys. J. 138, p.832.
Wilson, O.C.: 1968, Astrophys. J. 153, p.221.
Wilson, O.C.: 1978, Astrophys. J. 226, p.379.
Wilson, P.: 1981 in "The Physics of Sunspots" (ed. L.Cram and J.Thomas), (Sunspot: Sacramento Peak Observatory).
Vogt, S.S.: 1981 in "The Physics of Sunspots" (ed. L.Cram and J.Thomas), (Sunspot: Sacramento Peak Observatory).
Xu Zhen-tao and Jiang Yao-tiao: 1982, Chinese Astron. Astrophys. 6, p.84.
Zirin, H.: 1975, Astrophys. J. 199, p.L63.
Zirker, J.B.: 1977, "Coronal Holes and High Speed Wind Streams" (Boulder: Colorado Assoc. Uni. Press).

LIST OF POSTER PAPERS

I. MAGNETIC FIELDS IN STELLAR PHOTOSPHERES

II. ROTATION - ACTIVITY - CYCLE - AGE CONNECTION — OBSERVATIONS

III. THEORY OF STELLAR MAGNETIC FIELD GENERATION

IV. THE ROLE OF MAGNETIC FIELDS IN THE STRUCTURE AND ENERGY BALANCE OF STELLAR ATMOSPHERES

LIST OF PARTICIPANTS

AUSTRALIA	
Cram, L.E.	CSIRO, Lindfield, N.S.W.
Giovanelli, R.G.	CSIRO, Lindfield, N.S.W.
Saliba, G.J.	University of Sydney, N.S.W.
AUSTRIA	
Kömle, N.	Institut für Weltraumforschung, Graz
BELGIUM	
Koeckelenbergh, A.	Universitè Libre de Bruxelles
Scholiers, W.	Universitè Libre de Bruxelles
van Santvoort, J.	Universitè de Mons
CANADA	
Gaizauskas, V.	Herzberg Institute of Astrophysics, Ottawa
Gray, D.F.	University of Western Ontario, London, Ont.
DENMARK	
Nordlund, Å.	University Observatory, Copenhagen
Vennerström, S.	University Observatory, Copenhagen
FINLAND	
Tuominen, I.	University of Helsinki
Vilhu, O.	University of Helsinki
FRANCE	
Bel, N.	Observatoire de Meudon
Bommier, V.	Observatoire de Meudon
Bonnet, R.M.	LPSP, Verrières-le-Buisson
Crivellari, L.	Observatoire de Meudon
Cuny, Y.	Observatoire de Meudon
Didelon, P.	Observatoire de Strasbourg
Foing, B.	LPSP, Verrières-le-Buisson
Freire, R.	Observatoire de Strasbourg
Frisch, U.	Observatoire de Nice
Grappin, R.	Observatoire de Meudon
Hénoux, J.-C.	Observatoire de Meudon
Heyvaerts, J.	Institut d'Astrophysique, Paris
Koutchmy, S.	Institut d'Astrophysique, Paris
Léorat, J.	Observatoire de Meudon
Leroy, J.L.	Observatoire du Pic-du-Midi
Malherbe, J.-M.	Observatoire de Meudon

Megessier, C.	Observatoire de Paris, Paris
Mein, N.	Observatoire de Meudon
Mein, P.	Observatoire de Meudon
Muller, R.	Observatoire du Pic-du-Midi
Ribes-Nesme, E.	Observatoire de Meudon
Rozelot, J.-P.	Observatoire du Pic-du-Midi
Semel, M.	Observatoire de Meudon
Stellmacher, G.	Institut d'Astrophysique, Paris
GERMANY, F.R.	
Bues, I.D.	Dr. Remeis-Sternwarte, Bamberg
Galloway, D.J.	MPI für Astrophysik, Garching
Kaufmann, J.P.	Institut für Astronomie und Astrophysik, Berlin
Knölker, M.	Kiepenheuer-Institut, Freiburg
Moreno-Insertis, F.	MPI für Astrophysik, Garching
Saado, H.	Institut für Theoret. Geodäsie, Bonn
Schüssler, M.	Universitäts-Sternwarte, Göttingen
Soltau, D.	Kiepenheuer-Institut, Freiburg
Spruit, H.C.	MPI für Astrophysik, Garching
Stix, M.	Kiepenheuer-Institut, Freiburg
Tritakis, V.P.	MPI für Astrophysik, Garching
Ulmschneider, P.	Institut für Theoret. Astrophysik, Heidelberg
von der Lühe, O.	Kiepenheuer-Institut, Freiburg
Wittmann, A.	Obere Mühlenstrasse 7, Rosdorf
Wöhl, H.	Kiepenheuer-Institut, Freiburg
INDIA	
Gokhale, M.H.	Indian Institute of Astrophysics, Bangalore
Hasan, S.S.	Indian Institute of Astrophysics, Bangalore
ITALY	
Caccin, B.	Istituto di Fisica Sperimentale, Napoli
Chiuderi, C.	Osservatorio Astrofisico di Arcetri, Firenze
Galli, M.	Università di Lecce
Godoli, G.	Osservatorio Astrofisico di Arcetri, Firenze
Landi Degl'Innocenti, E.	Osservatorio Astrofisico di Arcetri, Firenze
Paternò, L.	Osservatorio Astrofisico di Catania
Russo, G.	Osservatorio Astronomico di Capodimonte, Napoli
Serio, S.	Istituto di Astronomia, Palermo
Severino, G.	Osservatorio Astronomico di Capodimonte, Napoli
Vaiana, G.S.	Istituto di Astronomia, Palermo
JAPAN	
Kaburaki, O.	Tohoku University, Sendai
Sakurai, K.	Kanagawa University, Yokohama
Sakurai, T.	University of Tokyo
Uchida, Y.	Tokyo Astronomical Observatory
NETHERLANDS	
Kuin, N.P.M.	Astronomical Institute, Utrecht
Kuperus, M.	Astronomical Institute, Utrecht
Martens, P.C.H.	Astronomical Institute, Utrecht

Mewe, R.	Space Research Laboratory, Utrecht
van Ballegooijen, A.A.	Astronomical Institute, Utrecht
Zwaan, C.	Astronomical Institute, Utrecht

NORWAY

Maltby, P.	University of Oslo

POLAND

Musielak, Z.	University of Gdansk

PORTUGAL

Campos, L.M.B.C.	Instituto Superior Tecnico, Lisboa
Lago, M.T.V.T.	Universidade do Porto

SPAIN

García de la Rosa, J.I.	Instituto de Astrofisica de Canarias, Tenerife
Herrero Davo, A.	Instituto de Astrofisica de Canarias, Tenerife
Vazquez, M.	Instituto de Astrofisica de Canarias, Tenerife

SWEDEN

Dravins, D.	Lund Observatory, Lund
Eriksson, K.	Astronomical Observatory, Uppsala
Linde, P.	Lund Observatory, Lund
Malmort, M.	Stockholm Observatory, Saltsjöbaden
Saxner, M.	Astronomical Observatory, Uppsala
Wyller, A.	Stockholm Observatory, Saltsjöbaden

SWITZERLAND

Barrow, C.H.	Institute of Astronomy, ETH Zurich
Benz, A.O.	Institute of Astronomy, ETH Zurich
Benz, W.	Observatoire de Genève, Sauverny
Bernold, T.	Institute of Astronomy, ETH Zurich
Bochsler, P.	University of Bern
Bürgi, A.	University of Bern
Deuel, W.	Institute of Astronomy, ETH Zurich
Dürst, J.	Langwies, Schönenberg
Eder, M.	Fehrenstrasse 23, Zurich
Huber, M.C.E.	Institute of Astronomy, ETH Zurich
Kämpfer, N.	University of Bern
Kindl, Ch.	Institute of Astronomy, ETH Zurich
Müller, E.A.	Observatoire de Genève, Sauverny
Nussbaumer, H.	Institute of Astronomy, ETH Zurich
Pauls, U.	Institute of Astronomy, ETH Zurich
Reufer, F.	Institute of Astronomy, ETH Zurich
Rüf, H.	Sihlweidstrasse 1, Zurich
Schmutz, W.	Institute of Astronomy, ETH Zurich
Solanki, S.	Institute of Astronomy, ETH Zurich
Spicer, D.S.	Institute of Astronomy, ETH Zurich
Stenflo, J.O.	Institute of Astronomy, ETH Zurich
Tozzi, G.P.	Institute of Astronomy, ETH Zurich
Wülser, J.-P.	University of Bern

UNITED KINGDOM

Arter, W.	Cambridge University
Aspden, H.	IBM, Hursley Laboratories, Winchester
Berman, N.M.	University of Manchester
Cattaneo, F.	Cambridge University
Hughes, D.W.	Cambridge University
McWhirter, R.W.P.	Rutherford and Appleton Laboratories
Mendoza, B.	University of Oxford
Mestel, L.	University of Sussex, Brighton
Moss, D.L.	University of Manchester
Proctor, M.R.E.	Cambridge University
Roberts, B.	University of St.Andrews
Roxburgh, I.W.	Queen Mary College, University of London
Sidher, S.D.	Imperial College, London
Weiss, N.O.	Cambridge University

U.S.A.

Baliunas, S.L.	Center for Astrophysics, Cambridge, MA
Basri, G.S.	University of California, Berkeley, CA
Boesgaard, A.M.	University of Hawaii, Honolulu, HI
Bonsack, W.K.	University of Hawaii, Honolulu, HI
DeVore, C.R.	Berkeley Research Associates, Springfield, VA
Dupree, A.K.	Center for Astrophysics, Cambridge, MA
Endal, A.S.	Applied Research Corporation, Landover, MD
Giampapa, M.S.	Center for Astrophysics, Cambridge, MA
Gibson, D.M.	New Mexico Inst. of Mining and Techn., Socorro, NM
Gilman, P.A.	NCAR–High Altitude Observatory, Boulder, CO
Golub, L.	Smithsonian Astrophysical Observatory, Cambridge, MA
Haisch, B.M.	Lockheed Palo Alto Research Laboratory, Palo Alto, CA
Hartmann, L.	Center for Astrophysics, Cambridge, MA
House, L.L.	NCAR–High Altitude Observatory, Boulder, CO
Howard, R.	Mt. Wilson and Las Campanas Obs., Pasadena, CA
Ionson, J.A.	NASA–Goddard Space Flight Center, Greenbelt, MD
Kalkofen, W.	Smithsonian Astrophysical Observatory, Cambridge, MA
Keil, S.L.	Sacramento Peak Observatory, Sunspot, NM
Lang, K.R.	Tufts University, Medford, MA
Linsky, J.L.	JILA, University of Colorado, Boulder, CO
Lites, B.W.	Sacramento Peak Observatory, Sunspot, NM
Livingston, W.C.	Kitt Peak National Observatory, Tucson, AZ
Low, B.C.	NCAR–High Altitude Observatory, Boulder, CO
Marcy, G.W.	Lick Observatory, Santa Cruz, CA
Mouschovias, T.Ch.	University of Illinois, Urbana, IL
Mullan, D.J.	Bartol Research Foundation, Newark, DE
Noyes, R.W.	Center for Astrophysics, Cambridge, MA
Peterson, R.	607 Marion Pl., Palo Alto, CA
Rosner, R.	Center for Astrophysics, Cambridge, MA
Scherrer, P.H.	Stanford University, CA
Sheeley, N.R.	Naval Research Laboratory, Washington, D.C.
Shields, G.	University of Texas, Austin, TX
Simon, T.	University of Hawaii, Honolulu, HI
Soderblom, D.R.	Center for Astrophysics, Cambridge, MA

Tarbell, T.D.	Lockheed Palo Alto Research Laboratory, Palo Alto, CA
Title, A.M.	Lockheed Palo Alto Research Laboratory, Palo Alto, CA
Vaughan, A.H.	Mt. Wilson and Las Campanas Obs., Pasadena, CA
Vogt, S.S.	Lick Observatory, Santa Cruz, CA
Walter, F.M.	JILA, University of Colorado, Boulder, CO
Wentzel, D.G.	University of Maryland, College Park, MD
Widing, K.G.	Naval Research Laboratory, Washington, D.C.
Worden, S.P.	University of California, Los Angeles, CA

U.S.S.R.

Khokhlova, V.L.	Astronomical Council, Moscow
Kotov, V.A.	Crimean Astrophysical Observatory
Kuklin, G.V.	SibIZMIR, Irkutsk
Severny, A.B.	Crimean Astrophysical Observatory

AUTHOR INDEX

*) "Solar Magnetic Fields" (Invited Review).
Abstract and manuscript not available.

SUBJECT INDEX